TIMESCALE

NIGEL CALDER

TIMESCALE

AN ATLAS OF THE FOURTH DIMENSION

CHATTO & WINDUS · THE HOGARTH PRESS · LONDON

Published in 1984 by
Chatto & Windus
The Hogarth Press
40 William IV Street
London WC2N 4DF

British Library Cataloguing in Publication Data

Calder, Nigel
Timescale
1. Fourth dimension
I. Title
530.1 1 QA699

ISBN 0-7011-3925-0

Printed in England by
Jolly & Barber Ltd, Rugby, Warwickshire

CONTENTS

Author's note

This book is a singleminded attempt by one fallible but tolerably well-informed person to weave new cloth from other people's yarns. Writing it has been a matter of learning from experts about discoveries and ideas in a range of specialist disciplines, and then of assessing and collating the information along a timescale. Where contradictions of opinion or apparent fact could not be reconciled, decisions in favor of one view or another have been guided by the available evidence, and by the wish to produce a coherent account of events, while avoiding the ifs and buts that quickly make knowledge opaque, like photochromic glass. Blunders may have been unavoidable, but what Samuel Johnson called "useful diligence" will perhaps prevail. Even if discoveries and reassessments ceased, the scope for finer resolution of dates and events, and for analysis of cause and effect, would remain in principle unlimited. Predetermining the length of the book led to a natural pause, when fresh themes could be pursued only by deletion of existing material, with consequent loss of balance or clarity.

More than a hundred scientists and scholars gave direct help in the book's preparation. In many instances, though by no means all, their work is cited in the reference index. Hundreds more contributed to the author's general education in science and public affairs during a quarter of a century of science writing. In short he is grateful to the entire academic community, and only he is to blame for any misunderstanding or misuse of material that was supplied in abundance.

His partner Peter Campbell skillfully shaped an unusual book, and without his guidance and encouragement it would never have been finished. Thanks are also due to members of the author's immediate family who undertook many tasks, both thoughtful and mechanical, to Hillian Durell and Robert Muir Wood, who assisted in research, and especially to Jean Lake, the author's secretary, who coped patiently with a succession of drafts.

Acknowledgement is due to the following for permission to reproduce illustrations: p. 10, Shell Photograph; p. 15, W. Hamilton, U.S. Geological Survey; p. 21, Royal Observatory, Edinburgh; p. 23, Energy, Mines and Resources, Canada; p. 27, Georg Gerster (John Hillelson Agency); p. 29, Laboratory of Tree-Ring Research, University of Arizona (Broder); p. 35, International Deep Sea Drilling Program (Site 552); p. 38, H. Hori and S. Osawa, Nagoya University; p. 45, Hunting (see below); p. 47, Science Museum, London; p. 53, National Geographic Society; p. 55, K. Apostolov, Royal Postgraduate Medical School; p. 57, M. J. Johnson, et. al, submitted to *Journal of Molecular Evolution*, 1982; p. 61, upper left, Ikon; upper right, Werner Forman; lower, Victoria and Albert Museum; p. 70, L. Pales; p. 72, R. Morse, reproduced with permission from *Scientific American*; p. 77, Royal Observatory, Edinburgh; p. 96, W. W. Stewart, reproduced with permission from *Nature*; p. 98, CERN; p. 103, Royal Observatory, Edinburgh; p. 104, P. Griboval; p. 105, Harvard College Observatory; p. 106, W. Herbst and G. E. Assousa, reproduced with permission from *Scientific American*; p. 107, NASA; p. 108, D. R. Lowe, reproduced with permission from *Nature*; p. 116, M. F. Glaessner; p. 117, J. E. Repetski; p. 119, Glasgow Parks and Recreation

Department; p. 121, 123, Bruce Coleman Agency; p. 124, upper, Geological Museum; p. 128, 129, Bruce Coleman Agency; p. 130, J. R. Thomasson; p. 134, GEOPIC; p. 136, J. J. Yunis and O. Prakash from *Science* Vol. 215, 1982, p. 1526, © 1982 by the American Association for the Advancement of Science; p. 138, M. H. Day; p. 140, 141, P. Kain and National Museums of Kenya; p. 143, Bruce Coleman Agency; p. 149, John Cleare; p. 150, Merseyside County Museums, Liverpool; p. 153, British Museum (Natural History); p. 154, M. H. Day; p. 160, Novosti; p. 161, Popperfoto; p. 164, upper and lower, J. Mellaart; p. 166, upper, Musée du Louvre, Paris; lower, M. Chuseville; p. 167, Ashmolean Museum, Oxford; p. 168, H. Blumner; p. 169, B. E. B. Fagg; p. 170, upper, Georg Gerster; lower, Brian Brake (both John Hillelson Agency); p. 171, British Museum; p. 172, upper, Edinburgh University Library; lower, British Library; p. 173, 174, British Museum; p. 176, Science Museum, London; p. 177, Ikon; p. 178, Mauritshuis, The Hague; p. 180, National Library of Australia, Canberra; p. 181, Science Museum, London; p. 182, Peter Newark's Western America; p. 183, Vickers Ltd; p. 184, Victoria and Albert Museum; p. 185, Robert Harding Picture Library; p. 186, Popperfoto; p. 188, upper, Sunday Times; lower, Ikon; p. 189, John Hillelson Agency; p. 190, CERN; p. 191, David Baker.

The Space Shuttle imaging radar (SIR–A) image on p. 45 was prepared from film supplied by Jet Propulsion Laboratory, Pasadena, by Hunting Geology and Geophysics Limited, Boreham Wood, Hertfordshire, England, who were joint official observers with the Royal Aircraft Establishment, Farnborough, to NASA's SIR–A investigation team, under a contract from the British Department of Industry.

Special sources for diagrams are as follows: p. 104, lower, after F. Hoyle; p. 107, after W. K. Hartmann; p. 111, 112, upper, after L. Margulis; p. 112, lower, after H. J. Hofmann and J. D. Aitken; p. 113, after D. J. Dunlop; p. 115, sutures after I. Gass; p. 118, after A. S. Roemer, and W. N. McFarland and others; p. 120, after D. M. Raup and J. J. Sepkoski; p. 122, upper after L. B. Halstead and G. Caselli; p. 124, upper after B. Tissot; p. 125, after C. J. Orth, et al.; p. 126, lower, data from V. Sarich; p. 127, after K. D. Rose; p. 128, after N. J. Shackleton and J. P. Kennett; p. 132, data from J. P. Kennett; p. 133, after P. R. Hooper; p. 137, data from V. Sarich; p. 139, data from L. G. Marshall; p. 140, lower, after R. F. Leakey; p. 142, geology from Goddard Space Flight Center; p. 143, after P. D. Gingerich; p. 144, Isernia data from M. Coltorti, et al.; general data after A. G. Sherratt and J. W. Lewthwaite; p. 145, after N. J. Shackleton and others; p. 146, simplified after R. L. Cann, et al.; p. 147, data after N. J. Shackleton; p. 152, after J. A. J. Gowlett; p. 153, after G. Woillard and W. Mook; p. 155, after D. Ninkovitch, et al.; p. 156, after W. Penfield and P. Lieberman; p. 157, after B. Kurtén; p. 158, after C. J. Jolly and F. Plog, and J. Pfeiffer, cutting-edge data after C. J. Jolly and F. Plog, and R. F. Leakey; p. 159, after B. Kurtén; p. 160, lower, after G. Woillard and W. Mook; p. 162, after P. S. Martin and A. Long; p. 163, plate boundaries after Goddard Space Flight Center; grasses after I. G. Simmons; p. 165, data from G. W. Beadle; p. 168, upper, after a photograph published by K. C. Chang; p. 175, tree-ring data after V. C. LaMarche, population data after J. D. Durand; p. 179, data after J. D. Durand; p. 186, after H. W. Bentley et al.; p. 187, after T. N. Barr; p. 207, as in caption; p. 217, upper, J. D. A. Piper; lower, after H. H. Read and J. V. Watson; p. 249, adapted from H. Hori and S. Osawa; p. 251, after Genetics Department, Stanford University Medical School; p. 252, as in accompanying text; p. 274, data from N. D. Newell; p. 276, after P. R. Vail, et al.

Diagrams were drawn by Shirley Wheeler and David Gifford
Colour maps by Thames Cartographic
Picture research by Ikon

OVERVIEW

Charts of Time

Anyone wishing to say what happened when should sympathize with early cartographers, who tried to specify what lay where. Mapmaking was a prehistoric art and a navigator-priest of the South Seas drew for the explorer Cook a chart showing seventy-four islands spanning thousands of kilometers of ocean. Hyperactive imaginations nevertheless confused the early attempts to visualize the world as a whole. A holy city was often the navel of the Earth, and plotting the resting place of Noah's ark seemed more important than placing the ports correctly. Ptolemy of Alexandria made an imperfect compilation that misled scholars and explorers for fifteen centuries, so that men in ships had to prove with their keels that wishful continents were missing and put the actual reef-girt land masses into position. The first cameras in space showed the entire planet looking just as the surveyors said it should.

Knowledge of the Earth's surface now approaches saturation. Spy satellites can spot where you parked your car, if that matters, while other satellites observe storms brewing in the mid-Pacific or monitor the cotton crop in Egypt. Laser pulses fired at the moon help to measure distances between continents to within centimeters. The South Pole is inhabited. Explorers in a deep-sea submersible have, it is true, come upon a lost world of bacteria and giant worms depending not on sunlight but on chemical energy that leaks from a rift in the ocean floor; zoology, at least, will never cease to surprise. But exceptions of that kind grow rarer and attention switches to interplanetary missions and cosmic space, now that the exploration of terrestrial space is broadly complete.

The exploration of time on Earth is, by contrast, beginning in earnest only now. In archeology and geology, reaching back in time often means going underground, because extinct animals and cultures tend to be buried under younger deposits, and the time machines of real life are the shovel and the drill. In the late 1960s American scientists pioneered an oceanic program of deep-sea drilling, now internationalized, that confirmed the new geological theory of plate tectonics and went on to explore 200 million years of oceanic history in splendid detail. In the 1970s the Russians began drilling wondrously deep holes through continental crust; the first, in the Kola Peninsula, was 2500 million years back in time when seven kilometers down. Such probes are muscular undertakings, though, and the shrewd time traveler has always allowed rivers or highway builders or oil-prospectors to do the heavy work wherever possible. But when the strata have been exposed, or the crumbled columns of a drilled core lies ready in its sections, discovering the age of a given layer is a more delicate matter.

A piece of the past comes to light, with the removal of a core sample from the drill pipe of an offshore platform, during a search for petroleum. The sample consists of ancient sediments under Europe's North Sea.

Chronographers, if I may use that term for those who put dates in sequence in a more comprehensive frame than earlier chronologists, are at roughly the stage of geographers after the first circumnavigators had returned to base. When Mercator laid out his chart of the world in 1569, he knew the size of the planet and the whereabouts of most of the principal land masses; important elements were nevertheless missing, and ignorance jogged his hand as he drew the coastlines. Exploration proceeded at only a few knots, and more than a century later Jonathan Swift could justly remark:

> So geographers, in Afric-maps,
> With savage-pictures fill their gaps;
> And o'er unhabitable downs
> Place elephants for want of towns.

In the late twentieth century, human beings think they know the approximate ages of the universe and the Earth, and when many significant events occurred, through to the present day. For this book, there seems no shortage of dates with which to decorate a timescale fifteen meters (fifty feet) long. And even though the time explorers turn up every day with new findings and breathtaking assertions, an Afric-map of past events may be better than nothing.

A sailor checks the foot of his chart for the date of the last correction: if it is more than a few weeks old he keeps a sharper lookout. Similarly, a scientist concerned with events far back in time scans with a critical eye the *anni domini* of cited references. A little integrated knowledge subject to continuous amendment is more reliable than a library's worth of complacent suppositions. Experts were recently teaching their students that the continents were fixed, that there were four ice ages, that Europeans invented cast iron, and so on. Exposure of errors gives hope of offering something better, but it also proclaims fallibility, and suitably documented corrections to any statement in this book will be welcomed, for assimilation in later editions.

Publications and conferences proliferate, summarizing segments of the past: world histories, world prehistories, theories of galactic evolution, and timetables for the first split second of the Big Bang. An International Geological Correlation Program is operational, and the Paleogeographic Atlas Project at the University of Chicago is systematically mapping geological, climatic, and other data onto a series of reconstructions of the positions of drifting continents. My objective is far more modest in detail, but less modest in scope, because this book pulls together and reconciles data about the past from many areas of research that deal in cosmic, geological, and human history.

Why bother? Let me offer motives in the order of ascending audacity. The first purpose is to present in an accessible form a selection of time-oriented data that are not always easy to find in other books. Second, people may be as curious to know where their atoms, their genes, and their languages come from as to study the recent chronicles of their nations, in which case history should start with the high-energy physics of the Big Bang and finish in the documented epoch of conventional history; the narrative then becomes a provisional scientific analog of old creation myths, and it may have similar interest for those who seek meaning in their lives. Third, close attention to sequence and timing helps to sort out causes and effects surrounding particular events and allows the reader to assess for himself the quality of the accounts on offer. Finally, science has done more than to fill in details of a story: it has set new standards of objectivity in the description of the past.

The heart of this book is the timescale and the accompanying narrative, on pages 98–191, where significant events that occurred between the origin of the universe and the present are put in order by the best available dates, and described and linked by concise explanations. The reference index provides a look-up facility, cites the sources of data, and sets out by topics certain timetables that have gone into the construction of the main timescale. Alternative narratives lurk there, for those who wish to browse. The present overview tours the sciences and considers the frame of mind appropriate to modern timescale making.

The comparison with mapmaking persists. Sextants and chronometers brought precision to navigation and marine surveying, and they restrained the flourishes of the cartographer, when the sailors told him firmly where his pen had to go. Scientific measurement and discovery similarly curb the freedom of self-expression in history. Geologists and archeologists have been among the first to find that they can no longer wave their arms and say what they fancy.

The search for mankind's roots in time has never proceeded more vigorously or productively than in this latter part of the twentieth century. The story is a dual one: of what happened in the past, and of current efforts to say what happened in the past. These two components fold together, because humans are prisoners of time and of the very processes that they describe, and subjectivity cannot be eradicated. Yet human beings alive today may be the very first creatures to have a roughly correct idea of how the game in which they find themselves embroiled has run its course so far. When they deployed their finest instruments and sharpest ideas for exploring the past, the results did not pander to preconceptions but delivered a succession of intellectual shocks.

The Physics of History

Beginning early in the twentieth century, physicists bulldozed through the gardens of knowledge, opening a freeway that cut across paths laid out with loving care by generations of specialists. Their authorization was a geiger counter: the physicists had found how to fix the ages of objects by the decay of radioactive materials they contained, and a particle counter helped to locate an event at its "absolute," not merely relative, date. That was only the start. By midcentury, sensitive magnetometers were detecting the fossil magnetism preserved in old rocks and providing evidence for a revolutionary version of the Earth's history. Methods of probing fine details of the atomic and chemical composition of materials transformed the study of minerals, meteorites, ancient climates, and the evolution of molecules. These techniques became, in their turn, powerful aids to dating.

The cries of protest were pitiful to hear. "This date is archeologically inacceptable," an eminent prehistorian declared when confronted with a radiocarbon result that made nonsense of his textbook. A famous earth scientist, defending the last rampart of his lifelong beliefs, called the physicist's new picture of the Earth "a hasty generalization of certain data whose significance has been monstrously overestimated." In field after field, dismayed experts saw their theories and narratives swept aside by measurements. "Unsubstantiated," the weathermen protested, when atoms of oxygen in sea-floor fossils announced the imminence of the next ice age. The discovery of impacts of giant meteorites disrupting life was greeted by the fossil hunters with a roar of "absolute nonsense."

Physics is the arrogant science. Its own theories are subject to sudden death: a couple of photographs showing stars microscopically misplaced destroyed the Newtonian theory of the universe cherished for two centuries. In no other area of learning are ideas so brightly burnished or the attacks upon them so abrasive. A woolly-minded physicist is soon looking for a woolier job, but sharp-minded heirs of Galileo have infiltrated other people's labs and taken charge. Chemistry became a branch of atomic physics. Physicists with X-ray machines broke into biology's innermost sanctum and emerged with the secret of life: the self-replicating structure of the DNA molecule, the stuff that genes are made of. ("Very little to do with real genetics, you know," an affronted biologist remarked at the time.) Compared with these strongholds of knowledge, the history of the world was a relatively easy conquest.

Newton taught physicists to distance themselves from their own planet, as if to observe it from far out in space. Their equations treated the Earth as a point of mass M orbiting around the sun or, when needs

What is the age of this precipice, which stands 700 meters tall, and 2200 meters high above sea level, in Victoria Land, Antarctica? Modern analyses and dating methods distinguish events widely separated in time. The light-colored features are fluvial sandstone, laid down gradually, layer upon layer, between 400 and 250 million years ago. But the broad dark band is younger than the rocks above it, being a tongue of lava injected between the layers about 165 million years ago. The whole assembly was uplifted comparatively recently to make the present mountains. Glaciers, like that in the foreground, fashioned the cliff face during the past 5 million years.

must, enlarged it into a ball for the study of its tides and its magnetism. Earthquake waves passing through the Earth's interior revealed to physicists the planet's inner structure and the molten core, the source of the magnetism. Thus the physics of the Earth was globally scaled from the outset. Surface decorations—the barely perceptible wrinkles of the Andes and Himalayas, or the thin film of life—were seen in an unsentimental perspective, and physicists could ask what global processes made the planet the way it is. Traditional geologists, on the other hand, might well have been flat-earthers: they studied individual provinces with little notion that processes and events on opposite sides of the world were directly related.

A flaw in physics might have barred its inroads into history if the human imagination were easily hobbled by philosophical niceties. Strictly speaking, physicists were disqualified from pronouncing on anything historical because their notions of time were plainly defective, from Newton's day until the late twentieth century. Indeed, two sorts of physics conflicted. In one of them, concerned with the motions of individual particles, interesting objects could in theory move forward or backward through time. The second kind of physics dealt with the random motion of many particles, manifest as heat; here time could run only in the one, sensible direction, but the creation of interesting objects seemed almost impossible.

In the physics of the first kind, space and time are laid out as a four-dimensional continuum. There is nothing at all mysterious about that: any event is specified by its location, within three dimensions of space, and the instant of its occurrence, along a single dimension of time. A relation between time and space is implied in the term light-year, which reckons a distance by the time taken by light to traverse it. Time as the fourth dimension is an idea older than relativity, although Einstein discovered that the rate of a clock can be affected by high-speed motion or the action of gravity, so that space and time become mixed together. These effects have little relevance for chronography, except at the origin of the universe. More significant is the fact that, if the general theory of relativity is correct, as it certainly seems to be, the laws of nature never vary. They are the same at all places and times, so that results of laboratory experiments here and now are directly applicable to the remote past of the Earth, and to the most distant galaxies. Perceptions of the past are not distorted by any change in the laws of nature: there is mistiness perhaps, but no funny business.

The equations that describe the motions of stars or atoms would remain correct if time ran backward. This antihistorical character of certain laws inspired the physicist Laplace to suppose that, in principle, the entire past and future could be deduced by computing backward or forward from the present motions of particles. Einstein shared this belief that distinctions between past and future were in some sense illusory. Serious efforts are in progress at last to bring mainstream physics into line with the facts of life, and relate the one-way reality of cosmic time to the fundamental character of the cosmic forces. Among subatomic particles, certain rare, weak interactions were discovered in the 1960s, which could not retrace their steps if time ran backward; this behavior is associated with a bias in the direction of spin of certain particles. At the frontiers of theory-making the chief task on the agenda is to reconcile Einstein's theory of gravity with the quantum theory of particles. The direction of time may turn out to be a feature of gravity itself, and linked to the laws of heat in the caldron of the Big Bang.

The alternative physics of heat can always tell past from future. It declares that disorder increases as time passes, so that the universe tends toward a lukewarm uniformity, and this increase in entropy, as it is called, gives due recognition to the direction of time. It is a useful law for many purposes, including the design of more efficient fuel-driven engines, yet it creates a false impression of where the world is heading. Living things and human brains, with their high degree of organization and information content, defy the alleged tendency to dull disorder. More subtle studies of the laws of heat, pursued by Ilya Prigogine in Belgium during the past quarter century, explore systems that have not settled down to equilibrium, with everything evened out and no changes pending. Amended laws then allow for order to appear out of chaos; examples include the neat pattern of rising and sinking water in a kettle heated from below, and the stationary whirlpools in a chaotic torrent. Moments of decision, bifurcation points when a system must adopt one pattern or another, begin to take effect in the new theories and experiments, and they bear more resemblance to historical events than do crisis-free descriptions of particles in motion.

Life on Earth is permissible because the universe has not yet reached its fatal equilibrium. But to say so transposes the question to another realm: how has the universe eluded an outcome so clearly required by the laws of heat? The answer seems to lie in the cosmic expansion. As the physicist Steven Frautschi observed:

> Far from approaching equilibrium, the expanding universe falls further and further behind. . . . This gives ample scope for interesting nonequilibrium structures to develop out of initial chaos, as has occurred in nature. [*Science*, Vol. 217, 1982]

The calculations involve the entropy of black holes and other features of cosmic economics. Suffice it to say that if objects billions of light-years off were not moving away at a high speed, nonequilibrium structures like you and me might not exist. As the fundamentals of physics begin, belatedly, to match the well-known processes of time and history, they express a unity in nature transcending anything that holy men ever contemplated.

An early intrusion of physics into Earth history was a blunder. In the late nineteenth century leading physicists were assuring fellow scientists that the sun and the Earth must be less than 50 million years old, on the grounds that they could not have stayed warm any longer. Geologists who reckoned, correctly, that greater intervals of time were needed to lay down the thicknesses of sediments that they observed were not pleased by such pronouncements, which also disappointed evolutionists who wanted as long a period as possible for the known changes in life to accumulate. The physicists' mistake was due to ignorance of certain nuclear reactions that allow the sun to burn for billions of years, and others that have kept the Earth's interior warm and mobile for a similar interval. Radioactive decay is responsible both for the prolonged activity of the Earth and for the means of determining its correct age; the discovery of radioactivity at the end of the nineteenth century opened the way to the physicists' takeover of the timescale.

Only a few decades have elapsed since the age of the planet settled down at around 4500 million years, and no chronographer should laugh at Archbishop Ussher of Armagh, who, in the seventeenth century, calculated that the world began six thousand years ago, in 4004 BC. His was a painstaking reckoning based upon the Bible, one of the few sources of data on prehistory then available, and Newton himself shared the belief in a very young Earth. Creationists in the United States still affirm that the Earth is about six thousand years old, making current science wrong by a factor of almost a million. It says something for the grasp that physics now has on history that creationists feel obliged to amend physics and to assert that heat and pressure have affected the rates of radioactive decay. In particular, they report variations in the radioactivity of cesium-133 and iron-57: this is just dust in the eyes of nonscientists, because those atomic species are not radioactive.

Creationists have another problem. Any object in the sky more than six thousand light-years away should not yet be visible if it was created only six thousand years ago, so the entire fabric of the wider universe, as elucidated by two centuries of galactic astronomy, has to be explained away. By contrast, one of the strengths of the scientific timescale is that the astronomical and earthly schedules now conform very well.

Havoc in the Sky

Natural philosophy made a U-turn in the mid-twentieth century. Telescopes had penetrated the realm of galaxies—vast collections of stars resembling our own Milky Way, scattered like ships in the ocean of space. Astronomers seemed to be plumbing a universe infinitely large and infinitely old, as foreseen by many sages. Even when the galaxies turned out to be dispersing as if from an explosion, one could envisage the continuous creation of new matter in the spaces left by the receding galaxies, and that was a theoretically respectable way of ensuring that the universe never changed very much. Then came the discovery in 1965, during experiments with microwave radio, that empty space is about three degrees above the absolute zero of temperature. The only plausible explanation is that there really was an explosion, and the universe is not incomparably older than the Earth. Several lines of evidence now converge on an age for the universe of 13,500 million years, and this inference more than any other gives point to efforts to construct a timescale: like Mercator's planet, the past is finite.

If the picture of the universe has changed and sharpened almost beyond recognition in the past few decades, it is chiefly because of an irruption of physics into astronomy—an event that did not fail to arouse resentment among astronomers. As it happened, optical astronomers were themselves well trained in physics, yet when telecommunications engineers and radar physicists first tuned in to the sky by accident, and discoveries followed thick and fast, many tried to stay aloof. Indeed, astronomers in the Netherlands, renowned for their early broad-mindedness about radio astronomy, are the exception that proves the rule. The revolution worked by radio telescopes was so comprehensive that by 1965 an astronomer was telling me with conspiratorial glee that the importance of a new class of objects he had discovered was that they were *not* radio sources. Meanwhile, theorists had injected the nuclear physics of element formation into the study of the private lives of the stars, and revolutionized that too. If old-time astronomers, in the nearest science, could feel dismay as physics engulfed their subject, imagine the bewilderment of geologists, biologists, and archeologists, when brash young physicists told the professors that, frankly, their studies had missed the point.

Astronomy is now reintegrated as a branch of high-energy physics where nature performs experiments far beyond the scope of man-made apparatus. Radio telescopes and other new instruments of the physicists have unveiled a universe of great violence. The serene night sky of the poets is loud with the birth pangs and death throes of stars, and with bleeping pulsars that are highly compressed remnants of exploded stars, or supernovas. Most of the pinpricks of light are giant nuclear reactors like the sun, busily transmuting the elements; some are distant quasars radiating energy at an astonishing rate by the action of massive black holes.

Fanciful star wars among sentient beings are tame compared with conflicts among the stars themselves, when vampire-like X-ray stars decant the plasma from their stellar companions. The detonation of one star can trigger the formation and detonation of others, in a procreative chain reaction of supernovas sweeping across the light-years. Entire galaxies have been torn, deformed, or sterilized by mishaps occurring on an unimaginable scale. Some of them shoot huge bolts of plasma into intergalactic space. By a seeming paradox, the more turmoil the universe reveals, the more plainly does life on Earth appear to be part of it and, incidentally, the more inviting space becomes for human colonization.

Human beings are the children of cosmic concussions. The Big Bang made hydrogen, the primeval fireball made helium; and thereafter a succession of little bangs, through generations of supernovas, yielded the carbon and other elements needed for life. That chemical prelude occurred against a ground bass from the center of the Milky Way, where a big black hole aspired to be a quasar. Eventually a supernova induced the birth of the sun and the Earth, which was not a gentle delivery, and the chemical elements arrived in packages, as comets and similar microplanets crashed into the embryonic Earth. Nor were the cosmic links severed: the sun's warmth stirred the fluids of the Earth, the gravity of moon and sun heaved the oceans in twice-daily tides, while the gravity of other planets toyed with Earth, slowly altering its orientation in space and the shape of its orbit, and so varying the climate. If the sun were not a relatively well-behaved and long-lived nuclear reactor, the Earth would be uninhabitable. Yet the calm blue of the daytime sky is just as deceptive about the state of the solar system as is the pretty night sky about the universe at large. Collisions between members of the sun's family of orbiting objects release violent energy on a scale that is petty by cosmic standards but disconcerting when they occur at the surface of the Earth.

Ancient fears aroused by comets were transformed by mathematical physics into a possible intersection of orbits, and Newton's friend Edmond Halley suggested that the Caspian Sea was the crater of a comet that struck the Earth and caused the biblical flood. Although his identifications were misguided, the idea was brilliantly correct, and subsequent discoveries of many more microplanets—small, dull asteroids as well as blazing comets—made astronomers fully aware of danger

Dusty debris of exploded stars adorns and obscures the interstellar scene in the Belt of Orion. The dark protuberance right of center is the Horsehead Nebula, and the vast cloud below it blots out the more distant stars. The "sunrise" effect beyond the Horsehead is due to emissions from streaks of hydrogen gas. Gravity recycles the dust and gas to make new stars, like the young, intensely bright star at the left. (The spikes and rings are effects of the telescope system.)

in the sky. Nearly three centuries after Halley's hypothesis, most geologists and evolutionary biologists still disregarded the possibility of major cosmic impacts disrupting life. Astronomers and physicists had to resort to the elaborate sciences of space research and nuclear chemistry to drive this elementary lesson home.

The exploration of the solar system by manned and unmanned spacecraft amounts to a rediscovery of this sector of the universe. The Moon, Mars, Venus, and other siblings of the Earth, seen at ground level or from close flybys, reveal similarities and differences that cast an intense new light on the character and history of our own planet. The giant craters on the moon proved to be the scars of impacts, and closeup pictures of Mercury and the moons of Jupiter and Saturn confirmed that such events were widespread. In modern theories of the origin of the solar system, cosmic impacts are simply a prolongation, at a slow rate, of the coalescence of small objects that built the planets.

These confirmations intensified interest in signs of impacts on the home planet. The Earth tends to heal its scars, but a number of fairly large craters have been identified. A multimegaton explosion in the forests of Siberia in 1908 is ascribed, with hindsight, to a very small piece of a comet hitting the Earth, while much older scatterings of glassy beads, called tektites, were plainly made by molten rock sent flying across wide areas by cosmic impacts. In some cases the impacts must have been equivalent to millions of megatons of explosives, far exceeding the present inventories of mankind's nuclear weapons. For some astronomers, indirect evidence of cosmic catastrophe has for long been obvious in the slaughter recorded by the wholesale disappearance of species from the fossil record. But the paleontologists—the professional fossil hunters—almost unanimously scorned the idea that mass extinctions could have anything to do with cosmic events, for which they preferred more homely explanations. Craters and astronomical statistics were insufficiently persuasive.

The contentions about impacts came to a head in 1979 and 1980, in connection with the extinction of the dinosaurs and many other groups of animals at the end of the Cretaceous geological period, 67 million years ago. A thin, worldwide layer of clay lying at the very top of the Cretaceous limestone turned out to be tainted with exotic chemical elements, almost unknown at the Earth's surface but present in meteorites falling from outer space. The investigators took samples from the clay, and from strata above and below it, and put them in a nuclear reactor, thus activating the rare elements for detection. They were not looking for a catastrophe but, on the contrary, were trying to determine the

Impact craters are produced by the unending succession of collisions between the Earth and giant meteorites. All of the examples shown are in Canada. At Manicouagan (upper left, Landsat image) a frozen circular lake delineates a crater seventy kilometers wide, formed 210 million years ago by the impact of an asteroid more than a kilometer in diameter. The two-kilometer crater at Holleford (upper right, aerial photograph) is approximately 550 million years old, and illustrates how erosion and infilling conceals many wounds. Deep Bay (lower left, aerial mosaic is a crater thirteen kilometers wide, of uncertain age. For New Quebec Crater, three kilometers wide and 5 million years old, the map (lower right) contours a lake 250 meters deep; a high rim surrounds it, like a crater on the moon.

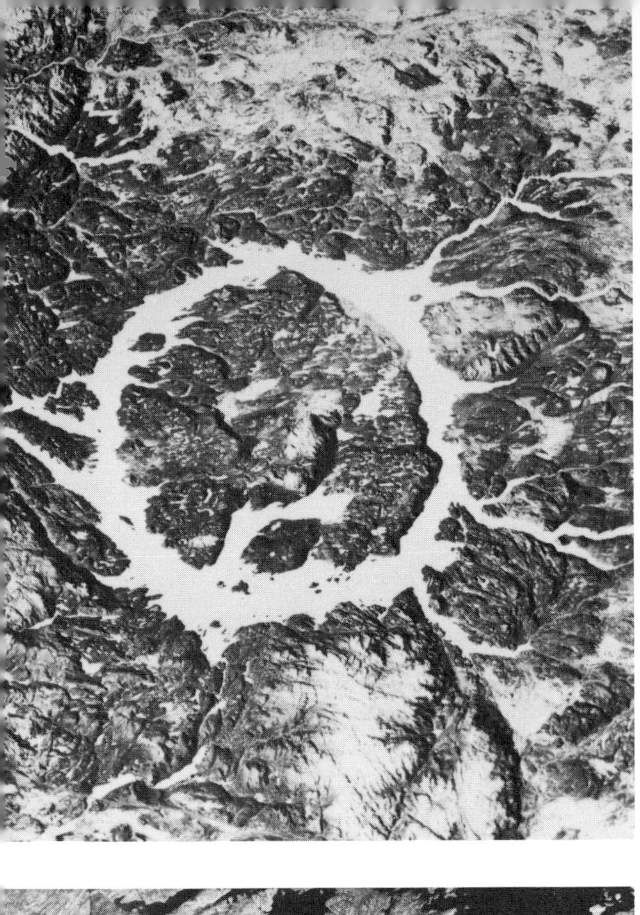

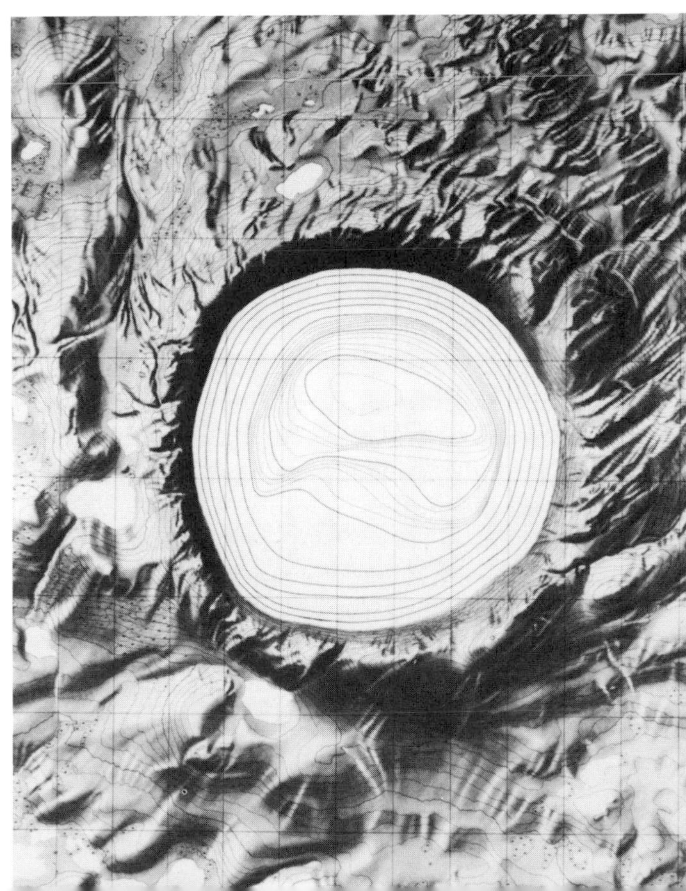

duration of a gradual turnover in life by relying on the gentle rain of meteoritic dust that continues at all times. The discovery of excessive quantities of rare elements in the clay, indicating a giant meteoritic impact, had therefore the instant legitimacy of the unexpected.

If the comet or asteroid that hit the Earth that day, 67 million years ago, had safely crossed the Earth's orbit an hour earlier, the dinosaurs would have survived; then the course of evolution would have been entirely different, and human existence unlikely. The story of life on Earth is not to be understood without reference to events of this kind, which repeatedly cleared the principal actors from the stage and gave understudies their chances. There have been, in fact, so many impacts that to mention any but the most consequential or best documented would become as tiresome as cataloging every war in human history. Major impacts probably occur every few hundred thousand years, on average, and stupendous ones every 50 million years or so. Some distinguished paleontologists remain mocking and resentful about this discovery, and their vehemence, laced with technical arguments, might be intimidating for a chronographer who had not heard similar protests before.

Radiometric Revolutions

More effort has gone into dating "the KBS tuff" in East Africa than any other sliver of the world. This layer of volcanic ash bears the initials of a geologist: it is the tuff at the Kay Behrensmeyer Site, lying to the east of Lake Turkana in Kenya. It contains ancient stone tools, and it overlies other strata that in 1972 yielded up one of the very earliest human skulls. Volcanic material is often suitable for age determinations that assess the radioactive decay of potassium-40 atoms into argon-40 atoms, and this layer promised to provide a benchmark date for the first human appearances, when geology begins to shade into archeology. Unfortunately, the KBS tuff consists of material that first fell elsewhere and was then swept up by running water and redeposited where it now lies. In the process, it became contaminated with older materials, and efforts to fix its age encountered awkward discrepancies.

Investigators visited the area again and again, took samples away to laboratories dedicated to atom counting, and obtained nominal dates. During the 1970s, different researchers using various methods made more than sixty measurements, but the resulting radiometric ages for the KBS tuff ranged from half a million to more than 200 million years. Early hunches that the correct date might lie between 2 and 3 million years ago gave the early inhabitants of Kenya some priority over others in East Africa. But by 1980 there was convergence: the best potassium-argon dates settled at 1.89 million years, and a quite different method, called fission-track dating, gave 1.87 million years. Geological comparisons showed that "tuff H_2" near the Omo River in Ethiopia was of the same age as Kenya's KBS tuff, and potassium-argon dating gave that 1.87 million years. One serious anomaly remained. Results with a method called argon-argon dating, akin to potassium-argon dating, still varied from one-half to two-and-a-half million years. Eventually, in 1981, a careful redetermination by the argon-argon method gave 1.88 million years, and the geologists and archeologists had their benchmark at last. The KBS tuff was around 1.88 million years old.

All methods of dating have drawbacks and inaccuracies, and each is limited in respect of its age range and the kinds of samples with which it can cope. For example, potassium-argon and fission-track dating both make use of material that was purged of argon gas and crystalline flaws during a volcanic event. Every natural radioactive clock used in dating was set to begin its countdown by some chemical or physical transformation of the material in question. The investigator has to decide precisely what event last reset his clock, and what confusions or contaminations may be lurking in his samples. Although dating measurements are made in well-equipped laboratories far from the scene, the experts always want to know a good deal about the context and history of each sample. Without shrewd and diligent explorations in the field, in tropical heat or polar cold, the clever atomic and magnetic equipment in laboratories would be relatively useless. The revolutions in knowledge worked by physics would have quickly petered out in error and confusion if alliances had not sprung up, marrying the new techniques to the traditional skills of geologists and archeologists.

Between AD 1500 and 1800, European scholars and surveyors gradually appreciated that the rocks were rich in fossils of prehistoric animals and plants, and that different layers of rocks, or strata, represented different stages of the Earth's history. They assigned names to eras, although dates eluded them, and the biblical conception of Noah's Flood dominated their search for explanations. From about 1800 onward, principles of stratigraphy developed by William Smith in England and George Cuvier in France became the basis of modern geology. These principles are: first, that in any one place, younger strata usually lie on top of older strata; second, when strata in different places contain similar fossils, they are of similar age. The analogous notion of a succession of Stone, Bronze, and Iron ages emerged at about the same time in archeology.

Around 1850, Jacob Worsaae in Denmark was the first to pay careful attention to levels when digging up antiquities, as a guide to prehistoric stages. Resemblances among tools, weapons, or pottery became the archeologists' equivalent of similarities among the fossils, and a means of correlating levels at different sites across a wide area. Where ancient cultures had writing, lists of kings and references to astronomical events provided clues for dating.

If sediments of mud or silt accumulate steadily on the bed of the sea or a lake over a long period, the thickness of sediments is then some sort of guide to the duration of the period. In the late nineteenth century geologists attempted to make their own age estimates from the maximum thicknesses of rocks known for each geological period. Answers, for the time elapsed since the appearance of abundant fossils, ranged from 700 million years (not a bad result) to a deferential 20 million years. Sedimentation rates are variable, and the process is often interrupted or undone by geological upheavals. As a guide to the past, sedimentation works best in tandem with other dating methods, when it provides some check on consistency and serves as a simple aid to interpolating dates for objects found between two layers of known ages.

When you can distinguish yearly layers of sediment by seasonal changes in color or composition, you can count them, although the reliability of the method may be disputed. Tree trunks often contain rings of annual growth, which again lend themselves to direct counting, and systematic use of tree rings for dating began with Andrew Douglass in Arizona in the early 1900s. Patterns of fluctuating growth during a run of years allow old wood to be matched with known sequences in newer wood, thus extending the count of years backward. The longest-living trees, matched to older remains, extend the dendrochronological timescale back more than eight thousand years. Variations in growth from year to year, when carefully interpreted, can give a vivid picture of changing climates in the places where the trees grew.

These were some of the methods available before the discovery of radioactivity revolutionized dating. By 1902 Ernest Rutherford and Frederick Soddy in Montreal had figured out that half the atoms in any sample will break up during a specifiable interval, the half-life, that varies from one species of radioactive atoms to another. The products of radioactive decay are atoms of a different chemical element. Most materials at the Earth's surface contain more or less harmless traces of radioactive substances, and the proportions of different kinds of atoms become clues to the age of a specimen. Seeing at once the possibilities of measuring the ages of rocks, Rutherford caused a sensation by announcing

Archeologists' labels mark successive floor levels at the ancient Mesopotamian city of Nippur, in present-day Iraq. New buildings arose on the rubble of the old, so the more recent levels stood progressively higher. Nippur was the holy city of the Sumerians; it yielded one of the earliest libraries of literary writings, about 5000 years old, and remained an important urban center for 3000 years.

E 24 floor 5
E 24 floor 6
E 24 floor 7
E 24 floor 8
E 24 floor 9
E 24 floor 10
E 24 floor 11
E 24 floor 12

that a particular piece of pitchblende was 700 million years old, far older than many people's conception of the age of the Earth at that time.

Rutherford was working with uranium, the atoms of which undergo a long series of transformations, through elements such as thorium and radium, before finishing up as atoms of lead of particular atomic weights. Comprehensive application of uranium-lead dating to geology came with the work of Arthur Holmes, who, in 1913, issued the first modern timetable for the past 600 million years. Since then, other slowly decaying atoms have been put to good use by the geologists. And fission-track dating, mentioned in connection with the KBS tuff, came in during the 1960s; this relies on naturally occurring fission. The splitting of uranium atoms produces fragments that tunnel through the surrounding material, creating tracks that can be spotted and counted, at least in glassy substances. On the whole, radioactive dating was reassuring and helpful to traditional geology. A detailed sequence of fossil-bearing geological periods and stages already existed, and dating confirmed it. The chief surprise was the discovery that this fossil-making era spanned only about one eighth of the Earth's history, but even as it delivered that news, radiometric dating offered a means of exploring the preceding eons.

More upsetting was the revolution in archeology brought about by radiocarbon, and the technique pioneered by William Libby in Chicago in 1949. Familiarity should not blunt the sense of wonder that one can take, say, a piece from a prehistoric rope, convert it into benzene, put it into a counter that registers emissions from traces of radioactive carbon, and find out how old it is. Once-living materials—wood, charcoal, bone, cloth—provide the samples, and radiocarbon dating tells when their original owners, plants and animals, died. Up to that time they were taking in carbon from the air or from their food, with roughly one radioactive atom among a million million stable carbon atoms. After death the radiocarbon gradually decayed with a half-life of 5730 years, well suited to most archeological purposes. Rays from space hitting the Earth's atmosphere manufactured the radiocarbon, but over the millennia the supply had fluctuated, and radiocarbon dates are misleading unless they are calibrated against items of known age. The long record of tree rings was ready-made for that task.

Radiocarbon's jolt to prehistory was partly due to the breathtaking opportunities it opened up, especially for comparing unrelated cultures worldwide. More immediate and emphatic were its contradictions of accepted ideas, when estimates of the ages of cultures just a few thousand years BC turned out in some cases to be wrong by two thousand years.

Rings of annual growth in a tree trunk provide the dendrochronologist with a direct count of elapsed years. Variations in ring widths tell of climatic fluctuations, and also make possible a matching of sequences from living trees to old timbers. There is no need to fell a tree; its rings can be sampled by drilling out a thin core through the bark.

Discoveries overturned preconceptions about comparative ages: intensive cultivation of crops was in progress on the banks of the Rhine sooner than beside the Nile; stone temples in Malta antedate the pyramids; a buried city in Turkey is older than any in Mesopotamia. Young prehistorians who scorned the "fine-arts-museum orientation" of traditional archeology were gleeful. Striving to understand the subsistence and social circumstances of prehistoric cultures, they had already taken to air photography and other new techniques, and radiocarbon accelerated the change in mentality. Computerized methods of analyzing the distribution and importance of settlements, and tracing catastrophic disclocations in complex trading systems, have followed in the wake of radiocarbon dating.

An old presumption was that hunter-gathering ancestors were feckless wanderers distinguished only for their cave paintings, until some bright spark suddenly invented agriculture in southwestern Asia—whence the light of civilization radiated all over the world. Radiocarbon rescued archeology from the circular reasoning that assigned relative dates according to how advanced a culture seemed to be. Hunter-gatherers were then seen to have used their brains purposefully to spread all over the planet more thoroughly than any subsequent group, and also to have invented pottery and other items previously credited to farmers. Agriculture evolved gradually, and originated independently in several regions. In examining this transition, and the reluctance of hunters to change their ways, the archeologists made common cause with anthropologists, who knew the ways of surviving hunter-gatherers and simple farming cultures.

A widening range of sciences helped to disperse the mists of prehistory. Metallurgists clarified the origin of the copper, bronze, and iron industries by reconstructing the difficulties and successes of smiths who knew no chemistry and had relatively cool furnaces, but who developed remarkable skills. Climatologists explained that modern humans first appeared in a mild interval during the most recent ice age, endured the worst of it, and then, when the ice melted, multiplied in complex cultures amid a continuously fluctuating climate. Techniques for tracing the climatic changes range from studies of ancient plant pollen and the thicknesses of tree rings, to scrutiny of records of the date of blossoming of Japanese cherries and of the quality of French wines from year to year. Among the prominent cultures thought to have succumbed to climatic disasters were those of the early Indus Valley farmers, and of the Mycenaeans and Hittites of the Mediterranean. Studies of prehistoric subsistence, metallurgy, and climatic fluctuations are all keyed to the radiocarbon timescale, and particle accelerators and laser beams now aid archeological research, in efforts to make radiocarbon dating more sensitive and extend its useful range.

Many archeologists remain innumerate in their use of discrepant dating systems, and one learns never to trust a date from an archeologist unless he explains how he arrived at it. As late as 1980, when James Mellaart, an archeologist of distinction, proposed in the journal *Antiquity* that radiocarbon dates should overrule doubtful quasi-historical chronologies even on the hallowed ground of Ur and Babylon, he was scolded in print by his peers. Nevertheless, measurement in the fourth dimension has been a powerful tamer of fancies. Archeology illustrates radiometric dating's most important quality: it is largely assumption-free, uncontaminated by preconceptions about whether ancient Egyptians were cleverer than ancient Bretons, or whether the taming of animals should have followed or preceded the cultivation of crops. Although technical considerations demand careful attention to archeological methods and the circumstances of a find, the numbers coming from a particle counter deliver a verdict of Olympian quality, to which all hypotheses ought to defer. Furthermore, the radioactive carbon that humans and their artifacts assimilated embeds them in the planet's atmosphere in life as surely as the soil surrounds their remains at death. They are not items of romance or legend, but part of the furniture of the planet, and no more suitable for fable making than ammonites or coal.

Fossil Compasses and Thermometers

Inferences from magnetism were as upsetting for geologists as radiocarbon dating was for the archeologists. When rocks form from volcanic material, or from sediments containing iron, they become imprinted with the direction of Earth's magnetic field prevailing at the time and place of their consolidation. During the 1950s physicists who had developed magnetic instruments of high sensitivity used them to sense the directions of magnetization in samples of old rocks from various continents. They found these fossil compasses misaligned, as if the rocks had formed at locations on the globe completely different from their present-day positions. The continents had plainly moved, just as heretics like Alfred Wegener and Arthur Holmes had been saying for half a century. Most geologists chose to ignore these results; so the physicists tried again.

They showed that the Earth had also switched around its north and south magnetic poles, repeatedly reversing the direction of its magnetism. From this finding flowed a new method of dating, magnetic stratigraphy, that identifies specific reversals in datable rocks, and spots the same events in other rocks around the world. A related discovery was a curious pattern of magnetic stripes detected when research ships towed instruments across the oceans. The pattern took the form of alternating zones of stronger and weaker magnetism, and in 1963 a graduate student in geophysics, Fred Vine, realized that the ocean floor had grown and spread, acting like a tape recorder while it did so. Its rocks had adopted the direction of the magnetic field at the time of their formation, and as a result the ocean floor consisted of bands of rock whose magnetism alternately opposed and reinforced the present-day magnetic field. If the ocean floor had spread, the continents must have moved to make room for it.

Many geologists still remained unconvinced, until a deep-sea drilling ship recovered samples showing the ocean floor to be unexpectedly young, and confirmed in detail the story deduced from the magnetic stripes. Meanwhile, other geophysicists came out with the brand-new theory of geology called plate tectonics. It was a matter of geometry on a globe, and of broken pieces of the Earth's outer shell jostling one another—here easing apart to widen an ocean, there pressing together with one plate overriding another and destroying the ocean floor. The venerable continents were then seen to be the playthings of juvenile oceanic plates that pushed them unceremoniously across the face of the globe. Plate tectonics answered overnight many questions that geologists had almost given up asking: why did earthquakes and volcanoes occur in particular places, and what caused the upheavals that created mountains and minerals?

Instant sophistication is a feature of human adaptability and makes it hard to remember what the earth sciences were like in 1960, when most people thought the continents were firmly rooted in the Earth—almost as hard as empathizing with the old conviction that the Earth itself was fixed at the center of the universe. Conservative geologists did not give up easily. Seeking to minimize the importance of plate tectonics, they alleged that the recent supercontinent called Pangaea had existed for billions of years until it suddenly decided to break up 200 million years ago. When that idea became untenable, a previous supercontinent was next supposed to have brooded over the world until about 800 million years ago. Now it seems that plate tectonics and continental drift have been fully functional for at least 2800 million years, and that four supercontinents have formed and fragmented during that interval.

In its pristine form, plate tectonics did not give a sufficient account of the geology of continents. Ocean floors are rigid, and they conform well to the geometry of plate tectonics; continents on the other hand are more crumbly. A second major step in modernizing geology was to describe how tensions due to plate action cause continents to stretch and sag, creating low-lying basins where sediments accumulate. The geophysicist Dan McKenzie, one of the originators of plate tectonics, was responsible for this further advance in geological theory in 1978. Stretching thins the crust of a continent and lets hot, deep-lying rocks well up underneath it. As these cool, the continent thickens again, but deep slices, near-vertical faults that formed rift valleys in the initial stages of continental stretching, remain as permanent scars. Subsequent compression of the continent can reactivate these faults and extrude individual blocks sideways, like lemon pips, or exalt former valleys to make new mountains.

The ice ages were the next target for the techniques of the physicists. The assault began in 1955, when Cesare Emiliani, a young Italian working in the United States, measured the atomic composition of small marine fossils recovered from the seabed. This was a matter, not of radioactivity, but of comparing stable atoms of different masses, the proportions of which were affected by global changes of climate. Emiliani at once counted considerably more than the four ice ages that had supposedly occurred in the current series. Twenty years elapsed before the picture of a quick-fire succession of glaciations won acceptance. By that time there were matching discoveries of multiple layers of ice-age dust on land, but the thermometer-like marine fossils remained the prime indicators. Their content of heavy oxygen (oxygen-18) varied according to their ages, in a rhythmic manner, indicating many glacial cycles in which great ice sheets formed and melted. For Nicholas Shackleton, in England, oxygen-18 became a means of weighing the ice in all the world's ice sheets, because the removal of water from the oceans to the ice sheets was responsible for altering the proportion of oxygen-18 in the sea, and in the animals that lived there. Other explorers of past climates studied the changing populations of heat-loving and cold-loving organisms, by sea or land, and these elaborated the emerging pattern of climatic change.

Dozens of theories about what caused the ice to come and go had been debated for a hundred years, but the new techniques swept most of them away and identified the real cause. Three powerful methods converged on the problem: the counts of oxygen atoms outlined the climatic fluctuations, a magnetic reversal some eight ice ages back offered a conspicuous time marker, and radiometric dating fixed the age of the reversal at about 700,000 years. It then became obvious that the pulse of the Earth's climate was governed by the moon and by the other planets, which, by their gravity, varied the orbit of the Earth and altered the tilt of its axis in a cyclical manner. This accorded with the Milankovitch theory, adumbrated in the nineteenth century but zealously advocated in the twentieth century by Milutin Milankovitch in Yugoslavia. The astronomical influences are calculable over long intervals of the past, and of the future too, so that they predict the course of the next ice age in some detail.

These findings also offered a new method of dating the past, because the oxygen-isotope and marine-fossil records can be "tuned" to the computable astronomical influences on the Earth's orbit and attitude, and the resulting orbital dating is the best framework for age determinations for most of the past half-million years. The Milankovitch rhythm also traces back, before the start of the current series of ice ages, over 3 million years ago. More generally, the study of climate now dovetails neatly with knowledge of geographic and oceanographic circumstances in the remoter past, when the continents were arranged differently and the Earth's climate was usually warmer than today. Present understanding of the ice ages thus relies on yet another comprehensive revolution that was virulently opposed, as were most of the others, by conservative experts.

Like the heroine in an old Hollywood movie, whose makeup never smudged even if she was in the clutches of cannibals or King Kong, scientific dignitaries and philosophers try to preserve an impression that all is polite, tidy, even logical, in the advancement of science. Advertising codes designed to protect the public from misleading claims do not apply in education or the popularization of science, and current knowledge is offered for assimilation by students and the public, as if science had stopped and was immune from correction. Research would soon become boring, of course, if that were the case, and clever young people would take their minds elsewhere. Although the chronographer might wish the multiple revolutions to cease at the time of writing, this convenience is rightly denied to him. He is required instead to guess where the next intellectual fracas will break out.

Offered then, as an example of an incipient revolution, is a puzzling picture of ancient fluctuations in the level of the sea, worldwide, discovered during the hunt for petroleum. Physicists working for oil companies are in the habit of setting off explosive charges or air guns, which act like miniature earthquakes and send seismic waves through the

The rhythm of the ice ages appears plainly in the uppermost sections of a core of sediments obtained by drilling in the bed of the North Atlantic. The light bands, rich in chalky fossils, correspond with warm interludes, and the dark bands with ice ages. The youngest sediments appear at the top of the segment on the left, and those at bottom right (twelve meters down the core) are about 640,000 years old. (The pattern continues deeper in the core, into sediments more than 2 million years old.) Rates of accumulation varied but astronomical calculation can date the transitions accurately, because the cycles of warmth and cold are governed by the Earth's antics in orbit.

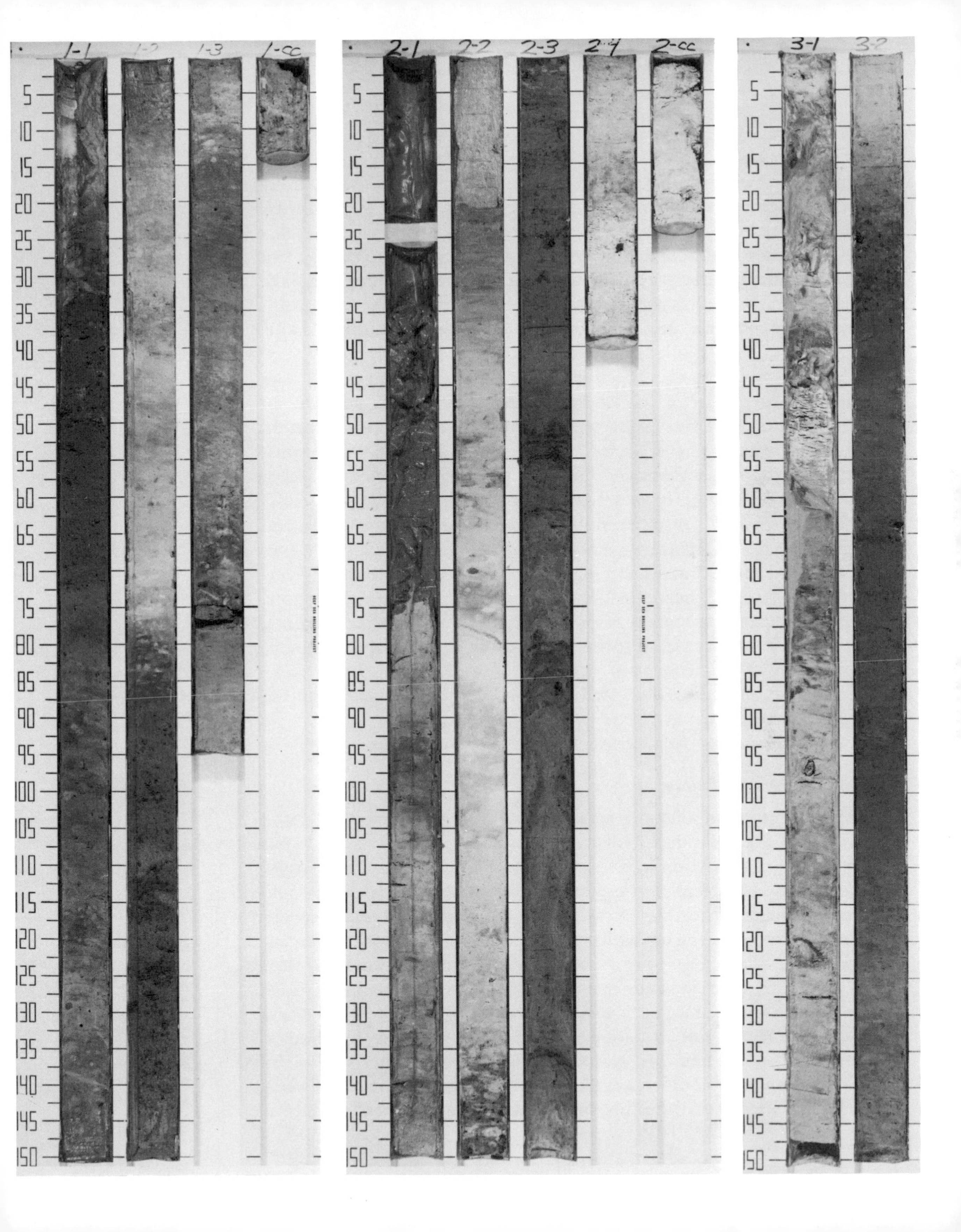
1-1
1-2
1-3
1-CC
2-1
2-2
2-3
2-4
2-CC
3-1
3-2

Earth's crust. Strata lying deep underground reflect the waves back to detectors at the surface. Conspicuous echoes correspond to occasions when the sea level fell, interrupting in many places the processes whereby sedimentary layers grew thicker under water. While keeping their raw data secure from competitors' eyes, oil-company scientists (initially Exxon's) issued chronologies of major sea-level changes, for the benefit of their academic colleagues. A bewildering number of quick and large drops in level have happened during the past 200 million years, unrelated to any obvious change in the Earth's geological regime. In some cases they coincided with notable changes among living things.

The only plausible hypothesis is geologically unpalatable: that long before the current series of ice ages began, many freak episodes of wholesale icing occurred. These would lower the sea level by locking up water in great ice sheets on land. If so, they struck suddenly in the midst of otherwise benign climatic conditions, and surreptitiously too. Except for certain well-known ice ages in the more remote geological past, geologists have detected few other hints of such episodes. The hypothetical quick-freeze events, if real, would have had far-reaching effects on life, on land, and also on the continental shelves, drained disastrously whenever the sea fell. Researchers are looking for possible connections between the sea-level falls, magnetic reversals, and cosmic impacts. These may all be wrapped together in a package of geological gelignite, and until the riddle of the seismic echoes is solved, many bets are off. The students of the fossil record of life's evolution may be due for another headache.

Museums versus Molecules

Chemical analysis of a famous fossil called Piltdown Man showed it to be a forgery, artfully composed of a human skull and an ape's jaw. That was in 1955, and with hindsight it was a capricious signal that the quiet life of the paleontological museums would never be quite the same again. A quarter of a century later, the professional fossil hunters were decidedly grouchy. The physicists told them the continents had moved, so they had to start reassessing the evolution of life on shifting platforms. The physicists went on to say that cosmic impacts caused mass extinctions in the fossil record, and for many fossil experts that was too much to swallow. Meanwhile, squabbles broke out within their own ranks, about how fossil organisms should be classified and arranged on the museum shelves; also about whether the evolution of new species was a gradual or a sudden process. The latter dispute even acquired a political tinge, when suggestions that evolution proceeded by fits and starts ("punctuated equilibria") were said to be Marxist in spirit. But most hateful of all was molecular evolution.

Whatever else was going on around them, paleontologists thought they could claim Darwin's idea of evolution by natural selection as their very own scientific principle, equal in grandeur and explanatory power to anything that physicists might have to offer. By this principle, living things evolve because those that best suit their circumstances leave the most surviving offspring. In the early 1970s inanimate self-reproducing molecules turned out to be capable of evolving in a test tube; natural selection became a physical law and Charles Darwin an honorary physicist. Moreover, the rise of molecular biology, marked in 1953 by the physicists' discovery of the nature of the genetic material, DNA, led within twenty years to wide-ranging studies of the evolution of biological molecules. These traced the story of life without resort to fossils, by comparing similar molecules in different living orgnanisms.

The key molecules of life, nucleic acids and proteins, consist of chains of various subunits arranged in a strict order, like letters in a coded message. Genes are nucleic-acid molecules that specify the manufacture of protein molecules. The sequence of subunits has been subject to mutations, or misprints, and these accumulated during the long history of life on Earth in the bacteria, plants, or animals that owned the molecules. Between closely related living organisms (say, a mouse and a rat) the molecular differences are few, but they are more numerous between distant relatives (a mouse and a fish). From molecular data computers can construct family trees that pinpoint the most recent common ancestor of any pair of species. Furthermore, the evolution of the molecules is a story in its own right, at least as important as the evolution of shapes of shells and bones.

Hemoglobin, for example, the red-colored oxygen-carrying pigment in human blood, consists of two kinds of molecular chains, called alpha and beta. Both of them are descended from another active molecule called cytochrome *c*, via an ancestral molecule that gave rise to myoglobin, the pigment of red meat, and also to leghemoglobin, which occurs in the root nodules of soybeans and other plants that fix nitrogen from the air. Making up proteins from pieces of existing ones has been an important process, and so has gene doubling, a mechanism that provides complete spare copies of genes. These are free to evolve without fatal consequences, and may thereby come up with quite new tricks, from the modification of old genes. About 1000 million years ago, gene doubling produced myoglobin and the ancestor of a peculiar hemo-

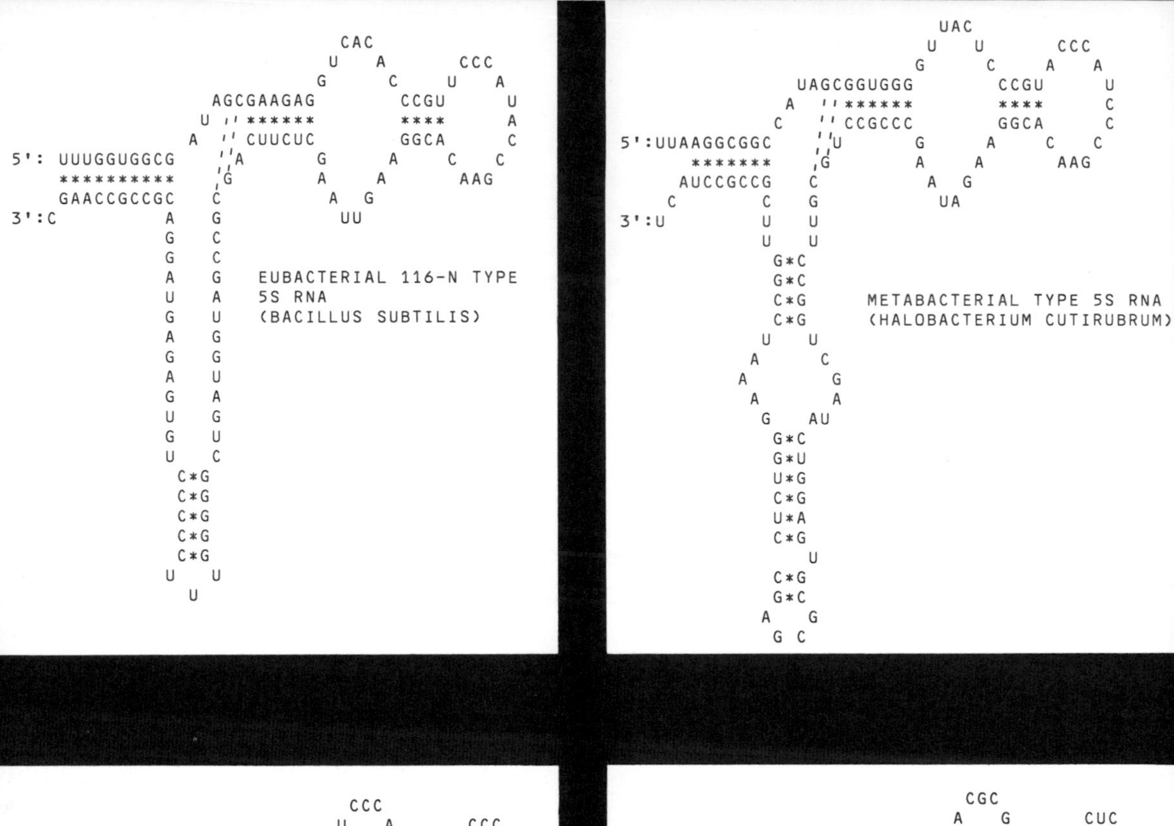
EUBACTERIAL 116-N TYPE
5S RNA
(BACILLUS SUBTILIS)
METABACTERIAL TYPE 5S RNA
(HALOBACTERIUM CUTIRUBRUM)

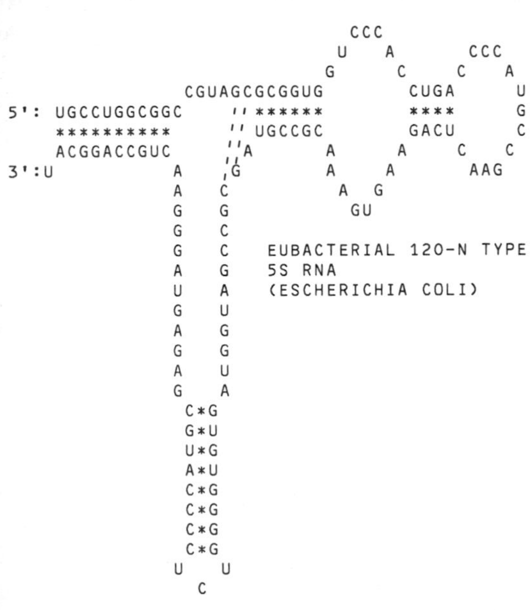
EUBACTERIAL 120-N TYPE
5S RNA
(ESCHERICHIA COLI)

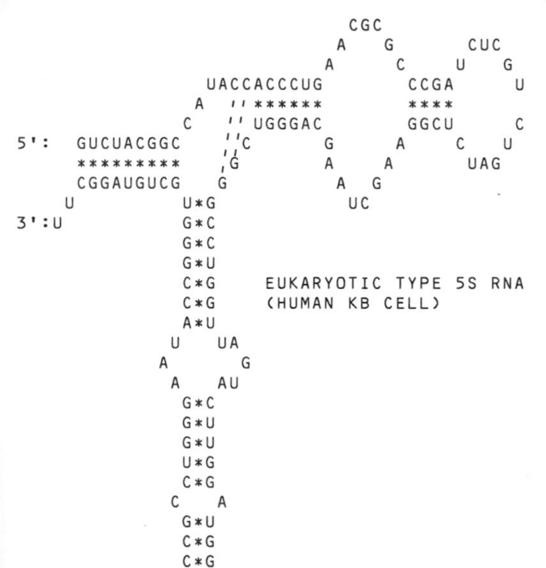
EUKARYOTIC TYPE 5S RNA
(HUMAN KB CELL)

globin found in jawless fishes. Further gene doubling in vertebrates, about 500 million years ago, made possible the evolution of the doubled (alpha plus beta) chains in modern hemoglobin molecules. Gamma hemoglobin, used by the human fetus to snatch oxygen from the mother's blood, evolved by stages from beta hemaglobin, in mammals.

The molecular evolutionist selects the material appropriate for the task in hand: some molecules that evolved extremely slowly can trace back even to the common ancestor of human beings and present-day bacteria, while others changed rapidly enough to illuminate human evolution over the past few thousand years. Molecular differences can be converted into a timescale by putting in the dates of one or two evolutionary events known from the fossil record, and assuming a steady "tick" of the molecular clocks. Apart from this calibration, molecular dating has little need for the fossils in museums, and it is particularly useful when fossils are scarce, for example in the early history of life, and in the evolution of those organisms, including human beings, which left few traces.

To say so is to invite the wrath of those biologists who regard the very idea of a molecular clock as an attack on cherished notions about how evolution proceeds. These biologists are "selectionists," who see themselves as custodians of a great Darwinian truth, that changes in organisms occur as a result of a positive pressure of natural selection encouraging the appearance and refinement of features that continually improve the organisms. Molecular dating assumes, on the contrary, that genetic mutations survive purely by chance, at a rate that is constant for a given molecule, without regard to the outward forms and adventures of organisms carrying the genes. That indifference is possible only if most evolutionary changes that survive are neutral in their effects, neither benefiting nor harming the organisms in which they occur.

The neutral theory of molecular evolution, developed particularly by Motoo Kimura in Japan from 1968 onward, provoked a prolonged battle between neutralists and selectionists. At times, the neutralists were treated like heretics in a theological dispute. All genetic mutations occur by chance, most mutations are harmful, and harmful mutations are soon eliminated by natural selection because lineages that possess them tend to die out. Selectionists and neutralists agree on these points. The difference of opinion concerns mutations that are not harmful, and may therefore survive. According to selectionists, these are mainly beneficial in effect, while neutralists say they are mainly neutral. If positive selection were at work, one might expect to see high rates of change when a novel molecule—a new variant of hemoglobin, for

Computer-generated diagrams of snake-like molecules illuminate the early history of life on Earth. When analyzed by molecular biologists in Japan, similar materials (5S ribonucleic acid) show kinship and contrasts in different living organisms. Human beings (lower right) are descendants of the metabacteria (upper right), peculiar microorganisms that split from the oldest bacterial lineage (upper left) about 2500 million years ago. The class of bacteria characterized as "gram-negative" (lower left) diverged from their bacterial relatives about 1800 million years ago. The letters A, C, G, and U refer to chemical units composing the molecules.

example—first appeared, and there was scope for refining it, to make it better suited to its task. The history of hemoglobin is said by some selectionists to show such accelerations, but the neutralists firmly deny them.

The arguments about the neutral theory are part of the general ferment of ideas concerning evolution stirred by recent research. The dearth of evidence in the fossil record for any transitions between one species and the next casts doubt on the idea of continuous refinement in evolution. It inspires the talk of punctuated equilibria, in which the origin of species is a sudden rather than a gradual process. And further discoveries about the genes provoke fresh thinking about evolution.

Genes are not mere computer tapes but participants in a micro-ecological drama within the cell. Careful reading of the sequences of chemical units in the genetic material of DNA reveals that genes are organized, or disorganized, into fragments interspersed with strands of other DNA. There are, indeed, so many dead "junk" genes and multiple maverick "selfish" genes that only a small part of the DNA in higher organisms actually codes for proteins, and the genes and pseudo-genes compete for survival. The DNA parcels illuminate the processes by which genes are switched on or off according to the needs of their owners. They also suggest mechanisms of large-scale evolutionary change. According to an emerging theory of "molecular drive," due to Gabriel Dover, the chief event in the creation of a new species is a rearrangement of genetic material within a population of organisms that have become separated from their relatives. One consequence is that any attempt to reunite with the relatives results in sterile matings. Furthermore, a change in the pathways of growth and development quickly produces a conspicuous change in the appearance of the new species, as compared with the old. Molecular drive thus fits comfortably with the idea of punctuated equilibria.

The processes of evolution and the origin of species are being illuminated as never before, and if the story is evidently not quite as Darwin had it, it is worth remembering that Darwin did not even know about genes, never mind the genetic code or the intricate events of molecular evolution. To treat Darwin's writings, brilliant though they still are, as holy writ is as optimistic as expecting Newton's *Principia* to account for the behavior of newly discovered subatomic particles. Selectionists made loyal efforts to reconcile modern genetics with their understanding of Darwin's ideas, but the trend of the discoveries was against them. In the outcome, natural selection seems to be on the whole conservative rather than creative, preserving efficient species with little change, and safeguarding their vital molecules. While selectionists see evolutionary changes as evidence of natural selection working like a creative sculptor, neutralists regard the same events as a relaxation of selection, with genes changing when, as it were, the sculptor has turned his back. Selection comes back into play once a molecule is "adaptive"—when it already has the features that make it worth preserving.

Enough has been said to caution anyone against lightly adopting molecular dating without pondering the evidence, and its theoretical, not to say philosophical, implications. Yet from a bare idea that was rank heresy when it first appeared, and was dismissed by many as self-evidently absurd, the neutral theory of molecular evolution has gathered remarkable strength and weathered many attempts to falsify it. The same can be said for the related proposition that molecular clocks may often be a better guide to the timing of events in evolution than the fossil record.

Fossil hunters spared no effort to try to prove that molecular dating was unreliable. Their chief attack was against the assumption that the rate of change of genes was largely independent of the character and circumstances of the organisms possessing them. The astonishing fact is that the assumption seems to be roughly correct, except in special circumstances when there is a physiological reason why the rate of evolution should vary—if, for example, a molecule changes its function. Among all the lineages of mammals, from bats to whales, over the past 100 million years, the discrepancies in the rate of change of molecules used in evolutionary studies are no greater than one would expect to arise by chance, in a steady but chance-driven process of mutation and survival. Such evidence is reassuring to those who use molecular differences between bacteria to probe evolutionary events of billions of years ago, at the dawn of life.

Most controversial of all the inferences of the molecular evolutionists was that of Victor Sarich and Allan Wilson, who asserted in 1967 that humans and chimps shared a common ancestor as recently as 4 or 5 million years ago. It seemed ludicrous; the paleontologists "knew" from their fossils that the lineages diverged 15 or 20 million years ago. But subsequent fossil discoveries all favored the much shorter timescale, and one supposedly prehuman animal turned out to have been a forerunner of the orangutan. An outstanding fossil hunter, Richard Leakey of the National Museums of Kenya, conceded in 1982, "I think the molecular people are closer to the truth than we've ever given them credit for."

How to be a Martian

Many accounts of Earth history deal with geological and evolutionary phases, and come to modern humans on the last page or two; equally, books on human prehistory or history may simply nod at evolutionary precursors for a few paragraphs, before getting down to the serious business of pots, kings, and battles. This division of attention reflects a change in the nature of the narrative. Humans are a distinctive life form: in their labors they are in effect a new geological and evolutionary force; in their chatter and information processing they rebel against entropy, leaving the world more ordered and complex than they found it. The nature of change alters, and becomes more purposeful. To join the two parts of the story as seamlessly as he can, the chronographer may decide to say as little as possible about pots, kings, and battles.

It is a matter of sitting, in effect, on Mars. By distancing himself from fascinating details of history, and from ethnic preoccupations, the observer may be better able to see the main currents in the life of our species. No reliable overarching theory, equivalent to continental drift or biological evolution, provides a ready framework for the human phase. Instead, there is an abundance of data, especially in the period of written documents, all of them of interest to one historian or another, and freighted with tales of endeavor and villainy of a kind scarcely available to archeologists, never mind paleontologists. The chronographer, though, may prefer to preserve the dispassionate tone of the earlier narrative and, from a Martian vantage point, look for the recent analogs of prehistoric population movements and adaptations.

Physicists, as mentioned earlier, began by treating the Earth as a point of mass M, and built up its features from there—a very different approach from looking out of one's window at the unemployed youngsters or the passing fire engine. If the Martian view seems less than human, that may be a requirement for seeing the planet's human crew in the same unsentimental frame as the mountains and forests. The good Martian scents transience, accepts that the future scene on the Earth may be very different from those of the present or the past, and has no special admiration for those who, like Jonathan Swift's Lilliputians, suffer death rather than break their eggs at the smaller end.

In the manner of Swift in *Gulliver's Travels*, let us cultivate detachment, perched on Mars with an interplanetary zoom lens that can pick out details on the Earth, right down to blades of grass, or else pull back to show the distant planet as a pebble in the universe. Shall we investigate that object, somewhat larger than Mars and quite different in color?

First to catch the eye are the white clouds of condensed water vapor swirling in the wind. They are products of the second striking feature of the Earth, the pools of water filling the ocean basins. Not everywhere: bright dollops of ice bury the southern continent, and glacial relics picked out by the zoom lens tell of ice recently melted from vast areas of the northern lands. We are observing the Earth during a comparatively warm interval.

At first glance the outlines of the dry surfaces of the planet are meaningless shapes, but then we discern how they once fitted together. A supercontinent has exploded and scattered. Where pieces are bulldozing over the floor of old oceans, volcanoes tower. Where fragments have collided afresh, the wreckage takes the form of mountain ranges. Plainly a very active planet, this, and the lineaments of ancient, worn-down mountains hint at a dance of land masses older than the last supercontinent.

Large green patches turn out to be natural solar-energy stations, where immobile forms of life, the plants, spread leaves to catch the energy of sunlight. (There is nothing of the sort on Mars.) Some green areas are bounded by improbable straight lines and right angles; particularly eye-catching are ribbons of green patchwork that follow the rivers, even through arid zones. Evidently certain cunning plants have taken over the choicest parts of the Earth's surface. A quick survey of plants confirms the suspicion that their apparent passivity is deceptive: they manipulate the animals. Flowers have recruited battalions of small six-legged and large two-legged animals to serve them. The reproductive self-interest of the plants requires that animals find their flowers attractive, their nectar and fruit tasty. Beside a bed of particularly elegant flowers, a biped stoops in a humble posture, carefully removing any plants except those that have charmed him to their service. And other bipeds swarm in large numbers around the green rectangles in the river valleys.

The master of the planet Earth is now identifiable: it is grass. (Lew Kowarski, a French physicist, hypothesized as much twenty years ago.) Grass first evolved its biped slaves, the human beings, as hunters on the grasslands, and then lured them into the cultivation of special grasses, including wheat, rice, and maize. With unstinted effort, the humans cleared trees and other plants out of the way, and they irrigated the kempt fields for the benefit of grass. To magnify the energy available to grass, beyond the immediate resources of sunlight and enslaved human muscle, farmers dutifully brought in oxen and manure. The most favored species—especially the cereals—were transplanted between the continents. It was greatly to the advantage of grass that humans multiplied their numbers without limit, and eventually recruited the energy of fossil fuels to pampering the grass with tractors and fertilizers. Now the grass waits patiently to be transplanted into outer space.

Watching grass waiting becomes a little monotonous even for a well-trained Martian, who may be better employed observing the humans at harvest time. They accept the grasses' bribes of food, but of course they put seeds aside—which is the object from the grasses' point of view. The farmers ship the food to markets and cities, and in exchange they receive trinkets and commands from other humans. A game is in progress among the slaves of grass, in which half of them manage to avoid having to tend the grass themselves, even though they cannot live without it. Inspection reveals various ways of succeeding in the game, for example by trinket making: a person can induce others to look after his share of the grass in exchange for pots, shirts, or automobiles. Or he can offer entertainment, teaching, doctoring, or moneylending. Another ploy is to invent machines and chemicals that groom the grass with relatively little human labor. That is of no disadvantage to the grass, as long as the human population increases and, with it, the total energy devoted to the multiplication of grass.

Free gifts come to those who manage the spiritual life of the community, and the priests redistribute some of the offerings to the neediest. But the most successful way to avoid muddying one's feet in the fields is by force. Individual earthlings declare themselves landowners or rulers of large areas of ground occupied by the grass, and demand that the farmers give a substantial part of their harvest to them. These involuntary gifts are called rents and taxes, and although violence is seldom needed to recover them, the owners and rulers can resort if need be to men with guns to enforce their claims. Of course, the men who make and carry the guns have to be fed too, along with their wives and families, and also a retinue of bureaucrats and lawyers, all supported by taxes.

This is called government. Its nature is concealed in most places because humans have short memories and prefer a quiet life. Manufacturers, traders, teachers, and others support the system because it provides a framework in which they, too, can go on eating without growing any food themselves. Only where farmers dispute the claims of landowners or rulers does the nature of the game become apparent, in bloody battles. To minimize provocation, and also to keep people alive and at work, rents and taxes are typically fixed a little below the level at which desperation sets in. As a Martian-like historian, William H.

Camel country at China's western gateway, the Kansu Corridor, appears in an image obtained by radar from a Space Shuttle. Geologists use variations in the radar brightness, due to surface roughness, to delineate large-scale features, including the immense fans of alluvium on the left and right of this picture. The valley slanting toward the lower right marks a major fault between two blocks of the Earth's crust. This district lies at the western extremity of the Great Wall of China, also visible by radar, and ancient roads across the adjacent desert converged at An-hsi, at the bottom of the picture. (Hunting/JPL/NASA.)

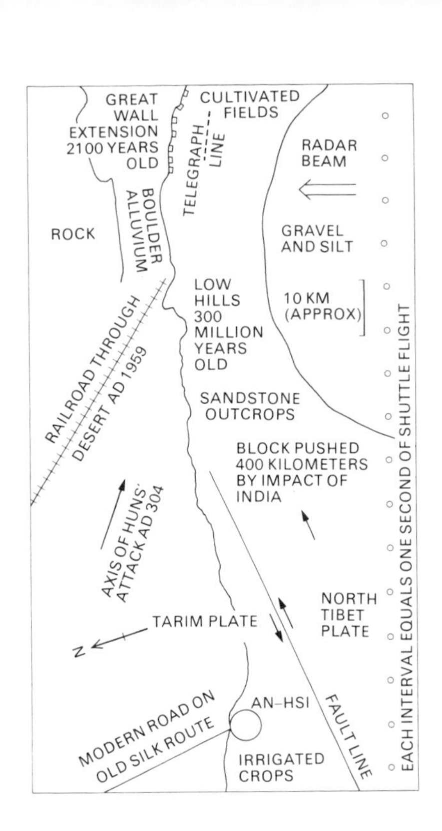

GREAT WALL EXTENSION 2100 YEARS OLD
CULTIVATED FIELDS
TELEGRAPH LINE
RADAR BEAM
ROCK
BOULDER ALLUVIUM
GRAVEL AND SILT
LOW HILLS 300 MILLION YEARS OLD
10 KM (APPROX)
RAILROAD THROUGH DESERT AD 1959
SANDSTONE OUTCROPS
BLOCK PUSHED 400 KILOMETERS BY IMPACT OF INDIA
AXIS OF HUNS' ATTACK AD 304
NORTH TIBET PLATE
TARIM PLATE
N
EACH INTERVAL EQUALS ONE SECOND OF SHUTTLE FLIGHT
AN-HSI
MODERN ROAD ON OLD SILK ROUTE
IRRIGATED CROPS
FAULT LINE

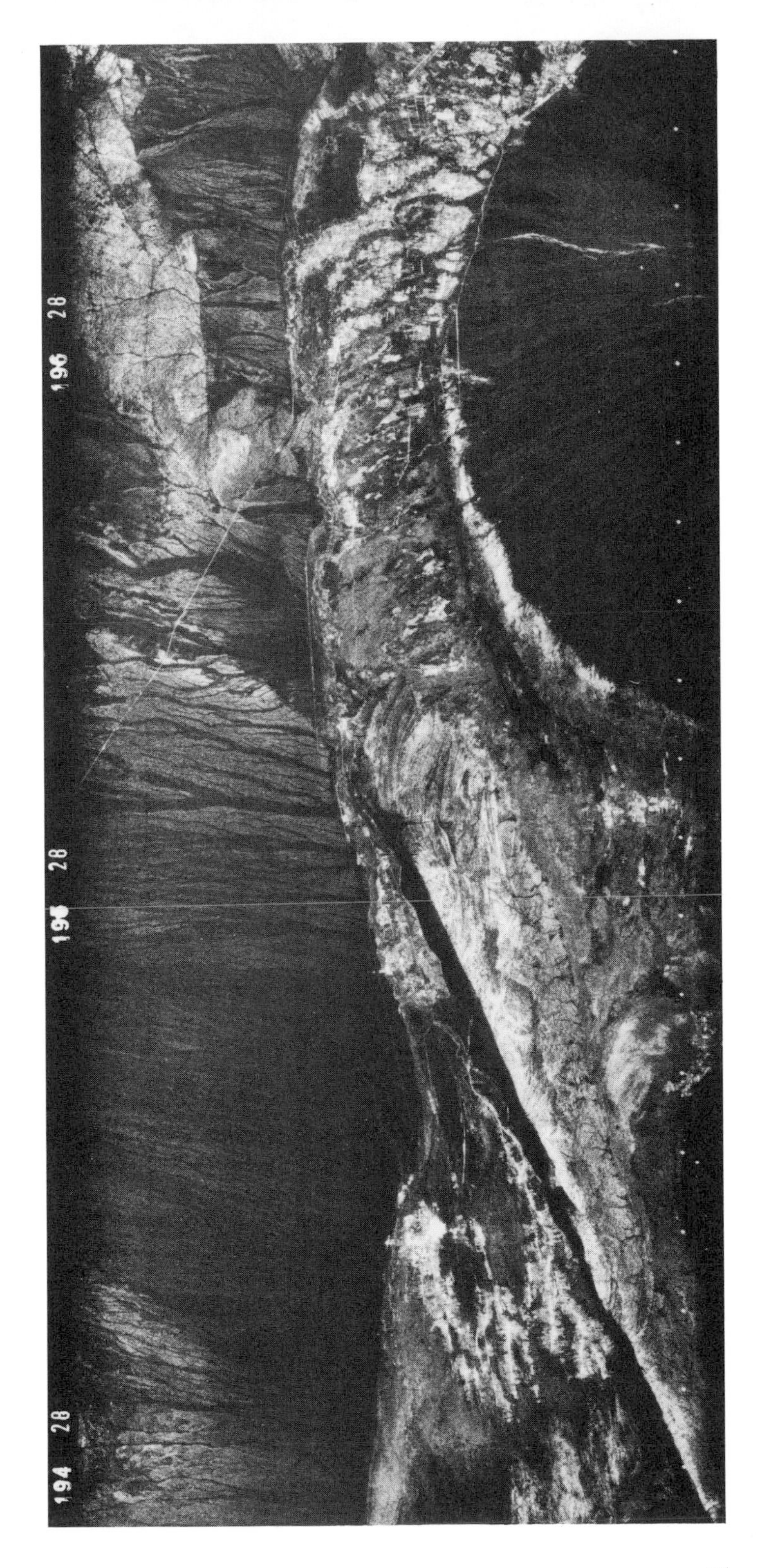

McNeill, has observed, it is the principle of all successful parasites. Malaria never quite exterminates its host species, nor the wise master his servants.

In the game of government, called politics, the aim is to secure control of tax revenues. Typically this is accomplished by force—by murder, revolution, military coup, or conquest. Individuals interested in the power, the treasure, and the control of the men with guns will say almost anything that helps them to acquire these prizes. The result is an unconscionable amount of rhetoric about rights of royalty or rights of workers, while the wish retrospectively to justify successful ploys in the power game colors the humans' accounts of their history. Critiques of the game, in the treatises on political philosophy from Plato, Machiavelli, Locke, Marx, and the rest, seem at the distance of Mars to be like so much paint on the guns.

The observant Martian sees local variations among the humans, whom he distinguishes according to where they live, what languages they speak, how they make ends meet, and whether they look well fed and healthy, or skinny and sick. The coloration of their skins is evidently an adaptation to the altitude of the sun, the amount of cloud, and other environmental factors. Nevertheless, it figures in the Martian first law of human biology: quality of nutrition is inversely proportional to skin pigmentation. At one end of the scale are half a billion pink-skinned people, soi-disant whites, who push food away, trying to control their waistlines, and at the other half a billion dark-skinned people who are seriously malnourished; exceptions in the form of fat blacks and scrawny whites show that this is not a genetic but a social phenomenon. Every two seconds a small child dies of poverty, and from Mars the planet Earth is just South Africa written large. A task for any Martian historian must be to explain how this state of affairs came about.

Food is prominent among the cargoes of ships seen moving in a stately fashion across the oceans, so evidently effort is not lacking to redress imbalances in nutrition, at least for those who can offer something in exchange. The most generous source of surplus grain turns out to be North America, where the care of grass is a slick, energy-intensive business, more highly mechanized than many industries. The conspicuous recipients of grain are not starving blacks but whites in northern Eurasia, in particular the Russians, who occupy the most spacious territories on Earth but seem incapable of feeding themselves as well as they would like. What is left over, when the northern whites have eaten their fill, nourishes the growth of tropical cities.

The wakes of tankers carrying crude oil make another pattern. The

This 35-horsepower harvester groomed the grasslands of Washington State, during a large expansion of territory devoted to cereal production in North America, about a hundred years ago.

whites, and not least the grain-rich Americans, have a great thirst for inanimate energy, slaked from oil fields that lie scattered for the most part around the tropics and subtropics, and especially in southwestern Asia. A chain then becomes apparent: Arabs send oil to Americans, who turn it into food that goes back to the Old World to save people from famine and enable Russians to eat a little meat. Masterly cooperation or worldwide fecklessness? The Martian notices the flash of guns near the southwestern Asian oil fields. He watches rainclouds drifting irresolutely across the North American grainlands, and wonders what fluctuations in climate might break the chain. With sonar vision he observes submarines ready to sink the life-giving ships, and others equipped with nuclear-tipped missiles for killing Americans and Russians in their homelands on a grand scale. Evidently, between the feeders and the fed, relations are less friendly than one might expect.

The puzzled Martian studies banners, badges, and cheering crowds, and soon discerns a characteristic of human beings: their instantaneous loyalty to any group to which they find themselves assigned. It is a means of finding an identity, enjoying social warmth, and joining in creative ventures. In scale, the groups can be anything between a children's gang and an intercontinental alliance. But this behavior also carries the seeds of conflict, because prizing one's own group means feeling in some sense superior to other groups. Political and spiritual leaders have evidently used group loyalty to organize the earthlings into rival nation-states and religious alignments. They have also invented a habit of war; this too cries out for retrospective investigation.

Shared beliefs help to knit a group together. These can range from a half-hearted hope that a certain football team may win its game to a conviction that one's ethnic group is especially favored by God. The substance of beliefs often seems arbitrary; what matters is that they should be distinctive, and shared by the group. History and beliefs, churned together by word magic, create the traditions and legends that underpin the culture of the group. These influences begin at the mother's knee, and scientists and scholars are not immune from them.

Many human histories appear biased when seen from Mars. Legitimate interest in individual groups and nations too easily becomes legitimation, justifying, for instance, dominance by the whites, on the basis of cultural descent from the Sumerians, Egyptians, Jews, Greeks, Romans, and the Europeans of the Enlightenment. Events and innovations that do not fall on that preselected world line are often devalued: stone structures appeared in Brittany too early to fit on it, and a noted archeologist writing in 1976 dismissed them as "megalomania," while she saved the word "magnificent" for the temples of Sumeria. In the latest edition of the *Encyclopaedia Britannica*, thirty pages devoted to the history of technology are vitiated by a disregard for work of the biologist-historian Joseph Needham showing that several key inventions originated in China and not in Europe as formerly believed. The same article gives special emphasis to innovations at the peak of Britain's industrial revolution and empire.

"Imperial" is nowadays a dirty word, but "progress" is not, and it is used in the manner of Dr. Pangloss to embrace whatever has led up to the prevailing situation, or has inspired current programs of revolution or reform. Karl Marx's error in imagining that the revolution would come in an industrialized country, rather than in agrarian Russia and China, was a consequence of the belief in step-by-step progress, via the latest fashions. Even the most skeptical historians seem barely able to distance themselves from the assumptions of their culture.

To give a minor example, everyone takes it for granted that reading and writing are blessings. In practice, illiterate people have achieved high levels of sophistication, for instance in reading the stars well enough to navigate safely to and fro across the Pacific. Skill in archery may have been as important as writing in shaping the course of history, and as for cultural progress, there has been no very obvious improvement on the oral literature transcribed by Homer. Modern science explicitly rebelled against the intellectual tyranny of books: the motto adopted by the Royal Society of London in 1663 was *Nullius in verba*, or "don't take anyone's word for it."

Writing evolved for tax gathering, and the Mesopotamian clay envelopes of 5600 years ago must have been even more alarming than the tax collectors' paper envelopes coming through the mail nowadays, because they depicted bound prisoners, presumably defaulters. Schools promptly created the meritocracy that, to this day, reckons the worth of young citizens by their facility in the cumbersome information technology displayed on the wafer of wood pulp now in your hands. Most humans have lived and died unable to read or write, and some bright individuals are dyslexic. New technologies may soon make the art as outmoded as oarsmanship for galleys. In short, the emphasis laid upon literacy by scholars who earn their living with written words appears self-serving.

The most loaded word of all is "civilization." Literally it has to do with populations concentrated in cities; technically, according to one modern-minded archeologist, a culture is called civilized if it scores two out of three on a checklist of (1) towns, (2) monumental buildings, and

(3) writing. But in scholarly as in common usage, civilization frequently means all that is best and desirable in the human condition. To speak of "the end of civilization as we know it" is not a neutral comment, and the fall of Rome is supposed to evoke horror, even among descendants of the Germanic peoples who gained by it. Tourists who wander among relics of ancient regimes are invited to admire grandiose monuments decorated with scenes of slaughter and enslavement, and take them as signs of the "greatness" of the kings concerned. Scholars from the political left and right see all this as prologue and progress toward the modern industrial nation-state, which awards grants for professors of archeology and history. If giant temples and tombs, and the successions of kings and battles, are steps along the way, so be it, and they are in any case much easier to identify than the detritus of "softer" and less propagandist cultures.

Achievements of the white people's civilization include scientific discoveries that this book in part celebrates. Yet its two leading nation-states, the United States and U.S.S.R., have recently suffered defeats in war by rural peoples of Vietnam and Afghanistan. Industrial economies squander nonrenewable resources, and several might founder if a single waterway—the Strait of Hormuz—were closed to supertankers carrying oil. At home, young citizens are often estranged from the systems and traditions of their societies. And with the whites living from minute to minute on the brink of nuclear annihilation, the Martian observer must at least entertain the possibility that they may soon go the way of Babylon and Monte Albán. At least he would not assume as readily as many earthlings do that when humans fly to the stars they will have white skins.

Restraint in the use of proper names is an aid to Martian detachment in speaking of human affairs. Not to name individuals, whether they be Julius Caesar or Albert Einstein, smooths the transition to documentary history from the prehistoric phase, in which the innovators and leaders are anonymous. It also avoids the arbitrariness of immortality that came from having one's name written down and repeated by historians, and it redresses regional imbalances. Many Western readers who have been schooled to think of Julius Caesar as a person of some importance on the world stage may not have heard of his grander counterpart Shih Huang-ti. Another device is to minimize the use of geographic names for nations, and to refer to "the Chinese" rather than to China. While admittedly making no distinction between the leaders and the led, this policy copes with migrations and expansions of populations, damps the poetic resonances of nation names, and cuts out reifications of the form "China attacked India," which seem like descriptions of continental drift.

A Martian chronographer has to be prepared to be interested in the minutiae of human life, such as the marriage customs of little-known tribes or the changing fortunes of the herring fishery, and be equally ready to skim over tracts of knowledge that many earthlings regard as highly important, concerning the Persian emperors, say, or World War II. At any time, several wars are in progress, making them as commonplace as the harvests of grass, while emperors are to human history as fossils of predatory dinosaurs are to evolution: when they first appear and eventually disappear is of great interest, but their identities and antics in the interim less so.

Among all the outpourings of human chroniclers, basic information is often the hardest to come by—about methods of grass tending, for example, or the booms and slumps in human populations. The usefulness of relatively high-grade data on the medieval Chinese population is limited by the absence of comparable information from other parts of the planet. The provincialism of many historians ensures that balanced global accounts of any topic are scarce. Nor can one expect to find fully warranted, uncontradicted information on any subject whatsoever.

After distancing himself to clear his head, the would-be Martian has to reconnect himself with the fun and grief of human existence. Coming down for breakfast on the planet Earth, he should note that his steel knife was invented by metallurgists on the island of Cyprus, three thousand years ago. The maize in the cornflakes traces back to mutant grass of Mexico, while the wheat in the toast is of southwestern Asian origin. The chicken eggs and citrus juice came first from Indochina, the coffee from Ethiopia via Arabia, and the sugar from New Guinea. The milk, butter, and cream are biochemical fossils from a six-thousand-year-old revolution of cow-and-plow that spread productivity and cruelty around the world. And as the Martian reads of some of the consequences in his newspaper, printed in Etruscan characters on Chinese tissue, he should have a sharp sense that "human life" is exactly that: a peculiar sort of biological system that the biological sciences can illuminate.

The Biology of History

In the era covered by documentary history, which begins with the invention of writing more than five thousand years ago, radiocarbon dating gradually loses its importance as the counting of years in historical documents becomes more widespread and reliable, and radiocarbon less precise than other sources of dates. In nonliterate regions, such as North America and the South Pacific before the arrival of the Europeans, radiocarbon continues to cast its special light on otherwise obscure events, but radiometric dates have caused no shocks in the mainstream of documentary history, which at first sight may seem safe from scientific molestation. Biology takes up the challenge before the physics of radioactivity fades out.

Where the crop plants and domesticated animals originated in the wild, how they changed under human management, and with what cultural and economic consequences—these are questions that nowadays provoke intricate and rewarding studies. When archeologists speak, for example, of crops being transferred from one area to another, they now take proper account of the need to adapt the growing season of the plants in question to the new environment. Demography, the study of human numbers, draws upon knowledge of human ecology and the biology of reproduction. Given present concerns about an exploding world population, it is useful to know that natural and contrived means of regulating the birth rate have always been a feature of human life; a salient discovery among hunter-gatherer women concerns the natural loss of fertility during prolonged breast-feeding.

Disease and immunity to disease provide a particularly striking commentary on history. Epidemics are the subject of an exact science that draws on a knowledge of pathogens, their animal reservoirs, and human immunological or genetic resistance to disease. Microbiology and immunology are firmly integrated with molecular biology, while mathematical theories tell how viruses require a certain population density in humans to take root and spread. Africa is identified as a continent supercharged with disease, carried by mosquitoes, tsetse flies, and water snails, where cultivation has increased the suffering. Elsewhere, in the light of epidemiology, cities appear as disease factories, which through most of history lured ambitious young people from the countryside and quickly killed many of them, thus making room for more newcomers. The pockmarked survivors were, though, very hardy, and their acquired immunity to disease gave them a military advantage over rural peoples, among whom the diseases were not endemic and who could be laid low by well-directed sneezes. Disease, not military skill, was often the arbiter in war.

Biomedical knowledge made possible a major clarification by William H. McNeill, already mentioned as a rare Martian among historians. In 1976 he reasoned that what was true of cities and farms applied also to kingdoms and continents: the rise of empires could be

No accident buried this colossal Olmec head that was disinterred at San Lorenzo, Mexico, in 1946. It was toppled on purpose 2900 years ago, by intruders rebelling against an authoritarian regime. The Olmec elite, arising 3250 years ago, employed fine sculptors and also developed trading and ceremonial systems, including a jaguar cult. These were imitated by later hierarchies in Middle America.

related to the export of old diseases, and their fall to the import of new or resurgent ones. A central puzzle of modern history, the defeat of fierce Amerindians by a handful of Europeans, was explained as a matter of unaccustomed viruses. What price glory, when it derives from the unplanned exploitation of viral nucleic acids?

Studies of the human era are experiencing foreshocks of another earthquake, the force of which comes from molecular biology. The earliest tremors began many years ago in work on blood groups, the inherited characteristics that determine whether or not blood can be safely donated from one human to another. Blood-group tests of peoples in different parts of the world reveal systematic differences. Among people living on one side of the Chimbu River in New Guinea, blood group B is twice as common as it is among people on the other side of the river, and the genetic boundary shows where hunter-gatherers (high B) halted an invasion by gardeners (low B) about five thousand years ago. This is a simple instance of how genetic analysis can help to recapture unrecorded events.

Nowadays there is no limit to the range of human genes that can be studied. Variant protein molecules, manufactured from variant genes, can be sorted simply by racing them, spurred by an electric voltage (physics again) along tracks made of jelly. Refined methods of studying the genes directly, by reading the coded sequences of the DNA, are also available. Human genetics is ripe for comprehensive application in history. The genes directly relate a person to his parents, his local gene pool, and his remote ancestors. In some respects they may be more reliable than chronicles or archives, even when these exist; for example, the covert interbreeding of populations does not escape detection. But the promise of enlightenment about the history of human populations is unwelcome to many.

One of the lesser but not unimportant blights for which Hitler and other racists are responsible is that many scientists and historians have adopted a frightened silence on the subject of human genetic differences. Liberal etiquette nowadays calls for finely judged schizophrenia that is simultaneously blind to color, religion, and geographic origins, and yet ever watchful about the rights of ethnic minorities. In accounts of prehistory one can sometimes spot where the scholar has censored his words, and even his thoughts, to minimize any suggestion of migrations or conquests by identifiable racial groups. But political squeamishness soon becomes intellectual perversity, and no help at all in efforts to understand why some human populations, or sections within them, came to be prosperous or downtrodden, and why the economic bias in

An electron microscope unmasks a fateful pathogen, the measles virus, and reveals the string-like nucleic acid of its genes. The immunity commonly acquired in childhood makes the disease less deadly than it was when it swept the Chinese and Roman empires 1800 years ago, or when Europeans first communicated measles to the Amerindians in the sixteenth century AD. The black bar on the photograph is for scaling.

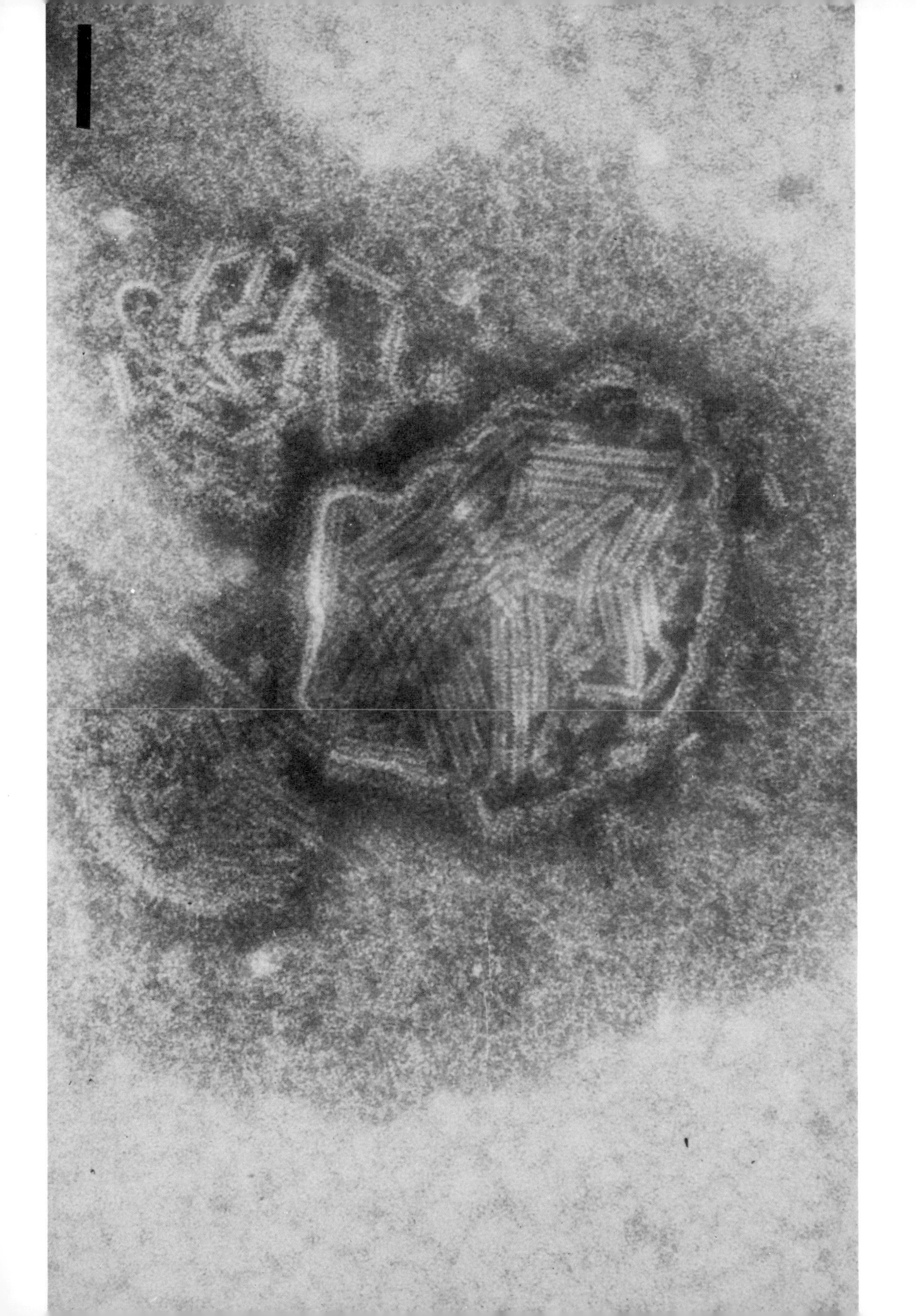

favor of white skins is visible from Mars. In any case, comparisons of gene frequencies are usually technical and nonemotive; agitators concerned about skin color or the shape of noses have not yet learned that a particular variant of a human acid-phosphatase gene is symptomatic of life in a cool climate.

More academic reasons for denying population movements arose in the aftermath of prehistory's radiocarbon trauma. When the new dating challenged the old idea that all good things diffused from southwestern Asia by showing that the innovative talents were at work in all parts of the world, the very idea of diffusion was discredited among some archeologists. Two quite distinct questions became thoroughly confused.

1. Was everything invented in Mesopotamia?
2. Do inventions tend to diffuse from their centers of origin, wherever they may be?

The old school wanted to say yes to both; young rebels were tempted to say no to both. Some innovations, for example the cultivation of crops, and writing, did indeed originate independently in more than one place. Equally, other inventions, for example the plow and the wheel, diffused rapidly from a single center of origin, so the correct answers seem to be no to Question 1 and yes to Question 2. As an open-minded archeologist complained:

> We are seeing the extreme reaction to an equally extreme past position. . . . Much present-day antidiffusionist scholarship seems to imply that nobody ever went anywhere during the Bronze Age. [James D. Muhly, *The Coming of the Age of Iron*, 1980]

One way for an invention to spread is for it to be taken to new localities by migrants. Documentary history is full of migrations, both peaceful and warlike, and in the spirit of Muhly's comment, one might add that to scorn similar movements in prehistory is to imply that people could not walk before they knew how to write.

In any case, biology delivers a clear-cut verdict: people moved. Genetic differences between populations are plainest for those that have lived for long periods in isolation, such as the Australian aborigines, the Arctic Eskimos, and the Khoisan of southern Africa. Most populations have been subject to migrations or to influxes of aliens, with consequent hybridization that makes analysis tricky—but ultimately rewarding, precisely because of its historical implications. For example, the dominant feature of European genetic geography is a set of genes that shows a gradation from the Balkans in the southeast to Scotland and Scandinavia in the northwest. The geneticist Luigi Cavalli-Sforza and his colleagues take it as evidence that the spread of farming from western

Genetic signatures of human beings appear when molecular fragments are separated by electrical means. Genes from the human cell components (mitochondria) have been cut up by means of an enzyme, producing fragments of different sizes. Applied to one end of a gel (the top, in this illustration), the molecules traveled through it, driven by a voltage applied for some hours between the two ends of the gel. The smallest molecules went fastest and therefore farthest. All had been made radioactive, so that they would reveal their final positions by fogging a photographic film. The pattern of fragments at the far left is the commonest for most of the human beings tested except the San Bushmen of southern Africa, among whom the second signature is most typical and the third and fifth also occur. The fourth pattern appears in a few people of Eurasian origin. (M. J. Johnson, et al.)

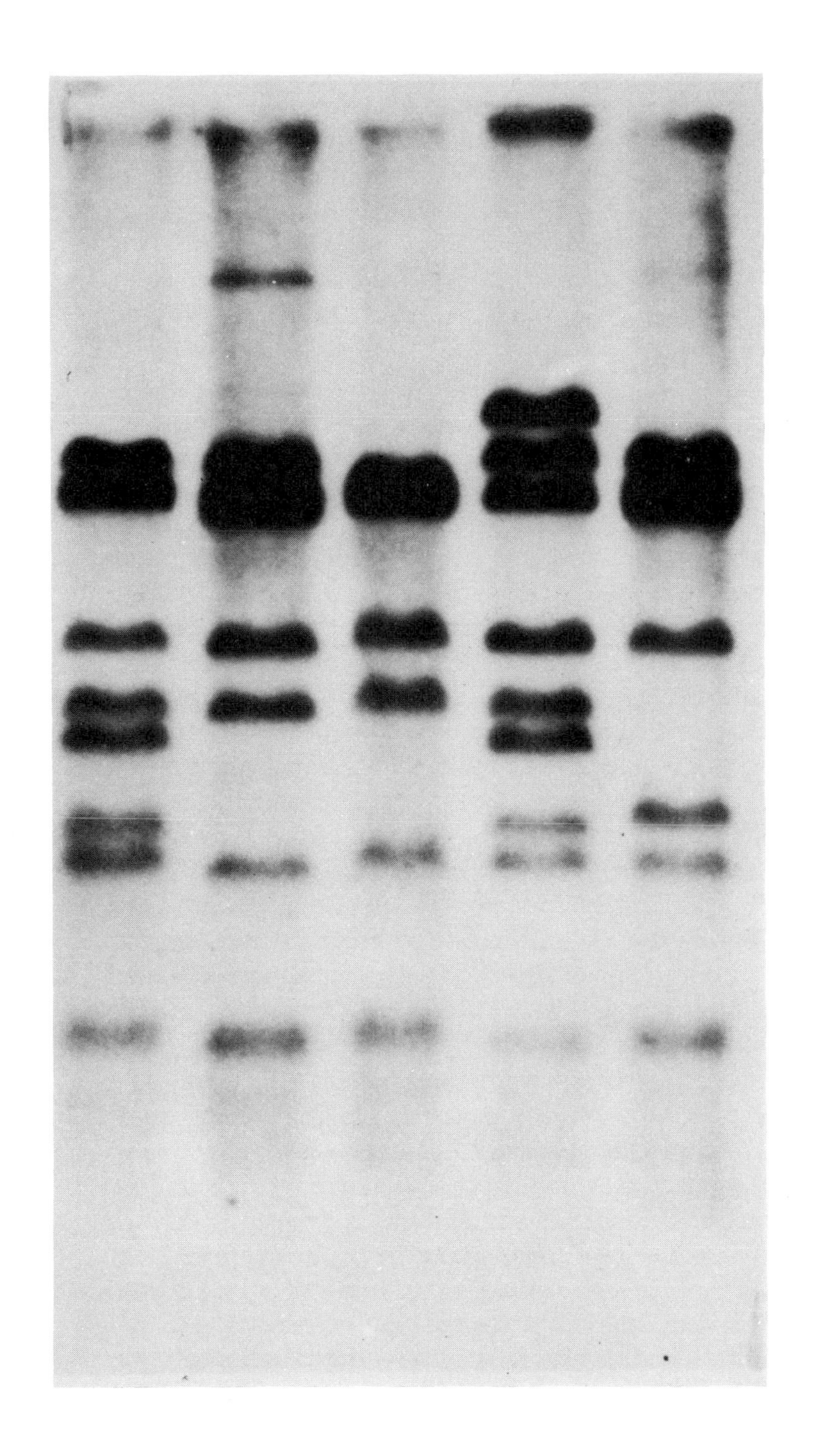

Asia across Europe involved, not an imitation of technique, but a migration of the farmers themselves, whose genes became increasingly diluted by interbreeding with indigenous hunter-gatherers. Examples of more subtle genetics in Europe include a notable Levantine admixture in the genes of Sardinians with known Phoenician-Carthaginian colonization; also a genetic influence in Icelanders attributable to hybridization with Irish slaves. Paris and London stand out as genetic islands different from their immediate surroundings because of the influx of people from the farther provinces of their respective countries.

By the boldest interpretations of genetic geography, modern humans may have emerged in south-central Asia—perhaps somewhere east or southeast of the Caspian Sea on the Kazakhstan-Baluchistan axis—because that is where the indigenous populations are most "intermediate" between genetic extremes of Africa, Australia, and the Americas. Genetic similarities and differences then follow naturally from geographical contacts or separation, taking account of the routes followed by the hunter-gatherers; for example, Amerindians proceeding to America via the Bering Strait or the Pacific Ocean are more closely related to the Chinese than to Europeans. Such findings all accord well with the archeological evidence of early human dispersal.

How can genetic data be fed into the narratives of prehistory and history? In one important episode it turns out to be straightforward. Western nutritionists have woken up to the fact that only a minority of the human species can digest fresh milk after infancy. Most, in common with nearly all adult mammals, are unable to tolerate the lactose sugar present in milk: it makes them ill. In certain humans, a mutation of genes controlling the production of the enzyme called lactase led to the persistence into adult life of the suckling baby's ability to digest milk.

The custom of milk drinking among older children and adults spread with the dispersal of peoples possessing the mutation—a dispersal that embraces much of the military and demographic history of the world. In prehistoric times the milk drinkers, who possessed plows as well as cows, moved into large areas of Asia, Europe, and northern Africa. The archeologist Andrew Sherratt elucidated the technological and political correlations of this breakout; they included, for instance, dominance of men over women. In recent centuries, milk drinkers went to the Americas and Australia. Many of them had light-colored skins, and political and economic ascendancy has recently belonged to that assortment of peoples whose chief common attribute is the milk-drinking mutation inherited from common ancestors of perhaps eight thousand years ago. A genetic marker as clear and consequential as this one may have to be counted as a rare bonus for the biology of history. It is unlikely to be repeated very often, or in so vivid a fashion; for other peoples and cultures, a more systematic approach will be needed.

Languages evolve independently, much as genes do, as a result of geographic separations of human groups, but they mutate at a faster rate than genes, and the kinship of language groups can be almost erased in ten thousand years. Nevertheless, languages are important markers for objective history, and it is a pity that historical linguistics is underdeveloped and spotty. Putting the present-day linguistic map of the world alongside the genetic map of supposedly "aboriginal" populations shows almost at a glance who has recently conquered whom. Hypothetical distributions of languages at earlier stages can be well matched both with genetics and with general historical data.

The mother-tongues for the greatest numbers of speakers are Mandarin Chinese and (a long way behind) English, but their patterns of distribution are entirely different. Speakers of Mandarin and related languages expanded and multiplied in a relatively small area of eastern Asia, while the English language became scattered around the world, along with closely related Indo-European languages that together surpass Chinese in numbers. This language group is associated, of course, with known migrations of Iranians, Indo-Aryans, and Europeans who swept across the globe; but the milk-drinking mutation also links it, more cryptically, with other linguistic groups, including the Semitic and the Uralic, represented in Arabic and Hungarian respectively.

Linguistics and genetics working together foil those prehistorians who would prefer to minimize migrations and invasions in their accounts; the maps make sense only in terms of peoples breaking out of core areas and spreading their languages or their genes, or both. The two markers are complementary, and comparisons can reveal different military and political conditions. An official language is a marker of political authority, which an invading minority may impose on an indigenous majority, even while its heredity is swamped by the preexisting gene pool. Conversely, less assertive migrants may adopt a local language while multiplying their genes.

A Martian account of human history should best start with the mosaics of languages and genes, mapped worldwide, and then turn to archeological and historical data to explain the patterns. First call should be made on other objective factors such as the natural environment, climatic change, demography, and disease; then on local advantages and disadvantages created by particular agricultural or industrial innovations, and by the sources and markets of items of trade. When all that has been fed into the story, there may be little left for kings and queens to do except find employment for their military retinues. Such a program for history has barely begun, but the shape envisaged for it explains why this book omits many conventional themes of history. Already, elements from the emergent archeobiology can be injected into a global narrative that uses data on how many people speak various languages, as a guide to relative success.

A proper linking of biology and history will also demand clearer thinking about genetic versus cultural evolution. The two processes often look alike: the wings of birds and of aircraft show comparable histories of experimentation, adaptation, and selection. Luigi Cavalli-Sforza, the pioneer of global human genetics, has started also to develop a scheme of cultural evolution in which elements of a culture develop and spread through populations like genetic mutations: new words, for instance. Can scientific ideas about cultural evolution lead to historical concepts worthy of radiocarbon dating and genetic and linguistic geography? If so, it may be the turn of historians to find their pet hypotheses savaged by measurement and calculation, and their freedom of self-expression circumscribed on the charts of time.

The big difference between genetic and cultural evolution is that human life is purposeful. The transition from apes to humans, and especially to modern humans, involves a switch from a Darwinian to a Lamarckian mode of evolution, wherein endeavor creates changes. Cultural evolution is then a matter of norms and goals, and these have enabled humans to outpace the cheetah, to plunge with the whales, and to fly with the birds. But norms and goals are ideas in people's heads. Can they be matched in a better than arm-waving fashion to the data of genetics and history?

Try poetry. Word magic is the artistic counterpart of thermonuclear fusion, life-giving and deadly dangerous. Scriptures show the power of poetry to shape history, and for that matter to shape the scriptures themselves—whether one thinks of the Jewish tale of Esther or the ballads of the Arab robber bands that preceded the Koran. The works of Homer and Shakespeare were preludes to aggressive breakouts of the Greeks and the English; the poetry helped to define both the language and the self-image of these peoples. The scene is not unlike the dance of the bees where one bee announces, by its wiggles and turns, the route to the honey, and all the others join in.

Human beings are not insects and only humans make potent sentences, but a biologically closer parallel is with the chimpanzee, almost unique as a species that has something of the human ability to see himself as

When Europeans irrupted around the world in recent centuries, African and Asian perceptions of the intruders were unflattering. Drunkenness is indicated by the drum from Zaire (upper right), in the form of a bottle pouring into a trader's brain. Hairiness is the theme of a Japanese print (upper left), where a woman married to a European sailor has given birth to a child that shocks the midwife. An artist in India depicted a British functionary traveling like a princeling (lower illustration).

others see him. The biologist Lawrence Slobodkin alluded to the chimpanzee's capacity for self-image when explaining why poetry is historically powerful. A chimp looking in a mirror will try to remove marks from his face that seem improper. Among human beings, even small children have enough self-awareness to act the roles of others. Indeed fantasizing is a part of human nature. Adult humans try to act out publicly shared fantasies, and that is one of the means by which they transcend their genetic heritage. Slobodkin comments:

> The fact that the northern Europeans may have prided themselves for centuries on their warriors and on their toughness and on their ability to drink and be generally rude says nothing one way or the other about the biological propensities of the northern Europeans compared to, say, the Bushmen. [*Perspectives in Ethology*, Vol. 3, 1978]

In purging history of much of its conventional poetry and rhetoric, the present book does violence to valued traditions that, in Slobodkin's sense, are properly part of the story even at the level of the science of human behavior. On the other hand, the poetry of shepherd kings, swashbuckling seafarers, and bomb-making mathematicians may be the death of us.

A further hint of how poetry may become comprehensible to science is the concept of "memes" advanced by another biologist, Richard Dawkins. A meme, for him, is a popular idea, whether a tune, a clothing fashion, a theory of evolution, or the idea of God, that replicates between human minds in very much the same fashion as genes replicate in the presence of suitable nourishment. Memes can also undergo mutations. Although scarcely developed, this theory is already quite compelling, because the existence and behavior of memes are evident as soon as they are pointed out. They are informational parasites, prospering or dying in brain tissue in accordance with their appeal and usefulness, and propagating from brain to brain by word of mouth, by demonstration, or by radio waves.

How much of human history and public behavior can be captured by the theory of memes remains to be seen. The most consequential memes may be slogans: proverbs, mottoes, religious formulas, political clichés, trader's street cries, and the buzzwords of the intellectual marketplace. They can serve programs good, bad, or indifferent. The Arabs' battle cry *Allah il Allah* changed the face of half the world, while the Royal Society's *Nullius in verba* encapsulated the modern scientific attitude. A slogan for the present book is *Let's pasteurize the past*, meaning, scald it clean of legend before the poetry of contending ethnic groups detonates the nuclear weapons and punctuates hope.

Causality Is Not Destiny

One way of studying events on the Earth around 2 billion years ago is to use radioactivity to figure out when ancient deposits of iron ore were formed. Another is to compare molecules in living bacteria. To expect anything better than mutual indifference or mismatch between such contrasting sources of data might seem too optimistic. And yet a decline in the rate of iron-ore formation coincides with evolutionary changes among the bacteria, as inferred by the ages of the last common ancestors of important groups of present-day microorganisms. Natural nuclear reactors appear on the Earth at about the same time, as a bonus for investigators. The common factor linking these phenomena seems to have been the appearance of free oxygen in the Earth's atmosphere and water. The oxygen, in its turn, can be related to earlier developments: a hardening of the Earth's crust and an increase in the abundance of living things. Dates and data converge, not grudgingly, but in enlightening ways, to afford a consistent and reasonable account of major events a very long time ago, reassuring to the chronographer.

That such accounts are possible at all, that natural law prevails and sequences of cause and effect are discernible, should not be taken for granted. Scientists always believed that stories of this quality would eventually emerge, but that was a matter of faith. Their day-to-day skepticism is, of course, the very opposite of faith, and raises objections against too glib a story. The chronographer cannot resolve the internal arguments of all the contributory branches of science and scholarship. He must cultivate his timescale, ensuring that evident causes precede evident effects. In the end, events must make sense, in much the same way as the lie of the land makes sense to a cartographer. This means that events ought to be linkable in a consistent and reasonable manner by concise causal descriptions. These should not be Just-So Stories that fill gaps in knowledge with ad hoc speculations, but should come either directly from the evidence, or else from well-developed theories.

A frequent wish to impose on many different events a single explanation—climatic change, for example, or the class struggle—is specialism gone mad. Confusions arise also from a presumption that cause and effect should follow one another like night and day, whereas in practice the outcome of a given cause is often unpredictable and conceivable effects may simply not occur. Otherwise, there would be no need for newspapers. The events of history are what did happen, in sharp distinction from what might have happened but did not.

Efforts of human beings to discover their place in nature involve self-reference or recursion. Humans are themselves pieces of the nature they explore, and in that respect their statements resemble the sentence: "The words in this sentence constitute a sentence." Even if no new discoveries were made about the past, historians would always wish to rearrange existing data to match the interests of their time. History is unavoidably a mental construct, a set of temporary hypotheses that reflect the perceptions, preoccupations, and analytical methods of those who write it. This book is no exception. In current science the theory of memes is itself a meme, an idea communicable from brain to brain. Mathematical theory suggests that an ultimate bar will prevent full comprehension of human nature by those observing it from the inside, but we are a very long way from any such limit.

Each of the assorted theories of human behavior that stalked through psychology classes in the twentieth century—Sigmund Freud's sex-crazed families, B. F. Skinner's robot-like people, Konrad Lorenz's killer apes, and so on—was sustained by prejudiced collation of the evidence; all are now clearly flawed. Similarly, historians, divided on whether ideas or economic realities are more important in shaping the course of events, offer data that play up a favored theory and minimize the opposition's. Polemical scholarship is unavoidable, because convictions about particular ideas are needed to spur the effort for research, but convictions affect the eyesight and people tend to see what they expect to see. A chronographer has to stand back from the enthusiasms and consider that any theory may be helpful, but not all of the time.

Time is a leveler, in more than one sense, and distinctions that seem impressive in a short timespan become reduced to averages in the long run. A hundred million years from now the great piles of the Himalayas will be worn down to gentle undulations, and wide oceans will have transformed themselves into new mountain chains; similarly, today's hiker can cross the hilly traces of an ancient ocean during an afternoon's stroll, and the oldest mountains are discernible only to a skilled geologist. Significant details of very ancient times include the origin and location of ores, and the effects of changing geography on the evolution of life, but those are matters partly for special, local investigations, and partly for global considerations of climate, sea level, and atmospheric changes. An accurate map of the world 200 million years ago is easier to prepare than a map for 2 billion years ago, and is also more useful.

A person has only two parents, but sixteen great-great-grandparents; going back a century in family time dilutes the individual significance of each ancestor. Some Americans like to claim descent from the Pilgrim Fathers, or from African chiefs, but after a dozen generations a single famous ancestor contributes in principle only one in four thousand of the genes possessed by an individual; cousin marriages delay the dilution but do not prevent it. Within a few centuries, details of family trees have less meaning than the statistics of the gene pool of miscellaneous ancestors.

Fossil hunters are apt to bridle if one claims a particular organism found in the rocks as an "ancestor," and their objection is not just pedantic. The chances of finding any individual animal truly ancestral to ourselves are negligible. Repeatedly, along the timescale, bottlenecks occurred when small isolated populations of animals evolved new tricks, while most of their relatives were dying off. Many lineages became extinct, and few fossils can have any living descendants at all. Vagueness strikes again, and fossils close to the prehuman line, whether fishes or early primates, have to be regarded as great-aunts rather than grandmothers. To say so is not to deny the most important genealogical point of all, that all human genes are descended in an unbroken, albeit much mutated, chain of inheritance going back 4 billion years. As a satirist wrote a hundred years ago:

> I can trace my ancestry back to a protoplasmal primordial atomic globule. Consequently my family pride is something inconceivable. [W. S. Gilbert, *The Mikado*, 1885]

Evolving human cultures present a similar picture of leveling. Nowadays a few nations are engaged in highly competitive developments of microcomputers and nuclear weaponry, yet it seems likely to be only a matter of time before any group except the smallest or poorest will have whatever it thinks it needs in the way of microcircuitry and H-bombs. The pioneering role of certain nation-states will be forgotten except by historians. Similarly, such innovations as the tamed camel, the wheel, writing, guns, and oceangoing ships gave advantages to the peoples who first possessed them; now they are part of the human repertoire, available to all, at least in principle, and few users stop to think where they came from. Anyone offering causal explanations should be aware of this historical entropy.

Philosophers used to fret about the various categories of causes—"material," "formal," "efficient," and "final"—but the geneticist J. B. S. Haldane pointed out that these distinctions were really a matter of the timespan of the processes examined when searching for candidate causes. He compared possible explanations for the song of a chaffinch:

> Muscles contract with the help of adenosine triphosphate.
> The brain sends patterned signals to muscles.
> Long spring days evoke hormones that act on the brain.
> The bird learned the song from other chaffinches.
> The bird's song attracts females and repels males.

All of these are correct answers, but they relate to processes of very different timespans: a split second for the molecular, a few seconds for the neurological, a season for the hormonal, centuries for the song dialect, and millions of years for the evolved function of birdsong. In a similar vein, a range of possible explanations, with timespans, can be offered for an event in human history, namely the first appearance of life on the moon in 1969:

The Apollo systems functioned correctly, and the astronauts would sooner have died than disobey their instructions (one hour).

The Americans wanted to surpass the Russians in space exploits (a decade).

The invention of liquid-fueled rockets made spaceflight feasible (half-century).

Gravitational physics specified attainable conditions for flying to the moon (three centuries).

The talkative human subspecies proved capable of extraordinary collective enterprises (45,000 years).

Toolmaking enabled animals to perform tricks beyond the limits of their anatomy (2 million years).

Animals included mobile organisms that ventured in an exploratory fashion to new places (700 million years).

Life began on the Earth and in due course vaulted to the moon, which was not very far (4 billion years).

Once the nuclear force had made carbon atoms, the electric force was capable in principle of assembling self-reproducing systems that would tend to enlarge their territories (13 billion years).

As in the case of the chaffinch song, there is no contradiction between these accounts, but they emphasize completely different items and processes, loosely linked at best. There is little connection between the nature of the gravitational force at long ranges and the human social behavior that organized the Apollo program. Nevertheless, each item is legitimate and illuminating, and ranks as a concise causal description of the arrival of life on the moon, depending on what timespan and tendencies one may wish to stress.

Too detached an observer may consider that the least impressive of the reasons offered for the flight to the moon is the matter of political history: that the Americans wanted to outperform the Russians. Readers on the moon five hundred years from now may be scarcely more interested in the twentieth-century East–West competition than modern Americans are in the ebb and flow of rivalries between Portugal and Spain at the time of Columbus. With hindsight the flights to the moon may look as if they were inevitable. Yet the political decision, a decade ahead of time, was in fact the crucial cause, and without it there would be no man-made litter on the moon—perhaps ever. It was a voluntary act of sentient beings almost wholly in charge of their own future. If it appears otherwise, that is an illusion created by historical entropy.

As physics obtrudes in the history of life, it introduces a certain prejudice about the nature of causes, because its traditional, atomistic approach is to seek simple principles underlying the complex untidiness of nature. An opposing tendency in current science is to investigate complex systems such as the human brain, a meadow, or a national economy, using computers to comprehend complexity, in the belief that reducing the system to its simplest parts would destroy holistic properties that one wishes to grasp. Complex systems have a will of their own, and their internal interactions can generate booms, slumps, and cycles of activity, independently of outside causes. Some archeologists, for example, seek to integrate the environmental, political and cultural circumstances of the anonymous peoples whom they study, in order to grasp more fully the richness of their lives, and to search for subtle, potentially catastrophic, dislocations of their economic systems. These admirable holistic studies can claim some affinity with the new thermodynamics mentioned earlier.

Yet simpleminded reductionism has succeeded far beyond expectation, and may well go on doing so. Who would have imagined, before the discovery of the genetic code, that all the variety of life was spelled out in a four-letter alphabet, with the origin of a cancer lying in a single misprint in a gene? Who was entitled to hope that the turmoil of the Big Bang would lend itself to precise description in the language of particle physics, or that the ragged mountains of the world would yield to the geometric analysis of plate tectonics? From an intricate computerized study of the ecology of North American grasslands, the most striking result was not the tabulated printouts showing how overgrazing reduced the birds that controlled the insects but the particular discovery that the diminutive mealybug was eating as much grass as all the cows. Events in complex systems may very well have simple causes, and where that assumption—the bias from physics—will turn out to be misleading, no one knows yet.

Entirely inappropriate for history are some other habits of thought of the physicists, especially the confidence with which they make predictions and the compulsory character of most of their physical laws. Even with hindsight, precious little of the course of evolution, or of human history, would appear to be predictable. The rationality of cause and effect is not at risk: circumstances keep changing, while genes and social systems keep mutating, and these provide explanations for the opportunistic behavior that has driven evolution and history along. But causality is not destiny, and evolution can no longer serve as the surrogate for a creative, purposeful God, as some biologists have wished.

So far from being a likely product of life, the appearance of clever human beings on the planet Earth was a matter of astounding luck, at the end of a dizzying succession of improbabilities.

Given enough time, improbable events are bound to happen, whether one speaks of very rare phenomena (comets colliding with the Earth) or freakish things (the evolution of frogs that nurture their tadpoles in their vocal sacs). Yet the character, timing, and effects of such occurrences are unforeseeable, even deniable by the usual criteria of prediction. Moreover, unexpected changes were at least as important as any gradual tendencies in the evolution of life. In human cultures, too, the most radical inventions and social innovations have been as unexpected as the discovery of America. In short, changes flow from inevitable improbabilities that by their very nature cannot be predicted.

Negative statements have positive force, provided they are emphatic enough. The second law of thermodynamics prohibits perpetual-motion machines; it is one of the chief stones in the edifice of modern physics, supporting entire towers of explanation and inference. From the rule that the speed of light, as measured, can never vary, Einstein wrested a new conception of the universe, and inferred real effects previously unimagined. A firm enough denial of the existence of any purpose, plan, or overall tendency, either in evolution or in alleged laws of history, may also have some predictive value. For instance, the chances of other forms of intelligent life in the Universe within communicable distances almost vanish—although the search for intelligent radio signals is then more necessary than ever to test the theory that there should be none. And anyone who calls upon his followers to pursue the natural or historical destiny of mankind is as much of a charlatan as the man who claims to run his car on water.

So far from diminishing the utility of history, a negative law should enhance it, by making people more receptive to the lessons of the past. Information in the fossil record about environmental catastrophes, for example, can help to cure a weakness of present-day ecological science, namely that one cannot safely perform large-scale experiments on the environment. Nature has carried out many trials in the past—of desertification, glaciation, decimation of species, variations in carbon-dioxide levels, and so on—but we shall not read the results clearly if we persist in thinking that we live in the best of all possible natural worlds, or that there is something planned, purposeful, or perfected about present relationships between organisms and environments. Similarly, past events in human history should be seen, not as inevitable steps leading toward our own glorious civilization, but as multiple social and political experiments played with a variety of outcomes, typically disagreeable.

How can all this be reconciled with strong intuitive impressions of directional change and social progress? By the principle of the ratchet. Genetic changes in living organisms cannot normally be reversed, and so biochemical and anatomical tricks tend to accumulate, along with many junk genes that serve no useful purpose but are still copied from generation to generation. During human history, material and social inventions amassed and information about them was not easily lost. Some Polynesians forgot about making pottery, by dint of voyaging thousands of kilometers over many generations. It will probably be harder for humans to forget the possibility of flying, or running protection rackets. Amid continuous change, the ratchet maintains directionality toward an accumulation of useful innovations and potentially harmful junk. As to which is which, that is a value judgment to be exercised item by item, but no one should mistake the totality of ratchet-governed change for progress in the only sense that matters: making a better world for the next generation.

To deny the inevitability of progress is not to rule it out but only to say that it has to be achieved by some combination of personal endeavor and the seizing of unexpected opportunities, not by reliance on abstract laws of biology or history. Many people are indeed far healthier, better informed, and in specifiable senses freer than their grandparents. Yet humane rationality is under threat, and if it is to be anything more than a bubble in the swamp of time, only its wits can save it.

Since modern science took up, in earnest innocence, the old Greek questions about the nature of matter, the ratchet has carried the species to the point where the nearest half-anticipated events are nuclear war and the creation of police states run by microelectronics. But it has also brought knowledge of the kind that may help to prevent such outcomes. At present the cutting edge of intellect is the physics that anatomizes the Big Bang and computes its products. It is good that the most powerful science has spilled over into fields of research telling where humans stand in space and time. If destiny's guiding hand has been retracted, humans must grow up and walk unaided. The very consternation caused by reassessments of the past arouses hope that humans in the nick of time will face the facts of life, with physics as the hormone of their species' puberty.

Those who repressed Galileo in the seventeenth century and the Soviet geneticists in the twentieth knew what they were doing. Science is anarchical in method, unpredictable in content, and subversive in effect. So too is the perspective of the fourth dimension, generated by

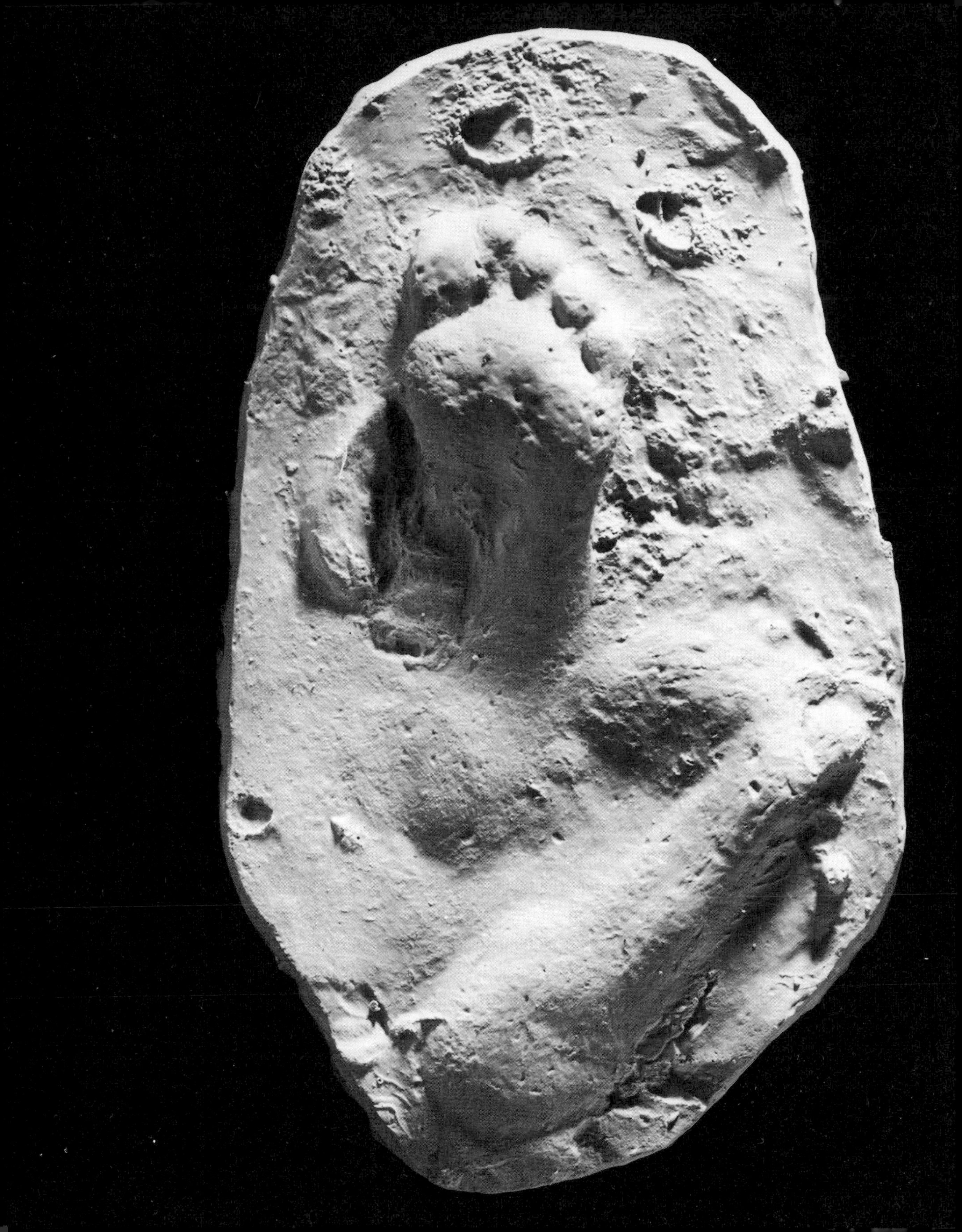

science: it tells of repeated triumphs of the underdog and is most plainly subversive where it clarifies the temporary nature of man-made institutions. Farming, cities, male domination, taxation, social class, chronic warfare, and many other features of life were all invented by human beings in specifiable periods and places, so they cannot be defended as "natural" but may in principle be disinvented or superseded. So too can science and technology. Yet along the spectrum of totally different worlds that present technological opportunity and human action could create, alternatives to nuclear war or computerized police states include invitations to life in space, or to the use on Earth of Santa Claus robots or Green Machines that satisfy all material needs. The more clear-headed human beings are, the better their chances of making sensible decisions.

Many accounts of the cosmos, of life on Earth, and of the "glories" of the human past are reverential, even priest-like, and large numbers—billions of years for example—serve too easily to overawe an audience. A sense of wonder is all very well, and certainly necessary for stirring curiosity even among professional researchers. But awe seems the wrong mood for digesting the findings of research, especially when they tell of tyrannosaurs and tyrants alike falling on the banana skins of history. Indeed, the first requirement for clearheadedness is to be rid of any vertigo that may be induced by peering into the abyss of time.

The Logarithmic Imagination

Physicists minding their own business—that is to say, the behavior of matter and energy—have been running experiments that give intimations of mortality for matter itself. Theories that relate the fundamental constituents of atoms and the cosmic forces in a grand unified scheme require that all atoms be eventually radioactive and prone to decay. The proton, the archetype of heavy matter, turns into an anti-electron, which then meets an electron and vanishes in a puff of gamma-rays. Stars and planets would not have survived if the rate were not extremely slow, but if you monitor a thousand tons of ordinary water, then, perhaps once a month at most, your instruments may detect a flash of light telling of the decay of a "stable" atom.

Some 15,000 years ago a high-spirited child, about eleven years old, deliberately pressed a foot into soft clay, to leave a mark in a cave at Niaux in France. The photograph shows a cast of the footprint. Léon Pales has found many traces of youngsters who explored deep inside caves that were inhabited by the large bears and hyenas of the ice age. Their daring typified the joy of life evident from the first appearance of modern human beings.

To measure that rate and trace the route by which the proton shuffles out of existence will keep experimentalists busy for years to come. Yet to learn that matter is temporary comes as a shock. Even though the average lifetime of matter is better than ten thousand billion billion billion years (1 followed by thirty-one 0's, or 10^{31} years in shorthand), it

may trigger a neurone of despair. Did it? If so, that reaction tells you that your mind is capable of traversing intervals of time far greater than anything covered by the timescale in this book. The quantities may be unimaginable in a strict arithmetical sense, yet the imagination can travel freely to the funeral of the universe.

Long before the sun loses any noticeable mass by this cosmic sleight-of-hand, it will grow steadily warmer and overtax the regulatory mechanisms that preserve mild conditions at the Earth's surface. In any case, when the sun ages it will swell up, first scorching the Earth and then engulfing it, some billions of years from now. Thereafter the whole universe may collapse and annihilate itself, perhaps 50 billion years hence, in the Big Crunch predicted in some versions of cosmology. The physicist Freeman Dyson pursues the alternative notion, of a universe that will go on expanding forever, and he is at pains to explain how intelligent beings could use black holes as sources of energy after the last star has guttered out. Observe the desire for an immortal culture.

This excursion into the far future will have served its purpose if it leaves the past seeming compact by comparison. Among the gross quantities of imaginable time, the universe itself is relatively young—a sprightly 13 billion years or so, and still warm from its creation. It gives other signs of youthful vigor. Many galaxies, including our own Milky Way, remain rich in gas for making new stars. Nothing to be seen in the sky is more than three times the age of Earth, and most of the bright stars are far younger. To accept that the Earth is some 4.5 billion years old may be less daunting to imaginations that have rehearsed longer timespans. And when geologists speak of an event occurring "very recently," just a few million years ago, I know they mean it. That age may be a thousand times the centuries since Moses, but you can stand on exposed rocks a thousand times older still. No geologist, or any astronomer or archeologist to my knowledge, loses a moment's sleep brooding about such numbers.

A cavity in a salt mine in Ohio was adapted to record the rate at which matter vanishes from the universe. Now filled with ten thousand tons of water, and shielded from cosmic confusions by 900 meters of overlying rock, the cavity contains more than two thousand detectors, which await flashes of light telling of the decay of supposedly stable atoms. It is a process occurring on timescales incomparably longer than the span of the universe so far.

To find comparisons that accommodate the extreme disproportions of time, and yet register in the imagination, is not easy. For many people, the most vivid experience of time and distance comes from a tiring walk. Broadway is the street that snakes for more than twenty kilometers along New York's Manhattan Island from end to end, and a person might take six hours to traverse it on foot, allowing for stop lights. Using that length of Broadway as a cosmic timescale, and setting the Big Bang at Baker Field in the north of Manhattan, the stroller has to proceed most of the way down the island, past the George Washington Bridge, Columbia University, and the New York Coliseum, before the

Sun and Earth come into existence at 48th Street, a few blocks before Times Square. When he has crossed Midtown to 34th Street near the Empire State Building, the Earth's oldest surviving rocks are showing up, carrying hints of early life. Every step he takes represents 400,000 years. While he plods on for another hour, through Madison Square, and past Greenwich Village, nothing larger than single-celled creatures is yet alive. The time walker's perseverence is rewarded with a wisp of seaweed at the Woolwich Tower.

Various kinds of marine animals, from jellyfish to jawless fishes, put in their first appearance between Fulton Street and Wall Street, but only after the Stock Exchange has been passed does the dry land become inhabited. Half a kilometer from the southern tip of Manhattan, our stroller enters Battery Park, just as the recent supercontinent of Pangaea is assembling itself; after a couple of hundred paces, it begins to break up again. Across most of Battery Park the dinosaurs are in their heyday, but they are wiped out a hundred meters from the end of the island. The earliest human beings come into existence just five paces from the water's edge, and the whole of history since the time of Christ is represented by the outermost three millimeters of the sea wall. The life of an individual is no more than a coat of paint on the railings.

A timescale laid out like a measuring rule, with equal distances along it corresponding to equal intervals of time, is useless when, within a cosmic frame, you want to say anything about Johnnies-come-lately. In order to accommodate a single succinct remark about modern human beings, who emerged around 45,000 years ago—"they talked a lot"—a narrative dealing evenhandedly with all earlier intervals of time would have to be as long as the Bible and spun out with an intolerable deal of facts about algae.

In any case, that is not how the brain thinks about widely variable quantities of anything. Two people haggling about whether a used car is worth $2000 or $3000 might settle on $2600: if either party suggested $2615.31 he would be considered mad. Units of measure, whether of money, of distance, or of time, go up in stages convenient for different purposes, by tens, hundreds, thousands . . . and what people consider to be round numbers change in a similar fashion.

To say "the assailant was about twenty" means that the police are looking for someone between eighteen and twenty-two. Similarly, a chronographic inquiry about what the world was like "around twenty thousand years ago" may mean roughly eighteen to twenty-two thousand years; whether for millions, or hundreds of millions of years, the "slop," the span of interest and of uncertainty, remains proportional to the magnitude. A method of scaling exists that is exactly right for keeping such matters in proportion, and the fact that it is called logarithmic is the only forbidding thing about it.

The logarithmic timescale used in this book is no more mysterious than the maneuver of an aircraft as it nears touchdown and flares out to avoid hitting the ground too hard. The rate of "descent" through time diminishes as one approaches the present, according to a strict but simple rule that a stipulated proportional change in ancient dates always corresponds to the same distance along the timescale. For example, any reduction in age by 90 percent—whether from 100 million to 10 million years ago, or from 10 million to 1 million years ago—is allocated the same room. Scientists use such scales to cope with phenomena spanning many orders of magnitude. The difference in thinking between people in general and the most mathematically minded is slight and is noticeable, if at all, in their choice of a round number between ten and a hundred. Many would say fifty, but a mathematician might prefer thirty, because it falls midway among the logarithms.

A minor snag is that a logarithmic scale becomes too sensitive at the shorter end, like an aircraft that flares out so smoothly that it skims with its wheels just above the runway, not making contact. Centuries ago become years ago, become days, minutes . . . and before you know it, you are into microseconds and nanoseconds, in hot pursuit of the present but never quite reaching it. For this reason, the target "present" has to remain somewhere in the future—a little beneath the runway, in effect—and it is for arithmetical not millenarian reasons this book sets the "present," before which dates are reckoned, at AD 2000.

A device that puts the cosmic eons in a nutshell, while letting the human story unravel, may seem to some linear minds a gross distortion. On the contrary, the logarithmic scale is a necessary aid to contemplating the past, and to seeing the sun, hills, trees, cows, and passing automobiles as items of very different yet commensurable ages. It liberates what the archeologist Andrew Sherratt has called the logarithmic imagination.

For a start, this timescale reconciles what is known with what one needs to know. Ask how the world came to be the way it is, and happenings of ten years ago will clamor for attention alongside events of 10 billion years ago, but the farther back in time one goes, the sparser, mistier, and more generalized the information becomes. When it matters that he should do so, the investigator can often resurrect details from the remote past, to detect a significant exploding star more than 4 billion years ago, follow the misfortunes of a trilobite species half a billion years ago, or plot weather maps for the storms that wrecked a Spanish armada four hundred years ago. But fine details are generally less important to comprehension in the older phases of the story.

In another important respect, the logarithmic timescale of human history comes closer to reality than the linear timescale, because it compensates to some extent for the growth of the human population. For example, many imagine that preliterate societies were technologically slow-witted compared, say, with the Chinese at their medieval peak or industrial nations today. This belief is falsified as soon as population size is taken into account. If an estimated 8 million people were alive ten thousand years ago, and we can credit them with two notable innovations every thousand years, or one per 4 billion man-years, that is equivalent to a major invention every year in the 1970s, when 4 billion people were alive. If ten inventions of the 1970s prove to be as tricky and consequential as the domestication of wheat, our generation will have done no better than par, for technological creativity.

A timescale that demands fewer entries per thousand years in the period before Christ, and more in the very recent period of high populations, therefore gives a fairer impression of continuing inventiveness. The underlying creativity may vary very little, because for every successful inventor there are dozens of disappointed ones, who confront a persistent conservatism in human society that ignores, actively resists, or even punishes them. The style and substance of innovations has of course changed. With new inventions building on the old, dissemination is faster and more comprehensive. The contemporary world still feels the effects of Victorian enthusiasm for change. Nevertheless, the "acceleration of history" in the nineteenth and twentieth centuries may be due as much to population growth as to the special creativity of modern science and technology.

Granted an undespairing mood, awareness of the abyss of time ought to make mortality less oppressive. The most self-centered person could not seriously wish to live for a billion years, and in the cosmic frame humans are not very different from the mayfly that lasts a day. On the other hand, thay have an advantage over their mammalian cousins. Mammals in general, regardless of size, normally live for about 800 million heartbeats; mice die at the same physiological age as elephants, but their hearts beat much faster and by the calendar their lives are short. Humans are exceptional and live about three time longer, by heart count, than they "should." As it happens, their brains are also about three times larger, in proportion to body size, than a mammalogist might expect, and there are good developmental reasons why a

Galaxies in collision tell of happenstances that govern the histories of entire assemblies of stars. Faintly visible in the photograph are long streaks of gas and stars wrenched from both galaxies by their mutual gravity, and giving the pair its name: the Antennae. Some galaxies are sterilized by encounters that rob them of the gas needed for making new stars.

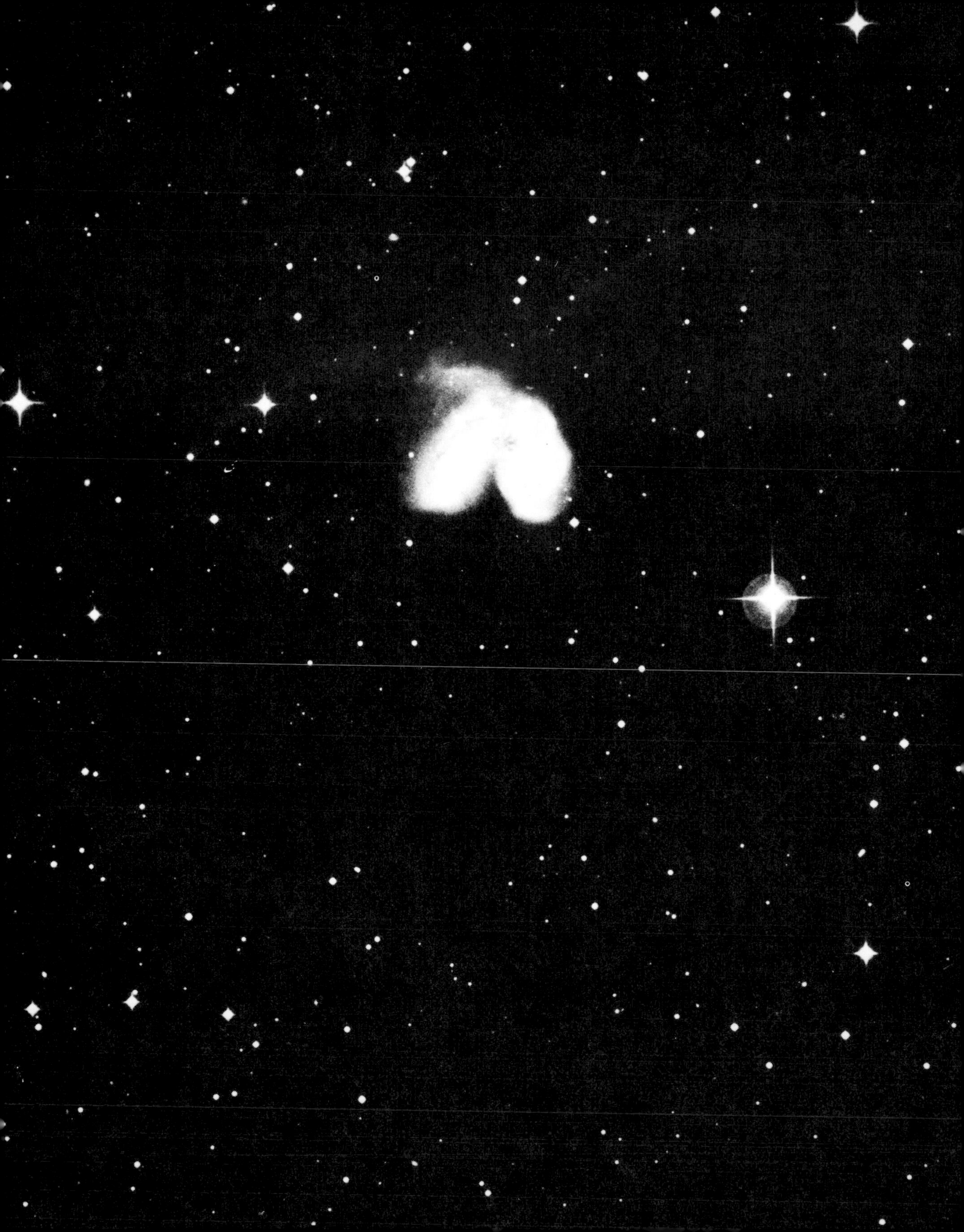

longer lifespan is needed for growing a large brain, stocking it with experience, and then exploiting it.

One use for a brain is to master time and contemplate ancient events attentively but not nervously. When excessive awe has dissipated and the logarithmic scaling has tamed time, the mind can ride it freely and the retrospects become comic in the end.

In caricatures of Mother Earth, where land masses ran riot, the first lords were colored slime, then upstart worms, then sprawling mammal-like reptiles, all for far longer intervals than humans have existed. In Darwin's paleontological estimation, ours is not so much the era of the risen ape as the Age of Barnacles. Holy mountains turn out to be wreckage of continental traffic accidents, while Chicago and Leningrad sit in the chairs of glaciers gone for lunch. All in all, the refurbished creation myth owes more to Groucho than to Karl Marx. It is a tale of hungry molecules making dinosaurs and remodeling them as ducks; also of cowboys who put to sea, quelled the world with a magnetic needle, and then wagered their genes against a mushroom cloud that knowledge was a Good Thing.

MAPS

Abbreviations

Myr = million years ago
yr = years ago (before AD 2000)

Sources of Data

Base maps of continental outlines and shelves follow computer-generated maps supplied by A. G. Smith, Department of Earth Sciences, University of Cambridge. Most maps employ Mercator Conformal projection. On pages 81 to 85, the base maps rely on paleomagnetic data available to 1979.

Basement rocks: p. 81, upper. These mainly follow the margins of Silurian continents as mapped by A. M. Ziegler, et al., *Tectonophysics*, Vol. 40, 1977, p. 13. Archean (pre-2500-Myr) regions are from J. F. Dewey, personal communication, 1981. Ocean rifts accord with a map in D. G. Smith, ed., *Cambridge Encyclopedia of Earth Sciences*, Cambridge University Press, 1982, supplemented from P. Molnar, et al., *Geophysical Journal of the Royal Astronomical Society*, Vol. 40, 1975, p. 383.

Continental positions: p. 81, lower. The maps (one hemisphere only) are Lambert equal-area projections for the Aldanian (Lower Cambrian), Frasnian, Ashgillian, Sakmarian, Rhaetian, and Santonian stages, as published by A. G. Smith in *Geologischen Rundschau*, Vol. 40, 1980, p. 91. Dates are adjusted.

550 Myr: p. 82. On A. G. Smith's Middle Cambrian base map (see above), the superimposed continental configurations and features draw also on A. M. Ziegler, et al., *Annual Reviews of Earth and Planetary Sciences*, Vol. 7, 1979, p. 473. The region of the Pan-African consolidation follows J. D. A. Piper, *Philosophical Transactions of the Royal Society*, Vol. A280, 1976, p. 469. Animals in the diagram accord with H. B. Whittington and S. C. Morris, *Scientific American*, Vol. 241, No. 1, 1979, p. 110, and other sources.

200 Myr: p. 84. A. G. Smith's base map (see above) is for the Pliensbachian stage of the Jurassic period, nominally 180 Myr, but 198 Myr by the timescale adopted here. In the colored modifications, North Iran and Tibet are relocated, following data of A. M. Celâl Sengor, *Nature*, Vol. 279, 1979, p. 590, and Japan and the Philippines are shown as part of the volcanic fringe of Asia. Kamchatka is transferred to Alaska, and Kolyma is let loose. Rift volcanism in Antarctica and southern Africa corresponds with K. G. Cox, *Nature*, Vol. 274, 1978, p. 47. Coastlines are as depicted by M. K. Howarth, in L. R. M. Cocks, ed., *The Evolving Earth*, British Museum (National History) and Cambridge University Press, 1981, and ocean currents are adapted from L. A. Frakes, *Climates Throughout Geologic Time*, Elsevier, 1979. Vegetation data follow P. D. W. Bernard, in A. Hallam, ed., *Atlas of Palaeobiogeography*, Elsevier, 1973. Desert areas accord with precipitation patterns inferred by J. T. Parrish, et al. (*Palaeogeography, Palaeoclimatology, Palaeoecology*, in press, 1982).

18,000 yr: p. 86. Data on ice sheets, sea ice, and sea-temperature anomalies follow *Seasonal Reconstructions of the Earth's Surface at the Last Glacial Maximum* by CLIMAP Project Members (A. McIntyre, leader LGM Project) as published by the Geological Society of America, 1982, MC No. 36. Vegetation areas are adapted from G. Kukla, *Natural History*, Vol. 85, 1976, p. 5 and CLIMAP albedo data. For archeological data see the reference index: breakouts (hunter-gatherers); crops.

human genetics: p. 88, upper. This computer-generated map was kindly supplied by L. L. Cavalli-Sforza of Stanford University. It was first published by A. Piazza, et al., *Proceedings of the National Academy of Sciences*, Vol. 78, 1981, p. 2638.

breakouts: p. 88, lower, and p. 89. These maps follow data and sources given in the reference index: breakouts.

5000 yr: p. 90. The regions of early cultivation, and the "cow" boundary in Africa, are adapted from maps due to J. W. Lewthwaite and A. G. Sherratt, in A. G. Sherratt, ed., *Cambridge Encyclopedia of Archaeology*, Cambridge University Press, 1980. The "cow and plow" contour accords with archeological and vegetational data. Languages are mapped to reconcile the hypothetical distribution of major human groups (Lewthwaite and Sherratt) with modern languages and known movements of populations.

500 yr (AD 1500): p. 92. Linguistic groupings are mapped according to contemporary data (see next map) adjusted to take account of known changes since the time represented by the map; see reference index: breakouts, especially 508 yr (AD 1492) onwards. The literacy contour is from R. I. Moore, ed., *The Hamlyn Historical Atlas*, Hamlyn, 1981; the principal cities and empires are adapted from G. Barraclough, ed., *The Times Atlas of World History*, Times Books, 1978. Tracks of voyages are from C. Platt, *The Atlas of Medieval Man*, Macmillan, 1979.

Late twentieth century AD: p. 94. Linguistic groupings are adapted and simplified from a map by B. Zaborski (Rand McNally) as published in P. Haggett, *Geography: A Modern Synthesis*, Harper and Row, 1979. The chief adjustments are in Africa, where groupings follow M. F. Goodman in *Encyclopaedia Britannica*, 1981. See also the reference index: language. The telephone contour is deduced from data in *The World's Telephones*, American Telephone and Telegraph Company (for 1978). The Intelsat data were supplied by British Telecom.

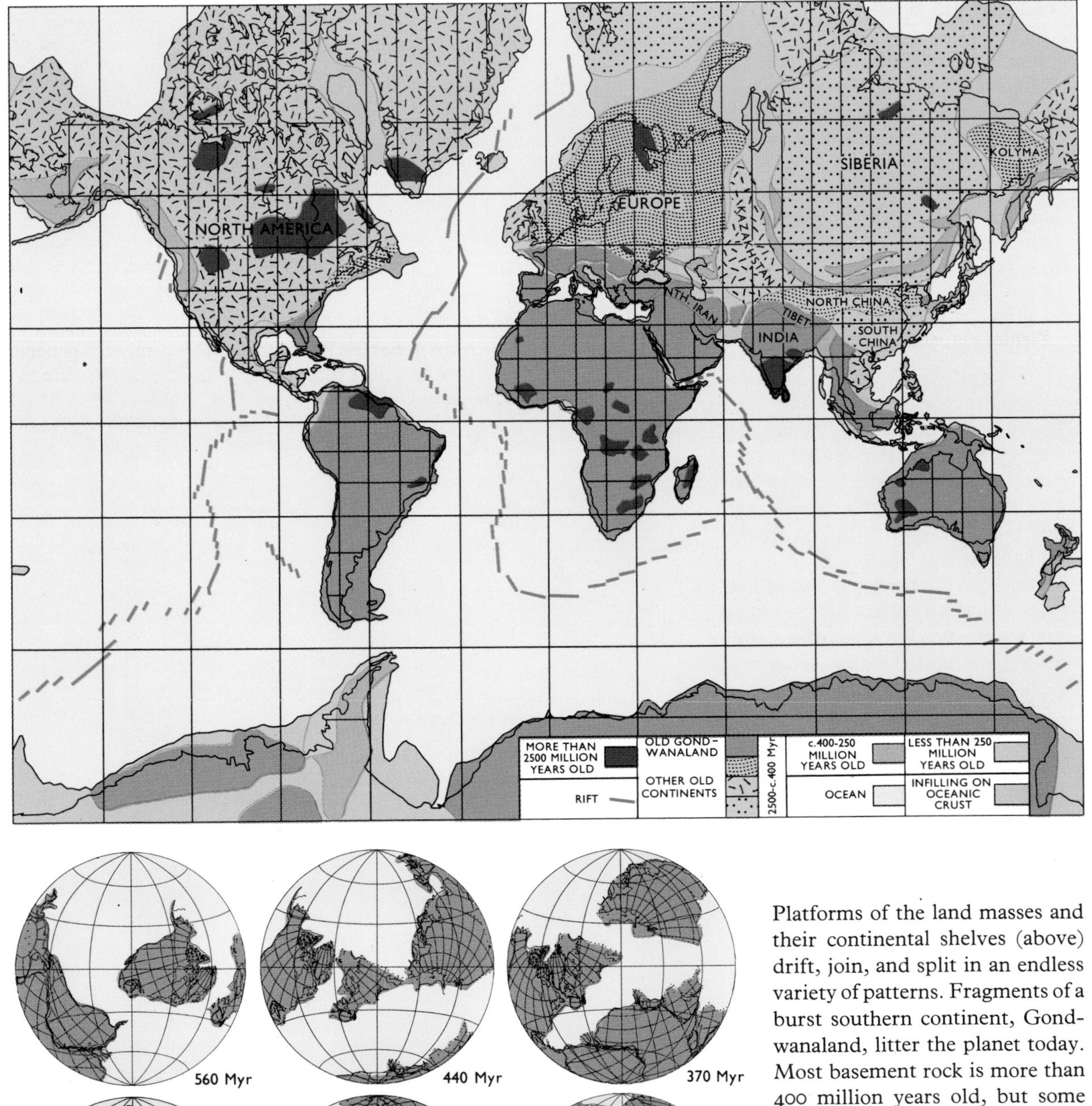

Platforms of the land masses and their continental shelves (above) drift, join, and split in an endless variety of patterns. Fragments of a burst southern continent, Gondwanaland, litter the planet today. Most basement rock is more than 400 million years old, but some has been newly added or reworked. Rifts manufacture fresh pavement for the oceans in the wake of moving continents, and present geography is temporary. The making and breaking of the supercontinent of Pangaea (small maps) dominated the Earth's recent history.

550 MILLION YEARS AGO
BARE LAND, TEEMING SEA

Antarctica basked in the tropics and Australia lay in the present latitudes of the United States, as provinces of the large continent of Gondwanaland, which was then cementing itself together in the spasm of mountain building called the Pan-African event. Other land masses were even more fragmented than they appear at first glance, because pieces of Asia were separated by oceans of unknown width. The base map, printed black, shows positions of major blocks deduced from fossil magnetism; their modern shorelines and continental shelves are depicted to aid recognition. Superimposed in color are continental masses that existed 550 million years ago, as inferred from other evidence.

Although the Earth was four billion years old, the land was still barren of life. On the flooded edges of the continents, marine animals were diversifying (diagram below), and many of them acquired hard jackets. These left, for the first time, a highly visible fossil record, in which oval-shaped trilobites were conspicuous during this Cambrian period of geological history. Soft-bodied animals were preserved in rare events, including the underwater mud slumps that entombed the animals of the Burgess Shale in British Columbia. There the trilobites were outnumbered by soft-bodied companions, among them many experimental forms that were heading for an evolutionary dead end.

Ancestors of bony animals were probably boneless chordates resembling tadpoles. These early forerunners of human beings have not been identified, but primitive chordates akin to the present-day amphioxus occur in the Burgess Shale.

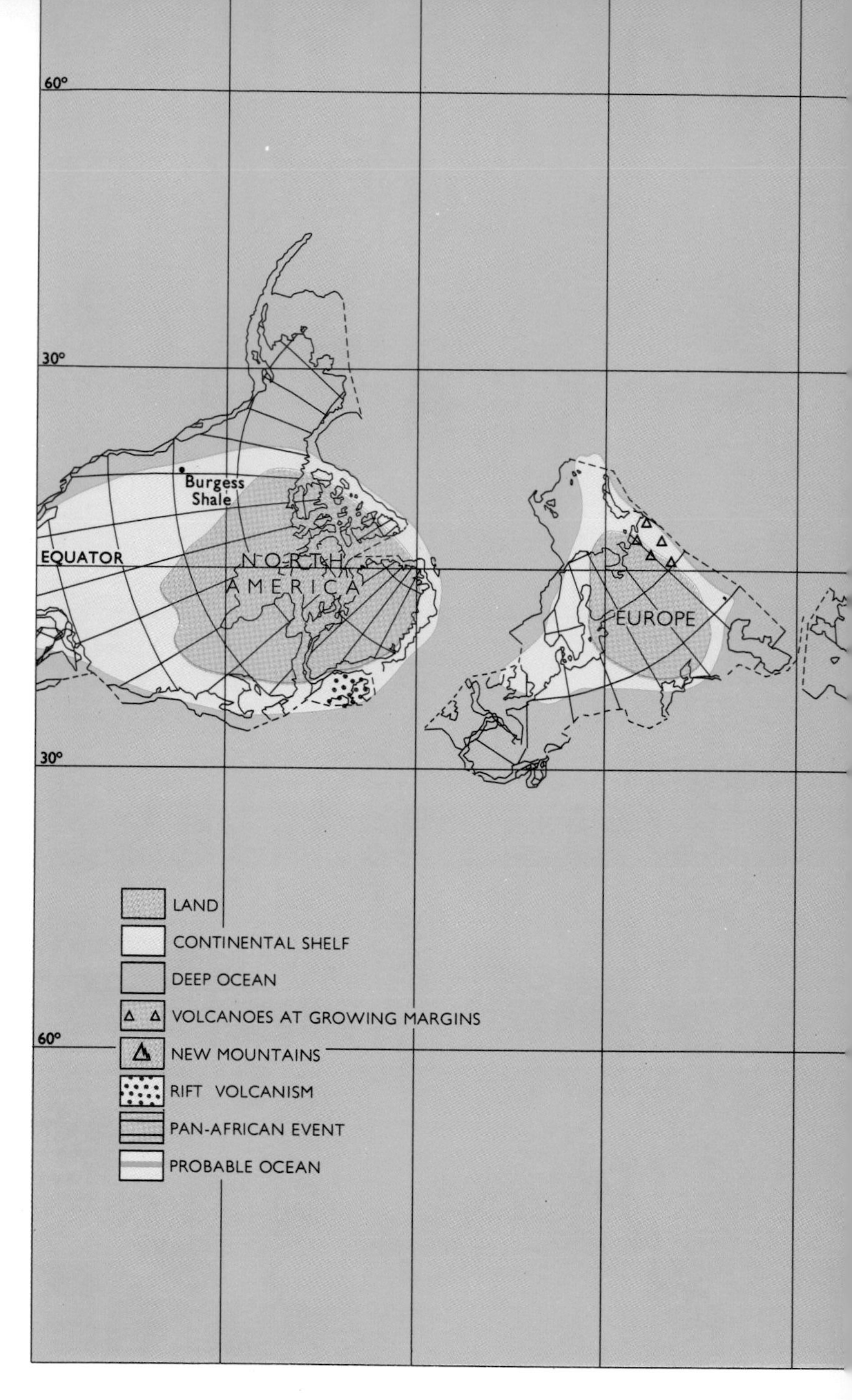

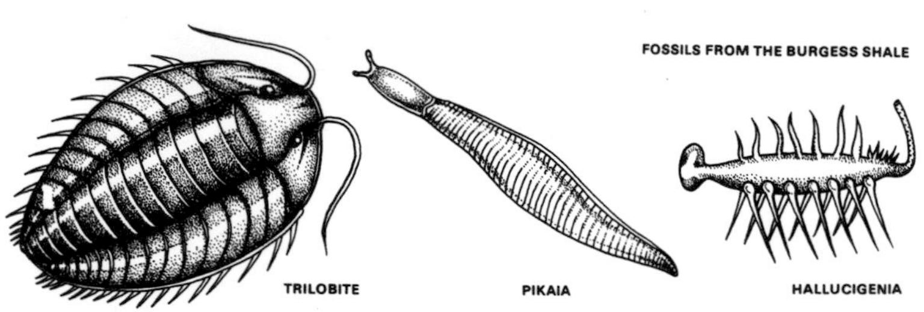

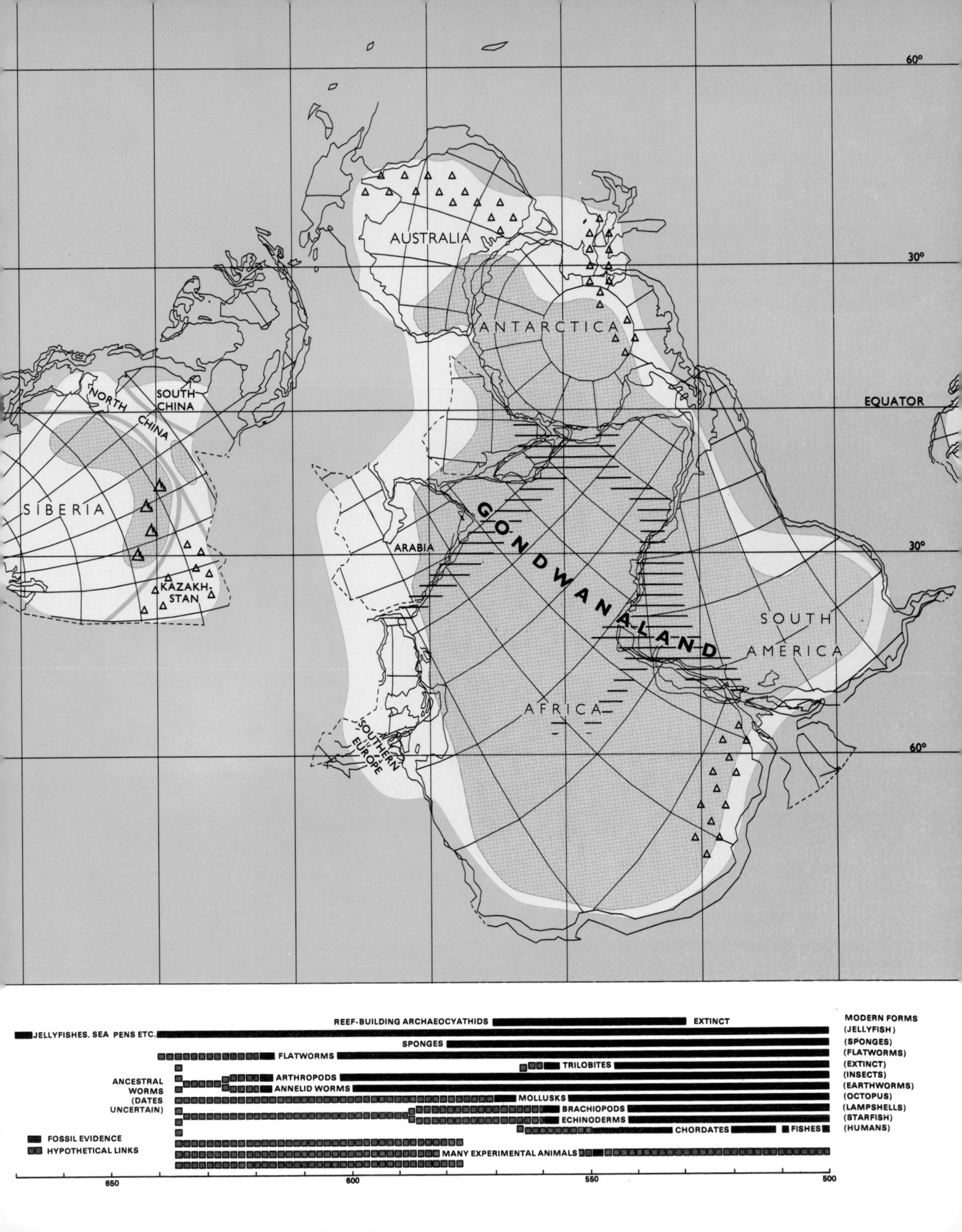

60°
30°
EQUATOR
30°
60°
AUSTRALIA
ANTARCTICA
SOUTH CHINA
NORTH CHINA
SIBERIA
KAZAKH-STAN
ARABIA
GONDWANALAND
SOUTH AMERICA
AFRICA
SOUTHERN EUROPE
REEF-BUILDING ARCHAEOCYATHIDS
EXTINCT
MODERN FORMS
JELLYFISHES, SEA PENS ETC.
(JELLYFISH)
SPONGES
(SPONGES)
FLATWORMS
(FLATWORMS)
TRILOBITES
(EXTINCT)
ARTHROPODS
(INSECTS)
ANCESTRAL WORMS (DATES UNCERTAIN)
ANNELID WORMS
(EARTHWORMS)
MOLLUSKS
(OCTOPUS)
BRACHIOPODS
(LAMPSHELLS)
ECHINODERMS
(STARFISH)
CHORDATES
FISHES
(HUMANS)
FOSSIL EVIDENCE
HYPOTHETICAL LINKS
MANY EXPERIMENTAL ANIMALS
650
600
550
500

200 MILLION YEARS AGO
THE DINOSAURS' EMPIRE

The land masses were gathered together in the latest of the world's supercontinents, Pangaea. Gondwanaland was joined to the assembly of northern continents called Lurasia, and in the east, where the Tethys Ocean filled a huge gulf, a cluster of microcontinents were attaching themselves to Asia, along lines of new mountains. The modern coasts and shelves of the major continents are shown in black; the probable land masses and shallow seas appear in color.

In these Early Jurassic times, the land was generally warm and green, with characteristic vegetation in various provinces. The pattern of the world's winds left three large desert areas deprived of rainfall. Two major impact craters, fairly close to the date of this map, may be associated with important turnovers in life. The supercontinent made a fitting arena for the giant dinosaurs, where meat eaters preyed on herds of plant eaters; the illustration below shows some animals of the Late Triassic and Early Jurassic. The dinosaurs displaced the mammal-like reptiles, whose surviving descendants included early mammals, all of them very small.

Where the floor of the oceans dived to destruction under the edges of the supercontinent, volcanoes were active, and arcs of volcanic islands piled up on the western margins of the Americas. The breakup of Pangaea began even as the last pieces were running into place, with the birth of the Atlantic Ocean between North America and Africa. Volcanic outpourings across southern Africa and Antarctica prefigured the eventually parting of those continents.

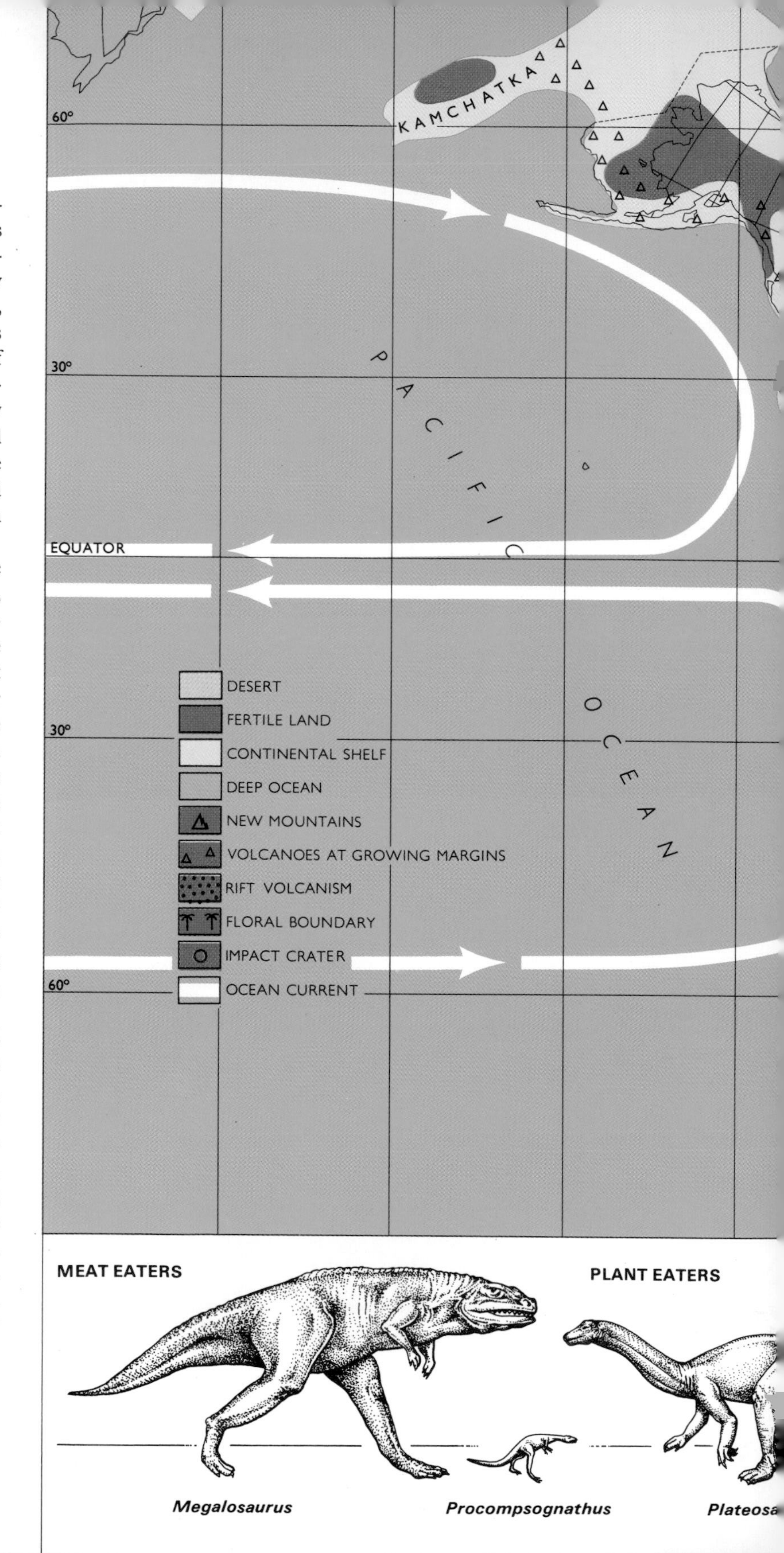

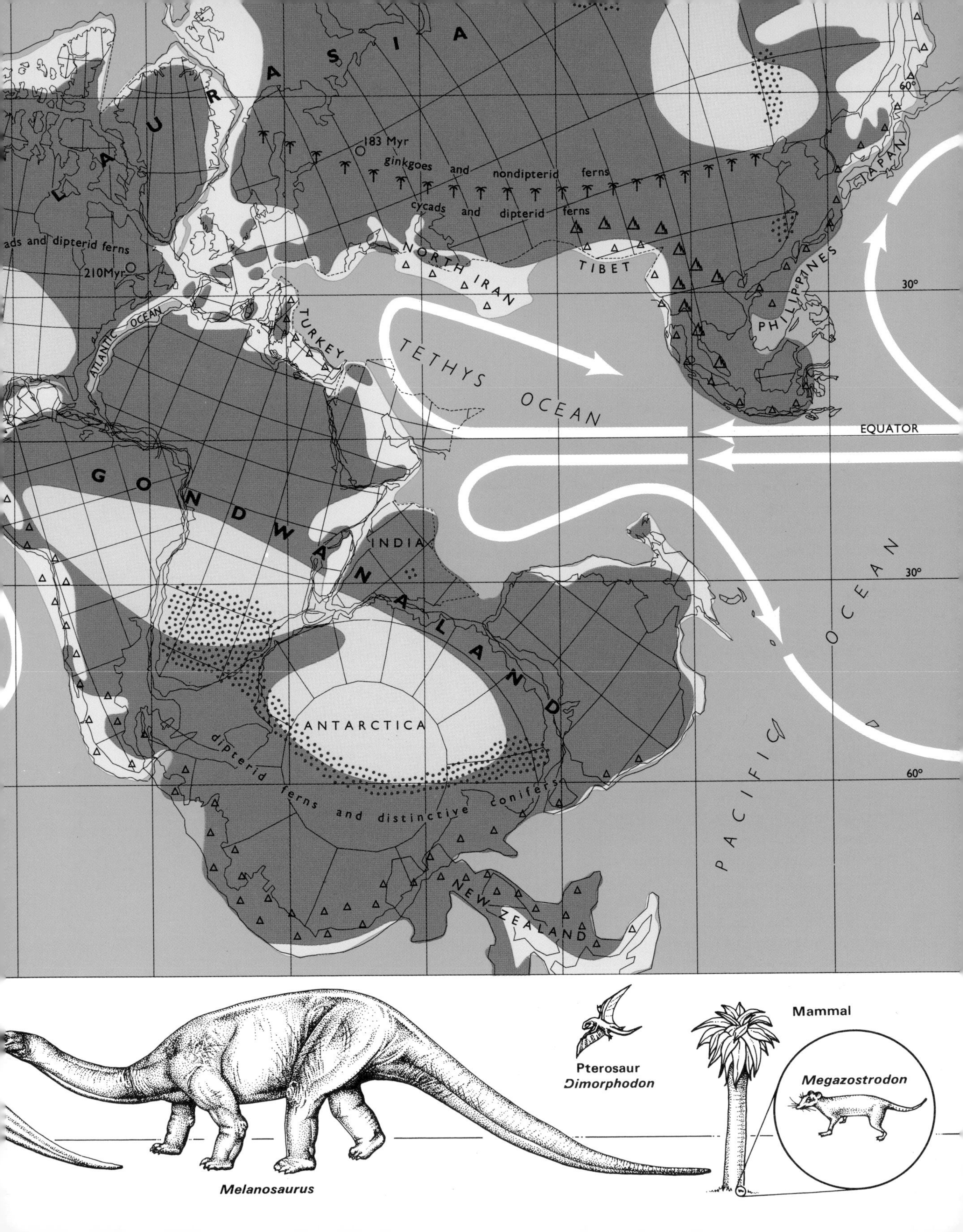

LAURASIA
183 Myr
ginkgoes and nondipterid ferns
cycads and dipterid ferns
ads and dipterid ferns
210Myr
NORTH IRAN
TIBET
PHILIPPINES
JAPAN
60°
30°
ATLANTIC OCEAN
TURKEY
TETHYS OCEAN
EQUATOR
GONDWANALAND
INDIA
30°
PACIFIC OCEAN
ANTARCTICA
dipterid ferns and distinctive conifers
60°
NEW ZEALAND
Melanosaurus
Pterosaur
Dimorphodon
Mammal
Megazostrodon

18,000 YEARS AGO
ICEBOUND HERITAGE

The continents were within a kilometer or so of their present positions, but the climate was very different when modern human beings took possession of the planet in the course of the most recent ice age. Their initial dispersal, beginning 40,000 years ago during a relatively mild respite, has left traces in most continents, including Australia, which could be reached only by crossing the sea.

The ice reached a maximum 18,000 years ago. Unmelted snow, piled in ice sheets thousands of meters thick, buried Canada and northwestern Eurasia. By robbing the oceans of water, the ice sheets lowered the sea level, changing shallow seas into land, for example in southeastern Asia and around the Bering Strait. Beside the drained Grand Banks of Newfoundland the sea surface of the Northern Atlantic was more than 12 degrees centigrade colder than at present, and the warm Gulf Stream tracked farther south.

In North America, some forests survived close to the ice sheet, and early inhabitants of Pennsylvania dwelt among oak trees and flying squirrels. In Europe the oak retreated to southern Italy, and reindeer lived beside the Mediterranean. Human beings did not merely survive their ordeal by climate, but made experiments in herding animals and raising crops.

Their descendants, living in warm conditions after the ice age, would form a misleading impression of their planet's climate. Glacial and drought-prone conditions, as depicted here, are typical of the present geological phase, and the ice will return unless human beings prevent it.

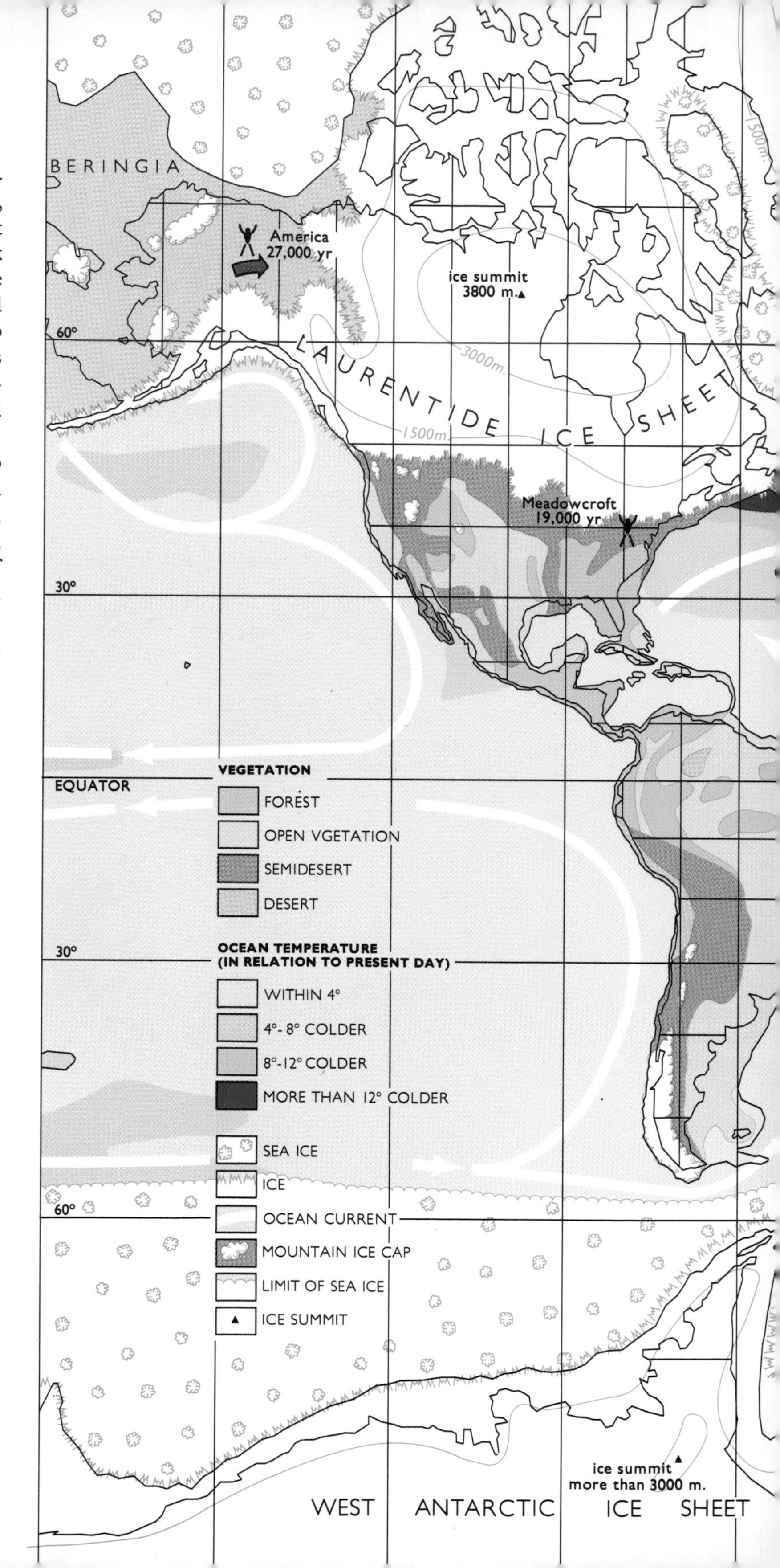

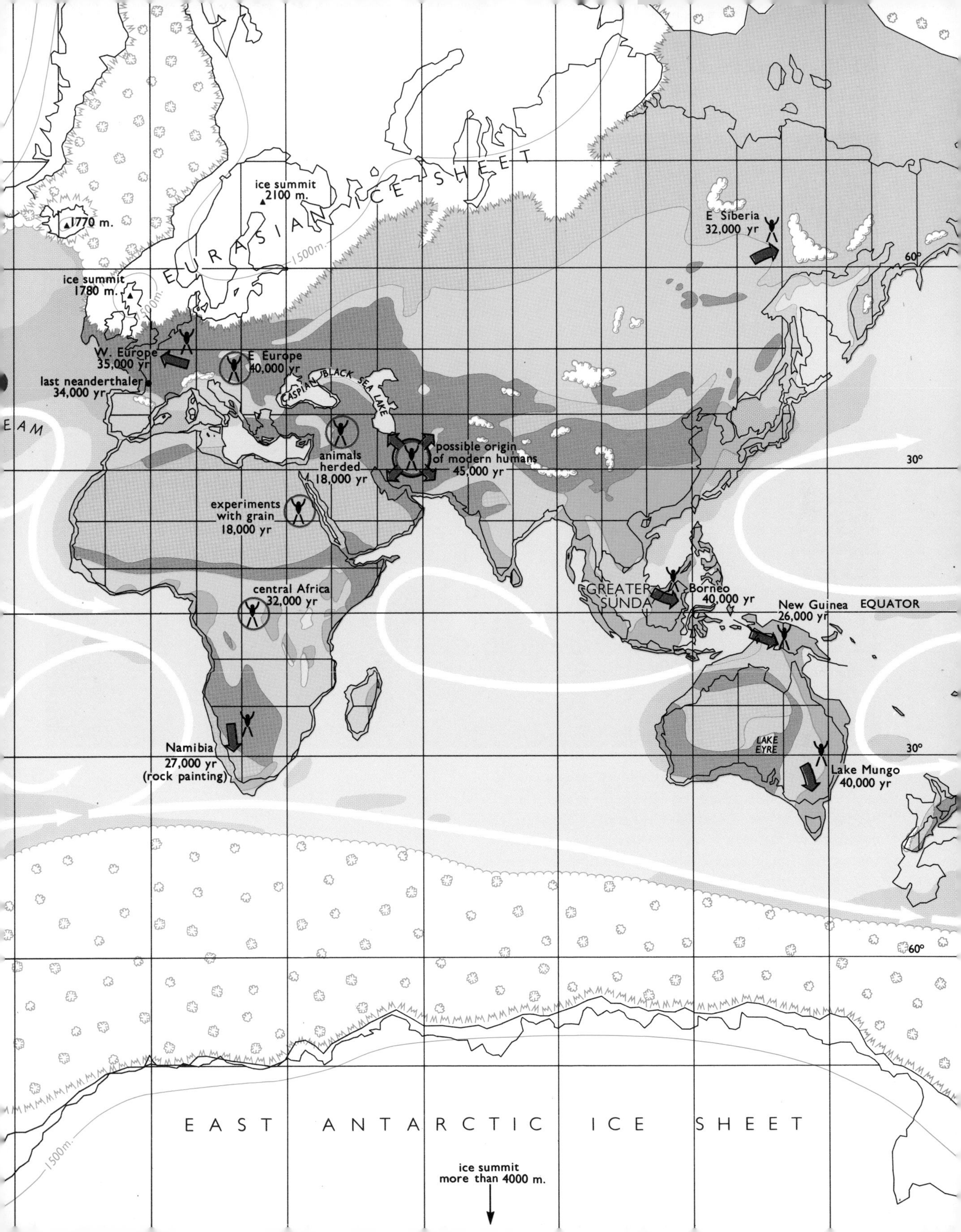

EURASIAN ICE SHEET
ice summit 2100 m.
1770 m.
ice summit 1780 m.
1500 m.
E Siberia 32,000 yr
60°
W. Europe 35,000 yr
E Europe 40,000 yr
last neanderthaler 34,000 yr
CASPIAN BLACK SEA LAKE
animals herded 18,000 yr
possible origin of modern humans 45,000 yr
30°
experiments with grain 18,000 yr
central Africa 32,000 yr
GREATER SUNDA
Borneo 40,000 yr
New Guinea 26,000 yr
EQUATOR
LAKE EYRE
Namibia 27,000 yr (rock painting)
Lake Mungo 40,000 yr
30°
60°
EAST ANTARCTIC ICE SHEET
ice summit more than 4000 m.

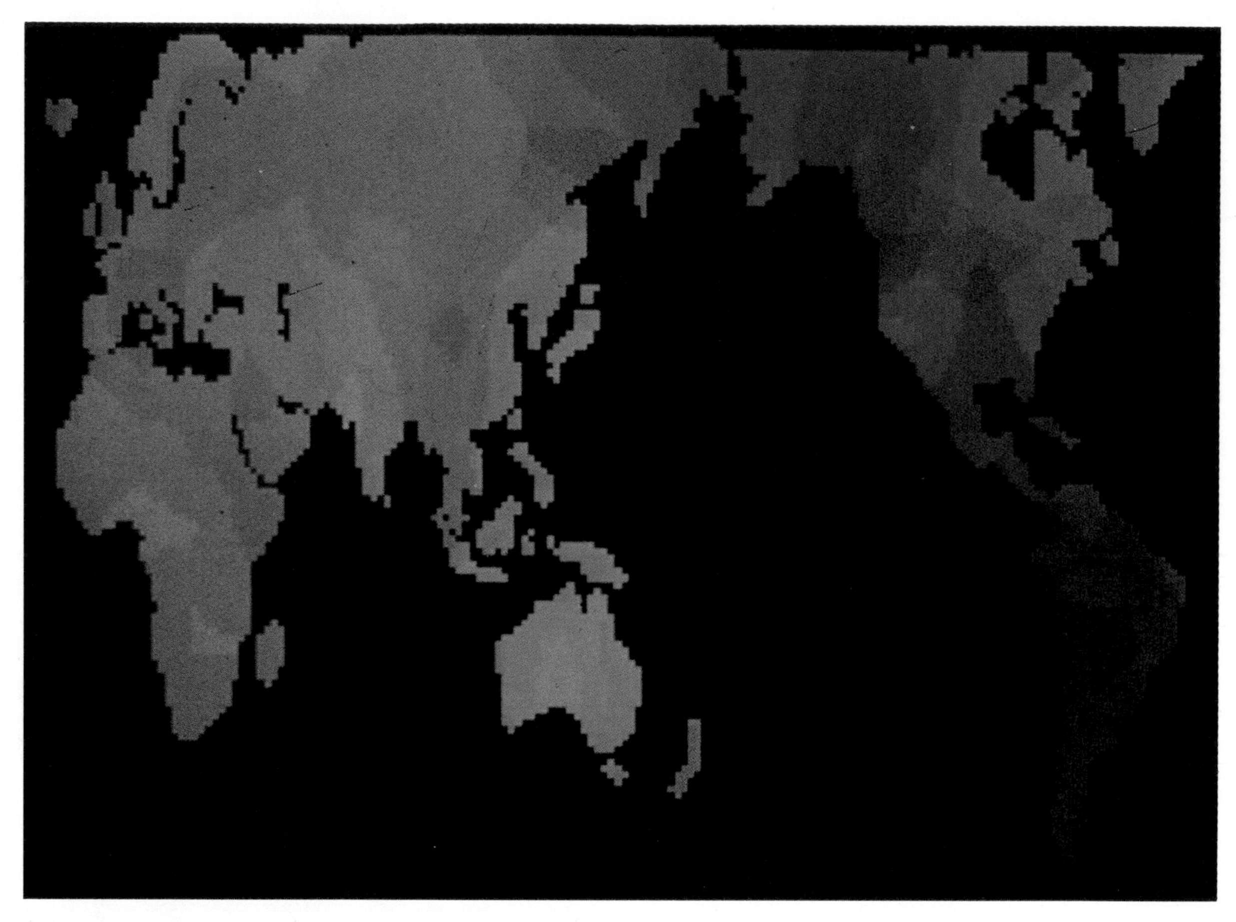

Genetic similarities and differences between supposedly aboriginal peoples, mapped above by Luigi Cavalli-Sforza's group at Stanford University, illustrate the human diversity that prevailed before the worldwide spread of Europeans. Earlier movements of peoples account for this genetic geography. The colors red, green, and blue denote three "clumps" of selected gene variants, and their relative intensities show how common each became in different parts of the world. The near-white region in south-central Asia is a possible place of origin of modern humans, who dispersed as shown in the map on the preceding pages. Subsequent breakouts of identifiable groups, some of which appear in the accompanying small maps, modulated the pattern. Linguistic and political consequences of breakouts were sometimes more obvious than their genetic effects, and languages are displayed in later maps.

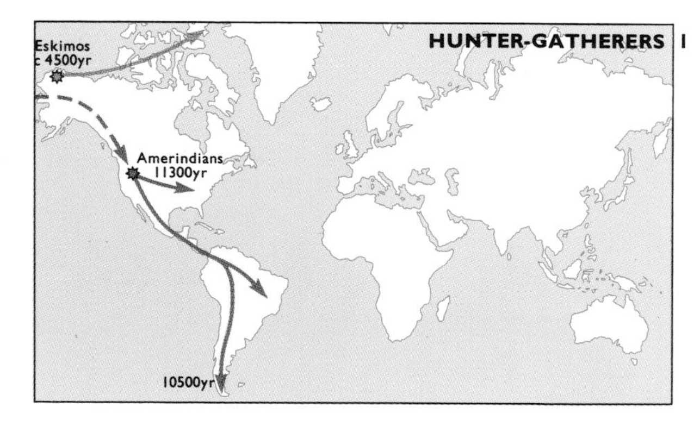

2

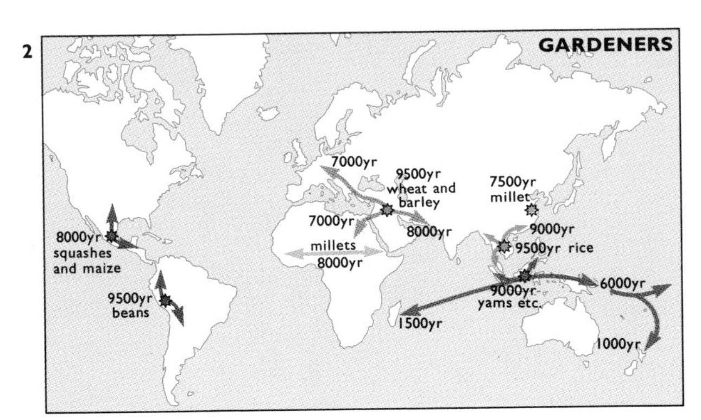
GARDENERS
7000yr
9500yr
wheat and
barley
7500yr
millet
7000yr
8000yr
9000yr
8000yr
squashes
and maize
millets
8000yr
9500yr rice
9000yr
yams etc.
6000yr
9500yr
beans
1500yr
1000yr

3

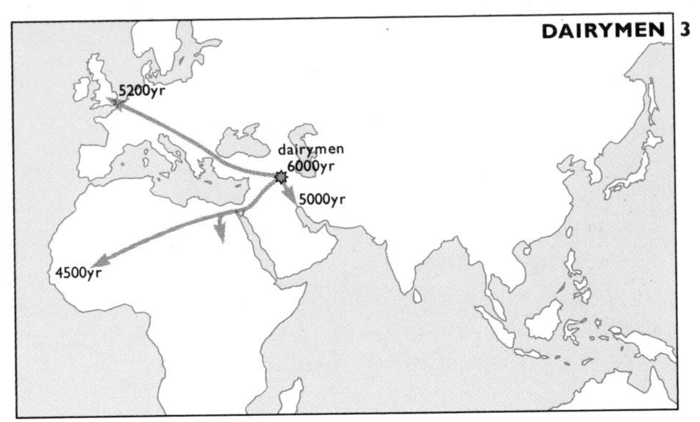
DAIRYMEN
5200yr
dairymen
6000yr
5000yr
4500yr

4

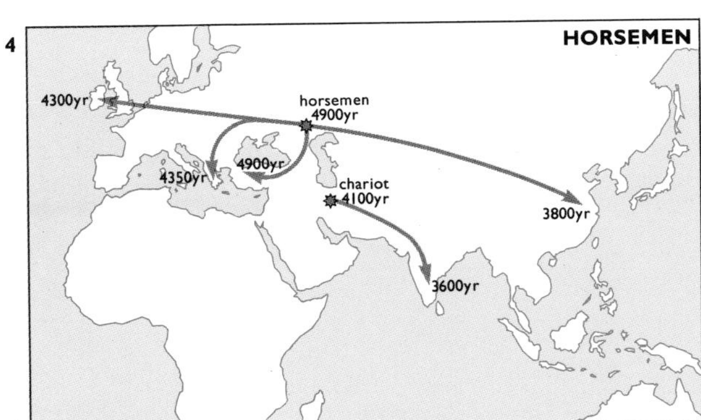
HORSEMEN
4300yr
horsemen
4900yr
4900yr
4350yr
chariot
4100yr
3800yr
3600yr

5

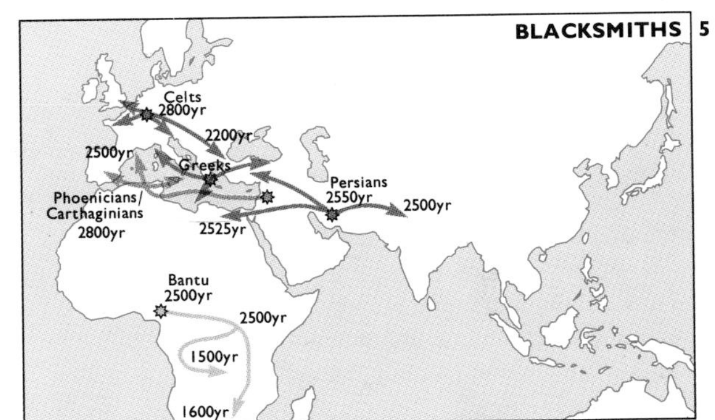
BLACKSMITHS
Celts
2800yr
2200yr
2500yr
Greeks
Persians
2550yr
2500yr
Phoenicians/
Carthaginians
2800yr
2525yr
Bantu
2500yr
2500yr
1500yr
1600yr

6

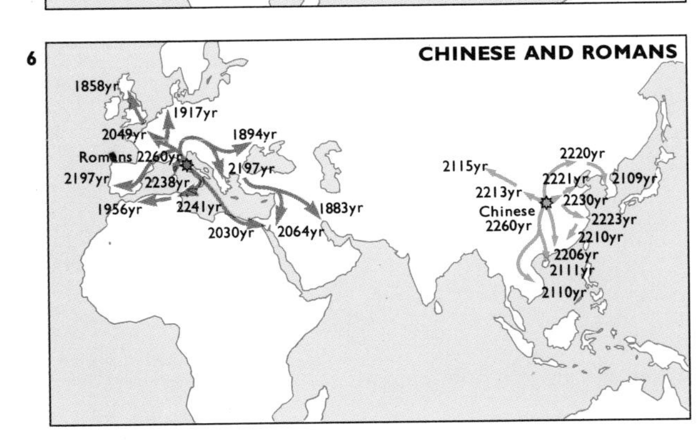
CHINESE AND ROMANS
1858yr
1917yr
2049yr
1894yr
Romans 2260yr
2197yr
2197yr
2238yr
1956yr
2241yr
2030yr
2064yr
1883yr
2115yr
2220yr
2221yr
2109yr
2213yr
2230yr
Chinese
2260yr
2223yr
2210yr
2206yr
2111yr
2110yr

7

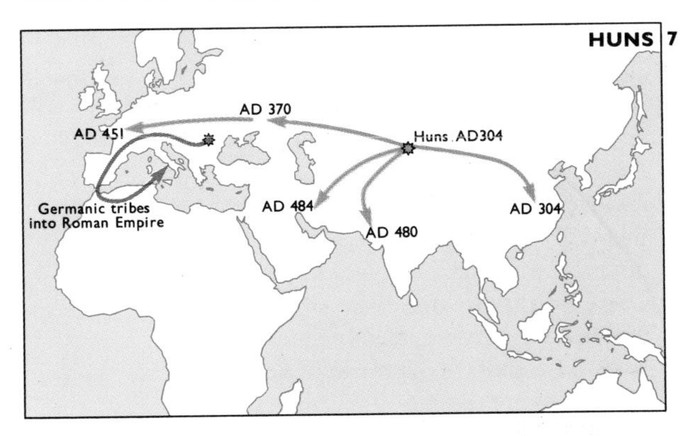
HUNS
AD 370
AD 451
Huns AD304
Germanic tribes
into Roman Empire
AD 484
AD 480
AD 304

8

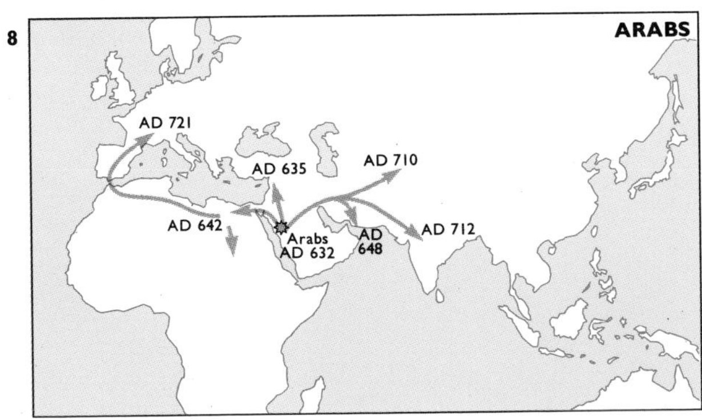
ARABS
AD 721
AD 635
AD 710
AD 642
Arabs
AD 632
AD
648
AD 712

9

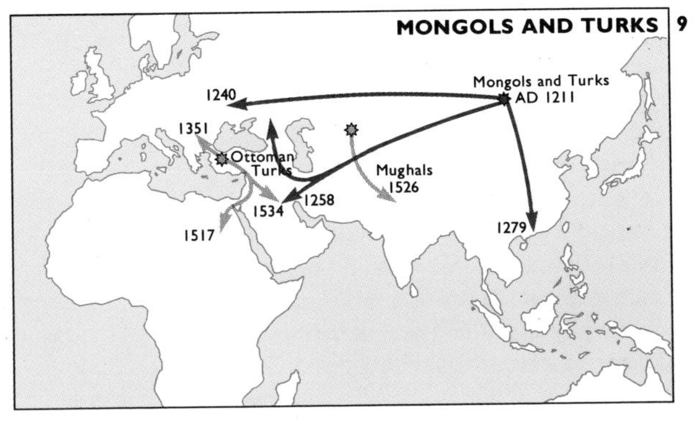
MONGOLS AND TURKS
Mongols and Turks
AD 1211
1240
1351
Ottoman
Turks
Mughals
1526
1534
1258
1517
1279

5000 YEARS AGO
EMERGENT COWBOYS

In a warm interlude, the Gulf Stream tracked northward. As the ice melted and the sea rose, independent cultivators in various parts of the world raised crops by simple hoe gardening; their populations grew, and spread to nearby areas.

A second agricultural revolution started in western Asia about 6000 years ago, with a potent combination of dairy cattle and ox-drawn plows. The new agriculture was linked with a genetic mutation that enabled certain human beings to drink milk in adulthood. These were paleskin speakers of "dairy" languages, here called Eurafrasian, ancestral to the Semitic (Afrasian), Indo-European, and Uralic language families. By 5000 years ago plows, or dairy cows, or both, had spread across northern Africa and Europe, and wheeled wagons were in routine use.

The first urbanized cultures of Kish and lower Sumer emerged in Mesopotamia. Writing began there 5500 years ago, but lack of written records elsewhere means that the language regions, distinguished by colors on the map, are hypothetical. They are inferred by combining genetic evidence with knowledge of the linguistic families that emerged subsequently in different parts of the world. Known migrations are taken into account (see previous and later maps) but it is assumed that major ethnic divisions of mankind were largely established by 5000 years ago. New Zealand and Malagasy (Madagascar) were still uninhabited then.

With the revolution of cow and plow came the social divisions that characterize Eurasian cultures, as between rich and poor, lords and serfs, and men and women. War became commonplace, and warriors set themselves up as kings.

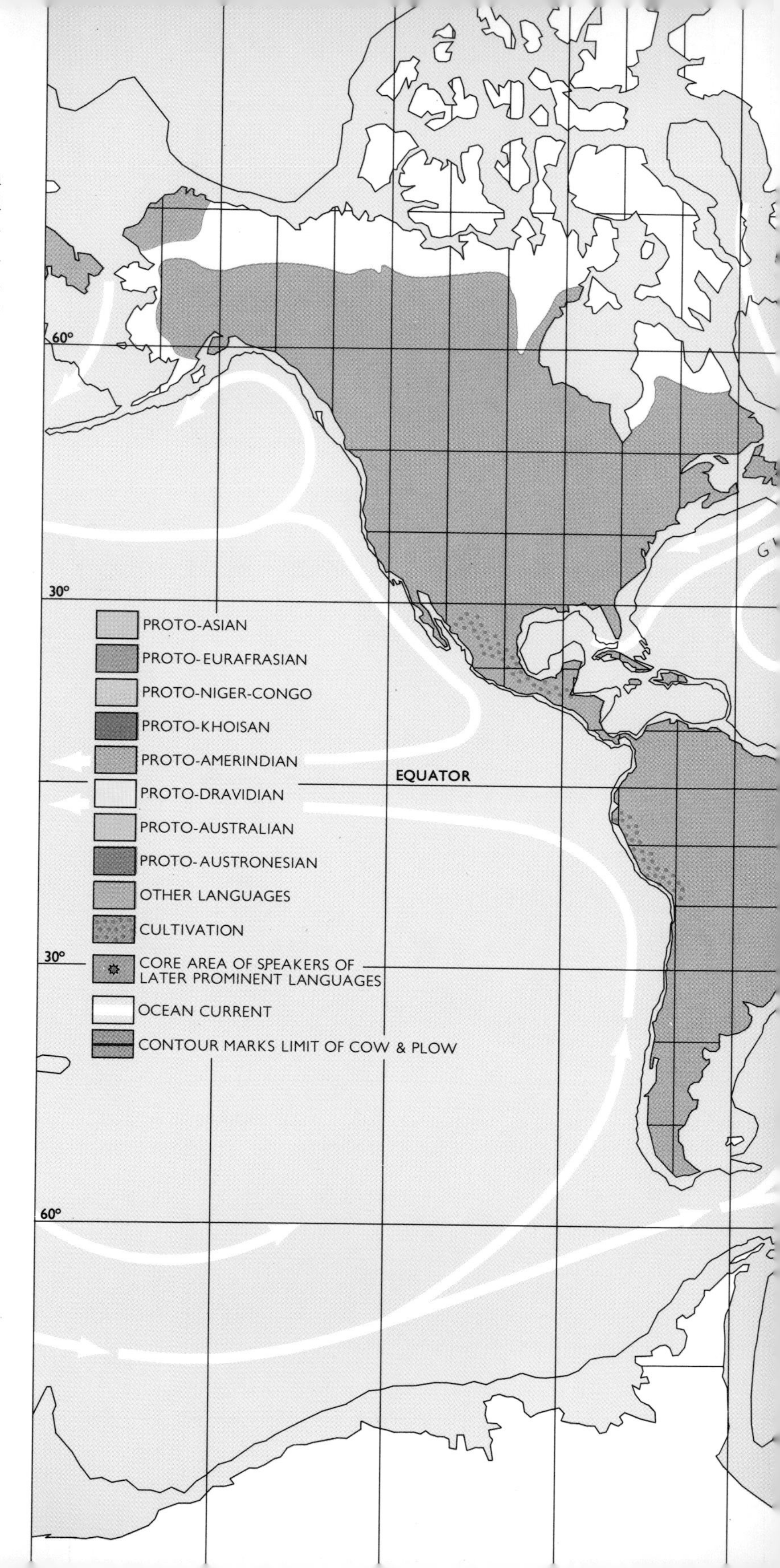

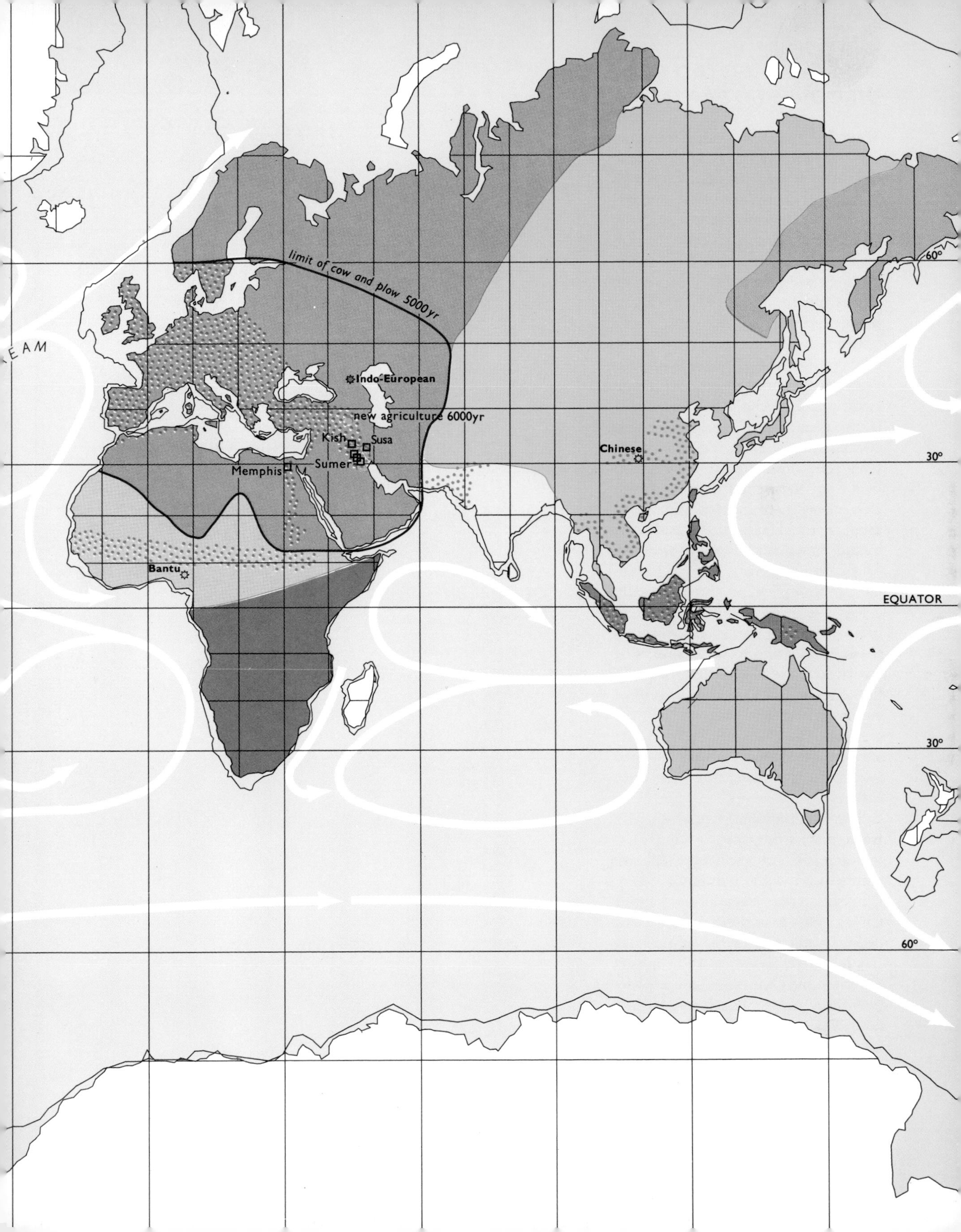

EAM
limit of cow and plow 5000yr
Indo-European
new agriculture 6000yr
Kish
Susa
Sumer
Memphis
Chinese
Bantu
EQUATOR
60°
30°
30°
60°

500 YEARS AGO (AD 1500) DISCOVERED EARTH

When Europeans in small ships began the voyages that were to alter the linguistic map of the planet, peoples in other continents were more worldly-wise. Literacy was widespread in the tropics and subtropics, which were largely knitted together by Islamic traders speaking Arabic and Persian. The Chinese of the Ming empire possessed the most advanced technologies.

From the proto-languages of 5000 years ago (previous map), the modern languages of the Old World had largely evolved, and they were distributed roughly as at present. Indo-European languages already occupied a broad belt from the British Isles, through Persia (Iran), to eastern India, but Uralic-speaking Hungarians had made inroads into central Europe. Speakers of Bantu languages of the Niger-Congo group were spread widely across Africa, and Austronesian gardeners had occupied Malagasy (Madagascar) and New Zealand. The assertive Turks who founded the Ottoman and Mughal empires were surviving representatives of the breakout of Mongols and other speakers of Altaic languages who had swept across Eurasia three centuries earlier.

The Amerindians were essentially isolated, but they had a population matching Europe's while the Chinese empire was twice as numerous. Large Chinese naval expeditions under Chêng-Ho demonstrated how sea power might master the world. But this initiative lapsed, and Europeans exploited the Chinese magnetic compass more comprehensively, as they headed eastward, westward, and right around the world.

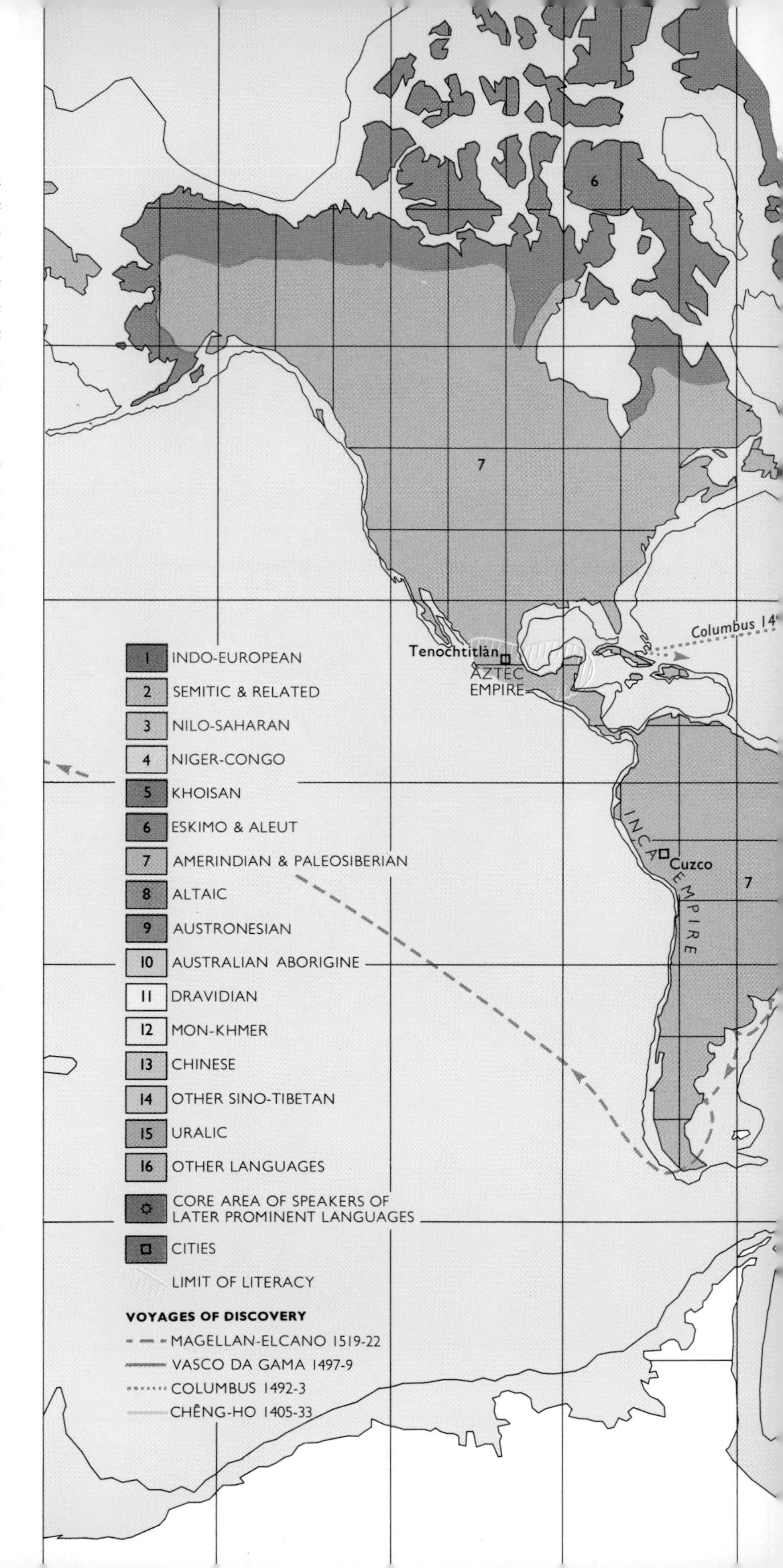

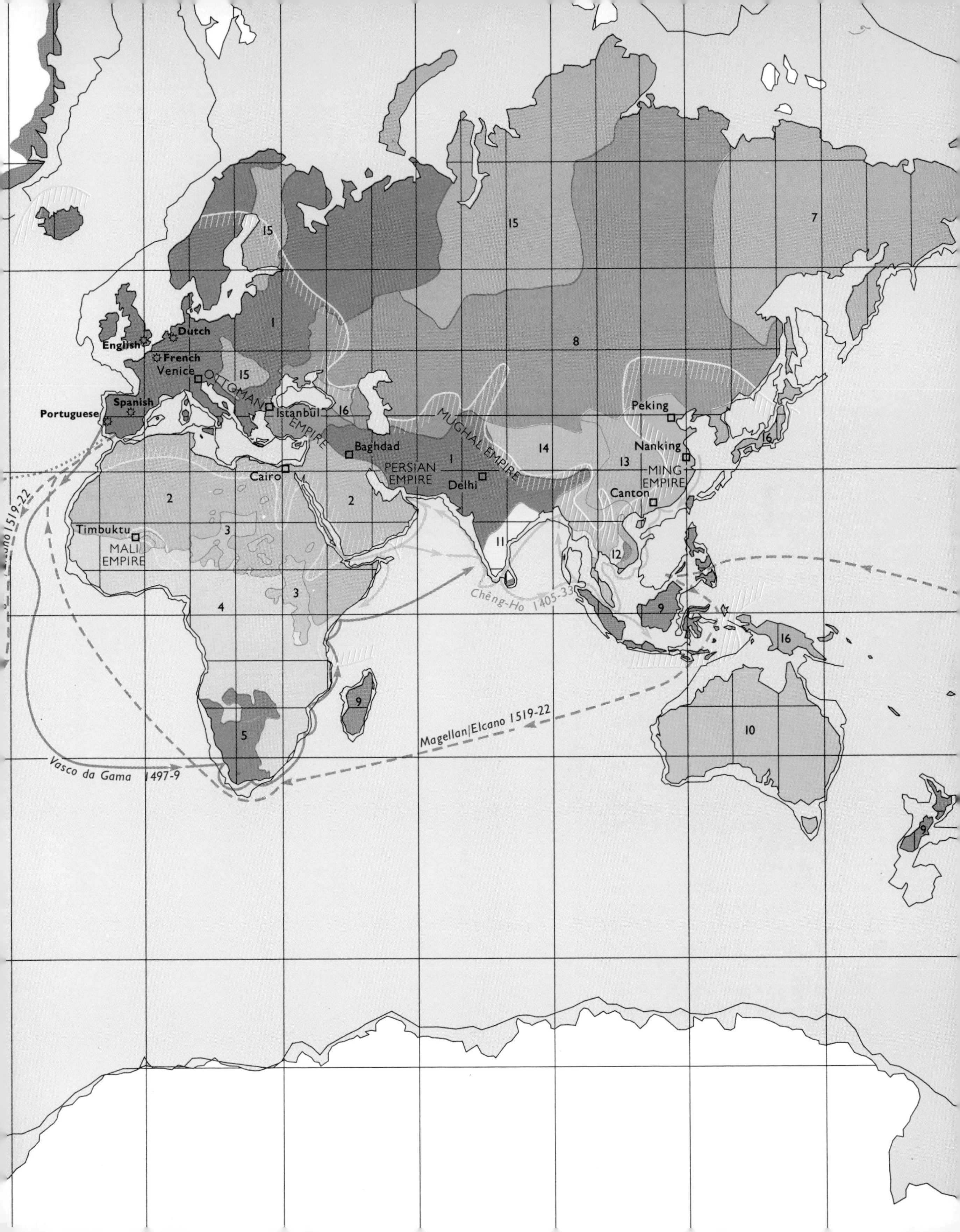

English
Dutch
French
Venice
Spanish
Portuguese
OTTOMAN EMPIRE
Istanbul
MUGHAL EMPIRE
Baghdad
PERSIAN EMPIRE
Delhi
Cairo
Peking
Nanking
MING EMPIRE
Canton
Timbuktu
MALI EMPIRE
Chêng-Ho 1405-33
Magellan/Elcano 1519-22
Vasco da Gama 1497-9
1
2
3
4
5
7
8
9
10
11
12
13
14
15
16

MODERN WORLD
ELECTRONICS IN ENGLISH

Indo-European languages dominated the world, after the European breakouts. The largest territorial gains were made by Russian in Asia, by Spanish and Portuguese in Middle and South America, and by English in North America and Australasia. The commercial power of the British in the nineteenth century and of the Americans in the twentieth helped to make English a global language. But the Chinese remained the most populous group, and trading successes gave new status to the speakers of Arabic and Japanese.

Electronics brought most parts of the planet into split-second communication with one another. A ring of satellites around the equator, revolving in step with the Earth's rotation, served an international system capable of handling torrents of data, words, and images. These told of a species divided by war and threat of war and by gross differences between rich and poor. In an era when communication counted for as much as tangible wealth, access to telephones was an apt indicator of which people shared in the unprecedented prosperity and personal freedom enjoyed by a minority of the species.

The existence of a hundred million tons of living humanity could be counted as a notable biological success for a predatory vertebrate animal, and the growing population was fed by a diminishing fraction of the workforce. On the other hand, nuclear weapons and an unreliable climate cast shadows over the species' future. The endless give and take of human history continued, with activity in California and Japan starting a shift in the economic focus of the world, from the Atlantic to the Pacific shores.

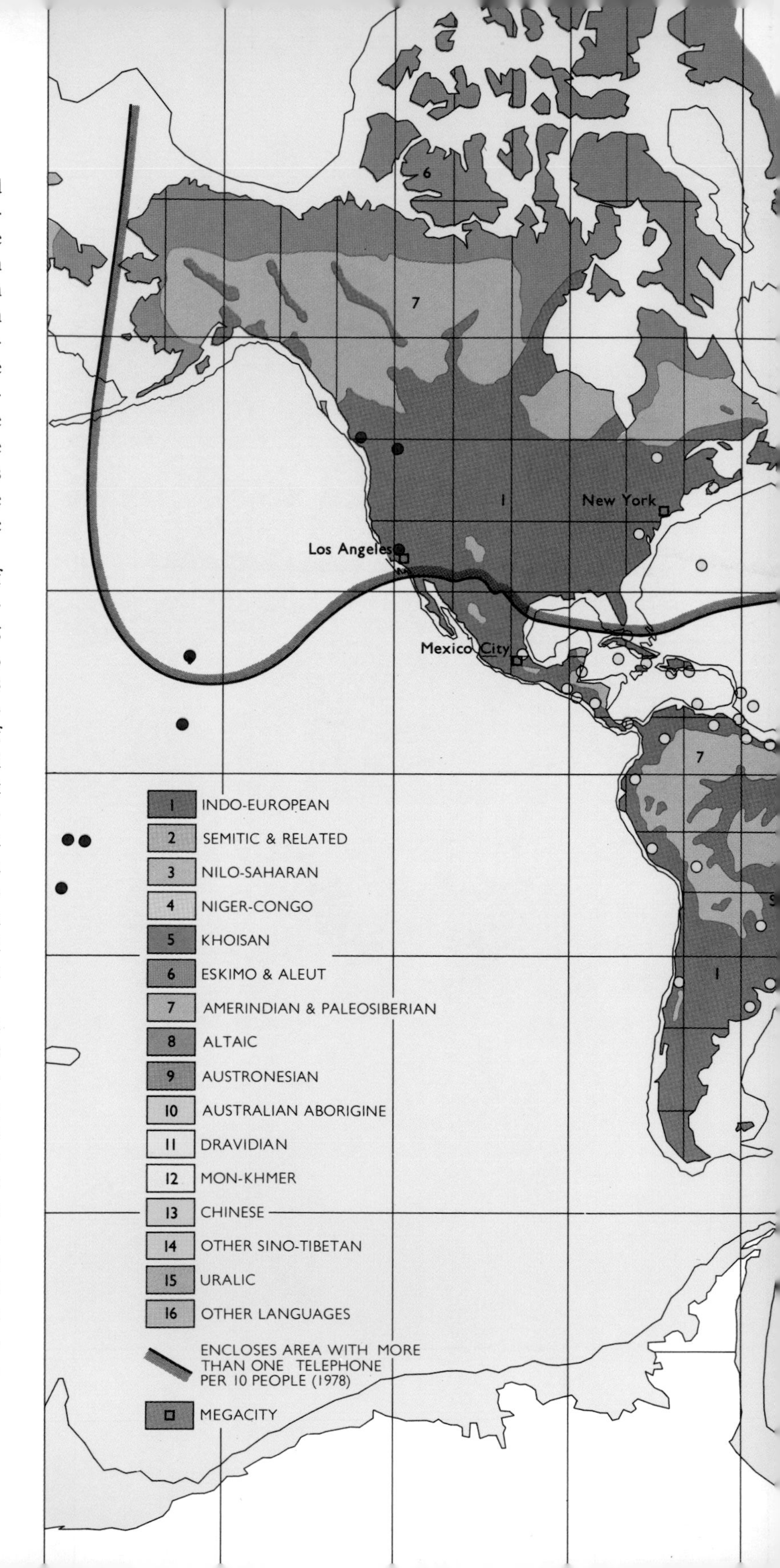

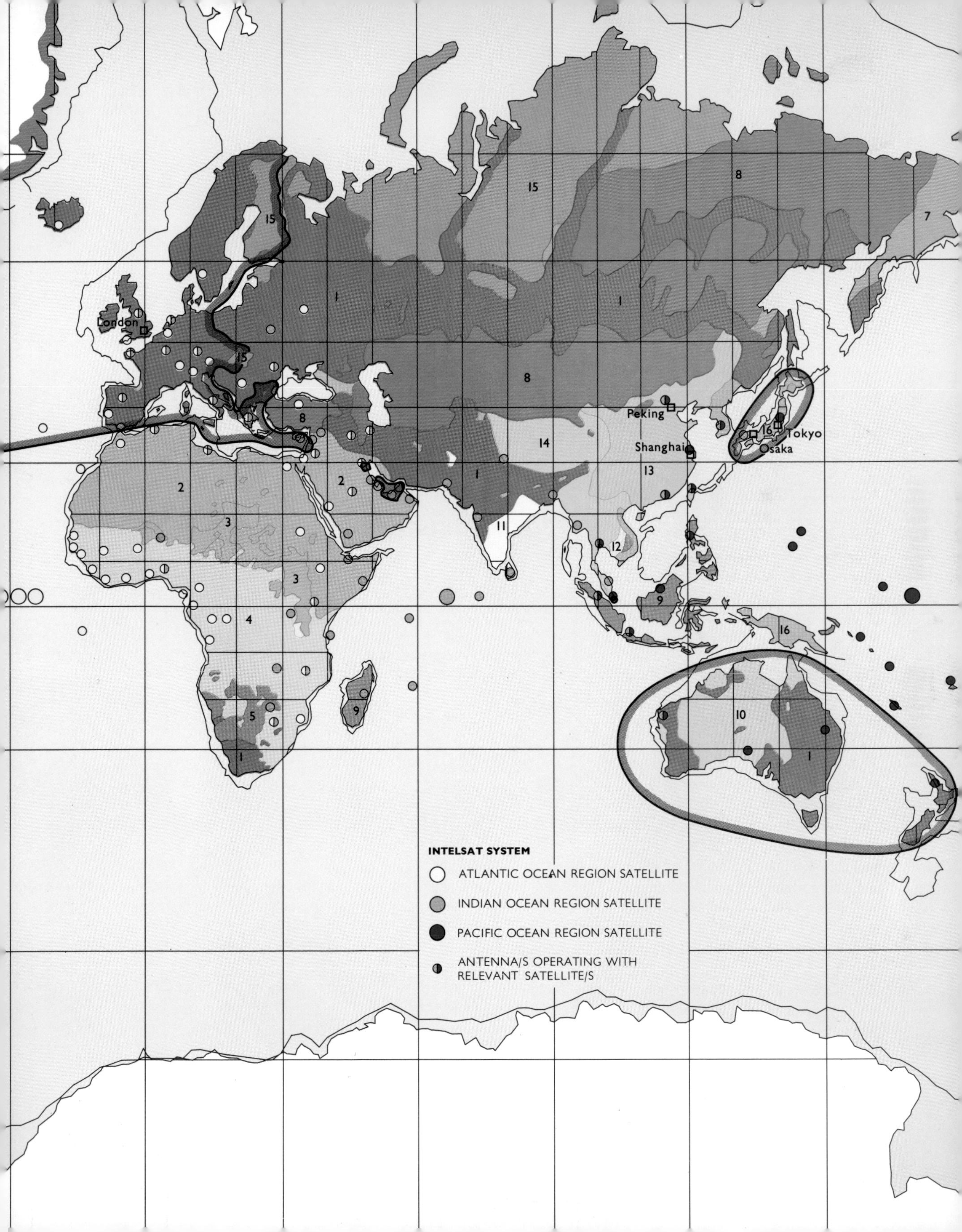
London
Peking
Shanghai
Tokyo
Osaka
INTELSAT SYSTEM
ATLANTIC OCEAN REGION SATELLITE
INDIAN OCEAN REGION SATELLITE
PACIFIC OCEAN REGION SATELLITE
ANTENNA/S OPERATING WITH RELEVANT SATELLITE/S

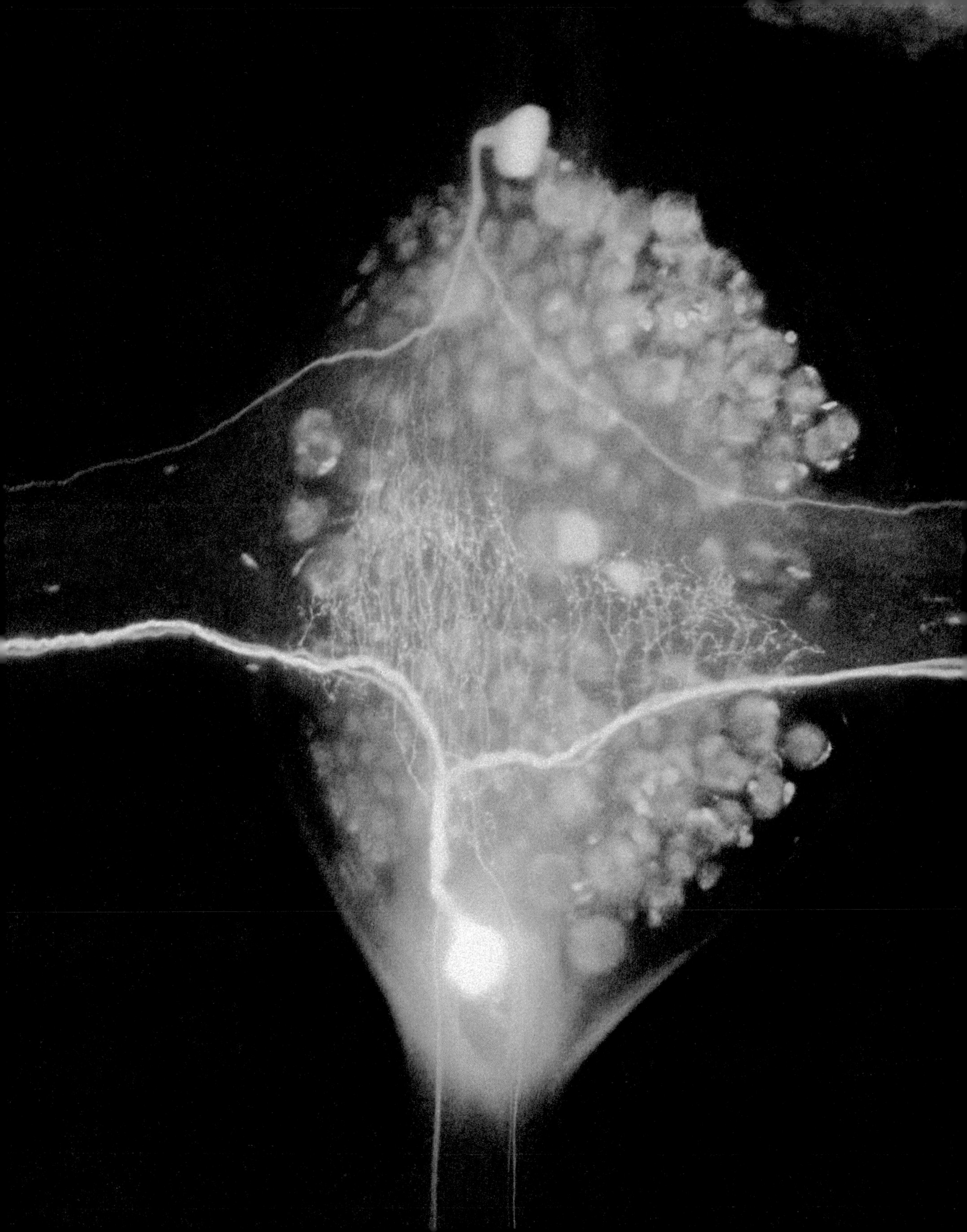

NARRATIVE AND TIMESCALE

The spiral summarizes *the main timescale, with each complete turn corresponding to a tenfold change in age. Starting near the middle (13,500 million years ago), the spiral passes clockwise through 10,000 million years ago, and around to 1000 million years ago, and so on, until the ninth turn reaches, at the top, 10 years before* AD *2000. Each tenfold turn is split into five segments, and these provide the segments of the main timescale, starting on p. 102.*

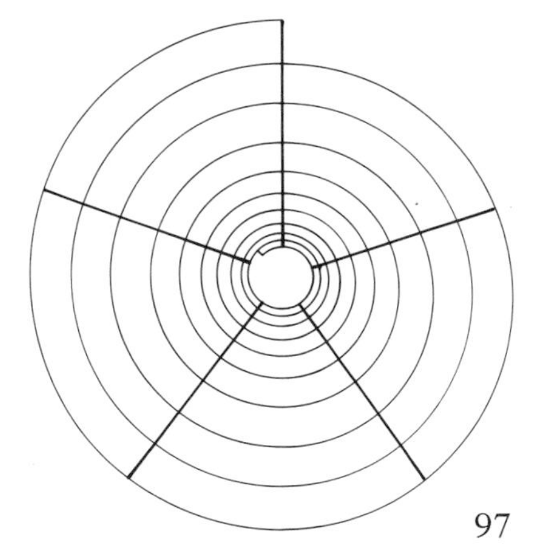

Nerve cells of a leech reflect the dawn of brainpower 620 million years ago, when annelid worms, the group to which the leech belongs, first registered in the fossil record. The cells shown here, greatly magnified, are from the nerve center, or ganglion, of a segment of the animal's body. A flourescent dye injected into the lower cell has spread into the upper one, making the connections luminous.

13,500 MILLION YEARS AGO

MELEE OF THE PARTICLES

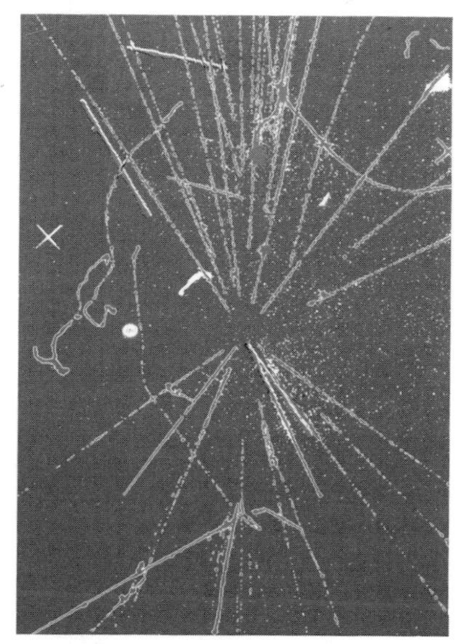

Tracks of energetic subatomic particles, transforming themselves in collisions between protons and antiprotons in a giant accelerator, recapture something of the conditions prevailing in the Big Bang, when the cosmic forces tried out every possible permutation of energy and matter.

Splitting a second. *The notation used on the timescale specifies unimaginable but computable shards of time. For example 10^{-43} second means one second divided by one followed by 43 zeroes, quite a large number: 10,000,000,000,000,000,000,-000,000,000,000,000,000,000,000. This may justify the use of the shorthand. In conversational terms, a billionth of a second (10^{-9}) is a nanosecond, and a millionth of a second (10^{-6}) is a microsecond.*

The beginning of time itself is the first entry on the timescale, because the event that created a growing volume of space, crammed with energy, also started the cosmic clock. The universe came into existence about 13,500 million years ago, out of a timeless foam. Other universes may be two-a-penny but, by definition, human beings can see and feel only their own. Time's progress was just one of many qualities of existence and unraveled themselves from the Big Bang in the first millionth of a second. At the outset, time was knitted with space so tightly as to be unrecognizable.

Spacetime looked vacant in the very first instant, and although its fabric shimmered with possibilities, it was smaller than a pinprick. The bubble of spacetime tentatively grew, and gravity was the first familiar force to establish itself, superseding the peculiar quantum gravity at work in the primeval foam. But the growth caused a hiatus, which allowed a temporary force to come into play. It was a grand repulsion, and it overrode the cohesive power of gravity. It also stoked spacetime with energy. The microuniverse inflated, very rapidly, to embrace a much larger volume of spacetime, the size of a tennis ball. By then it contained enough energy to build billions of galaxies of stars and hurl them across billions of light-years. But the energy took an intangible, obverse form, like that of a hole waiting for things to fall in it, and the universe remained momentarily dark and empty.

The fabric of spacetime yielded to strains of unimaginable intensity, and the tennis-ball universe filled with heat, as the pent-up energy made particles. The onset of the Hot Big Bang might seem dumbfounding, but physicists are eloquent on the subject. Besides gravity, another force was latent: an electronuclear force restricting the character of possible particles and threatening to impose a sterile perfection on the contents of the universe. But when spacetime cracked, so did the monolithic order of the electronuclear force, which broke into three versatile cosmic forces: the electric force, the nuclear force, and the so-called weak force, agent of radioactivity. A variety of particles of matter, and of antimatter too, was mass-produced in vast numbers.

The particles were frozen energy, parceled out in accordance with a law of creation, $E=mc^2$, that demanded a great deal of energy to make one small electron. The family of nimble electrons and their relatives coexisted with a family of much heavier quarks. The quarks are not abstruse entities (fingernails are made of them, and

beginning of time
▼ zero

normal gravity
10^{-43} second ▼

grand repulsion
10^{-38} ▼

Hot Big Bang; quark era
▼ 10^{-35} second

10^{-40} second

so is everything else), but nowadays they shyly inhabit only the nuclei of atoms. In the quark era they appeared in public as the chief thickener of a scalding soup. That the universe adopted a limited range of subatomic particles, neither all identical nor all different, tells of a delicate compromise with imperfection. Strict symmetry remained between positive and negative electric charges on the quarks and electrons, so keeping matter electrically neutral. On the other hand, the production of matter and antimatter was not exactly equal, which was just as well, or they would have destroyed each other entirely.

Space was bright with radiant energy. Distinctive particles, made of mixtures of matter and antimatter, purveyed the cosmic forces, and conspicuous among them were inherently speedy particles like light, carriers of the electric force. Light enjoyed a special relationship with spacetime, such that its progress through space was a fundamental measure of the passage of time. But the universe was packed with electrically charged particles that reflected and scattered the radiant energy, making space opaque—like a piece of the sun but unimaginably heavy. With only the present quota of matter, it would have been impenetrable enough, but in fact there was sufficient material for millions of universes.

Matter and energy were wholly interchangeable, and the frenzied contents of the quark soup were being continually destroyed and replenished. The weak force enlivened the scene by promiscuously altering the characters of individual particles. The first restraint on the free-for-all began when the universe was the size of a giant star: carriers of the weak force became scarce, and although the force continued to transmute matter, in radioactive decay, it operated much more slowly. Next, on the stroke of the microsecond, the nuclear force tamed the wild quarks. They had cooled and slowed down sufficiently for their mutual attractions to become irresistible, and the quarks set themselves up in *ménages à trois*, threesomes free of tension as long as no one tried to leave. Thus were protons and neutrons made, of three quarks apiece. These less unfamiliar particles of nuclear matter came into existence in what, by earthly standards, was still an extremely hot mass of matter and energy, although wider than the solar system by that time.

Protons and antiprotons dominated the scene for a while, until the universe became too cool to create them anymore. Only the slight imbalance in numbers then prevented the total destruction of heavy

10^{-35} SECOND

EXPLODING UNIVERSE

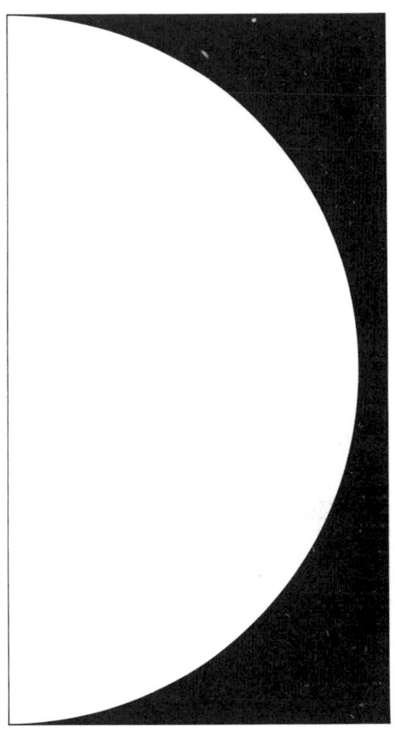

The contents of the entire observable universe filled a volume only ten centimeters in diameter, at the onset of the Hot Big Bang.

We were quarks. *Lineal kinship of modern humans with objects and organisms in the remote past is noted for each segment of the narrative. Here too, formal names appear for certain segments of the timescale. Thus 10^{-43} second is the Planck time; Hot Big Bang is a technical term. These events occurred at an estimated 13,500 million years ago, but the timing of the initial events, reckoned from the start, is largely independent of the question of the age of the universe.*

weak force weakened
▼ 10^{-11} second

proton era
▼ 10^{-6} second

10^{-20} second

10^{-10} second

ONE MICROSECOND
MATTER AND ANTIMATTER

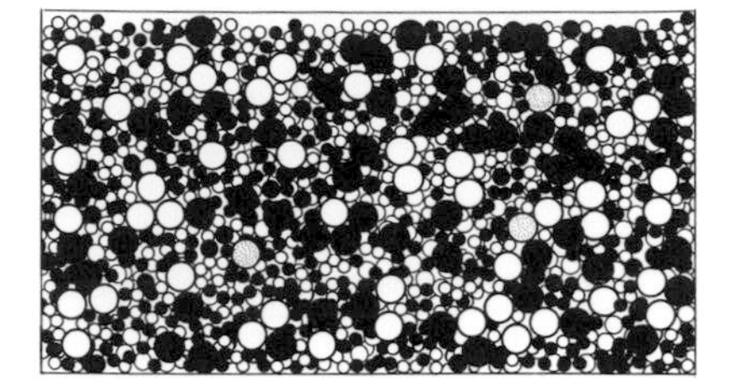

ONE HUNDRED MICROSECONDS
PROTONS FROZEN OUT

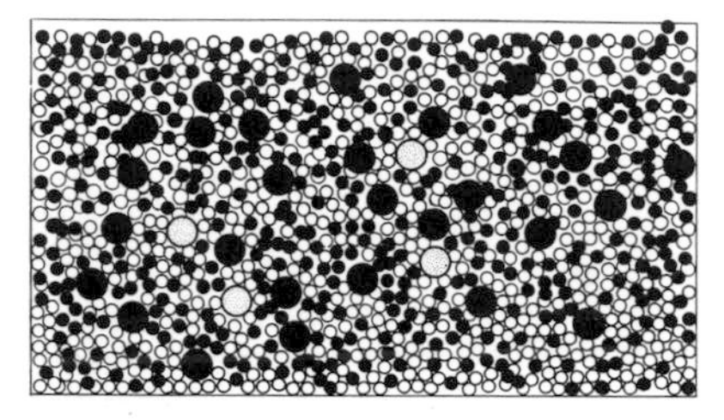

ONE SECOND
ELECTRONS FROZEN OUT

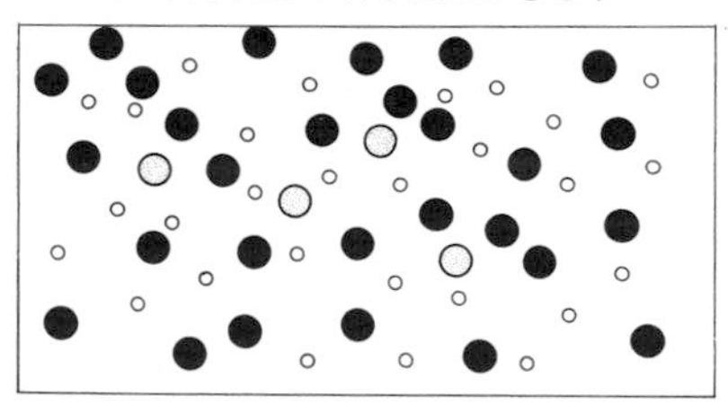

PROTONS ●
NEUTRONS ○
ELECTRONS ∘

The Big Bang made both matter and antimatter in abundance, most of which vanished in mutual annihilation when the temperature fell below what was needed to create new particles and antiparticles. A slight imbalance in numbers left a residue of matter: first the heavy protons and neutrons, and then an appropriate number of lightweight electrons.

matter, when the antiprotons wiped out almost all of the protons in a process of mutual annihilation. This carnage, when the universe was about a hundred microseconds old, left the electrons, antielectrons, and their kin still freely creatable, as the most abundant form of matter in the universe. Their moment of reckoning came after one second had elapsed. Annihilation by antielectrons began to reduce the electrons to much more modest numbers—exactly equal, in fact, to the number of protons. Radiant energy then predominated for a while, in a glare of gamma rays.

Through these changes of regime, the universe seemed to be rehearsing its orchestra, section by section. One company of particles was dismissed: neutrinos, ghostly electrons lacking electric charges, were capable of interacting with other matter only by the enfeebled weak force, and as a result they were condemned to rush about the universe forever, passing obliviously through stars and planets, accomplishing little. From the other available particles of matter, the cosmic forces were capable of building stars and rocks and living things, some of which would eventually use electrons to paint faces on television screens and nuclear matter to menace one another. But all that would take time. At one second after the beginning of time, the contents of the known universe filled a volume three light-years wide, and the pace of events slowed. The chief hint of what might be pending was the way the particles glittered with ungratified force.

A thermonuclear explosion racked the entire universe in the next notable event of its early history. The fireball, forty light-years wide, was a mild affair compared with what had gone before. It occurred when the universe was about three minutes old and was cool enough, at a billion degrees, to allow the nuclear force to bind protons and neutrons together, releasing energy in the process. Almost a quarter of all the heavy particles joined together in foursomes (a pair of protons and a pair of neutrons) to make nuclei of helium atoms; no atoms formed in this era, because the electrons needed to complete them were still far too agitated. Protons, later to serve as the nuclei of hydrogen atoms, outnumbered the helium nuclei by more than ten to one.

This matter might have scattered diffusely into a gassy desert as the expansion continued. If the universe was to do more than break wind, its material had to become sufficiently clumped to let the cosmic forces make interesting objects. Gravity was the force best

proton era 10^{-6} ▼ | **electron era** ▼ 10^{-4} second | **gamma-ray era; neutrinos loose** 1 second ▼ | **helium making** ▼ 3 minutes

1 second | 1 day

able to reach out and scoop material together, but it was racing against the disruptive expansion of the universe and needed a start. Pressure waves, most familiar as sounds, create zones of relatively high density in any material through which they pass, and waves of this kind were the origin of clumping. Local disorderliness in the young universe, beginning when particles first formed, persisted as vibrations, because matter was as taut as a drum: gravity tried to compress it while pressure tried to explode it. The vibrations made some regions a little denser than the rest, and the hum that pervaded the cosmos sang of galaxies to come.

The melody of the Big Bang is a matter of dispute, with some cosmomusicologists favoring bass notes that eventually summoned up giant clusters of galaxies, and others preferring the treble that descanted on far smaller clusters of stars. Given a little harmony, clumping at different scales was entirely possible. When the universe was ten years old, the bass had descended to a pitch corresponding to the mass of an individual galaxy, making that a reasonable moment to log the seeding of the galaxies. Visibility remained poor: all views were obscured by the fog of electrons that scattered or absorbed radiant energy.

Ten thousand years of expansion cooled the radiant energy to the point where the stock of matter outweighed it; since that transition, most of the accessible energy of the universe has been frozen in particles of matter. Some three hundred thousand years passed, while space grew to about one billionth of its present volume and cooled to a few thousand degrees, about as hot and as bright as the visible surface of the sun. The electrons were then cool enough for the electric force to snare. The protons and helium nuclei took possession of them to make complete atoms of hydrogen and helium. Matter began to assume its familiar atomic form for the first time.

The fog cleared: purged of free electrons, the universe became highly transparent by the time it was 1 million years old. The all-pervading light from that era has passed freely through space ever since, but it dimmed as the universe expanded, shifting from an intense red glow, at a million years, to invisible infrared, until nowadays it takes the form of very cool microwave background radiation detectable only with special radio antennas. It remains a firm signature of the Big Bang.

The luminous stuff of the universe is marshaled in billions of galaxies, each consisting of billions of stars, and galaxy making was

THREE MINUTES
HELIUM NUCLEI FORMED

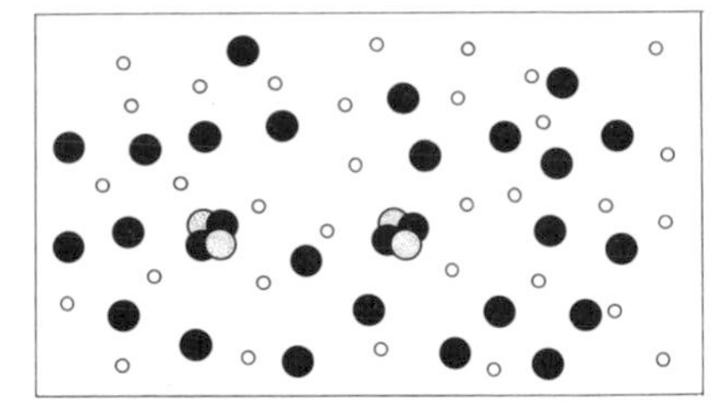

300,000 YEARS
ATOMS FORMED

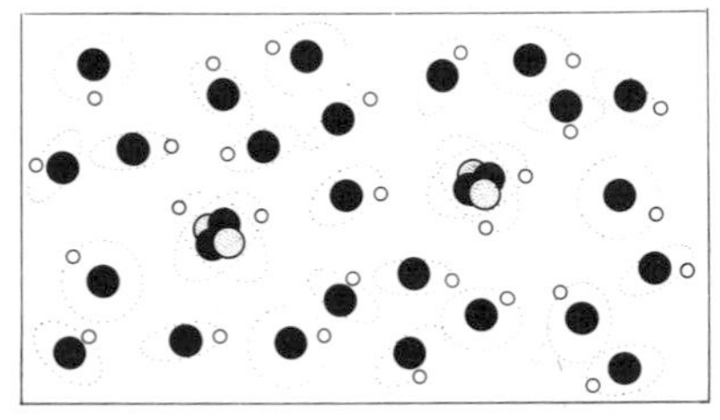

A thermonuclear explosion next converted the neutrons plus some protons into helium nuclei. When the electrons were cool enough, protons and helium nuclei trapped them, thereby making atoms of hydrogen and helium that drifted in space.

We were hydrogen. *A more formal name for the proton era is the Hadronic era, while the electron era is the Leptonic era (hadrons being nucleons and mesons, and leptons being electrons, heavy electrons, and neutrinos). The gamma-ray era is the Radiation era; helium making is also known as nucleosynthesis. The events occurred about 13,500 million years ago; even one million years is a very small fraction of the age of the universe.*

galaxies seeded
▼ 10 years

matter predominant
10,000 years ▼

atoms forming
▼ 300,000 years

transparent universe
▼ 1,000,000 years

10,000 | years

12,500 MILLION YEARS AGO
BIRTH OF GALAXIES

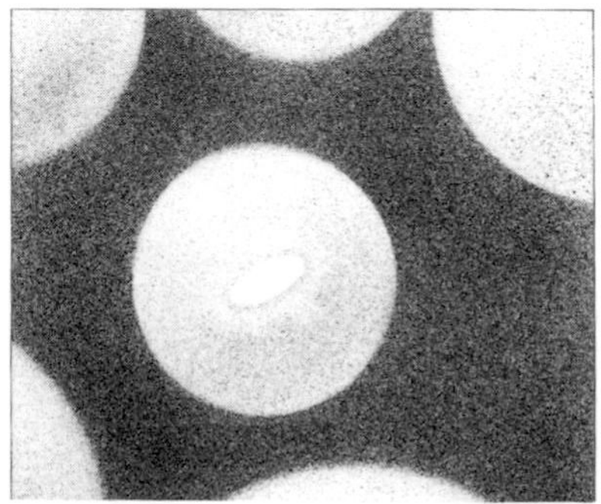

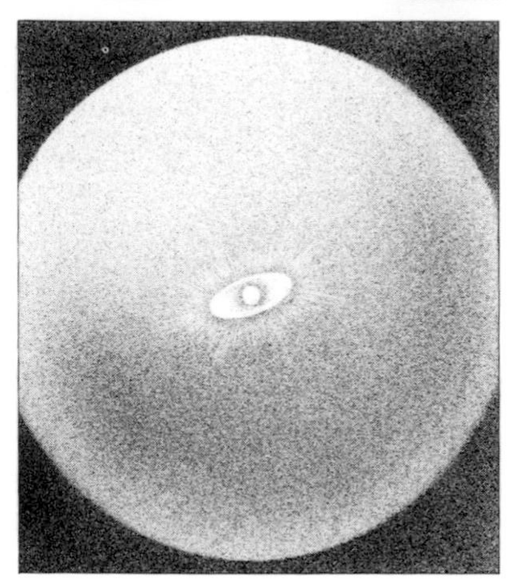

As the universe expanded, clumps of gas separated to make galaxies. The gas made the galaxies luminous by condensing into stars, and massive condensations in the very centers of some galaxies formed black holes that caused the intense outbursts of energy called quasars.

the chief cosmic event after the Big Bang. If the bass notes predominated among the vibrations of the early universe, the galaxies may have formed in giant clusters, like the nearby Virgo cluster containing thousands of galaxies. A higher pitch would account for much smaller packets of matter, including waif clouds of gas adrift in intergalactic space, and relatively small globular clusters of stars that contain some of the oldest stars known and occupy haloes around the galaxies. Another possible product of early clumping might be an undiscovered crop of black holes.

If the Big Bang occurred 13,500 million years ago, the major galaxies probably burst into light about a billion years later. The galaxies were made from clumps of gas, and each star was a microclump, a ball of highly compressed gas, ignited by gravity heating the core of the star to a very high temperature so that it could then run as a thermonuclear reactor. It converted hydrogen and helium into heavier chemical elements, and released copious energy in the process. Large stars burned quickly and exploded cataclysmically as supernovas, scattering newly fashioned elements into the galactic gas. Some stars destroyed themselves within a million years of their creation, so the accumulation of elements began at virtually the same time as galaxies formed. More slowly, gas gravitated toward the centers of the galaxies, where even greater violence impended.

The limpid universe offers earthlings a time traveler's view of events, because light makes long journeys across the abyss. The nearby bright star Sirius is seen not as it is but as it was nine years ago, because light from the star has taken that long to reach the Earth; it lies nine light-years away. And the particles of light from quasars, small, remote, but incredibly energetic beacons, arrive at the Earth tonight telling of commotions more than 10 billion years ago, associated with the birth of galaxies. It is as if a Greek runner were to burst breathless into the office of a twentieth-century historian and announce the battle of Marathon.

The past tense is appropriate for writing of the quasars; the oldest of them lies 12,200 million light-years away, and the youngest were active a billion years ago. Their first appearance signaled the emergence of organized galaxies, because quasars were outpourings of energy from the vicinity of massive black holes lying in the very centers of some galaxies. In a black hole, matter is so highly condensed, and gravity so strong, that no light can escape from it. Anything falling in is irretrievable, and its mass merely adds to the gravitational vigor of the black hole. Although such an object is inherently black, matter approaching it can be accelerated sufficiently violently to become extremely hot and bright, as in a quasar.

Material concentrated in mid-galaxy made a minotaur-like black hole that grew fat on a diet of gas. The black hole's tidal force ripped passing stars apart and reduced them to a disk of gas that swirled toward the hole, howling with radiant energy before oblivion. If the black hole became gross enough, some billions of times more massive than the sun, it could swallow its stars whole. Black-hole power made each quasar incomparably brighter than the sum of all the stars of the galaxy in which it lurked. Once it had digested most of the vulnerable matter in its vicinity, a quasar would peter out. Although there are no very recent quasars in sight, many galaxies are swept by lesser central outbursts, as if their black holes still take an occasional snack.

Quasars became numerous 11,000 million years ago, reporting that the centers of many galaxies were well developed. These central regions were also the chief settings for self-destruction among the stars, and the strewing of heavy elements, but a lesser boom in element making occurred far and wide, throughout every galaxy. With generous supplies of gas making giant stars in abundance, the young and crowded universe would have been a brilliant sight. Since then it has gone on enlarging itself, and the galaxies have scattered.

Astronomers honor the Milky Way Galaxy with a capital *G* because they live there. The sun and its attendant planets did not exist early in the Galaxy's history, but knowing their cosmic address helps in following the events that led to their appearance. On the outskirts of the Virgo cluster of galaxies lies the Local Group, consisting of three major galaxies, M31 (the Andromeda nebula), M33, and the Milky Way, together with a score of lesser galaxies. The sun resides in the suburbs of the Milky Way Galaxy, more than halfway from the center toward the edge of the star mass. The central black hole that lurks beyond the constellation of Sagittarius was probably never a quasar. The Galaxy contains about 100 billion extant stars, but also the fossils of many others, and for human beings the chief function of the Milky Way has been not to beautify the night sky but to smash stars.

Excluding the primeval hydrogen and helium, the chemical elements now composing the sun and its planets have an average age of

12,200 MILLION YEARS AGO

OLDEST QUASAR

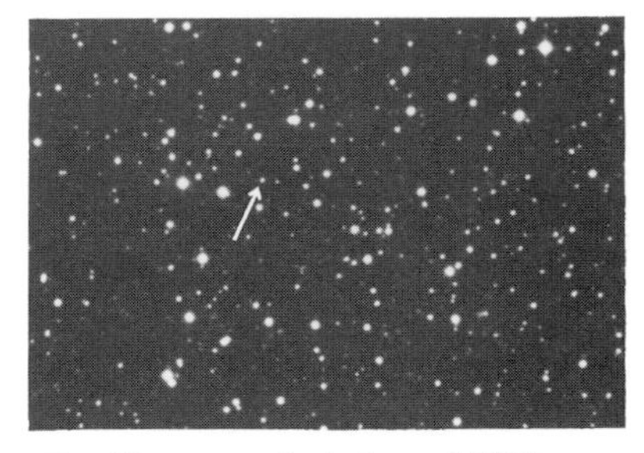

The object austerely designated PKS 2000 + 330 (arrowed) lies an estimated 12,200 million light-years away, and is the most distant object known. Allowing for its remoteness, it is the brightest entity in the universe. (UK Schmidt Telescope.)

We were hydrogen. *The frame of time is set by the expansion of the universe, with the degree of redshift (z), the increase in wavelength in the light of distant objects, providing the guide to dates. The universe became fully transparent about a million years after the Big Bang; that corresponds, on the timescale adopted here, to a redshift of z = 600, and for 10,000 million years ago the redshift is z = 1.5 (see the reference index: cosmological timescale).*

Abbreviation: *Myr = million years ago.*

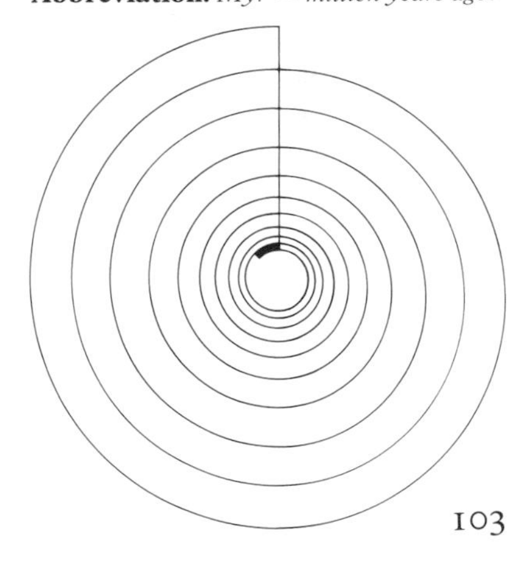

galaxies forming ▼ 12,500

oldest quasar ▼ 12,200 Myr

element-making boom; many quasars 11,000 Myr ▼

solar-elements (average age) 10,000 Myr ▼

12,000 | 11,000 | 10,000

10,000 MILLION YEARS AGO
ELEMENT MAKER

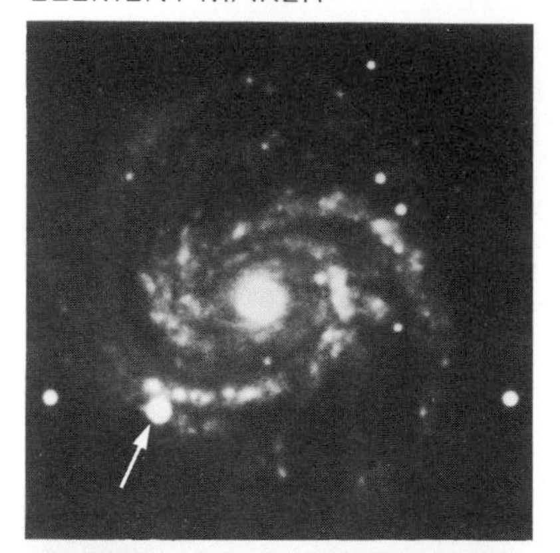

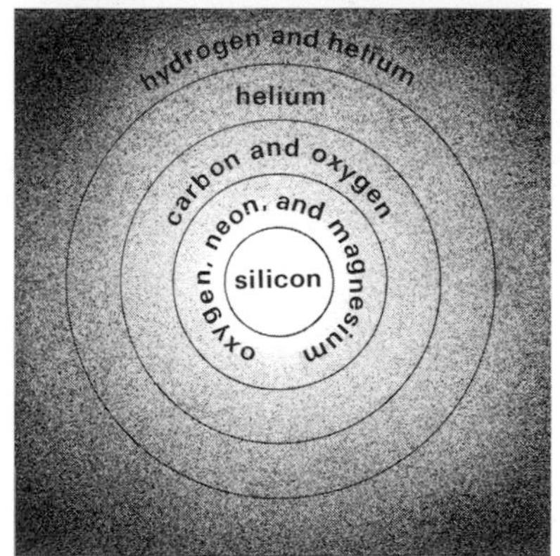

The chemical elements were made in giant exploding stars like the one seen here in another galaxy. (SN 1979 supernova in galaxy M100; electrograph taken by P. Griboval with the Griboval Electrographic Camera at the McDonald Observatory 76 cm telescope.) The diagram shows the core of a star twenty times the mass of the sun, nearing the end of its life, and burning elements heavier than hydrogen, up the scale of masses. At the very center, silicon-28 burns to make nickel-56, which decays radioactively into iron-56, the heaviest of the common elements.

about 10,000 million years. They are all the products of the big stars that burned hot, aged quickly, and expired violently, before the solar system came into being. Life is chemistry, and for anyone wishing to comprehend his or her existence, this stellar alchemy is one of the keys to the mystery. The human body is built from the ashes of stars of the Milky Way.

The Galaxy developed a flattened shape, with most of its stars and the residual gas between them orbiting around the center in a disk-like formation. The disk appears edge on as the broad, star-packed ribbon across the night sky for which the Milky Way is named. Less obvious to the naked eye are the Galaxy's elegant spiral arms, which would give it the outward appearance of a whirlpool, like the nearby Andromeda nebula and many other galaxies. The spiral arms are mobile, creative zones, always freshly stocked with bright-blue stars as shock waves travel through the Galaxy, commanding clouds of gas to collapse and make the stars. At any given time the spiral arms occupy those zones that happen to be experiencing the passing pressure, and chain reactions of exploding stars reinforce the shock waves. Such is the nature of the galactic manufacturing plant that made the elements; its spiral arms are now stocked with elements younger, on average, than those that built the Earth.

The region of the Galaxy where the sun and Earth would one day appear felt the shock of a spiral arm every 100 million years. It emerged from each encounter adorned with massive new stars, whose careers culminated after just a few million years in the star-smashing supernovas. These blasted most of the contents of the stars through interstellar space, and left a fossil residue in the form of bleeping pulsars or small black holes. Steps in the process are plain among present-day stars: Spica in Virgo is an example of a massive blue star in its prime, the red giant Betelgeuse in Orion is an exhausted star due to explode soon, and the Crab nebula in Taurus consists of debris scattered from a recent supernova, complete with a pulsar in its midst. Over billions of years stars like those were ostentatious performers on the galactic stage, as the nuclear force and the weak force contested with gravity for control of events in the hot cores of the massive stars.

To begin with, as in a star of any size, hydrogen burned in the thermonuclear fashion and produced helium—nothing new, and chemically worthless. But as the supplies of hydrogen fuel in the

solar elements (average age)
▼ 10,000 Myr

spiral-arm elements (average age)
▼ 9000 Myr

star's core began to run out, gravity tightened its grip and the core became progressively hotter, enabling heavier elements to serve as fuel. Helium nuclei joined together to make carbon, the very stuff of life, and oxygen too. Some of the carbon then burned to produce magnesium, while oxygen made silicon and sulfur, and finally, at a temperature of several billion degrees, silicon burned to make iron. Thus were the most common elements formed, along with many important by-products such as nitrogen.

That was not all. When the core of the star had exhausted its heavy fuels, and was unable any longer to oppose the force of gravity by nuclear-generated heat, it collapsed fatally, to become either a small, dense pulsar or an even more concentrated black hole. This last spasm released copious energy that not only blew the outer layers of the star into space but drenched them with nuclear particles. Neutrons transmuted some of the material into elements heavier than iron (silver and gold for example) and made many radioactive elements besides. For a few weeks the exploding star burned brighter than a billion suns. With every pass of the spiral arms, perhaps eighty times in 8 billion years, the flaring supernovas stocked the local region of the Galaxy with ever-greater quantities of elements. In the gaps between the stars, chemical reactions proceeded, as the electric force married different elements to make water, silicates, alcohol, and dozens of more exotic molecules, until the resulting clouds of ice and dust locally blacked out the view of the stars.

Concussive though these events were at the scale of stars, they continued only because the Milky Way Galaxy as a whole enjoyed a lasting tranquillity. This was nothing to be taken for granted. Self-engendered explosions continued in other galaxies: in addition to exhibiting the small, intense quasars, the stormy centers of galaxies also shot twin jets of particles in opposite directions into intergalactic space, creating enormous double lobes that broadcast strong radio emissions. But worse misadventures befell some of those mighty assemblies of stars. The universe had only a quarter of its present volume 6700 million years ago, and in crowded conditions in clusters a large galaxy could gobble up smaller ones, or else disrupt them by strong tides to make irregular galaxies. During outright collisions between galaxies of comparable sizes, the widely spaced stars passed between one another safely enough, but the interstellar gas clouds collided like winds. They were brought up

7000 MILLION YEARS AGO

GALACTIC DUST CLOUD

Chemistry began with the release of interesting elements from the supernovas. Reacting together, the elements made grains of dust that gathered in giant clouds, blotting out the stars. (Harvard College Observatory.)

We were carbon. *In the cosmological time frame, 10,000 million years ago corresponds to a redshift z = 1.5, and 6300 million years ago to a redshift of z = 0.4. In terrestrial terms these ages are, respectively, roughly twice and 1.4 times the age of the Earth.*

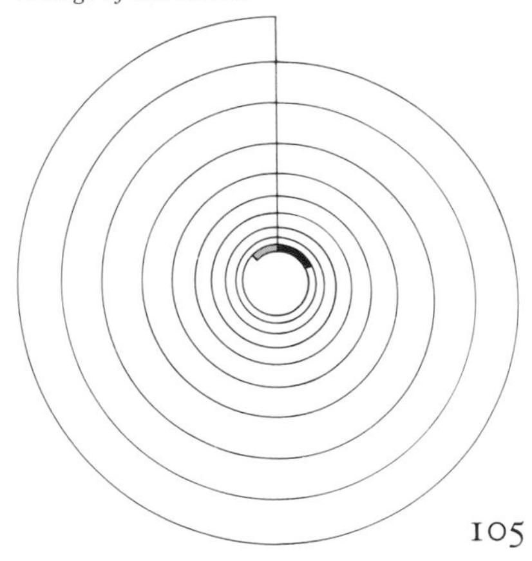

quarter-size universe
6700 Myr ▼

8000 | 7500 | 7000 | 6500

4550 MILLION YEARS AGO
STELLAR CHAIN REACTION

Amid the expanding shell of debris from a recently exploded star, new stars are forming. The bright and the dark patches immediately to the right of the curved, bright shell contain young stars. A similar event 4550 million years ago triggered the formation of the Sun and its family of planets.

short, and left behind when the galaxies went on their way. In the most densely packed clusters, where the risk of galactic collisions was greatest, this loss of gas became a runaway process around 6000 milion years ago. Gas already stripped from galaxies became itself a powerful stripping agent, and many galaxies were sterilized at that time. Robbed of gas, they were no longer able to breed new stars.

In its foxhole with relatively harmless galaxies of the Local Group, the Milky Way escaped that fate, and its supernovas continued to fling their ashes into fertile gas clouds. One such cloud, darkened by dust, was the maternal mass that gave birth to the sun and the Earth. Many dying stars had stocked it, but two supernovas, separated by 100 million years, were implicated more directly in the origin of the solar system. About 4650 million years ago, the star-making shock wave passed through the presolar cloud, without causing it to collapse, but creating a big, short-lived star nearby. Within a few million years the star blew up and spattered the presolar cloud with a highly radioactive charge of elements. Just over 4550 million years ago, the next pass of the spiral arm once more failed to make the sun, but fashioned another massive star in the vicinity. This time the blast wave from the supernova induced the implosion of the presolar cloud, in a violent act of midwifery.

The cloud shrank inward, slowly at first and then faster and faster. Much of the gas fell right to the center of gravity and made the sun in about ten million years. The inrush became sufficiently intense to achieve, in the very core of the sun, a temperature high enough to trigger the nuclear reactions of hydrogen by which it still burns. A flattened cloud of gas, ice, and radioactive dust swirled around the newborn sun, looking like a giant version of Saturn's rings.

The materials coagulated into microplanets that grew by collision and mutual attraction, and assembled the planets. Jupiter and Saturn, the big ones, gathered large quantities of hydrogen and hellium, mimicking the stars. Beyond Saturn lumps of ice amassed, in snowball fashion, to make Uranus and Neptune. From the inner regions of the ring cloud, the sun drove out the lighter gases and ices, leaving stony and ironbound materials that built petty planets: Mercury, Venus, Earth, and Mars. By 4000 million years ago the universe at large had half its present volume, but attention now focuses on one small planet, and what could be done with a mass of atoms held together by gravity.

We were tar. *Events 4500 million years ago correspond, cosmically, to a redshift of* $z = 0.3$. *On the Earth, no rocks survive for this period, but it is sometimes designated the Hadean, or "hellish" era.*

sterilization of many galaxies 6000 Myr ▼

6500	Myr		6000					5500			

The energy of the collisions that formed the Earth, combined with intense radioactive heat from its constituents, softened the planet's interior. About 50 million years elapsed in assembling five sixths of the Earth's substance, and during the interval molten iron sank into the core of the planet, while lighter rocks rose. Completing the Earth took a further 50 million years, and the new outer layers remained cooler, although they spouted numerous volcanoes. The moon was an embryonic planet captured by the Earth, perhaps 4500 million years ago, in the midst of the tumult. Collisions with microplanets did not cease even when the Earth was essentially complete, 4450 million years ago. Snowball comets and stony asteroids continued to rain down, and the moon preserves craters and great basins dug by impacts during that long episode. The Earth, with its stronger gravity, must have experienced fiercer pummeling, with every scrap of its surface being involved in more than one major impact.

The Earth acquired an atmosphere of steam and carbon dioxide gas, coming partly in burps from the planet's interior and partly from the cargoes of frozen vapors carried by impacting comets. Thunderstorms drenched the world, and when the surface was cool enough, perhaps as early as 4450 million years ago, liquid water gathered in basins created by the biggest impacts; these were the first seas. In contrast, the atmosphere of Venus was a pressure cooker in which the temperature soared, while Mars became a refrigerator where surface water froze. The adolescent sun was feebler than it is today, and the Earth's own seas might have frozen forever but for the warming effect of carbon dioxide gas and thin cirrus clouds. As the sunshine grew stronger, spontaneous adjustments to the cloud cover helped to maintain a mild climate.

The seas and pools were flavored with chemicals, making the infusion that students of the origin of life call the soup. Whether "volcano soup" or "comet soup" is the the more apt name for the menu is moot: materials vented from the Earth contributed ingredients to the soup, but so did comets and tarry meteoric bodies, known to contain carbon compounds. Among all the products of exploding stars, carbon showed an unusual talent for building elaborate and subtle molecules. These included proteins, and also nucleic acids, the self-copying forerunners of the genes of heredity. Pools of concentrated soup, on the flanks of volcanic islands, supported a ceaseless roulette among spontaneously generated chemi-

4300 MILLION YEARS AGO

FOSSIL MOON

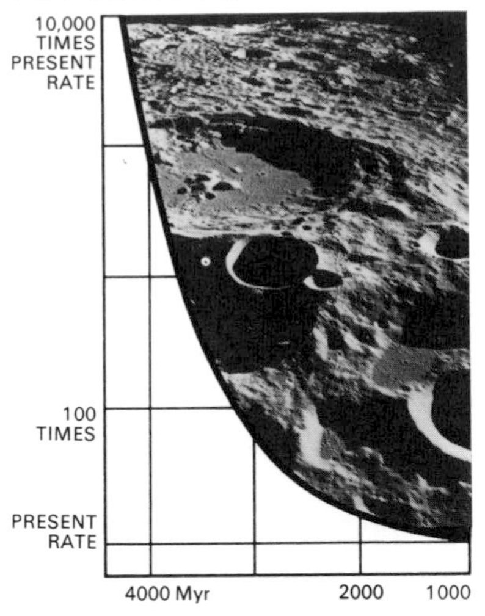

The cratered moon, photographed from Apollo-11, *gives an impression of what the Earth was like 4300 million years ago – except for the lack of volcanoes, clouds, and ocean water. The oldest surfaces of the moon are saturated with impact craters made by giant meteorites. Studies of the moon and other objects of the solar system lead to estimates of the rate of major impacts on the Earth, during and since the early heavy bombardment by cosmic objects.*

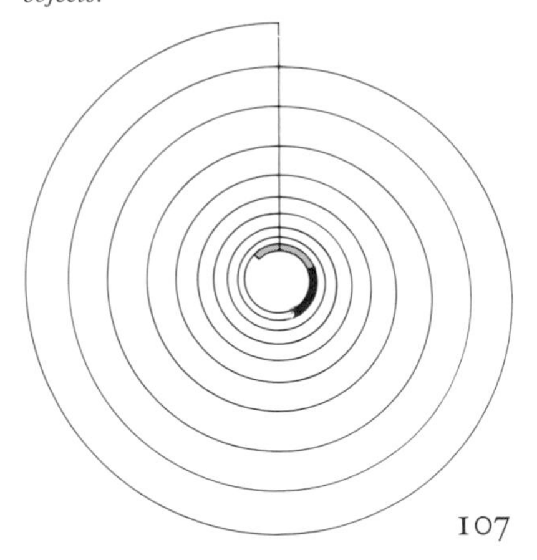

early presolar supernova 4650 Myr ▼

solar system ▼ 4550

Earth complete; storm of cosmic impacts ▼ 4450 Myr

life on Earth 4000 Myr ▼

5000 | 4500 | 4000

3500 MILLION YEARS AGO
LUMPS OF LIFE

Bacterial colonies made conical humps a few centimeters wide, on an ancient shore now embedded in northwestern Australia. Called stromatolites, objects like these were conspicuous for billions of years around the early continents.

cals. Some molecules, well engineered and suited to their chemical environments, made copies of themselves more rapidly than others. Even before life existed, selective evolution was at work among the molecules.

Life began before the heavy cosmic bombardment abated, and geological action has erased any record of the event, save only for what can be inferred from differences among the most ancient genes of living bacteria. These indicate an origin at least 4000 million years ago. Hungry molecules competing for the soup fared better when they helped one another, and a winning combination of interacting molecules was the sticky grandmother of all life on Earth. Certain genes prescribed the manufacture of key proteins, enzymes, which in turn favored the reproduction of the genes. The recipe was copied in billions of microscopic packets, the first bacteria. They still fed on the soup; had they failed to evolve better tricks than that, they would have starved when the soup ran out.

Fairly faithful copying of genes and proteins was essential if organisms were to function properly, yet mutations were indispensable if they were to evolve. Misprints in genes, the inherited instructions for life, caused alterations in proteins, so that individual organisms and strains came to differ from one another, sometimes creatively. By 3900 million years ago, if not earlier, some bacteria had acquired the means of absorbing sunlight with pigments, as plants do today. They used the free supply of solar energy for photosynthesis, to build sugars from carbon dioxide, then the commonest gas in the atmosphere.

The cosmic bombardment eased about 3900 million years ago. The oldest surviving rocks on the Earth's surface, 3800 million years of age, include bands of ironstone that bacteria helped to make; they are the first mark of life. The ancient rocks occur in Greenland, and bear witness to a planet bathed in a sea in which sediments could form. The world was an ocean, punctured by high rims of impact craters and by many volcanoes that bled black, heavy basalt onto the floor of the ocean. Of continental material only a few scraps survive, in western Greenland, Labrador, southern Minnesota, northern Norway, Zimbabwe, southern India, and western Australia.

Among 3500-million-year-old rocks of the Pilbara region of Australia, and Zimbabwe in Africa, the oldest emphatic signs of life are the fossilized remains of living reefs called stromatolites, made by slimy mats of billions of pigmented, plant-like bacteria. Grains of silt were cemented in the mats, and new mats grew on top of the old, thus building durable mounds. For more than a billion years thereafter, the only living things that any sentient observer might have spotted with the naked eye were the humps of the stromatolites, perched in the shallow tidal water on the world's beaches.

Gold belts, known also as greenstone belts because of the color of reworked lava, are common among old continental rocks, especially 2900 million years of age or younger. The ocean floor renewed itself when basaltic slabs cooled, lost their buoyancy, and sank back into the main body of the Earth. The grinding action melted the rocks and liberated underground bubbles of ore rich in gold, silver, copper, and zinc; uplift and erosion have since brought them close to the surface.

Abundant supplies of radioactive heat within the planet stirred the rocks and refurbished the surface. The moon, being smaller, cooled faster and developed a thick crust that choked off all activity near the surface, making it defunct, geologically speaking, by 3000 million years ago. The small planet Mercury, nearest to the sun, became inactive at roughly the same time.

The Earth, by contrast, cooled just enough to become more creative. The outer shell of the planet had hardened by 2800 million years ago into a number of rigid yet mobile plates that jostled one another in the processes of plate tectonics that still operate today. Where two plates rifted apart, basalt welled up along a mid-ocean rift, filling the gap and enlarging the plates. Where two plates approached, one of them had to sink under the edge of the other, making a long, deep trench in the ocean floor and grinding its way fiercely into the solid Earth. Beyond the trench a line of volcanoes burst forth, and beneath them buoyant granite solidified to make the unsinkable stuff of continents, the slag of the Earth.

In a spurt of continent building, 2800 to 2600 million years ago, the Earth created two thirds of its land masses. They included portions of every present-day continent, although their names, Baltica, western Africa, and so on, say nothing about their locations long ago, because plate movements have shuffled them disrespectfully about the globe. When two continents ran together, they welded themselves along a range of new mountains. For instance, the microcontinents of Zimbabwe and Kapvaal, now pieces of Africa, collided about 2600 million years ago and heaved old gold

2800 MILLION YEARS AGO
MAKING THE LAND

Heavy basalt rock of the ocean floor, released at mid-ocean rifts, moved sideways and eventually burrowed back into the Earth at ocean trenches. The grinding threw up lighter, continental rocks, in arcs of volcanic islands. These collided to make wider masses, which then continued to grow. An episode of vigorous continent building began 2800 million years ago.

We were bacteria. *The Archean era runs from 3800 million years (the age of the oldest rocks) to 2500 million years ago. A general name for the entire early history of the Earth is the Precambrian, ending 570 million years ago at the start of the Cambrian period (see the reference index: geological timescale).*

See also map p. 81, including oldest rocks.

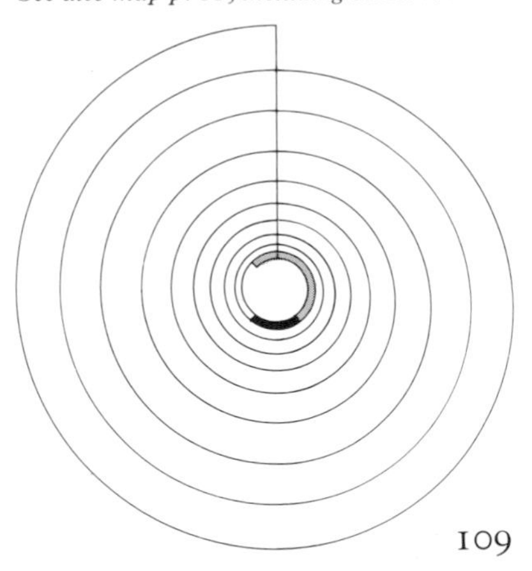

moon and Mercury defunct 3000 Myr ▼
gold boom ▼ 2900 Myr
continent making ▼ 2800 Myr
iron boom 2500 Myr ▼

3000 2500

2300 MILLION YEARS AGO
COLLISION AND BREAKUP

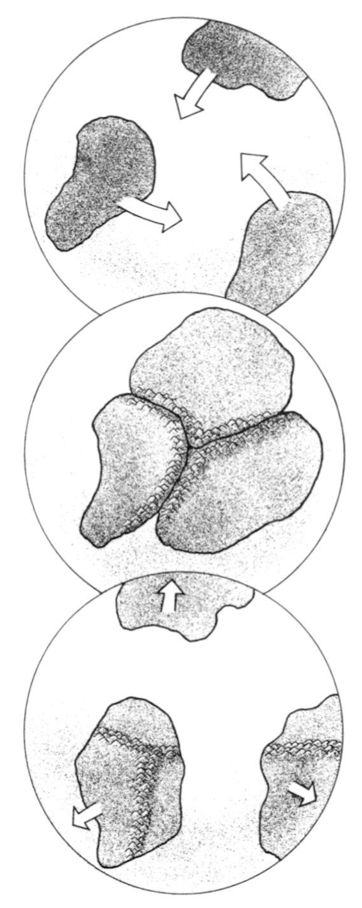

A succession of supercontinents, of which the first existed about 2300 million years ago, was the natural consequence of the continuous motion of continents on the confined surface of the globe. Pauses in mountain building corresponded with periods when a supercontinent existed, or was in the early phase of breakup.

We were naked cells. *The Proterozoic ("pioneering life") era began 2500 million years ago. The Early Proterozoic lasted until 1600 million years ago.*

belts into the air. Rivers that ran through the new Limpopo mountains carried glistening silt down to a lake, where the gold settled into the extravagantly rich deposits now buried in the Witwatersrand of South Africa.

Shallow seas on the continental shelves created the setting where peculiar bacteria turned by stages into fishes. This transformation began when certain bacteria in a fit of genetic absentmindedness lost their jackets and were left naked within a soft membrane instead of the usual sturdy cell wall. Bacteria called thermoplasma remain in that peculiar state today, and human beings are descended from similar organisms that first appeared 2500 million years ago. Whatever the inconveniences of nudity, the bacteria could wriggle better, and by 2100 milllion years ago some had acquired the capacity to engulf other bacteria in the folds of their membranes and admit them whole as guests in their interiors. Playing host paid off, because the planet was being poisoned.

Plant-like bacteria had responded with mindless zeal to the grant of large areas of shallows where they could lie on the bottom and sunbathe. Stromatolite reefs proliferated along the shores of the ancient continents, and as the bacteria grew, they consumed carbon dioxide and released oxygen. Free oxygen had been virtually absent since the world began: when any living or nonliving chemical process released it, other chemistry quickly mopped up the oxygen to form deposits of sulfates or compounds of iron. Oxygen made in abundance by the cohorts of stromatolites precipitated iron from seawater on a grand scale, laying down more than 90 percent of the world's minable iron, between 2500 and 1800 million years ago.

Carbon dioxide in the atmosphere had helped to keep the planet warm and humid, but the bacteria stole so much of the gas for their own purposes that they caused the first ice ages, beginning 2300 million years ago. A distinct lull in the creation of new continental rock at that time meant that the new land masses drifting around the globe had coalesced for a while, calling a halt to most geological activity. The resulting supercontinent, here called Kenora, was the earliest of four. Its geography is scarcely to be guessed at, but glaciers scoured pieces of Africa and North America.

In a bizarre side effect, the oxygen mobilized uranium compounds in the water, and bacteria stockpiled them in ores. Fissile uranium-235 was much commoner than it is today, and rich ores became natural nuclear reactors. In western Africa 2000 million

supercontinent of Kenora; ice ages 2300 Myr ▼
mountain-building pulse 2200 Myr ▼
host cells 2100 Myr ▼

2500 Myr | 2400 | 2300 | 2200 | 2100

years ago, chain reactions consumed several tons of uranium-235 in one small district. But the resulting mess of radioactive fission products was by no means the chief problem for the bacteria at that time. About 1800 million years ago the surface of the land rusted, and the resulting red beds tell of free oxygen escaping from the living sea into the atmosphere, as the oxygen-making bacteria overwhelmed the planet's chemical defenses.

The oxygen revolution was like pouring bleach into the seawater. The chemical vigor of oxygen that nowadays helps to energize animal life made it poisonous for all of the oldest forms of life. Stung by the oxygen, some bacteria died, and some survived only in dark, airless places. In a flurry of evolution, other bacteria found the means to endure and then to exploit the oxygen. Among them was a group of purple photosynthetic bacteria that lived in relatively deep water, where sunlight was faint and blue, and they were colored to suit. A naked bacterium then hit upon a labor-saving response to the oxygen crisis: instead of evolving the necessary biochemical machinery for itself, it welcomed on board a small bacterium that was already competent with oxygen. A mutant descendant of the purple bacteria that had lost its capacity to grow by sunlight took up residence inside its host and managed the oxygen in exchange for nourishment.

From this symbiotic chimera a lineage of host bacteria evolved that improved their housekeeping, thereby keeping the guests in order and incidentally inventing modern cells of the kind used by all animals and plants. In bacteria, the genes slopped about; the modernized cells arranged their genes on string-like chromosomes, creating a library for them in a nucleus in the cell, and neatly shared duplicated chromosomes between daughter cells when they were ready to divide. But still to be found in the cells of the human body are the oxygen-handling guests, now called mitochondria, with a private system of heredity tracing back to the mutant purple bacteria of a poisoned sea. The first owners of modern cells 1700 million years ago were simple, single-celled molds, or fungi, and also single-celled proto-animals distinguished only by greater mobility; a billion years were to elapse before they became true, many-celled animals.

The bombardment of the planet by asteroids and comets from outer space had not entirely ceased, and there are traces of major impacts around 1900 million years ago. One apparently delivered or

1700 MILLION YEARS AGO

MODERNIZED CELLS

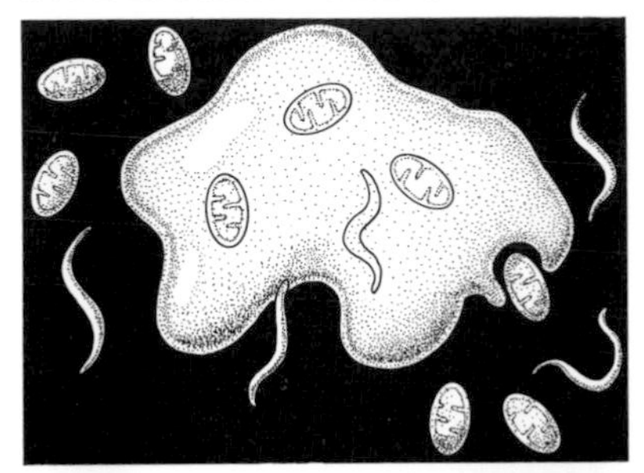

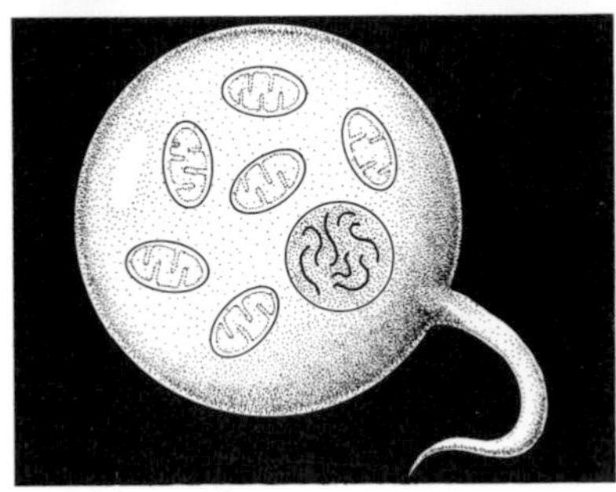

Oval bacteria settling in a larger bacterium became oxygen specialists (mitochondria), while the host gathered its genes into a cell nucleus. Whip-like appendages may have originated as swimming bacteria.

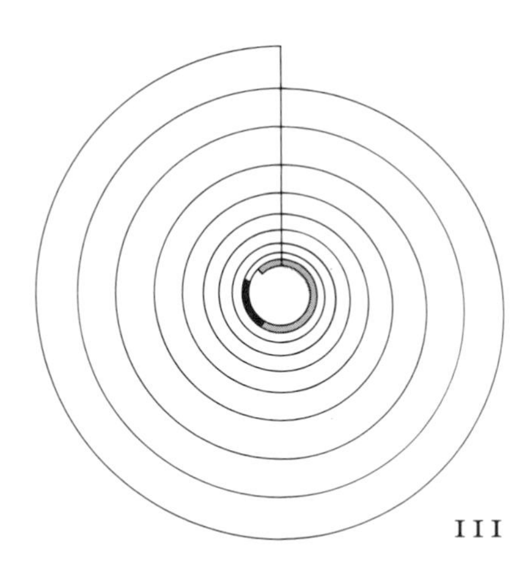

uranium reactors ▼ 2000 Myr

metals from space ▼ 1900 Myr

oxygen revolution; purple bacteria ▼ 1800 Myr

proto-animals ▼ 1700 Myr

blue-green algae 1600 Myr ▼

2000 | 1900 | 1800 | 1700 | 1600

1500 MILLION YEARS AGO
PROTO-PLANTS

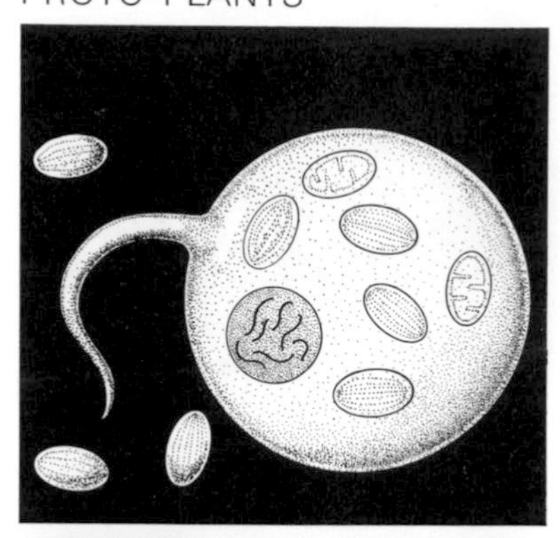

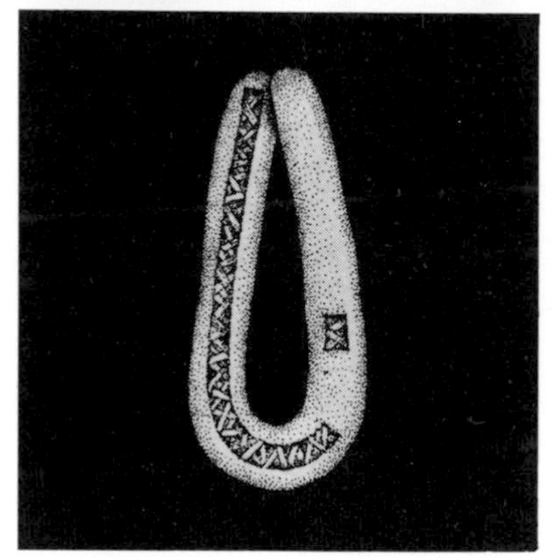

The assimilation of bacteria that grew by sunlight converted some modern cells into true algae. By 1000 million years ago, seaweed composed of many cells existed; in the lower illustration a fossil ribbon of Tawuia *from Canada is shown at its actual size.*

We were protozoa. *The Middle Proterozoic began 1600 million years ago and continued until 900 million years ago. In an alternative scheme based on stromatolite variations, the Early Riphean began 1700 million years ago, and the Middle Riphean lasted from 1350 to 950 million years ago.*

unearthed the world's largest deposits of chromium and platinum, in South Africa, and a giant iron-nickel meteorite falling in Ontario endowed Canada with the most productive nickel mines. In collisions of a much slower kind, between continents, widespread mountain-building occurred between 2200 and 1500 million years ago. The first supercontinent having broken up, the pieces were reassembling in new configurations, with core areas of North America, Africa, and Australia consolidating themselves at that time. Their locations and orientations still bore not the slightest relation to their present situations, but one broad block of granite now straddles the Amazon basin in South America. A pause in the formation and reworking of continental rocks 1500 million years ago marked the completion of the second supercontinent, Amazonia.

Evolutionary changes early in the Earth's history may seem tediously slow to humans unschooled in the intricacies of biochemistry or genetic control systems. In underwater gardens, plant-like bacteria grew and reproduced, and bacteria with screw-like tails nosed through the water as the only organisms other than humans that invented the wheel. The molds and proto-animals were not conspicuous, and the most dramatic sight was of well-nourished bacteria dividing and doubling their numbers every twenty minutes; the most methodical was the scavenging of dead plant-like bacteria by other bacteria which putrefied them into stromatolite cement.

Local appearances were deceptive, because the myriads of bacteria had longstanding networks of global trade, in which waste products of one species became raw materials for another. The natural transport systems of winds and ocean currents carried molecular freight: carbon dioxide, methane, nitrogen, ammonia, hydrogen sulfide, methyl chloride. Bacterial action also created solid sediments of limestone, phosphate, metal ore, and the like. When the global environment changed, for example when completion of a supercontinent lowered the sea level and drained the continental margins, the bacteria adjusted their economies to weather the recession and keep the Earth healthy. (This silent, business-like corporation of bacteria continues operating to the present, running great chemical industries in the mud of estuaries.) Oxygen, drifting high into the atmosphere, was drawing a screen of ozone across the sky, which protected exposed organisms from solar ultraviolet rays

blue-green algae
▼ 1600 Myr

supercontinent of Amazonia; proto-plants
▼ 1500 Myr

mountain-building pulse
▼ 1400 Myr

1600 Myr | 1500 | 1400 | 13

when the tide went out or the pool dried; but there is no evidence that any of them was yet fit to take up residence on dry land. The bacteria had no need for higher organisms, and when these did appear, some of the bacteria would attack them, as diseases, trying to putrefy them alive.

Along the beaches newcomers joined the ranks of the stromatolite-builders: the blue-green algae. These were not true algae but cyan-colored bacteria that grew by sunlight. The blue-greens made their appearance about 1600 million years ago and became one of the most durable forms of bacterial life, still flourishing opportunistically in the modern world. The increasing success of the photosynthetic bacteria made more oxygen available in the atmosphere and the water, and the oxygen loaded the evolutionary dice in favor of the modern cells that relied on it for their relatively sophisticated modes of life.

Certain proto-animals took photosynthetic bacteria aboard, as further guests, and these conferred the power of growing by sunlight on their hosts, thus turning them into single-celled proto-plants—true algae of various colors. Some of the blue-green bacteria formed a particularly successful partnership, which gave rise in due course to true plants. Fossil specks larger than usual, the remains of bloated proto-plants, appeared in great numbers to register this union in rocks 1500 million years old. Descendants of the blue-green bacteria inhabit the cells of modern plants, as the chloroplasts that color the foliage and paint the landscapes green.

The roster of diminutive ancestors for three kingdoms of higher organisms was thus complete by 1500 million years ago: fungi, proto-animals, and proto-plants. The next step was to organize many of the modern cells within a single organism—not a mere colony like a stromatolite, but a coherent system of cooperative cells. The plants were apparently the first to make this innovation and the earliest known elaborate, multicelled organisms on the planet were pieces of microseaweed, around a millimeter long, appearing in rocks 1300 million years of age. More demonstrative of the shape of things to come were pieces of seaweed several centimeters long, showing up in Canada and vaguely dating to about 1000 million years ago.

The interval 1400 to 800 million years ago was a further notable period of mountain building in many parts of the world. In one collision 1150 million years ago Grenvillia, a microcontinent two

1150 MILLION YEARS AGO
AMERICA ENLARGED

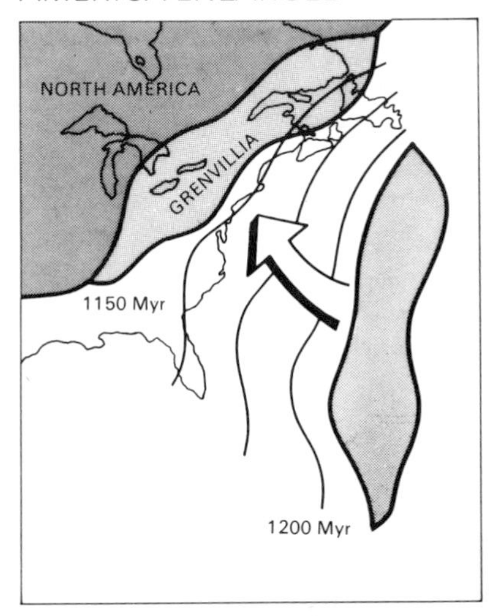

Much of the territory of the eastern United States and Canada was added to the core of North America during the Grenville mountain-building event. At that time North America was rotated from its present orientation and lay in northern latitudes.

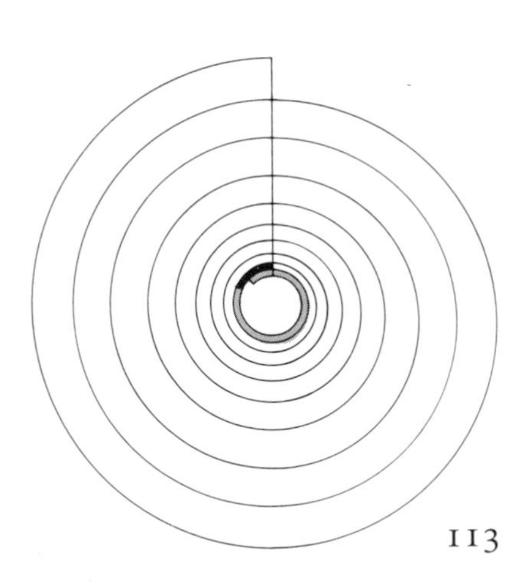

plants
ɔ Myr

Grenvillia hit North America
▼ 1150 Myr

seaweed
▼ 1000 Myr

1200 1100 1000

650 MILLION YEARS AGO
ORIGAMI ANIMAL

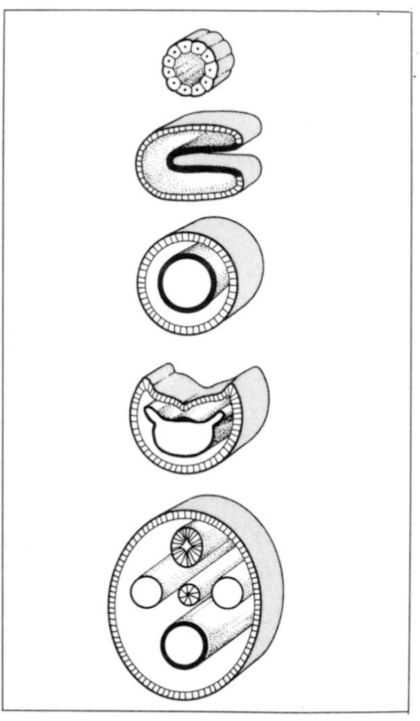

Sheets of cells, folded and involuted to form body cavities, gave rise to various hollow-bodied worms, prototypes of different kinds of higher animals. In the case illustrated here, the folding fashions a primitive chordate. Human embryos, during their earliest development, still imitate the contortions of their remote ancestors.

thousand kilometers long and five hundred kilometers broad, attached itself to the eastern edge of North America, enlarging the continent. The mountains produced by that collision have long since eroded away. The same can be said of other mountains built during this episode, straddling the Zaire-Zambia border in the heart of Africa, where they contain the largest deposits of high-grade copper, about 850 million years old. A three-thousand-kilometer chain of similar age runs across Asia and passes through the present region of Lake Baikal.

By contrast with the restless Earth, Mars expired. Geological action at the surface of this planetary neighbor culminated in the giant volcano of Olympus Mons, more than twenty kilometers tall, and then virtually ceased, as the frozen crust thickened. The date of 1000 million years ago for the fossilization of Mars is approximate. Of the main objects of the inner solar system, that left only Venus and the Earth in a geologically active state. Venus possesses continents and dry "ocean" beds, but its atmosphere is hot, heavy, and sulfuric–no place for life of the kind that developed on the Earth.

Here the supercontinent of Baikalia was completed by 800 million years ago, and promptly fell apart. Even as it did so, at least half the world's land, including the present southern continents, became incorporated in a single mass. Beginning around 650 million years ago, 200 million years of mountain building, across India, much of Africa, Arabia, Brazil, Antarctica, and southern Australia, consolidated them, as a portion of the next supercontinent. This land mass dominated the mobile geography of the next half-billion years: Gondwanaland.

The pieces of the continental jigsaw were in strange fragments, liaisons, and orientations, at the time when true animals first appeared. The stalwart but unpromising stromatolite builders attained a peak of diversity, despite two long-lasting ice-age episodes of 950 and 770 million years ago. But when the ice returned about 670 million years ago, it brought cold unmatched in the whole history of the world, and most kinds of algal microplants were wiped out.

The payoff for evolving an animal composed of large numbers of living cells was that the assemblage could sample larger volumes of water or seabed mud than single-celled proto-animals could do, in the search for food. Furthermore, until that age of scarcity, the sea may have been too rich in microscopic life, making it as unfit for aspiring animals as a stagnant pond. Perhaps 1000 million years ago, perhaps later in the run up to the invention of true animals, sex began among the proto-animals. Bacteria had long been in the habit of injecting genes into one another in a casual kind of sex life; but this was not directly linked to reproduction, during which a cell simply divided, making two individuals from one. Sex of the modern kind, making one individual from two, involved further refinement of the genetic machinery of modern cells, to parcel the gene-carrying chromosomes into germ cells. Male and female germ cells had then to unite to reestablish a full complement of chromosomes and genes for a new individual.

Why any particular organism should have found it worthwhile to go through such troublesome procedures merely to reproduce itself is one of the theoretical biologists' favorite puzzles. But evolution was certainly accelerated, and its mechanisms refined, by the sexual shuffling of genes and by the demarcation of a species as a mating population, distinct from other species with which it did not interbreed. All in all, the picture is of a hungry planet ready for novelty. The traveling continents were so disposed as to accumulate great thicknesses of snow, building ice sheets that chilled the world. The surest reservoir of food lay in the graveyards of the seabed mud, where the remains of microscopic plants settled after death, and the biotechnical means of reaching it were under development. After 3 billion years, life had become subtle enough for the hierarchy of genes controlling genes to find ways of differentiating animal cells into muscle, skin, nerve, and so on, according to where they lived within a many-celled body.

More than one group of proto-animals invented true animals. Jellyfish and quill-like organisms, called sea pens, were the first abundant animals that left fossil imprints (in Newfoundland about 670 million years ago). Jellyfish have prospered ever since, but more distinguished groups of animals may all trace back to flatworms, ribbons of cells with simple nervous systems that wriggled along the seabed in search of food. The most decisive changes in the history of animals then involved folding and curling ribbons of cells to make hollow bodies with internal organs. Pumped full of water, such bodies became more rigid and helped the resulting worms to burrow in the seabed. These superworms were the ancestors of all animals with hollow bodies—including human beings, who are descended from an odd kind of marine worm with tentacles. Entire divisions, or phyla, of the animal kingdom derive from different

650 MILLION YEARS AGO
OLD GONDWANALAND

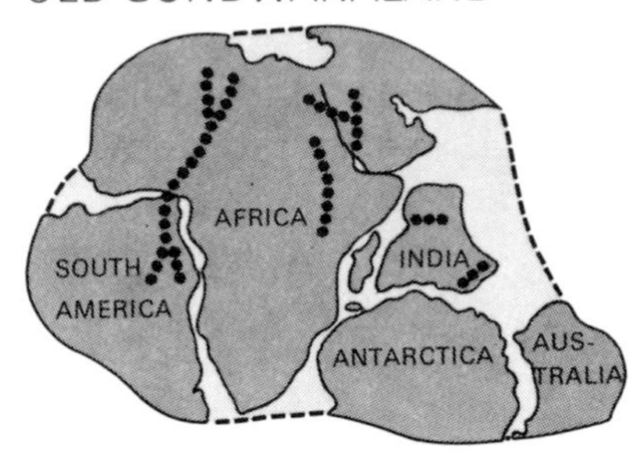

The land mass that split into the present southern continents was itself composed of older blocks. These coalesced along the internal lines that mark traces of ancient oceans, destroyed in the amalgamation.

We were protozoa. *The Late Proterozoic is designated as beginning 900 million years ago, although this transition may well relate to the onset of glaciation, taken here to be 950 million years ago, which is also the date at which the Late Riphean commenced. An era or period designated the Vendian began 680 million years ago, although this too can be keyed to an onset of glaciation, given here at 670 million years ago.*

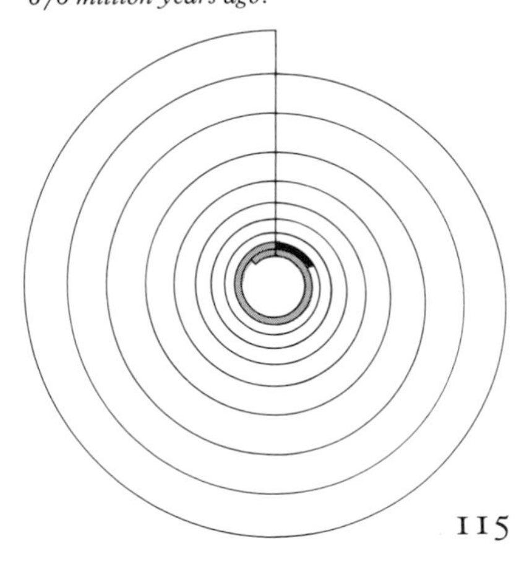

supercontinent of Baikalia ▼ 800 Myr
ice ages ▼ 770 myr
ice ages; animals 670 Myr ▼
mountain building; Gondwanaland ▼ 650 Myr

800 | 750 | 700 | 650

620 MILLION YEARS AGO
THREE-ARMED ANIMAL

The fossil of an early complex animal, here shown actual size, reveals three feeding tentacles on a disk-shaped body. Tribrachidium *foreshadowed similar five-armed animals, the echinoderms. This specimen is from the Flinders Ranges of southern Australia.*

We became odd fishes. *The Proterozoic era gave way to the Phanerozoic ("visible life") era, which began 570 million years ago and continues to the present. The first phase was the Paleozoic ("old life") era. Successive geological periods commenced as follows: Cambrian 570, Ordovician 520 or 500, Silurian 435, and Devonian 410 million years ago (see the reference index: geological timescale).*

See maps pp. 81, 82.

designs for worms. Animals with bodies built in segments, like those of earthworms and insects, lived on the shores of Australia and Russia, perhaps 620 million years ago. They were annelids and arthropods, akin to today's earthworms and insects, and they already possessed brain-like knots of nerve cells. All had soft bodies.

Many animals developed shells and other rigid jackets that enabled them to grow larger, resist sea waves and predators, and incidentally leave much more obvious fossils—thus initiating the era of "visible life." Different kinds of marine animals embarked on this consolidation, independently and at different times. Trials began with phosphate shells around 600 million years ago, but carbonate, especially in the form of calcite, became the most favored material for making hard parts by 570 million years ago. Even then the shelly animals coexisted with abundant soft animals reminiscent of the earliest forms. Sponge-like animals, some very small ancestors of mollusks and lampshells, and bizarre creatures called hyolithids all evolved further, but most of the early shelly animals were freaks that left no descendants.

From this phase of experimentation, the mollusks were the first conspicuously successful animals to emerge about 570 million years ago—forerunners of snails and octopuses. Cup-shaped, sponge-like animals called archaeocyathids were the earliest of a long succession of animals that left indelible marks on the planet in the form of limestone banks and reefs, grown in tropical waters. From about 560 million years on, the mobile masters of the undersea were the small trilobites with segmented bodies in oval calcite shells. They walked on multiple jointed legs and some evolved into burrowing and swimming forms; they also included the first animals with eyes. Similar-looking predatory animals, known as chelicerates, had pincers that became fearsome in sea scorpions.

The trilobites were not left to amble in peace among the archaeocyathids. Between 550 and 500 million years ago, in the latter part of the Cambrian geological period, about four successive disruptions occurred. The pioneering reef builders were exterminated, while the trilobites, repeatedly evolving into diverse species, were repeatedly choked off, and left with only a few hardy survivors. By 490 million years ago, when life had settled down for a while, a variety of new reef builders had appeared, and the trilobites had acquired many more hard-bodied neighbors in the form of new shelly animals.

worms; arthropods; brains 620 ▼
early hard animals 600 Myr ▼
many hard animals 570 Myr ▼
Cambrian disruptions 550 Myr ▼

630 Myr | 600 | 550

In the midst of the disorders, the first bony animals evolved as primitive fishes, founders of the vertebrate line that gave rise to the most ostentatious animals, including humans. Precociously sexy tadpoles were the probable forerunners of the fishes; that is to say, tadpole-like larvae of a sedentary marine animal became capable of reproduction while swimming, without first changing into their traditional adult form. They adopted phosphate scales for protection against the pincers of the chelicerates, and these, the first bony traces, show up 510 million years ago. A long interval elapsed before fishes developed bony jaws 425 million years ago.

The main geographical theme was the dispersal of pieces of the former supercontinent, creating many Noah's arks, on the continental shelves of which the sea animals could venture baroque experiments in form, as they cruised now through the tropics, now through colder regions. About 460 million years ago, North America was engaged in a slow collision with the microcontinent of Baltica, the core of what is now Europe. An ocean called Iapetus was converted into a chain of mountains, pieces of which are now to be found in Greenland, Norway, Scotland, Ireland, and northeastern North America. Scotland, for instance, was united with England in this event. Toward the end of the Ordovician period, 440 million years ago, many fishes were destroyed in a catastrophe that again afflicted the trilobites and many other marine animals. This event coincided with ice ages in which Africa, then straddling the South Pole, was extensively glaciated.

Earth entered a long warm phase of so-called greenhouse conditions 430 million years ago. Mid-ocean ridges rose hot and bulky, pushing the continents around, and their groundswell displaced much salt water onto the continental margins. At a time of high sea level and affluent reefs, the modern group of ray-finned fishes appeared 415 million years ago. These became the most successful of bony animals, boasting 30,000 species at present. By 400 million years ago certain predatory fishes of another sort, with fleshy, muscular fins, had converted portions of their guts into the first vertebrate lungs capable of breathing air. Among them were the ancestors of the bony land animals.

For billions of years life had been confined to the water. Rivers flowed through fields of rubble, across continents wider than Asia and as barren as Mars, until the first plants appeared on land 425 million years ago. These were soon followed by the first terrestrial

510 MILLION YEARS AGO

THE INVENTION OF BONE

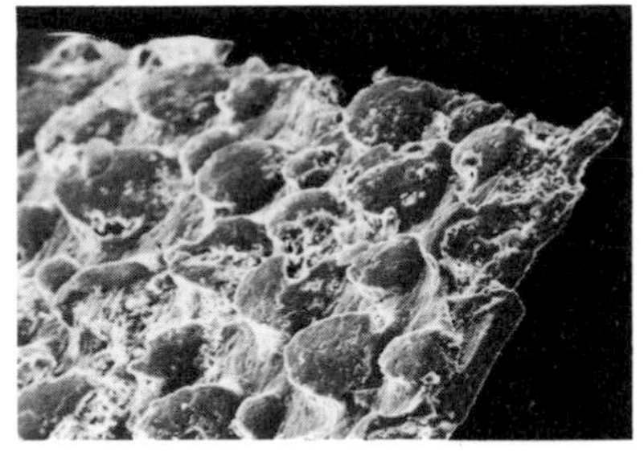

Small scales made of the bony material, calcium phosphate, protected primitive fishes from scorpion-like predators. These scales from Wyoming, shown greatly magnified, are the oldest traces of the vertebrates, the large group of bony animals that includes fishes and human beings.

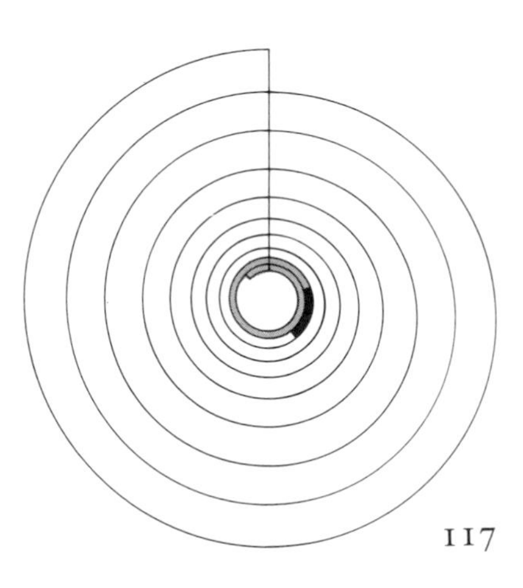

vertebrate animals ▼ 510 Myr

Europe hit North America 460 Myr ▼

Ordovician catastrophe; ice ages 440 ▼

jawed fishes; life ashore ▼ 425 myr

500 | 450 | 400

380 MILLION YEARS AGO
LIMBED FISH

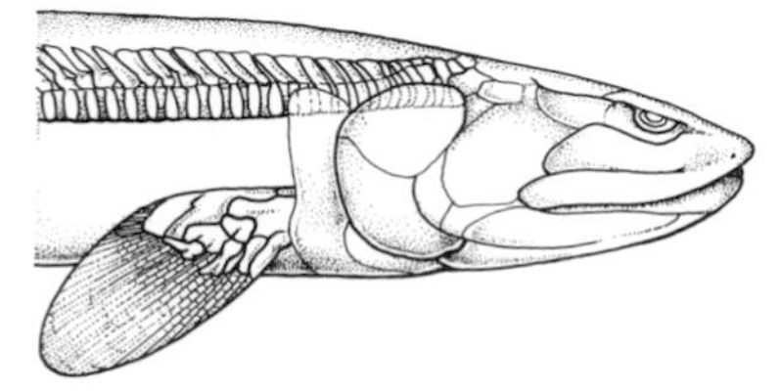

370 MILLION YEARS AGO
EARLY AMPHIBIAN

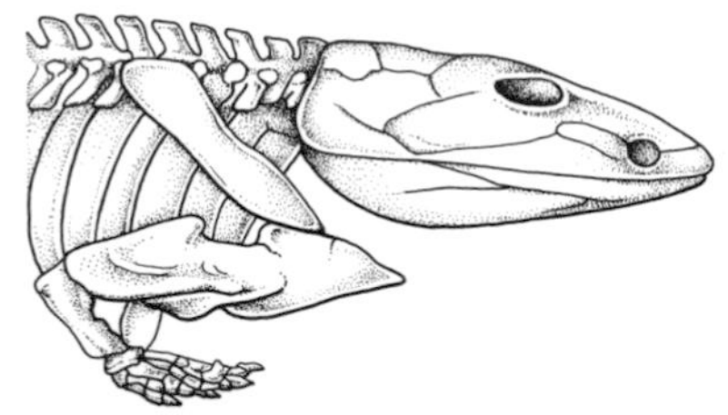

310 MILLION YEARS AGO
EARLY REPTILE

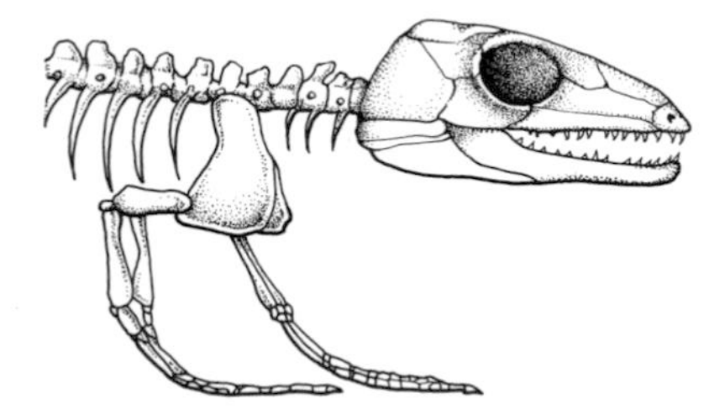

270 MILLION YEARS AGO
MAMMAL-LIKE REPTILE

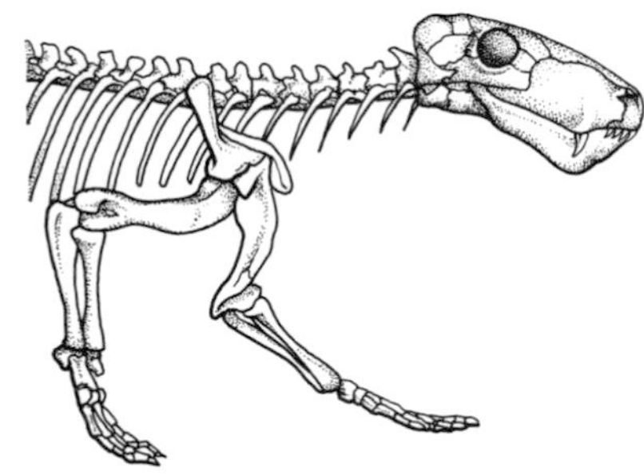

Bony land animals converted fleshy fins to limbs of increasing refinement.

animals, probably millipedes. The plants did not at first shake off their seaweed origins, but continued to reproduce in water, from spores and sperm. They could, though, support their own weight and avoid drying out in the air, and after small beginnings the plants formed trees and forests by about 370 million years ago. Some of the land plants had by then evolved landworthy seeds, and the very durable conifers date from 350 million years ago. All of the earliest animals ashore were arthropods, arising from segmented marine animals that acquired desiccationproof cases and means of breathing the air. By 395 million years ago millipedes, mites, and the first insects (springtails) were eating the plants, and spiders and scorpions were eating them.

Catastrophe struck the seabed 370 million years ago, when immense tidal waves scoured the shores and wiped out the limestone reefs of the world. These tsunamis were far greater than any produced by earthquakes, and were presumably due to a cosmic object hitting the ocean; there are craters in Sweden and Canada that might date from the same event. The catastrophe terminated the Frasnian stage of the Devonian geological period. Trilobites and shelly animals of the seashore took heavy punishment, while swimming animals, including fishes and the parvenu sharks, were relatively safe.

With tongue in cheek one could say that the Frasnian tidal wave washed our fishy grandparents onto terra firma. Certainly that event was followed at once (geologically speaking) by the first amphibians. Scaly creatures waddled through the steamy tropical swamps of Greenland looking like fishes on legs—as indeed they were, with fleshy fins adapted into four limbs. They could feed on insects, but few amphibians took to the land very seriously; they seemed happier in the water which, like the first land plants, they still needed for their breeding. The early amphibians filled the intermediate generations between fishes and the reptiles while the warm "greenhouse" phase gave way to a cooler "icehouse" phase, around 360 million years ago.

The main theater of action, geologically as well as in evolution, was in the tropics, where the wedded continents of North America and Europe began a long-drawn-out collision with Gondwanaland, the large southern land mass. About 350 million years ago, the South American corner of Gondwanaland butted Euramerica. It dragged down margins of the land and then crushed up

insects 395 Myr ▼

Frasnian catastrophe; amphibians; trees 370 Myr ▼

Euramerica hit Gondwanaland 350 Myr ▼

winged insects 330 Myr ▼

400 Myr ... 350

the Hercynian ranges that extend east-west across the middle of Europe and the main Appalachian Mountains of the eastern United States. Hemmed in were the swampy forests of Euramerica, running from Poland, through Germany and England to Pennsylvania and Kentucky. Giant horsetails, thirty meters tall, were the dinosaurs of the plant kingdom, and by 330 million years ago dragonflies and other winged insects were flying in their shade. The tectonic squeeze interred fallen trees and began making very rich coalfields 320 million years ago.

Reptiles evolved about 313 million years ago from amphibian ancestors. They had jaws that were less trap-like and more powerful than an amphibian's, but their supreme invention was the modern egg, with its leathery or calcified shell, and subtle membranes that handled the breathing and waste disposal of the embryo. Thus equipped, the reptiles were able to grow larger and to move about more freely, across the plains and forests of the continents. The first large animals to walk on land were members of a group of reptiles that emerged 310 million years ago, the pelycosaurs. They had mammal-like qualities and included forerunners of the mammals; some attained weights of over two hundred kilograms.

The collision between Gondwanaland and Euramerica took some 60 million years to complete, and before it had finished Siberia rammed Europe from the east, about 300 million years ago, to build the Ural Mountains. Gondwanaland extended across the South Pole, and about 290 million years ago it developed ice sheets that smothered large areas of the present southern continents, including India and Australia. The ice ages lowered the sea level, and ended the main Euramerican coal-making period. But when they relented other continental collisions were in progress, in Siberia and China, where coal making reached an all-time peak around 270 million years ago.

Some reptiles took to the sea, as the small nothosaurs of 278 million years ago, harassing the fishes and rehearsing the roles of later reptilian sea monsters. On land, meat-eating pelycosaurs, some with fins on their backs, preyed on their plant-eating relatives. In a turnover among living things 256 million years ago, the large pelycosaurs became extinct and were replaced by a second wave of mammal-like reptiles. For about 10 million years the sturdy therapsids, some of which weighed more than a ton, loomed large across a nearly completed supercontinent.

320 MILLION YEARS AGO

COAL MAKERS

Petrified tree stumps in Glasgow, Scotland, give an unusual glimpse of large club mosses that flourished in tropical forests of Euramerica, during a coal-making boom of 320 million years ago.

We became reptiles. *The Devonian period lasted until 360 million years ago, after which the Carboniferous ran until 290 million years ago, when the Permian began. In North America the Carboniferous is often designated as two periods: the Mississippian (early) and Pennsylvanian (late), with the division at about 325 million years. This was still the Paleozoic era.*

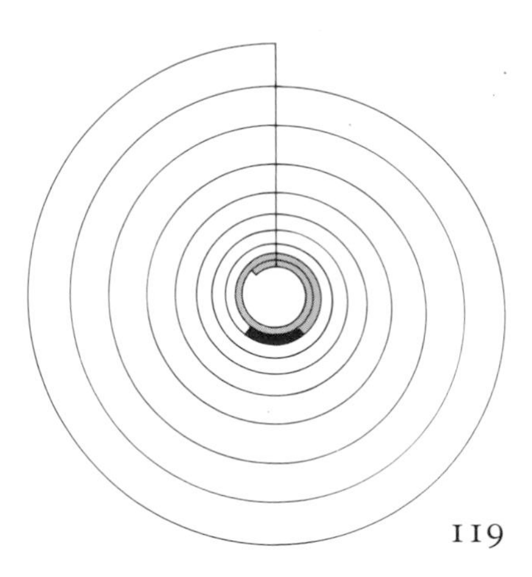

reptiles 313 ▼ | **pelycosaurs** ▼ 310 | **Siberia hit Europe** ▼ 300 Myr | **ice ages** ▼ 290 Myr | **coal peak** 260 Myr ▼ | **therapsids** ▼ 256 Myr

300 | 250

245 MILLION YEARS AGO

THE WORST CATASTROPHE

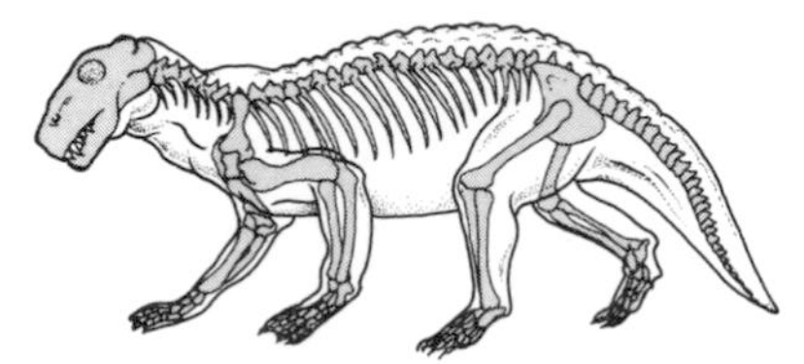

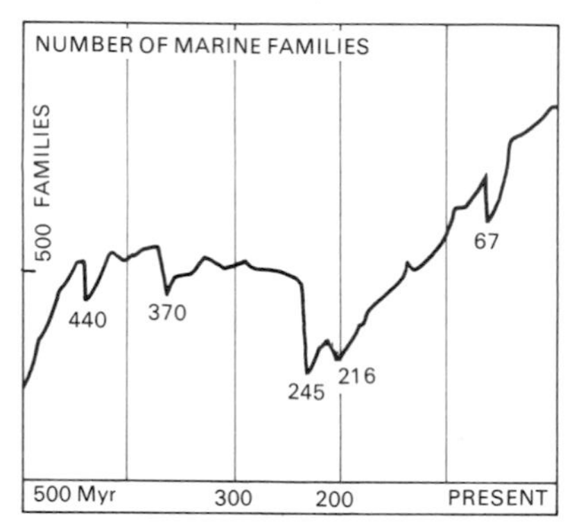

On land, the predatory mammal-like reptile, Lycaenops, *was among the conspicuous casualties of the Permian terminal catastrophe. In the sea the diversity of vertebrate and invertebrate animals was halved. The graph counts surviving families of marine species; if one reckons individual species, the event was even more disastrous. Four other mass extinctions, less severe, are also evident.*

Strange contrasts in the pace of events recur in this synopsis of life. A phrase disposes of a billion days on which the lopsided planet winked as it spun, showing first its watery, then its continental, face to the sun. Mountains grew and withered, plants deployed strange foliage, bizarre reptiles strutted and threatened, spiders scuttled, and all to the beat of a leisurely drum. Then, for a moment, time stood still as every alert animal on Earth sensed death.

The catastrophe of 245 million years ago was the most deadly disaster on record. At the end of the Permian geological period, 96 percent of all species of marine animals were annihilated, and the reefs and seabeds were scoured to sterility. On land, all of the large mammal-like reptiles perished. According to one gradualistic theory, the assembly of the supercontinent of Pangaea, accompanied by a fall in sea level, was responsible. A better explanation, comparable with other accidents that are known to have happened, may be a head-on collision between the Earth and a massive snowball, a comet proceeding, as Halley's Comet does, the wrong way around the sun.

When living things recovered, they looked so different that the Permian terminal catastrophe is one of the chief boundaries of geology. In the aftermath, certain mammal-like reptiles called lystrosaurs browsed untroubled by competitors and predators. During the next 10 million years many new animals evolved, challenging the incumbents. The first modern corals appeared on the shores, the coiled, jet-propelled mollusks called ammonites rose to prominence, and superpredators in the sea took the form of fish-like reptiles, ichthyosaurs more than ten meters long. Novel amphibians of the modern frog-like kind ventured on to the land. Plants called bennettitales sported the first flowers, more primitive than those of modern flowering plants yet luring insects to do the work of fertilization. A third wave of comparatively small mammal-like reptiles, the cynodonts, filled vacancies left by their annihilated predecessors.

We became mammals. *The Permian period ended 245 million years ago, when the Triassic began. This transition also marks the end of the Paleozoic era and the beginning of the Mesozoic ("middle life") era. An alternative name for the Permian terminal event is the Permo-Triassic catastrophe. The Triassic gave way to the Jurassic 208 million years ago.*

See map p. 84.

Reptilian meat eaters of different stock, two meters in length and slender, heralded another revolution. They had long hind legs on which they trotted like birds 235 million years ago, balanced by their long tails; their forelimbs became short arms. These were the first small representatives of the terrible reptiles, the dinosaurs, and long-legged beasts of various lineages gradually conquered the supercontinent. By 228 million years ago the last of the big pre-

Permian terminal catastrophe 245 Myr ▼ | **cyno-donts** 240 ▼ | **dinosaurs; flowers** ▼ 235 Myr | **pterosaurs** 225 Myr ▼ | **Pangaea** 220 Myr ▼ | **Norian catas-trophe; mammals** ▼ 216 Myr

250 Myr | 240 | 230 | 220 | 210

dinosaur meat eaters were preying on the first plant-eating dinosaurs. Giant dinosaurs, their weights reckoned in tons, appeared 225 million years ago, with the first pterosaurs, winged reptiles, flying or gliding overhead. The issue of supremacy was settled by a catastrophe at the end of the Norian stage in the Late Triassic period, 216 million years ago. It was not as severe as the Permian terminal event, but marine life was decimated and many reptiles were extinguished, both in the sea and on land. A large impact crater in Quebec province, of approximately the right date, may be related to the Norian catastrophe. After it, the dinosaurs were everywhere, in herds of plant eaters shadowed by meat eaters.

The mammal-like reptiles were not quite extinct, and a few had gone to ground as animals as small as mice. From this stock the true mammals arose, including remote ancestors of human beings. Small animals lose body heat rapidly, so these first mammals were no doubt furry, but their most characteristic invention was milk for the young, produced from re-engineered sweat glands. The mammals also owned relatively large brains, whose chief task for the next 150 million years would be figuring out how to elude the claws and jaws of dinosaurs and pterosaurs.

The northern and southern parts of the supercontinent slid slowly past each other from 260 to 220 million years ago, eventually setting Africa instead of South America against the eastern side of North America. The microcontinent of South China collided with eastern Asia and welded itself to North China along the line of the Chinling Mountains. By that time, 230 million years ago, South Tibet and Indochina were also running into place, and Pangaea came fleetingly to the configuration from which the present dispositions of the continents derive. A gap opened between western Africa and North America, at the birth of the Atlantic Ocean, about 210 million years ago. The Palisades of New York's Hudson River, 195 million years old, are made of basaltic lava from that great split; in related rifts across Europe's North Sea, and beside the eastern edge of Mexico, oil accumulated during a worldwide boom in petroleum making around 170 million years ago. It began as the Earth's climate entered a new "greenhouse" phase, better for oil than coal. Volcanic outpourings in southern Africa foreshadowed other continental splits.

The brachiosaurs, long-necked plant eaters, emerged after a change in regime among the dinosaurs 175 million years ago.

240 MILLION YEARS AGO

NURSERY OF JAWS

Young crocodiles still mimic the hatching of dinosaurs, to which they are distantly related. The crocodiles themselves evolved about 230 million years ago.

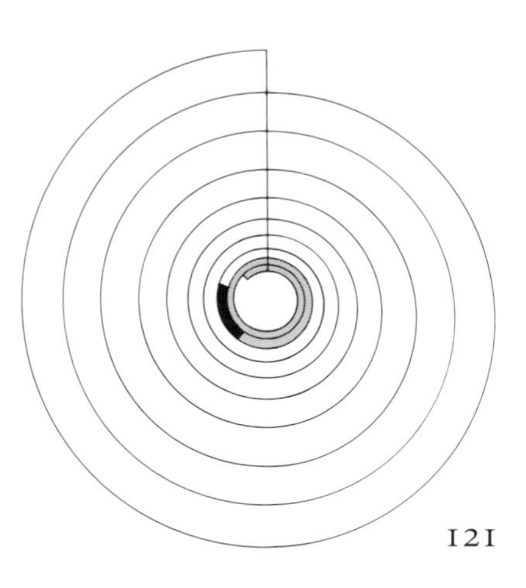

brachiosaurs
175 Myr ▼

petroleum boom
▼ 170 Myr

200 | 190 | 180 | 170 | 160

150 MILLION YEARS AGO
DINOSAUR INTO BIRD

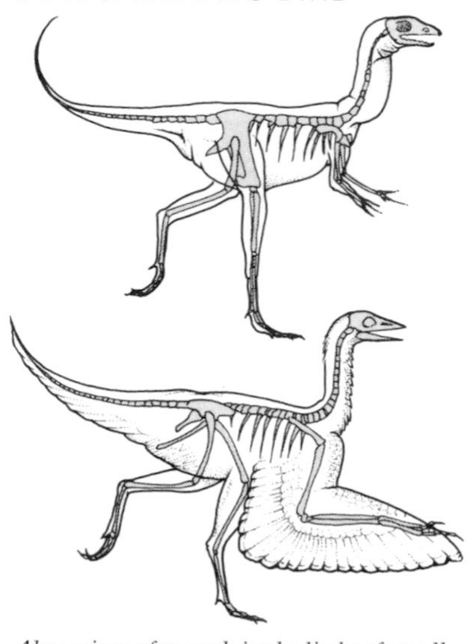

Alterations of growth in the limbs of small dinosaurs may at first have seemed like gross malformations, but they created the first birds, thus adding the last main branch to the family tree of bony animals.

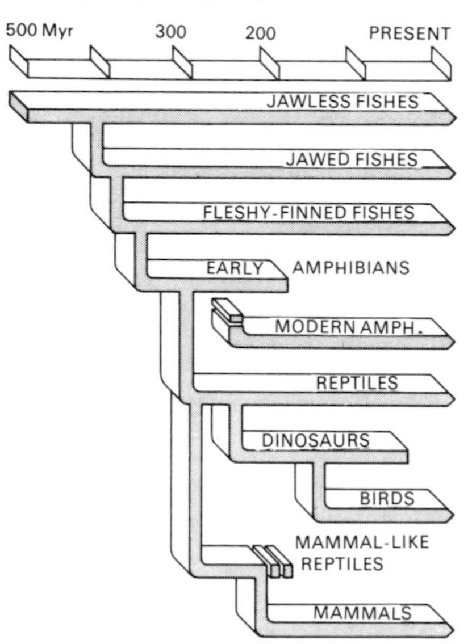

Ranging from forty to eighty, and perhaps even two hundred tons, the brachiosaurs were the largest animals ever to walk on land. Large body size in the dinosaurs maintained a high body temperature with little waste of energy, and also offered a certain protection against hungry jaws. The interplay between big plant eaters and big meat eaters was a long series of retakes of the same scene with different actors. From the six-meter megalosaurs of 200 million years ago to the twelve-meter tyrannosaurs of 70 million years ago, predators operated on two legs and took their occasional meals out of the herds that fed on the vegetation. Plant eaters with distinctive hip design produced the stegosaurs, distinguished by their curious back plates, 164 million years ago.

Defying the trend, some dinosaur species were no bigger than small birds. Like the mammals, they had more difficulty in keeping warm: one solution was feathers, and these opened the way to the evolution of birds, as descendants of the dinosaurs. Bird-like fossils of *Archaeopteryx* date from around 150 million years ago. It was little more than a glider, and may not have been a true forerunner of eagles and sparrows, but *Archaeopteryx* represented an evolutionary ploy that, within 20 million years, had produced more sensible-looking water birds.

A turnover of dinosaur groups began 145 million years ago, when the stegosaurs and other large plant eaters died out at the close of the Kimmeridgian stage of the Jurassic period; the brachiosaurs and others carried on, but reefs, fishes, and marine reptiles suffered badly, with the ichthyosaurs disappearing forever at that time. New dinosaur groups established themselves 133 million years ago: tall iguanodons, the heavy-headed pachycephalosaurs and panoplosaurs, and, as novel predators, nimble beasts called deinonychids, with sickle claws on their feet. The start of that long-playing revolution, 145 million years ago, coincided with a minor fall in sea-level, and a major fall at 133 million years ago drained large areas of the continental margins. Unrecorded climatic excursions may have occurred during those 12 million years of the Kimmeridgian turnover. And although the next major alteration in the Earth's environment was to take a very different form, in a colorful transformation of the realm of plants, changes of climate may again have played a part.

Modern flowering plants are the most successful vegetation in the world, boasting hundreds of thousands of species. Called angio-

birds
150 Myr ▼

Kimmeridgian turnover
▼ 145 Myr

sperms or encased-seed plants, because of the arrangement of seeds made graphic in peapods, they probably arose in Africa from the more primitive flowering bennettitales. They came out of obscurity and conquered the planet in a comprehensive floral revolution, formerly regarded as "an abominable mystery," but now well defined. Leaves and pollen of the first angiosperms date from 123 million years ago, and the plants began a phase of biotechnical improvements and limited geographical spreading.

Then, 117 million years ago, another environmental crisis occurred, marked by a major fall in sea level and the extinction of the mighty brachiosaurs. The flowering plants thereafter took off, diversifying rapidly, displacing the previously successful conifers from many settings, and driving the old-fashioned cycads, ginkgoes, and bennettitales to near-extinction. Early upstarts included magnolia-like plants, plane trees, and water lilies. The flowering plants entered into trade with insects and mammals. Insects multiplied on the nectar and other bribes offered by the plants in return for cross-pollination, while mammals and birds were well suited to keeping insects under control.

Against this botanical background, a burst of evolution produced modern mammals. The earliest mammals probably laid eggs, like the platypus today, but marsupials resembling opossums had evolved by 125 million years ago. A marsupial is born at a very early stage of development and the fetus then grows in a feeding pouch. More modern mammals possess a placenta and carry their young for much longer before birth. What seems to have been the first placental mammal showed up 114 million years ago in Mongolia, feeding on the insect offerings of the floral revolution.

The race was on, to stock the southern continents with the new mammals before juvenile oceans and rising sea levels broke the links with the old heartlands of Pangaea. By 140 million years ago, Eurasia and North America, still linked together, were edging away from the southern continent of Gondwanaland and creating an open sea around the equator. Gondwanaland itself was disintegrating by 130 million years ago, when incipient oceans had etched the outline of Africa. The marsupial mammals—opossums—marched via the Americas and Antarctica to the still-attached Australia. The flowering plants crossed over too, before rising sea levels created water barriers across that route, but modern mammals missed the boat and Antarctica-Australia remained a marsupial province. They

120 MILLION YEARS AGO
ANCIENT MARSUPIAL

Opossums living today bear a close resemblance to the earliest mammals that nourished their young in pouches. This is Marmosa, *from South America.*

We invented placentas. *The Jurassic period ended, and the Cretaceous period began 138 million years ago. This was still the Mesozoic era.*

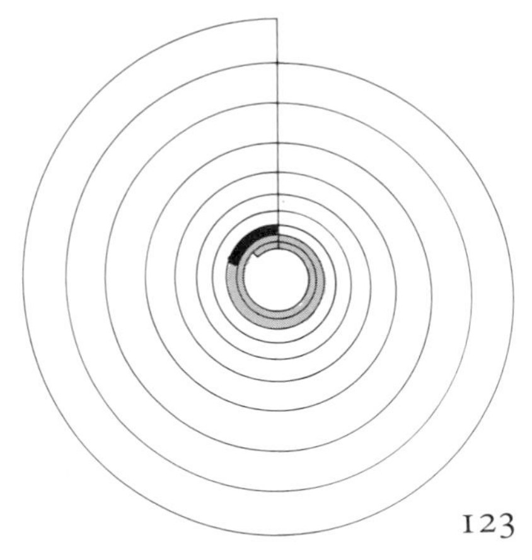

marsupial mammals ▼ 125 Myr | **brachiosaurs extinct** 117 Myr ▼ | **floral revolution; modern mammals** ▼ 114 Myr | **India quit Antarctica** 100 Myr ▼

120 | 110 | 100

90 MILLION YEARS AGO
PETROLEUM AND CHALK

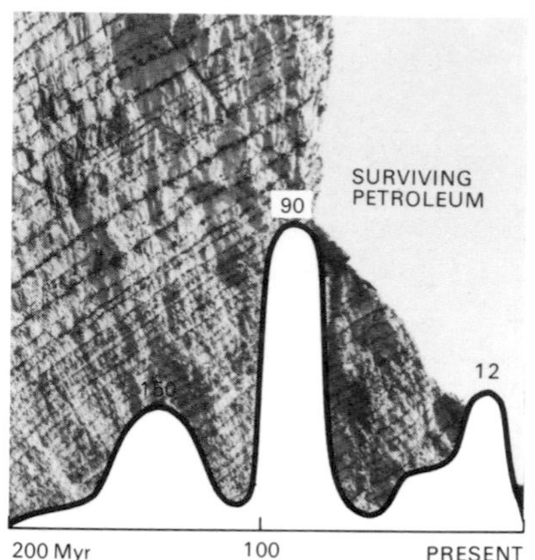

On flooded shelves of continents, most of the world's oil reserves formed at around this time, as did extensive deposits of chalk, conspicuous in southern England. The dark bands of nodules in the chalk reflect regular climatic cycles, similar to those governing the current ice ages.

70 MILLION YEARS AGO
THE LAST MONSTER

Tyrannosaurus rex, *twelve meters long and with formidable jaws and claws, was a characteristic predator in the phase immediately before the extinction of the dinosaurs.*

reached South America, though, before it became fully isolated, 85 million years ago, and gave rise to armadillos and other curious mammals sharing that continent with the marsupials.

The placenta by which a modern mammalian mother nourishes her fetus was an invention just as radical in its private way as the wings of the birds. By preventing immunological quarrels, in which the fetus might look like destructible foreign tissue for the mother's antibodies, it also moderated the hazards of inbreeding and allowed modern mammals to lead more sociable lives than the marsupials. They remained small, but a mark of modest success was their radiation into different lineages. By genetic evidence, ancestors of hoofed mammals, rodents, and modern carnivores went their separate ways between 100 and 85 million years ago. The flowering plants continued to interact with the mammals. They offered fruit to buy seed dispersal, and mammals climbing trees in search of insects or leaves found the fruit and nuts tasty. This food nurtured the primates, the group of mammals that now includes monkeys, apes, and human beings. Preprimates parted from other mammals an estimated 95 million years ago, and premonkeys split from lemur-like prosimians about 70 million years ago.

The microcontinent of India, a fragment of Gondwanaland, was on the loose 100 million years ago, after breaking from Antarctica. The geological tumescence of worldwide rifting and ocean making 93 million years ago pushed the sea level toward an all-time high, creating huge shallow seas that covered one third of the total areas of the continents. A broad seaway running from Alaska to the Gulf of Mexico divided North America. Single-celled carbonate organisms called coccoliths multiplied amazingly and plastered the flooded continents with vast quantities of chalk, 93 to 67 million years ago, earning for the Cretaceous geological period its name, which means chalky. The white cliffs of Dover, in southeastern England, are typical tombs of the coccoliths.

Fertile surface waters delivered a rain of organic sediments that created stagnant conditions at the bottom of the shallow seas, ideal for forming hydrocarbons. More than half the world's known oil reserves date from around this time, when the Gulf of Mexico, Venezuela, and North Africa, for example, lay under tropical seas. But where the torn edge of old Gondwanaland slanted across the equator, at what is now Arabia and southern Iran, 93 to 85 million years ago, petroleum accumulated at the greatest rate ever, creating

India quit Antarctica
100 Myr ▼

Cenomanian turnover; preprimates
95 Myr ▼

continents flooded; petroleum peak
▼ 93 Myr

South America quit Africa
85 Myr ▼

100 Myr | 95 | 90 | 85

the richest oil fields of them all. The sea level was past its peak by 70 million years ago, and the Earth's climate began gradually switching from "greenhouse" to "icehouse" conditions, better for coal than oil. Western North America experienced a new boom in coal making.

On fragmented continents reduced to still smaller islands by flooding, the dinosaurs were neither as big nor as varied as they had been. In the west of North America, the iguanodons had died out 95 million years ago, to the accompaniment of a brief but sharp sea level fall in the Cenomanian stage, but successful duck-billed hadrosaurs replaced them 89 million years ago, and the horned ceratopsids came in somewhat later. Tyrannosaurs were then the top predators. By 69 million years ago the diminutive mammalian forerunners of carnivores and hoofed plant eaters were registering in the fossil record, and the earliest known primate, *Purgatorius*, was living in Montana and watching from the safety of the trees as the tyrannosaurs passed by.

One day 67 million years ago a blinding flash filled the sky, and thereafter the world belonged to *Purgatorius* and its mammalian cousins. The object from outer space that blasted the Earth may have struck the Pacific Ocean; plants on land evidently suffered worse in western North America than in the east. Whether it was a comet or a giant asteroid, and whether it killed quickly or slowly, by shock wave, poison, heat, cold, the obscuration of the sun, or prolonged ecological disruption, the impact dusted the entire planet with exotic elements. For about ten thousand years afterward the oceans were dead.

The Cretaceous terminal catastrophe was not the worst of its kind, nor the last, but the lists of casualties and survivors have special poignancy. The sea level dipped, reefs were obliterated, and most species of marine plants and animals perished. Ammonites and reptilian sea monsters were wiped out. On land, the pterosaurs and every dinosaur expired, leaving only the lurking crocodiles to remind the world of what reptilian jaws could do, and the birds as feathered relics of the dinosaurs themselves. Small reptiles scraped through, and so did some of the small furry mammals.

The Earth's magnetism had gone flip-crazy, and frequent magnetic reversals characterized the new era. Between 84 and 72 million years ago there had been only one reversal; during the following 10 million years the Earth swapped its magnetic poles around at least

67 MILLION YEARS AGO

ATOMS OF DISASTER

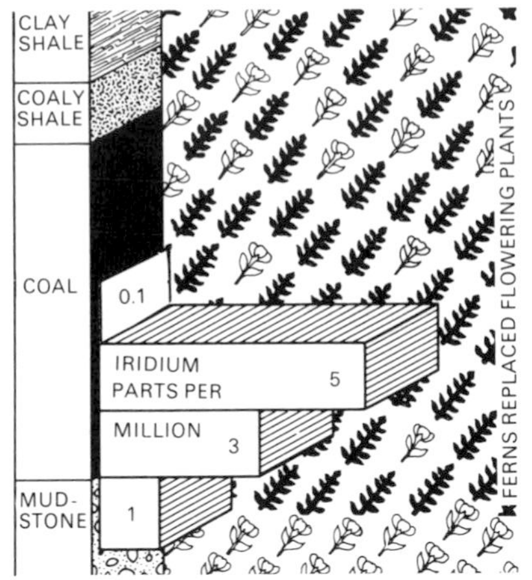

Unusual concentrations of iridium atoms, a rare material of extraterrestrial origin, occur all over the world at the level in the geological record when the dinosaurs were extinguished. This detailed record, where the iridium coincides with a narrow band of coal and with a marked change in vegetation, is from freshwater sediments in northern New Mexico.

We became primates. *The Cretaceous period and the Mesozoic era ended 67 million years ago, when the Tertiary period of the Cenozoic ("new life") era began.*

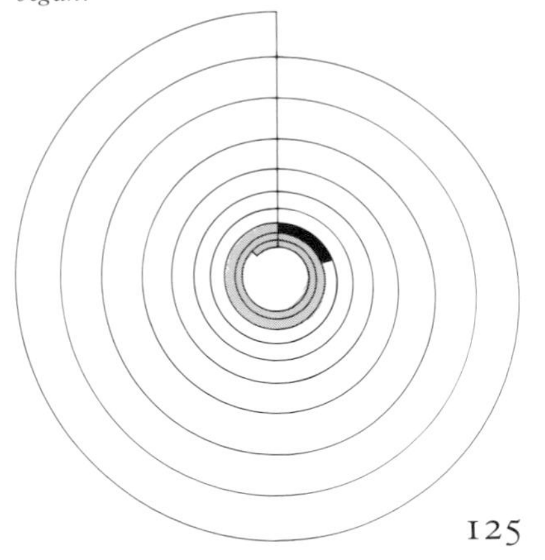

coal boom; preprimates 70 Myr ▼

fossil primates ▼ 69 Myr

Cretaceous terminal catastrophe; dinosaurs extinct ▼ 67 Myr

80 75 70 65

55 MILLION YEARS AGO
CLIMBING TO SUCCESS

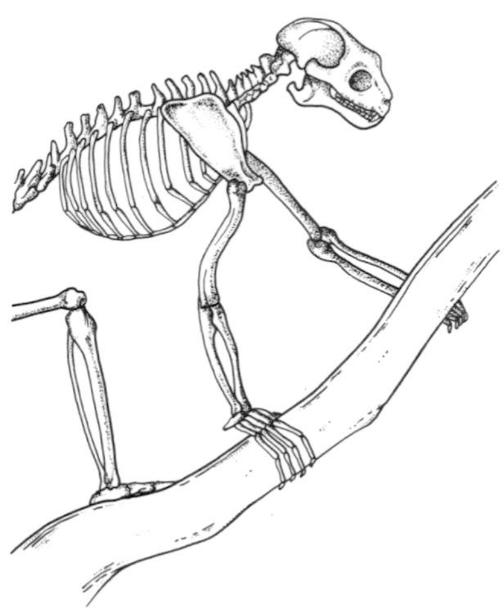

Hands with opposable thumbs, well suited to grasping, helped the primates to gain mastery of the trees, most conspicuously in the lineage of premonkeys, monkeys, and apes.

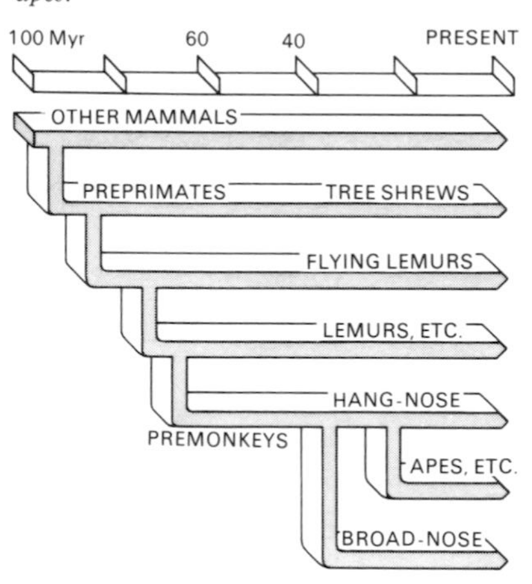

sixteen times. Since then, the rate has increased, by fits and starts, to roughly forty reversals in the most recent 10 million years. Uncertainty surrounds possible links between magnetic reversals, falling sea levels, and effects on life—and between all these phenomena and the cosmic weather of impacting comets and asteroids. At any rate, the overall trend to more frequent magnetic reversals paralleled a trend to a colder world.

The world belonged to the mammals and birds, and a summary of the past 60 million years, as brisk as for a comparable period in the times of earliest animals, might be as follows. Continental breakup gave way to reassembly and associated mountain building, the sea level fell, and the climate cooled. Mammals gradually took the places of the great defunct reptiles by sea (notably as whales) and by land (for example, as elephants and large cats). The flowering plants invented grasses that virtually asked to be grazed. During a short warm spell in the midst of a new series of ice ages, some talkative two-legged primates used their free hands to master plants and other animals, and built spaceships that carried the first of their kind, still chattering away, as far as the moon. Some of the details follow.

As usual after a major catastrophe, millions of years elapsed between the annihilation of the dinosaurs and the evolution of novel kinds of animals to replace them. The first ventures were unimpressive. In the northern continents and Africa, marsupial mammals faded. Among placental mammals, archaic carnivores prospered for a while, and the early insectivores and primates were perpetuated as shrew-like and lemur-like animals. Archaic hoofed mammals grew in size as stopgap replacements for plant-eating dinosaurs, and came to look like sheep as big as rhinos, but their line is long extinct, as is that of the early carnivores. After an environmental disturbance indicated by a major fall in sea level 62 million years ago, which closed the North American seaway, the first modern carnivores appeared, the ancestors of lions and bears.

The Americas were two separate continents drifting westward, belching volcanoes and piling up volcanic islands along their leading edges. This had been going on long before Pangaea started to break up 200 million years ago, but in the west of North America the activity reached one climax around 60 million years ago, when strata rich in dinosaur bones were crumpled in an early version of the Rockies. The event coincided with North America's departure from Europe, as Greenland slid northward past Canada and the rift of the northern Atlantic opened between Greenland and Scotland-Norway 60 million years ago. Overland connections between the northern continents were not permanently cut because North America had already collided with the eastern end of Asia, and the region of the Bering Strait beside Alaska provided an occasional route for animals. So, while South America nurtured its own peculiar populations of marsupial and primitive placental mammals, North America remained in the mainstream of evolution. The rise of the mammals began to look spectacular by about 55 million years ago, at a stage that saw the advent of rodents, bats, early whales, and the vanguard of two modern armies of hoofed animals, including horses, then the size of dogs. In the trees, premonkeys had evolved the forward-looking eyes and grasping hands that were to typify later primates.

Antarctica, installed at the South Pole, shed Australia 50 million years ago, and that island continent then drifted with its marsupial crew toward warmer latitudes. The humble opposums sired whole orders of large marsupials, matching many of the roles of placental mammals elsewhere; while plant-eating kangaroos and koalas were eccentric in appearance and habits, marsupial meat eaters were uncannily like cats and dogs. And 45 million years ago, wayward India rammed Eurasia amidships, in Tibet, at what by geological standards was a high speed, a hundred kilometers per million years. Large continental blocks were squeezed sideways, notably into China, during the collision. Older mountain chains in nearby Asia were rejuvenated. As possibilities of that kind were exhausted, India, traveling more slowly, burrowed under Tibet and raised it high into the air.

Elephant-like animals the size of pigs appeared about 50 million years ago, although these proboscidians and their early successors were confined to Africa. By 40 million years ago with the emergence of rabbits, and whales and insectivores of modern kinds, the cast list of the various orders of mammals was essentially complete. Their descendants were going to need their fur coats. The climate was benign until 50 million years ago, when the oceans cooled by several degrees Centigrade. Thereafter, global temperatures went up and down, but mainly down. An underlying reason was a rearrangement of the continents interfering with ocean currents that redistributed the warmth of the world. The planet was in an increasingly

55 MILLION YEARS AGO

FORERUNNER OF COWS

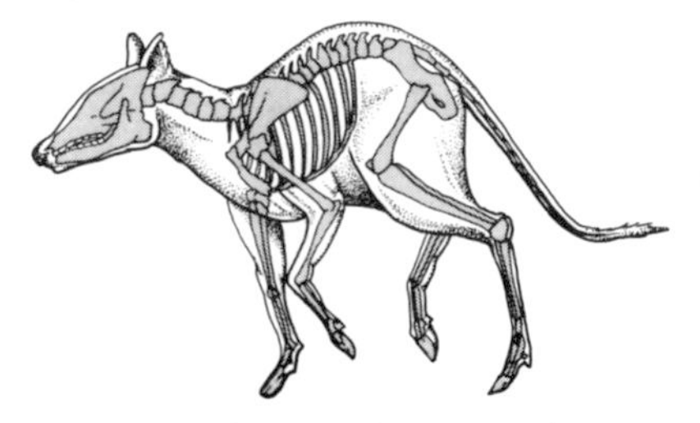

Rabbit-sized Diacodexis *scampered or hopped in North America and Europe 55 million years ago. It was close to the ancestral line of camels, and of the even-toed ruminants that include giraffes and antelopes. Cattle, sheep, and goats are in turn descended from antelopes.*

We were premonkeys. *The Paleocene, the first epoch of the Tertiary period, lasted from 67 to 55 million years ago. The Eocene epoch followed it, and lasted until 37 million years ago.*

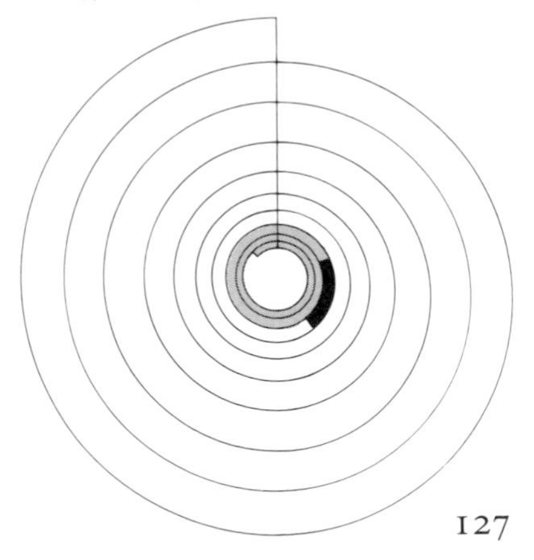

Australia quit Antarctica ▼ 50 Myr

India hit Eurasia ▼ 45 Myr

more new mammals 40 Myr ▼

50 | 45 | 40

37 MILLION YEARS AGO
TUMBLING TEMPERATURES

As the world's climate lurched toward colder conditions, changes in the atomic composition of small marine animals recorded the event, as one of a succession of steps down a staircase leading to the ice ages. Glaciers in Antarctica were spawning icebergs more than 25 million years ago, as evidenced by stony debris that they dropped when they melted.

vulnerable condition, climatically speaking, when it took the second pronounced step toward cooler conditions.

Another cosmic impact occurred 37 million years ago, and flung droplets of molten rock halfway around the planet. It caused widespread changes in life, at the Eocene terminal turnover. A related effect was the invention, or reinvention, of winter. Before the event, the difference in temperature between January and July had been quite moderate; after it, short days at high latitudes brought frosts. Vegetation changed drastically to adapt to the wintry conditions, while glaciers grew in Antarctica. Pack ice formed on the polar seas, and chilled water sank and circulated at the bottom of the ocean far from the polar regions, killing many marine animals and protoanimals that lived there. Another sign of colder oceans was widespread deposition of carbonates on deep ocean floors, where previously they would have dissolved. The development of cold bottom water initiated a regime of deep-water circulations, which continues to operate as a major feature of present-day oceans.

By genetic and fossil evidence the water birds diversified rapidly at this time, and penguins were appropriate newcomers. (Loons, albatrosses, herons, and penguins all shared a common ancestor approximately 40 million years ago.) The early whales suffered badly in an impoverished ocean, and on land many old-fashioned species of mammals died out, among them primitive insectivores, carnivores, and hoofed animals. On the other hand, hoofed animals called titanotheres and indricotheres grew to giant sizes, and the extinctions encouraged a further modernization of the mammals with the appearance, 35 million years ago, of the first true cats, dogs, and rhinos. By 30 million years ago, early pigs and bears had made their debut.

In South America, the toxodonts and mylodonts evolved; these became rhino-like plant eaters and the chief of the giant sloths. The most mysterious appearance was that of certain primates, forerunners of the broad-nose (platyrrhine) New World monkeys. By estimates from genetic differences, these last shared a common ancestor with the hang-nose (catarrhine) Old World monkeys and apes about 36 million years ago, and the earliest known fossil primate in South America turns up, on cue, in beds dating from immediately after the Eocene terminal turnover. This conformity of genetic and fossil evidence leaves no room for doubt that primates first arrived in that continent about 35 million years ago, together with some alien rodents. At that time South America was supposed to be separated from the other land masses by ever-widening oceans, and the general picture of evolutionary isolation is clear enough among the other animals of South America. So how did the ancestors of New World monkeys reach South America? Answers envisaging a pregnant primate Columbus on a raft crewed by rodents should be treated with due caution.

Japan was a piece of the Asian mainland until about 30 million years ago. Then the floor of the Pacific Ocean, diving to volcanic destruction under the margin of Asia, began to wrench it loose, opening a miniature ocean basin behind it, the Japan Sea. In East Africa, the ground rose in a dome-like fashion and cracked open, producing rift valleys strewn with volcanoes and lakes, and Arabia began to break from Africa along the line of the Red Sea. The mountains of the dome fed the Nile River. In another small but consequential adjustment to geography, a sea gap opened between Antarctica and South America about 35 million years ago, and between Antarctica and the southernmost Australian continental shelf about 30 million years ago. The waters around Antarctica were then free from encumbrances, and the cold and vigorous Circumantarctic Current, or West Wind Drift, began to flow around it, cutting Antarctica off from any warming water from the tropics. As a result, Antarctica accumulated ice more thoroughly, and the sea level fell drastically about 29 million years ago. Although the ice may not have constituted a permanent major ice sheet, its presence near the South Pole helped to confirm the trend to chilly conditions.

The lowering of the sea level 29 million years ago drained much of flooded Eurasia, making for easier movements of animals between the northern continents. Mammalian novelties in Africa included quite modern-looking primates of the Old World: animals with monkey-like tails and ape-like jaws, apparently the hang-nose forerunners that monkeys, apes, and humans had in common. They lived in Egyptian forests about 29 million years ago, when Africa was still isolated from Eurasia. The first of the modern whalebone whales, which fed by filtering small food from the sea, evolved in the southern oceans at about that time. The new regime was favorable for the whales because the Circumantarctic Current became highly productive of filterable food by 25 million years ago. As the gap between Australia and Antarctica widened, and the current

35 MILLION YEARS AGO

A LIVING FOSSIL

The two-horned Dicerorhinus, *living in Malaysia today, is virtually indistinguishable from fossilized predecessors among the early rhinoceroses.*

We became monkeys. *In the middle of the Tertiary period, the Eocene epoch gave way to the Oligocene 37 million years ago.*

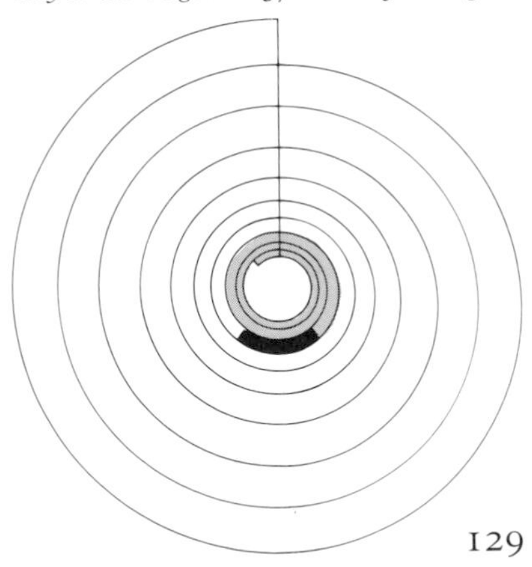

Circumantarctic Current; Japan quit Eurasia; Arabia quit Africa 30 ▼

sea-level fall; hang-nose primates ▼ 29 Myr

whales growing ▼ 25 Myr

30 | 25

24 MILLION YEARS AGO
PIONEERING GRASS

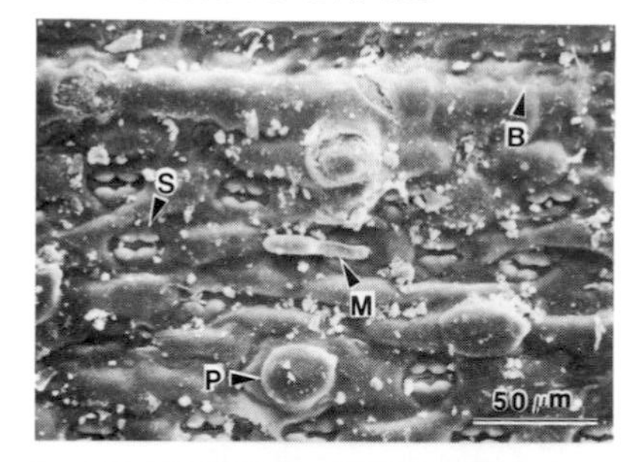

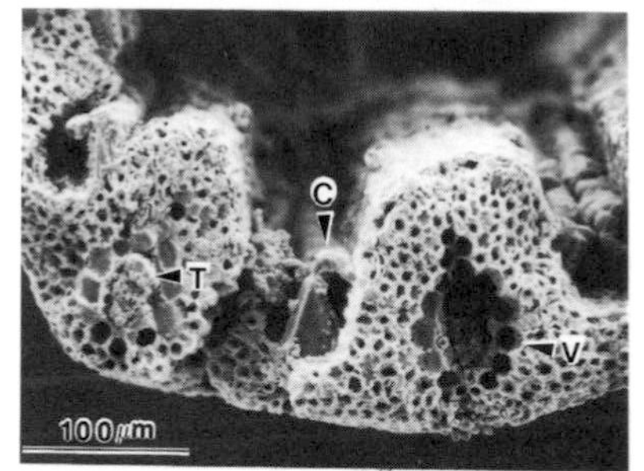

Remarkably old fossils of grass leaves (Early Miocene, North America) show characteristic features under the microscope. The outer surface, in the upper illustration, contains small prickles (P), microhairs (M), pores (stomata, S), and clover-shaped silica bodies (B). The lower illustration is a cross section of a leaf, with vascular tissue (T), a sheath of soft material (V), and bulliform, or blister-like cells (C). The scale bars are annotated in micrometers; 100 µm is a tenth of a millimeter.

grew greater, the whales progressively increased in size to become the largest marine animals of all time. Some carnivores also took up an aquatic life, becoming seals.

The climatic deterioration paused about 25 million years ago, and the ensuing 10 million years were relatively mild. There was also a distinct lull in the rate of magnetic reversals around 20 million years ago. The climate, though, remained cooler than it had been in the days of the dinosaurs, and drier too, with the result that vegetation was sparser in some areas of the world. Opportunistic flowering plants carpeted the treeless ground with grass. A world without grass may seem hard to imagine, yet grass is one of the most recent of botanical innovations. It evolved from bamboo-like plants about 24 million years ago, and in parts of southeastern North America a scene soon unfolded, which was not unlike the African savanna today, with large herds occupying a grassy plain. The potential of the grasses as world-transforming plants was not fully apparent until more regions became semiarid, about 15 million years ago.

Open country was characteristically windswept, so grass could rely on the breezes to scatter its pollen, thus becoming independent of insects and other animals, and able to dispense with fancy flowers. Turf grasses went a stage further and propagated underground. In times of drought, vegetation was extremely vulnerable to fires started by lightning or volcanoes and fanned by the wind. Grass evolved a way of growing, not from the tip of the leaf, which was the normal way, but from the bottom, so that fire could pass and leave the tender growing regions unscathed.

This method of recovery from fire had the important consequence that plant-eating mammals could graze without defoliating the landscape. Evolving into thousands of species, grass became a blessing as the climate went downhill, and it enabled armies of animals to survive in the ever-larger drought-prone regions; in return, the animals tended to destroy any bushes that might try to intrude in the territory favored by the grass. Plant-eating mammals responded with evolutionary enthusiasm to the gift of grass. Among the adaptations were especially long teeth, which could be worn away by tough and abrasive grass and still remain serviceable. But speed became important too, because on the grasslands there was nowhere to hide from the meat eaters. No other plants have more significance for human life than the grasses.

whales growing 25 Myr ▼

grass ▼ 24 Myr

apes and monkeys split 21 Myr ▼

25 Myr | 24 | 23 | 22 | 21

Like the apes, human beings are monkeys without tails. and the joint ancestors of humans and modern great apes were probably represented among the early tailless apes called dryopithecines, living in Africa 20 million years ago. From about the same time, also in Africa, comes the first sure trace of a fully-fledged monkey. It is only a tooth, but it fits well with genetic comparisons indicating that monkeys and apes shared common ancestors until 21 million years ago, and then diverged. Also appearing 20 million years ago were the deer, which descendants of the dryopithecines would prize as venison.

The Arabian microcontinent was still joined to Africa, and it extended north to Syria and east across the Arabian-Persian Gulf. It was at the height of a collision with Eurasia 20 million years ago. This event pushed up mountain ranges in southern Iran and Turkey, around the fringe of what came to be known as the Fertile Crescent, and rain on the mountains nourished the Euphrates and Tigris Rivers of Mesopotamia. A fall in sea level confirmed an overland connection between Africa and Eurasia, with Arabia as the bridge. Continental movements and changes in sea level had for long enriched the story of the mammals by repeatedly opening and closing intercontinental links, like doors in a French farce. The reunion of Africa and Eurasia 20 million years ago pooled the novel mammals of the Old World. From Africa, apes and elephant-like animals invaded the northern continents, while antique cats and horses were in the stream of traffic going the other way. The encounter between grass coming from the north and the hoofed animals of the south was especially promising, because it was in Africa that antelopes evolved 19 million years ago, the bovid ancestors of notable grazers that now include cattle and sheep.

The deployment of the continents looked by this time like a freehand sketch of a modern map of the world, to which finishing touches were being applied. The volcanic ridges that were manufacturing the Atlantic Ocean stood in part above the surface, creating Iceland as a basalt island larger than today. With Asia and Africa already locked in the east, near-closure between Spain and North Africa created the narrow Mediterranean Sea, cluttered with mobile islands. The most significant discrepancy, compared with the present arrangements, remained the watery gap between North and South America.

The 10-million-year mild interval ended in renewed cooling,

20 MILLION YEARS AGO
ORIGINAL APE

Dryopithecines lost their tails, spread widely in Africa and Europe, and founded the lineage of apes and humans.

We became apes. *The Oligocene epoch of the Tertiary period ended 25 million years ago, and the Miocene began.*

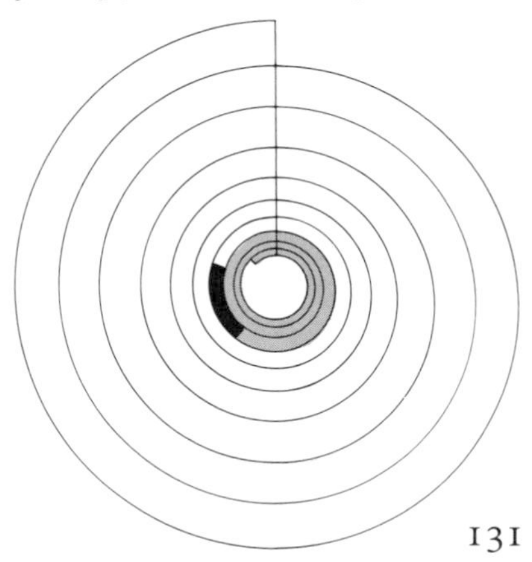

Old World interchange ▼ 20 Myr

early antelopes ▼ 19 Myr

Oregon eruptions 16 Myr ▼

20 | 19 | 18 | 17 | 16

14 MILLION YEARS AGO
PACIFIC PLUME

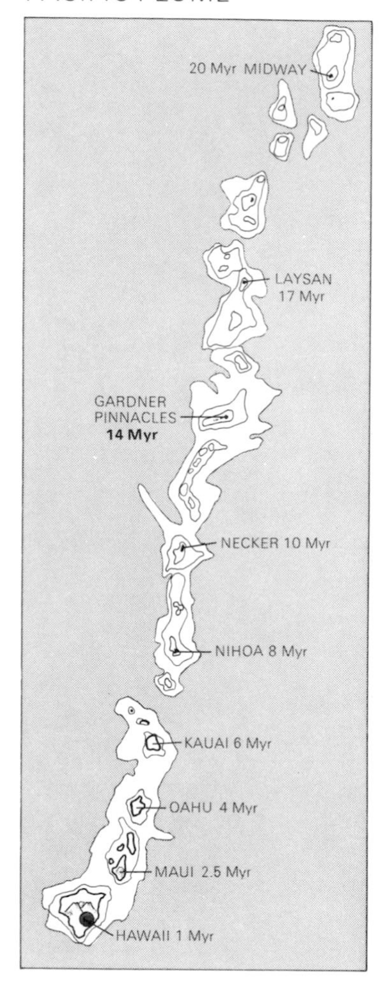

The islands of the Hawaiian chain have been created by a succession of volcanoes punching through the floor of the Pacific Ocean, as the plate slid northwestward over a chimney of molten rock rising from deep in the Earth's interior. The Hawaii plume joined in the chorus of volcanic activity 14 million years ago, creating the island of which the Gardner Pinnacles are the eroded remnants.

when a number of events occurred around 15 million years ago. A cosmic object of moderate mass fell on Europe; Antarctica went into permanent deep freeze beneath the great eastern Antarctic ice sheet, which has persisted ever since; in parts of the tropics, forests faded; and intensive volcanic activity occurred worldwide. During this flurry of cosmic, climatic, and geothermal business, the Earth switched its magnetic poles thirteen times in 3 million years.

About 16.5 million years ago, western North America began a paroxysm of volcanic outbursts that smothered Oregon and surrounding areas in the piles of hot basaltic lava that built the Columbia River plateau. This was the high point of a 30-million-year episode that continued to the recent eruptions of Mount St. Helens, due to North America overriding the floor of the Pacific Ocean. In the Oregon event, the digesting of an ocean rift, a seamount, or a deep-rooted hot spot may account for the great volume of basalt. Hot springs now located farther east, in Yellowstone Park in Wyoming, seem to be a mild continuation of that encounter.

The cosmic impact memorialized in glassy debris strewn across Czechoslovakia, and in a crater in Germany twenty-eight kilometers wide, occurred 15 million years ago. A minor fall in sea level ensued, and for about a million years the climate oscillated violently between cold and mild. As the impact may be implicated in these and other events, it gives its date to a disruption—the Miocene disruption, adopting the name of the prevailing geological stage.

The eastern Antarctic ice sheet began to grow decisively about 14 million years ago, and it may have attained roughly its present dimensions, from the pole to the sea, by about 11 million years ago. By far the larger of the two main Antarctic ice sheets, it has never again melted away since that time. Paradoxically, its creation may have depended on a supply of warm water, to provide the necessary snow for piling up to make the ice sheet. Otherwise it is hard to understand the delay between the isolation of Antarctica 30 million years ago and the formation of the full ice sheet. A possible explanation involves the submergence of much of the barrier that crosses the northern Atlantic Ocean, the Iceland-Faroes ridge; a consequent major change in oceanic circulation caused warm water to well up near Antarctica. In the oceans at large, these events ushered in the present era, in which the bottom water has remained close to freezing.

During an earlier cooling, around 30 million years ago, great

Oregon eruptions
▼ 16 Myr

Miocene disruption
▼ 15 Myr

East Antarctic deep freeze; volcanic chorus
▼ 14 Myr

volcanic outpourings had occurred in Indonesia, the Philippines, and elsewhere. A remarkable worldwide chorus of volcanic activity began about 14 million years ago. An obvious suspicion is that volcanic dust reduced the sunshine at the Earth's surface and so encouraged cooling and the formation of ice. That may well be correct, and yet be a distraction from a more profound connection, namely, that icing promoted volcanoes. The root causes of volcanism lie with the creation or destruction of great tectonic plates of the Earth's outer shell, driven by subterranean processes that are indifferent to the weather at the surface. Nevertheless, the timing and magnitude of volcanic eruptions may be influenced by the sea level, if the pressure of water helps to bottle up lava and steam. Ice sheets accumulating on land rob the oceans of water, and the consequent drop in sea level may liberate volcanoes. Whether or not this explanation is correct, the Hawaiian hot spot, Central America, and other regions were all engaged in frequent eruptions around 14 million years ago.

During the global upheavals, the land mammals lost much of their diversity, with the number of species groups (genera) falling by 30 percent. Most of the early dryopithecine apes became extinct, but at least one line must have survived, which led to modern apes and humans. Symptomatic of evolution among the primates, were apes with man-like teeth that evolved about 15 million years ago. These creatures, called ramapithecines, deceived a generation of fossil hunters, who took them as evidence that ancestors of human beings had by this time separated from the apes. They turned out to be forerunners of the orangutans, which diverged from their relatives about 10 million years ago. These apes are now confined to southeastern Asia, but the ancestral ramapithecines and sivapithecines flourished for about 7 million years, scattered in Africa, Europe, the Indian subcontinent, and China.

Grass prospered in the cool world, and about 11 million years ago a burst of evolution of grazing mammals began. Between Africa and Eurasia, the Mediterranean Sea was being squeezed by the slow processes of continental drift and plate tectonics that continue still, causing frequent earthquakes in Turkey, Greece, Yugoslavia, Italy, Algeria, and far out into the Atlantic, as the Mediterranean gradually changes into the mountain range that will complete the Africa-Eurasia collision, say 20 million years from now, when many sun-kissed beaches will be ski slopes. Protruding pieces of land created

14 MILLION YEARS AGO
THE DARK FLOOD

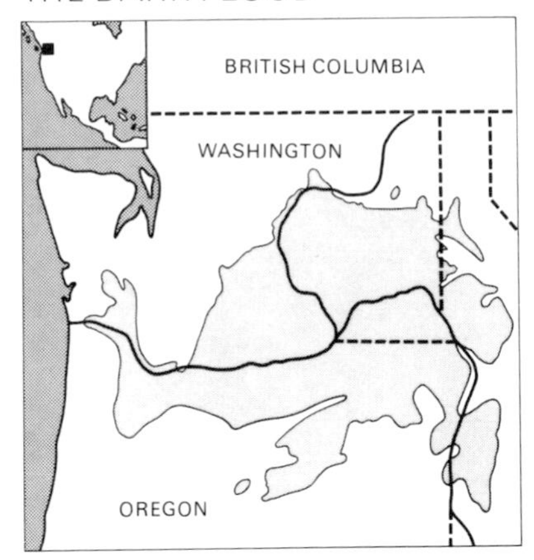

By 14 million years ago, after more than 100,000 cubic kilometers of lava had poured from vents near the junction of Oregon, Idaho, and Washington, the Columbia River basalts stretched as far as the Pacific coast.

We were apes. *The Miocene epoch of the Tertiary period continued.*

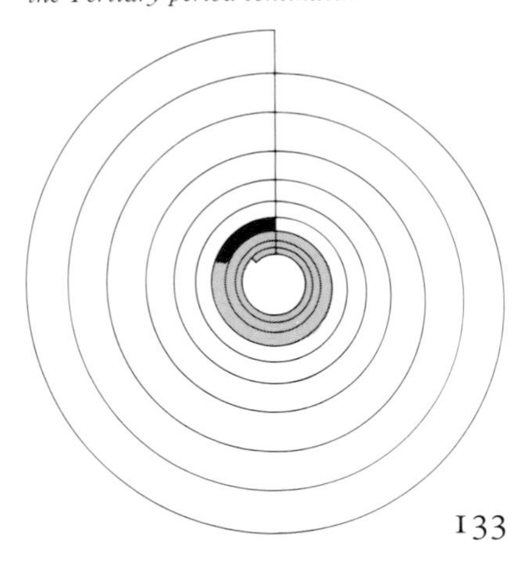

surge in grazing animals 11 Myr ▼

orangutans 10 Myr ▼

12 | 11 | 10

10 MILLION YEARS AGO

EUROPEAN COLLISION

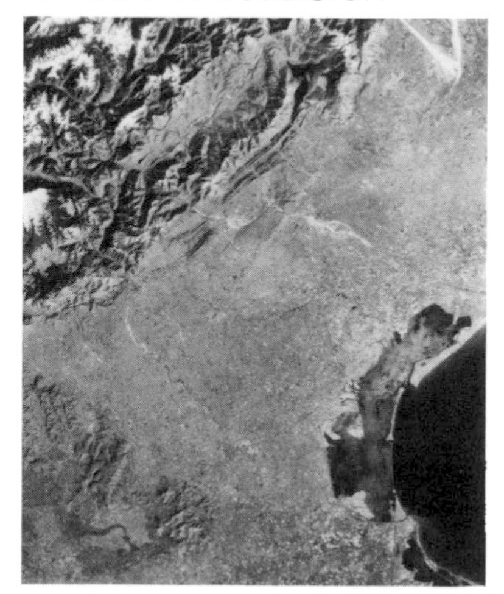

The crust of Italy's Venezia province has buckled, because the margin of Switzerland has burrowed under it, producing mountains to the north and a subsiding, sediment-filled basin to the south. As the pressure continues, the city of Venice is sinking into the sea. (Landsat image.)

early points of contact, where the impact of the opposing continents was concentrated.

Africa pushed Italy firmly into the heart of Europe. There had been long preliminaries, creating, for example, mountains in Switzerland as high as the Himalayas about 30 million years ago, but the present thrust began in earnest 10 million years ago. Italy rode up over the margin of Switzerland, and the resulting wreckage is the modern Swiss Alps, still rising by a millimeter or so each year. The intense pressure, felt through much of Europe, reactivated ancient faults in northern France and southern Britain, where the white chalk laid down at the time of the dinosaurs bowed upward into the air.

The late Alpine event was the most recent of the multidirectional collisions that made Europe a checkerboard of private valleys, promontories, and islands. But this collision was only one of the wounds in Eurasia caused by shrapnel from the slow-motion explosion of Gondwanaland, as impacting microcontinents pushed up mountains and plateaus all the way from Spain to Burma. Great rivers flowed from them into down-thrust plains—Rhine, Danube, Indus, Ganges, and others. In the undisturbed north, Eurasia was a flatland of plains and plateaus stretching for seven thousand kilometers, from the Rhine in the west to the Lena River in Siberia.

At both ends of the southern Eurasian collision zone, pieces spilled out sideways for want of room. Turkey edged westward, creating the volcanic islands of the Aegean Sea. A similar wedge of Tibet, squeezed eastward by India, still pressed into the heart of China, resurrecting crisscross mountains and creating another checkerboard realm that mirrored Europe; for Rhine and Danube read Yellow River and Yangtze. Burma and Indochina, elbowed by India, acquired mountain ranges like splayed fingers, gloved in forests; the Mekong and other great rivers issued between them. Southeastern Asia was an arc of fire, consuming the basalt of the Indo-Australian plate. The shifting continents and the cooling climate created lands of extreme contrasts, on scattered and battered continents quite different from Pangaea.

Africa, the ancient core plateau of Gondwanaland, stretched in both directions from the forested equator into the zones where warm moist air from the tropics sank cloudlessly over parched lands. Fragments of Gondwanaland littered the southern hemisphere: warm Malagasy (Madagascar); icy Antarctica straddling the

We were apes. *This was still the Miocene epoch of the Tertiary period.*

orangutans; Swiss Alps
▼ 10 Myr

northern glaciers
▼ 9 Myr

10	Myr				9.5						9					8.5		

pole; Australia, distinctively dry; and volcanic New Zealand being drawn out in concertina fashion between sliding plates. South America was a mobile Africa, and the mountains piling up on its leading edge created a vertical geography where a condor could proceed from tropical to polar landscapes simply by soaring.

The western side of North America, too, was tangling with a jumble of ridges and plates on the floor of the Pacific Ocean, and it was experiencing, as a result, a complex combination of tugging and pushing, stretching and uplifting. Inland, tilted blocks of a stretched crust began producing, about 8 million years ago, the characteristic Basin and Range formations of the western United States. These came to include the Sierra Nevada as the grandest range, and Death Valley as one of the basins, in the desert that formed in the lee of the new mountains. Farther inland the old Rockies, which had been worn to stumps, were being rejuvenated to attain their present elevations.

The fault along which the Pacific Ocean floor slid past the continent was about to jump ashore as the earthquake-prone San Andreas fault running through California. From Arctic islands to the forested volcanoes of Central America, North America was a grandiose continent, offering mountain ranges, deserts, great river valleys, and high, dry plains, in parcels of a million square kilometers, very different from fussy little Europe. To complete this nearly modern geography, plates in contention peppered the oceans with small volcanic islands.

Modern cats (the first of the species group *Felis*) and the first of the elephant family evolved about 8 and 7 million years ago respectively, while the chilling of the planet resumed. Mountain glaciers crept through northern landscapes for the first time in the present cold era, in Alaska about 9 million years ago. By about 6.6 million years ago, glaciations spread to southern America, and a new ice sheet buried the islands of western Antarctica. The amount of ice in the world was increasing rapidly, and the ocean surface dropped farther, by about forty meters. That fall contributed to unusual events in the Mediterranean.

A current flows continually from the Atlantic into the Mediterranean, to replace the water lost by evaporation from that warm, land-locked sea. The convergence of northern Africa and Spain threatened to choke the passage from the Atlantic, and 6.3 million years ago, aided by the falling sea level, a dam consisting of the

10 MILLION YEARS AGO

MOUNTAINS ALL THE WAY

Southern continents, drifting northward, were pushing up mountains across Eurasia, from Spain to eastern India, and the effects were felt in northern Europe and China.

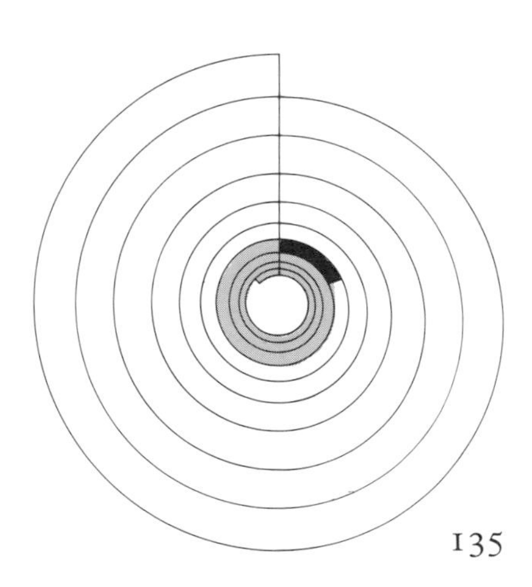

Northern American uplifts; modern cats ▼ 8 Myr

elephants 7 Myr ▼

West Antarctic deep freeze 6.6 Myr ▼

Mediterranean dryout 6.3 ▼

8 | 7.5 | 7 | 6.5

5 MILLION YEARS AGO
FAMILY TIES

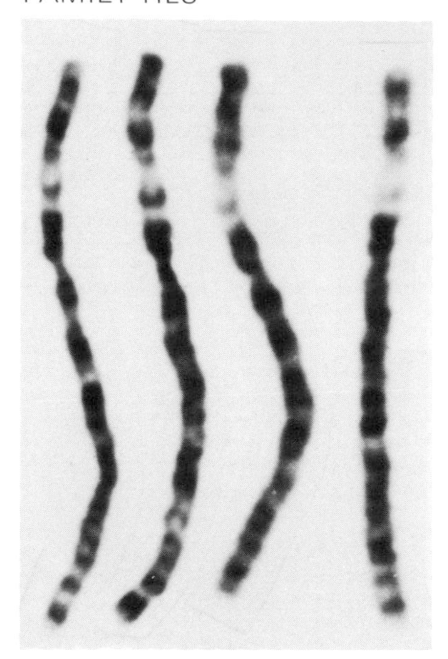

Equivalent gene-carrying chromosomes of a human being (left), a chimpanzee, a gorilla, and an orangutan confirm their close relationships.

southernmost hills of Spain cut off the water supply. Within a thousand years the Mediterranean dried out, leaving deep holes in the Earth, spattered with salt lakes that occupied the abyssal plains and trenches of the ocean floor, two kilometers or more below sea level. The grotesque scenery would be hard to imagine if a small-scale version did not exist today, in Afar, a dried-out piece of the ocean floor uplifted for examination at a corner of the Red Sea, in Djibouti in Africa. In searing heat, amid volcanic rocks and salt lakes, a few nomads live, as if on the wrong planet.

As the sea evaporated, the French Rhône, the Eyptian Nile, and other rivers feeding the chasm became waterfalls that cut deep canyons. The greatest waterfalls were salty, and occurred at the western end when the Atlantic spilled over the dam, replenishing the Mediterranean. That happened not once but many times. Every twenty thousand years or so the basin filled and dried again, until the deposits of salt became kilometers thick. Similar mishaps occurred in other episodes during the abolition or creation of oceans elsewhere, but this was unusually repetitive, and the world's oceans were losing significant amounts of salt. The Atlantic waterfall at Gibraltar finally excavated a channel deep enough to refill the Mediterranean permanently 5.3 million years ago, and ring down the curtain on an oceanic scandal.

The long-drawn-out collision between India and Asia came to a climax in the crumpling that built the Himalayas, beginning 5 million years ago. Far away, on the western side of South America, the Andes Mountains had been building and wearing away, in successive pulses dating back more than 200 million years, as the continent was pushed westward by the growing Atlantic, and plowed into the Pacific. The present eminence of the Andes goes back to 4.5 million years ago, when they began to rise from two thousand meters to four thousand. Rainfall diminished in the lee of the mountains, and forests thinned out in the Amazon basin. New species of mammals evolved there, in an interaction of geology, climate, and life that was typical of what was going on worldwide.

While changing climates extinguished many species, they also accelerated the origination of new species, in some groups faster than the extinction rate. With variegated geography overlain by ever-sharpening contrasts due to polar and mountain chills, and inequitable gifts of monsoon rains, the Earth offered a wider choice of real estate than ever before in its long history, propitious for the

Mediterranean dryout
▼ 6.3 Myr

modern dogs
▼ 6.0 Myr

Mediterranean normalized
5.3 Myr ▼

appearance of new species. Modern dogs of the species group *Canis* showed up 6 million years ago, and modern camels, bears, and pigs less than 2 million years later. Thus, around 5 million years ago, in a surge of evolution, the mammals began to make good the losses of 10 million years earlier, as the animals adapted to the contrasting habitats in an opportunistic and specialized fashion.

Yet life came up with another, quite different response to the geographic and climatic challenges. In one primate line, an evolutionary adaptation to changing circumstances became an adaptation to every circumstance. The eventual outcome was an animal so unspecialized in its habits that it could live in any setting whatsoever, by consuming plants and meat, rocks and trees, taking up stones and sticks to serve as teeth and claws, and burning firewood. Eventually it would be eating iron, coal, and uranium, and drinking petroleum, so as to wear the muscles of a giant. This human novelty, with all its attributes of limbs and tongue, was not evolved in one purposeful leap. Instead, millions of years of chance mutations and refinements, of evolutionary strides and long consolidations, brought it at last to an insecure success.

About 5 million years ago, by the evidence of genetic differences, the great apes of Africa separated into parties that were the ancestors of gorillas, chimpanzees, and apemen. Natural barriers in the form of rivers and rift-valley lakes may have helped to break the sexual contacts that would have prevented independent evolution. The parties occupied different environments: dense forest for the gorillas, more open forest for the chimps, and for the ancestors of apemen, sparse forest in eastern Africa that was becoming grassland. Our predecessors did not so much leave the trees as the trees deserted them. As the climate became drier, any monkeys or apes in eastern Africa had to evolve or perish. Some monkeys became baboons, about 4 million years ago, adapted to open country and using trees only for refuge. Apes became apemen at about the same time, although early signs of their existence are scanty: two teeth and a piece of jawbone from three different sites in Kenya, vaguely dated between 6 and 4 million years ago. The earliest reasonably vivid remains of apemen, the australopithecines, date from 4 million years ago. The creatures lived in Ethiopia, near the three-way split of Afar, where eastern Africa and Arabia were being torn from Africa.

The fragmentary fossils include a thigh bone plainly different

5 MILLION YEARS AGO

PARTING OF THE WAYS

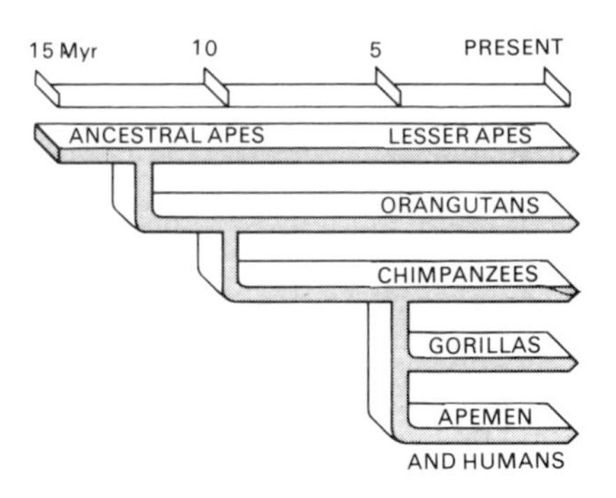

The family tree, based on molecular evidence, shows a common ancestor of humans and the great apes of Africa, living 5 million years ago, at most.

We became apemen. *The Miocene ended 5.3 million years ago, making way for the Pliocene, the last epoch of the Tertiary period.*

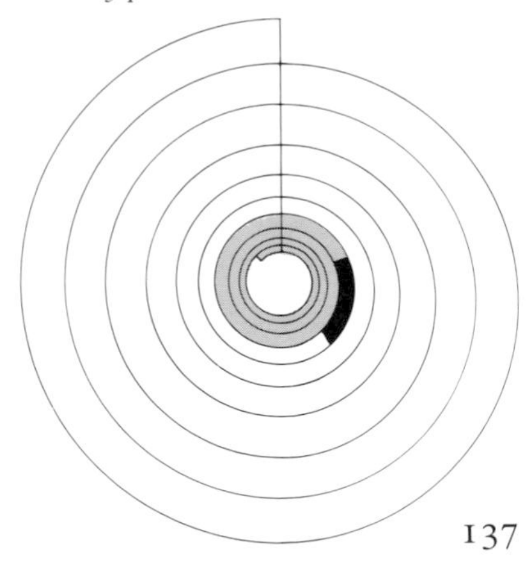

Himalayan uplift; apes and apeman split
▼ 5.0 Myr

Andean uplift; modern camels, bears, and pigs
▼ 4.5 Myr

baboons
4.0 Myr ▼

5 | 4.5 | 4

3.7 MILLION YEARS AGO
DAINTY FOOT

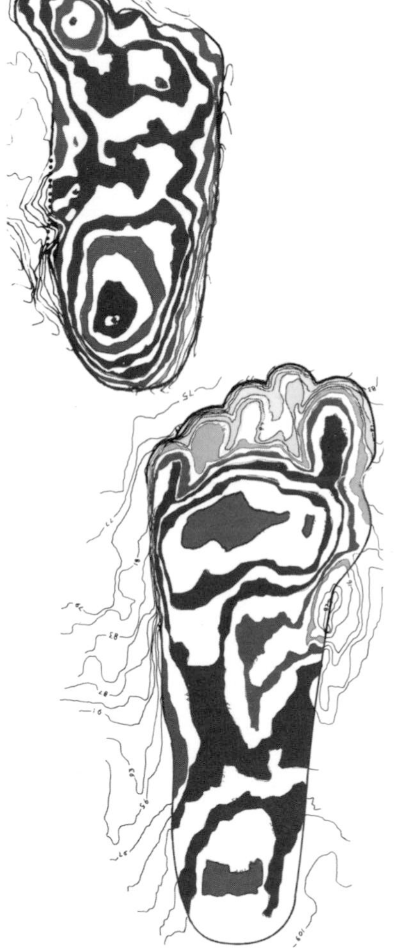

The fossil footprint of a juvenile apeman (above), contoured by photogrammetry, is compared with the footprint of a human adult from 15,000 years ago. Upright walking evolved long before the enlargement of the brain began.

We were apemen. *The Pliocene epoch of the Tertiary period continued.*

from an ape's, and show that the apemen used their hind legs. This balancing act was demonstrated, farther down the African rift 3.7 million years ago, when a volcano scattered ash of an unusual plaster-like composition across an almost bare Tanzanian plain. As it cooled, animals walked across it; rain then set the plaster, and more volcanic ash covered it and preserved it. Unearthed, the Footprint Tuff shows thousands of prints that a modern animal tracker can readily identify: elephants, giraffes, baboons, hares, guinea fowl, and so on. Some less familiar paw marks may be those of a large sabertooth cat, now extinct. And among these prints are a few that resemble human footprints, made by apemen who walked fully upright with a short stride across the fresh lava.

The first major evolutionary innovation that led to the emergence of human beings was therefore the re-engineering of the hind limbs and the perfection of the perilous balancing act of walking. It occurred long before the appearance of other human talents, in animals with ape-sized brains and short legs, who were ancestors both of later australopithecines and of humans. Upright walking set their hands free. Although diligent searches yield no sign of stone tools, the australopithecines may well have used sticks in the manner of chimpanzees to deter other animals or probe for food. The first species of australopithecines existed for more than a million years across a wide area, as an eccentric but not momentous addition to the fauna of eastern Africa.

While the apemen were qualifying as pedestrians, animals on which some of their descendants would eventually ride appeared in North America 3.7 million years ago, as horses of the modern species group called *Equus*. From the remote dog-size forerunners of 55 million years ago, the horse had evolved through a succession of intermediate forms into a sturdy, fast-running, grazing animal; in the wild it was not as elegant as stud farmers would eventually make it. The first primitive cattle evolved in Asia at about the same time. During these events, falls in sea level intermittently drained the shallow Bering Strait between Siberia and Alaska, and bears, for example, trudged from Asia into North America. Horses entering Eurasia 2.5 million years ago became zebras. Their path into Africa was constricted by the Red Sea opening between Arabia and Africa, and first paved with oceanic crust 3.5 million years ago.

An arc of volcanic islands curved south and east from Mexico, and it was building the isthmus that now bridges the gap between

North and South America. The whales, the most quick-witted animals of the time, may have noticed the change in geography, as the water route between the Pacific and Atlantic became dammed about 3.9 million years ago. One effect was to invigorate the warm current running into the northern Atlantic Ocean and across to Europe—the Gulf Stream, soon to serve as the world's greatest snow factory. The cooling of the world that had been going on for the better part of 100 million years culminated 3.25 million years ago in the onset of the current long-playing series of ice ages.

That was when the amount of ice in the world first exceeded the present stock of the Antarctic and Greenland ice sheets. From then on, warm spells like the one in progress now were the exception rather than the rule, and for much of the time many favorite tracts of land have looked like Greenland or the Gobi desert today—if not smothered with ice, then dry and cold. A belt of short grass fit for sprinters stretched for seven thousand kilometers from Europe to China, and on it, as a symptom of the modern climate, giant cheetahs appeared, the fastest beasts on legs.

Glaciers in the Andes were only one of the problems for the denizens of South America that had lived in isolation for many millions of years. For the giant armadillos, anteaters, and sloths, for marsupials resembling sabertooth cats, and for many other mammals decidedly strange to cosmopolitan eyes, the reunion with North America was a testing time. It was not quite instantaneous, because a few waifs had made improbable passages in both directions, along the islands of uncompleted Central America: sloths, for example, perhaps too slothful to jump off logs that drifted out to sea, turned up in Florida as early as 9 million years ago, and raccoons from the north were in South America 6 million years ago. When the Panama bridge became fully open to animal traffic around 3.0 million years ago, northern mammals marched across it and extinguished many of their South American counterparts, if not by preying on them, then by competing for the same food. But the great American Interchange was by no means a wholesale massacre of southern animals. Northern immigrants soon equaled them in diversity, on their own territory, but that was due more to an overall increase in diversity than to a catastrophic loss of southern species. Camel-like creatures, forerunners of llamas, arrived from the north, while opossums and other southern animals successfully invaded North America.

3.0 MILLION YEARS AGO

WINNERS AND LOSERS

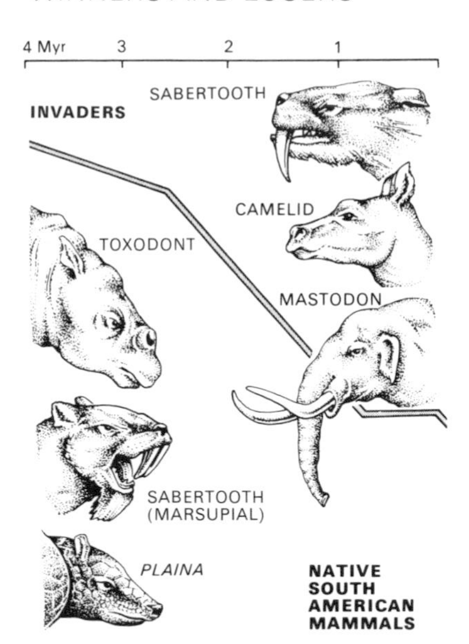

When the Americas were reunited, invaders from the north prospered in South America, until they made up half of all the species groups (genera) of mammals in that continent. A number of southern animals, including the sabertooth marsupial and Plaina *became extinct, but toxodonts held on.*

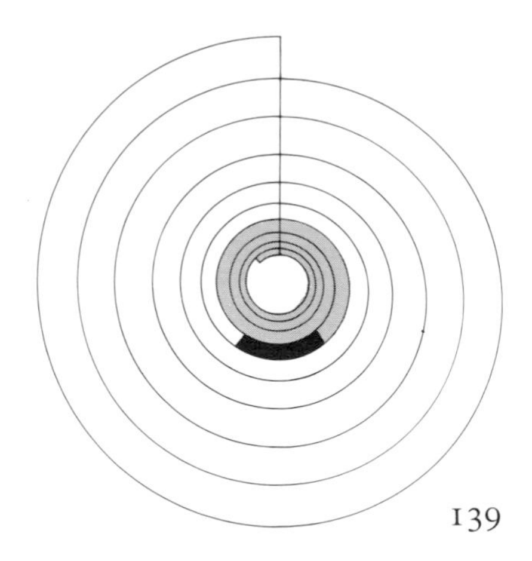

cheetahs
▼ 3.2 Myr

American interchange
▼ 3.0 Myr

3

2.5

2.0 MILLION YEARS AGO
FIRST HUMAN SPECIES

The most complete skull of Homo habilis, *the earliest human species, comes from Kenya. Older stone tools suggest that the species may have existed 2.4 million years ago.*

2.0 MILLION YEARS AGO
HANDLING TOOLS

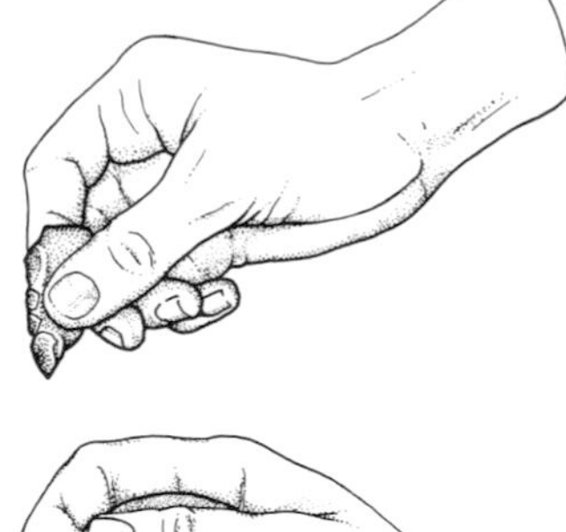

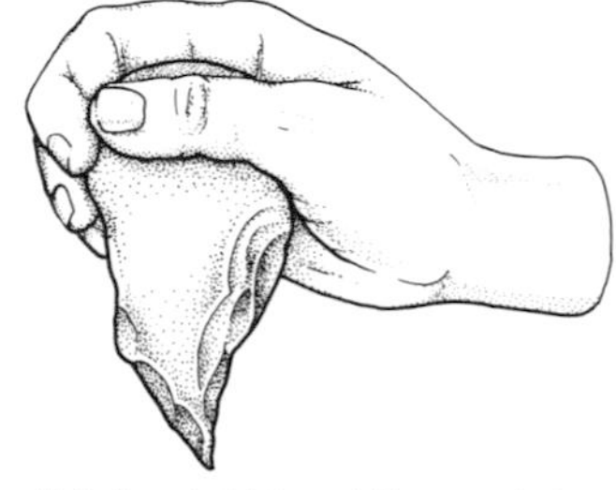

Small flakes, held in a delicate grip by Homo habilis, *could cut through the skin and flesh of a dead animal. Later (1.9 million years ago) a hand ax carried in the palm made a characteristic hunting weapon for* Homo erectus.

Settling to a rhythm, the ice ages gripped and relaxed in cycles of roughly ninety thousand years. They were not very severe to begin with, but they intensified noticeably 2.4 million years ago. Probably it was more than a coincidence that the first small stone tools, made by an unknown human hand, date from eastern Africa at that time. More heavily built apemen, robust australopithecines, appeared in the rift valleys 2.2 million years ago. There the new climatic regime expressed itself in droughts. A dramatic decrease of rainfall in eastern Africa occurred 2 million years ago, just when the true human beings first registered in the fossil record: *Homo habilis*.

They were mutant apemen, with overgrown heads and an unprecedented ability to enlarge their brains after birth. Although they were simpleminded folk by modern standards, they had a new trick for survival when food was scarce: a razor-sharp flake of stone, grasped in a well-engineered, well-controlled hand, could cut through the skin of a large animal and so reveal the meat. It was scarcely a weapon, and the earliest humans were scavengers rather than hunters of large animals. They competed with dogs and vultures for the pickings of beasts killed by the big cats, or dead from other causes, and one recumbent hippo or giraffe represented far more meat than could be obtained in a week's scrabbling after small game. Besides the small stone flakes that served as knives, there were cobbles with pieces knocked off, and they too may have been tools; cobbles served in any case as ready-made hammers.

Although *Homo habilis* had a brain only half the size of a modern human's, these people had taken an emphatic evolutionary step in the direction of exploiting gray matter. Fundamental to human social life is the sharing of food among relatives and friends, and it began with *Homo habilis*, who would bring chunks of meat on the bone from far afield, as if home for a family meal. The first humans evidently needed to balance their diets with a variety of plant and animal foods, thus unconsciously correcting for biochemical defects which left them incapable of making certain vitamins in their own tissues. This requirement helped to force humans to remain unspecialized and adaptable in their ways of life. On the other hand, food sharing allowed different individuals, different age groups, and different sexes to concentrate on tasks that suited them best.

Specialization boosted productivity and expertise. The characteristic division of labor among humans made men the zoologists and women the botanists: one sex supplied most of the meat, the

deeper icing; stone tools
2.4 Myr ▼

robust apemen
2.2 Myr ▼

2.5 | Myr | 2.4 | 2.3 | 2.2 | 2.1

other fruit, nuts, and salads, together with some small game. These sex roles are universal among surviving hunter-gatherers today, and they had to be adopted very early in the human story, to make possible the prolonged nursing and maternal care that were necessary for the human brain to grow postnatally. Once this practical step had been taken, later species of humans were able to acquire more powerful brains, even slower to mature.

Some representatives of *Homo habilis* must have wandered to Asia, because that was where the next step in human evolution occurred, in response to changes in the living environment. In the older landscapes, trees and shrubs fed lumbering browsers, such as rhinos, and these in turn provided meat for predators, notably sabertooth cats. As forests changed to open savannah, sleek grazing animals flourished at the expense of the browsers. They were too fleet-footed for the sabertooth cats to catch, and lions and leopards eventually replaced their feline cousins about 1.8 million years ago. Scavengers like *Homo habilis* lived cautiously alongside the predators.

While vacancies still existed for new predators, during the decline of the sabertooth cats, human beings of a second species established themselves. Designated *Homo erectus*, they were hunters rather than scavengers, and compensated by patient cunning for their lack of speed. Possible very early fossils of *Homo erectus* are known from Indonesia, and by one interpretation (not uncontested) the new people showed up in southwestern Asia near the Sea of Galilee, 1.9 million years ago. Their most characteristic tool and weapon was the hand ax, a stone implement about the same shape and size as a human hand. Rounded for clasping in the palm, it was sharpened at the other end like a saber tooth.

Mammals in general entered a phase of amazingly rapid evolution. For every genus (species group) of mammals that existed before the Pleistocene epoch started 1.8 million years ago, almost three new ones have come into existence since then. Apart from the modern big cats, early arrivals included bison, sheep, and the modern species of wild hogs. Evolutionary innovation was at its most feverish in the first one million years of the Pleistocene, and there was not room enough for so many species. Many of them, both old and new, died out, and each species lasted only a million years on average. Yet it was a turnover in which extinctions of pre-existing animals did not provoke, but followed, the origination of

1.6 MILLION YEARS AGO

SECOND HUMAN SPECIES

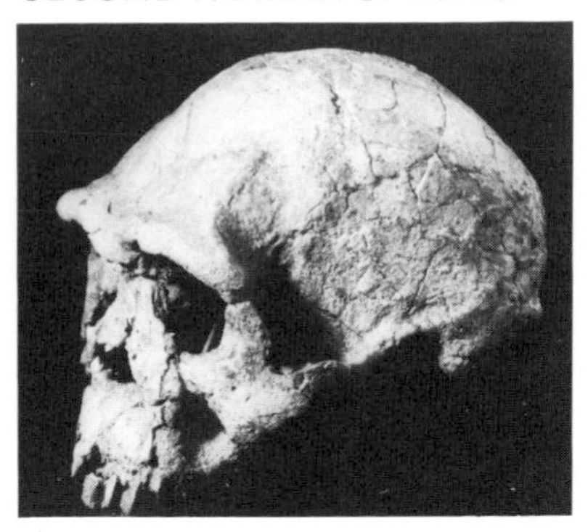

Also from Kenya comes this early skull of Homo erectus. *For this species there are indications of an origin outside Africa.*

We became human. *Archeology begins with earliest stone tools of Oldowan style 2.4 million years ago, launching the Early Paleolithic period. Industries associated with hand axes, appearing 1.9 million years ago, are called Acheulian. Geologically, the Tertiary period and the Pliocene epoch ended 1.8 million years ago, and the Quaternary period and its Pleistocene epoch began.*

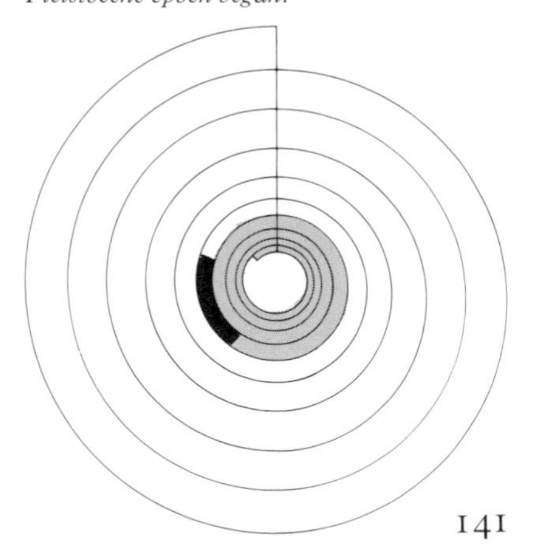

***Homo habilis*; scavenging** ▼ 2.0 Myr

***Homo erectus*; hunting** ▼ 1.9 Myr

modern big cats; bison; sheep; wild hogs ▼ 1.8 Myr

***Homo erectus* in Africa** 1.6 Myr ▼

2 | 1.9 | 1.8 | 1.7 | 1.6

1.6 MILLION YEARS AGO
RIFT VALLEY ROUTE

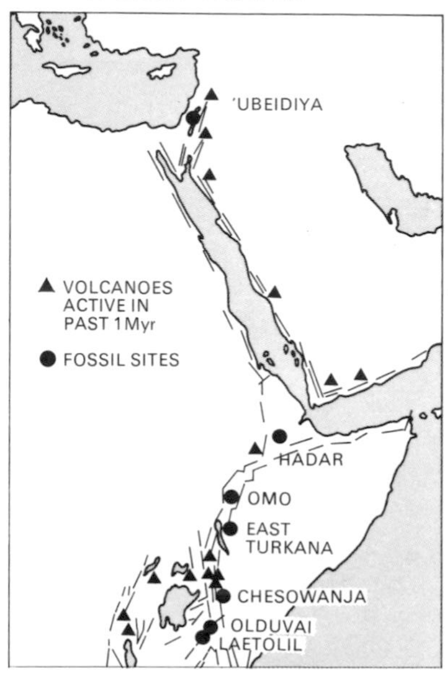

Movements in the Earth's surface, in and around East Africa, created rift valleys, peppered with volcanoes, where important fossils of apemen and early human beings are preserved. Homo habilis *and* Homo erectus *are known from Olduvai and East Turkana.* Homo erectus *occurs at 'Ubeidiya, and evidence for his use of fire comes from Chesowanja.*

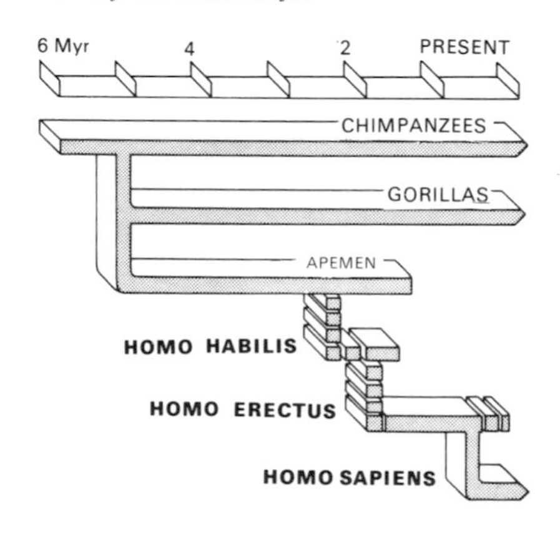

other species, and the diversity of mammals peaked around one million years ago.

Small groups of the earliest humans, *Homo habilis*, modestly held the stage in East Africa for just a few hundred thousand years, alongside their apemen cousins, until *Homo erectus* appeared in the rift valleys 1.6 million years ago and replaced them. The apemen, always more numerous than *Homo habilis*, survived the change in regime. The invaders were taller in stature, bigger in brain, as well as more active in hunting large animals. The symmetry of the hand axes was a sign of a higher mentality, and a Tanzanian factory for making them from a supply of choice stone is one of the earliest traces of the new order.

Stone implements last well, yet they give only a partial impression of early technology. Under a microscope, various stone tools found in Kenya, 1.5 million years old, revealed traces of wear characteristic of tasks for which they were used: butchering meat, cutting the stems of plants (probably grasses), and shaping wood. The evident use of stone tools to make wooden ones that seldom survive in the archeological record leaves one guessing at what this missing equipment was, although wooden spears and digging sticks are preserved from far younger (100,000-year-old) sites.

Human babies are born with an ape-sized brain, which then grows to attain its full size during a long childhood. Between the australopithecines and modern humans the brain trebled in size, a formidable increase even when part of it is discounted by an increase in stature. More important than mere swollen heads were alterations to the format of the brain. The greatest enlargements, compared with the apes, have been at the sides of the brain, in the parietal regions most closely involved in problem-solving skills and, latterly, in language.

When handedness began—the typical right-handedness in modern humans—is also obscure, although it should in principle be discoverable from tools, which were more conveniently grasped in one hand or the other. By the crossover wiring of the human nervous system, the left side of the brain was responsible for the right hand, and with handedness, that side became more skilled in the control of movements and other sequential actions that eventually included speech, leaving more general and spatial tasks to the right side of the brain. One present-day hint that handedness was primarily a female phenomenon appears in the way infant girls tend

***Homo erectus* in Africa; *Homo habilis* extinct**
▼ 1.6 Myr

firemaking; apemen extinct; human standstill 1.4 Myr ▼

to develop handedness and speech more quickly and decisively than boys. The custom of carrying a baby on the left side, where it is comforted by the maternal heartbeat, sets the mother's right hand free, and that may have been sufficient reason for the evolution of predominant right-handedness in humans. Creating in effect two brains under the one skull, the specialization implied by handedness was a quantum jump in the braininess of human beings. It tells also of feedback between optional behavior, such as baby carrying and tool using, and physiological change; humans were half-consciously engineering their own evolution. To say that handedness arose in *Homo erectus* is at present only a guess.

The control of fire was the next major step in technology, and there are indications that *Homo erectus* was using fire purposefully a little more than 1.4 million years ago. Charcoal does not survive in a tropical environment, but at a site in Kenya, pieces of burned clay are intermingled with remains of animals. The clay might have been heated by natural causes, but the fossilized scene gives the impression of a meal eaten around a man-made campfire. Apeman meat was on the menu, by the simplest interpretation of australopithecine bones found in the vicinity of the tools of *Homo erectus*. With a diet consisting mainly of fruit and nuts protected by tough skins, the apemen were scarcely competing with the human beings for resources, so if *Homo erectus* killed them it was for nourishment, and not in rivalry or spite. Nevertheless, about 1.4 million years ago, after six thousand centuries of coexistence, first with *habilis* and then with *erectus*, the apemen vanished from the scene. By the time their career drew to a close, the walking apes had existed for longer than human beings have lasted so far.

Once fire was mastered, human life showed little change for more than a million years. A small band of *Homo erectus* typically operated from a base camp beside a lake or a stream, in warm, open country. Hunting meat, gathering plant food, and making tools filled a few hours of the day, while expeditions of tens of kilometers fetched high-quality stone for the tools, and red minerals presumably used as cosmetics. With the changing seasons, the band may have varied its technologies, and shifted its base. But although human affairs now provide the chief thread for the narrative, *Homo erectus* was confined at first to warm latitudes. Elsewhere, animals continued their evolution unaffected by people. They were, on the other hand, much influenced by the alternations of warm interludes and ice

1.0 MILLION YEARS AGO

MAMMALS GALORE

A host of newcomers amongst the mammals pushed the number of genera (species groups) up to an all-time peak, before subsequent extinctions began to reduce the numbers. The leopard (evolved 1.8 million years ago) is a notable survivor, although some of its contemporaries among the cats died out.

We were *Homo erectus*. *The Early Pleistocene of the Quaternary period continued. Developed Oldowan industries appeared, and coexisted with the Acheulian, still in the Early Paleolithic.*

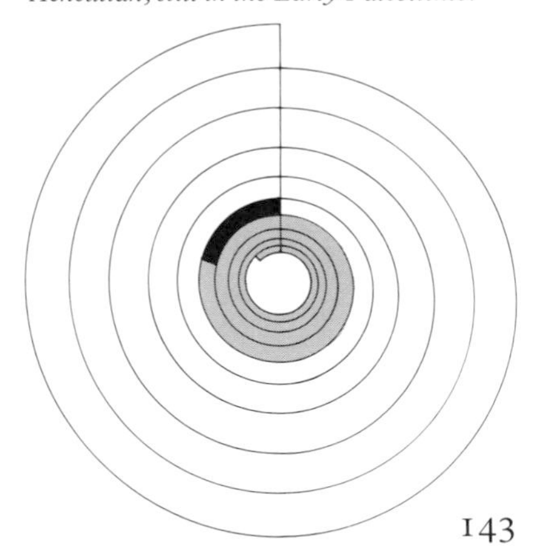

mammalian peak
1.0 Myr ▼

1.2 1.1 1

750,000 YEARS AGO
RIVAL TRADITIONS

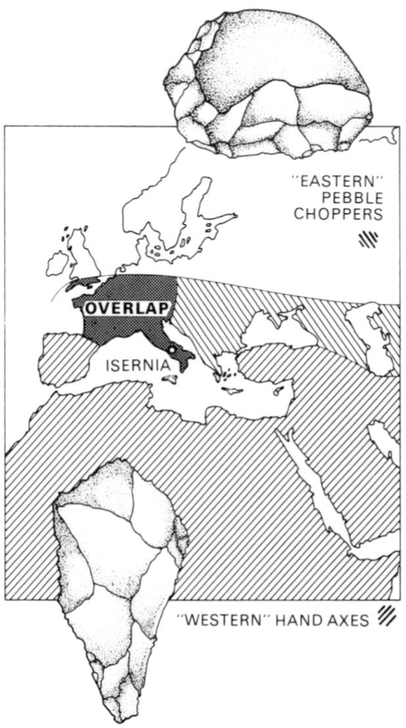

The earliest settlers in Europe used pebble choppers, like that depicted from Isernia in Italy. This stone technology was mainly associated with inhabitants of eastern Asia. Hand axes in the southwestern Asian and African tradition appeared later in Europe.

ages; these became more emphatic 800,000 years ago. About 1,000,000 years ago the Earth was entering a period of exceptionally intense and widespread volcanic activity. Also perturbing was a substantial object that fell from the sky and struck the Earth 730,000 years ago, leaving glassy tektites strewn on the floor of the ocean near Australia and ashore in Indonesia. At that time the Earth switched its magnetic poles around, to their present "normal" orientation. From the time of this mid-Pleistocene disruption onward, the diversity of mammals declined.

The early humans probably had no inkling of the magnetic reversal, but for their descendants in laboratories 730,000 years later, who could detect it in many deposits, it became a beacon illuminating an era that had been inscrutable and hard to date. A little before the magnetic benchmark, people broke out of the tropics into the chillier realms of ice-age Eurasia, in one of the outstanding events in the early human story. Previously, humans had been mainly restricted to Africa and southeastern Asia. One well-attested move into northerly climes was to central Italy, where people with chopper tools (not hand axes) were living in the hills of Molise about 750,000 years ago. They fed on the products of big-game hunting, including bison, rhino, and elephant meat; greater skill in hunting large animals may have helped these tropical primates to expand their range. Large lions made the northward trek from Africa into Europe at much the same time as the humans, and they passed newly evolved brown bears heading south.

In yet another turn of the climatic screw, year-round pack ice became normal in the Arctic Ocean. Northern China was very cold when *Homo erectus* lit fires in the caves at Choukoutien; here too is evidence of big-game hunting—bison, horse, rhino, and so on. The dating is insecure, but 650,000 years ago corresponds with a glaciation falling in the middle range of widely varying estimates. The skulls of "Peking Men" were all broken open, perhaps to extract the brains in routine cannibalism. Animals that were to play particular roles in the human story put in appearances in these Chinese deposits. One was the wolf, ancestor of the domestic dog. The rat, later to develop a parasitical relationship with urbanized and seafaring humans, was present at Choukoutien, and so was the marmot, the ground-dwelling squirrel that would be hunted for its fur and become a reservoir of the Black Death, transmissible to rats and humans.

We were *Homo erectus*. *The Late Pleistocene began 730,000 years ago, with the start of the present Brunhes geomagnetic epoch. The Early Paleolithic industries continued.*

Abbreviation: *yr = years ago.*

mammalian peak; volcanic chorus
▼ 1,000,000 yr

100,000 | yr | 950,000 | 900,000 | 850,000

The commanding inference from all the scrappy fossils and tools is that by 750,000 years ago human beings had wandered into Europe, the small continent that played a special part in forcing them to smarten up or perish. In interglacial episodes, Europe was invitingly warm because of the Gulf Stream, the ocean current from the American subtropics that drifted across the North Atlantic and bathed northwestern Europe in warm water. The Gulf Stream helped to launch each ice age, by feeding moisture and snowstorms to the incipient ice sheets, but when the ice became extensive the Gulf Stream tracked farther south, toward northern Africa, leaving much of Europe out in the cold. Climatic fluctuations were thus magnified in Europe, and barriers of mountains and sea compounded the effects on living species, by blocking most avenues of escape to the south. It was easy, too, for small populations to become isolated—a propitious situation for a new species to evolve.

A group of *Homo erectus* arriving in Europe from Africa or Asia evolved into *Homo sapiens*, the species to which modern humans belong. That, at any rate, is the simplest interpretation of the evidence, but in view of the controversies and confusions, described by a leading anthropologist as "the muddle in the middle," it might well be apt to call the people of this time the muddlemen. According to a conventional opinion, all these northerners were *Homo erectus*, meaning in this context the species that first appeared in Asia 1.9 million years ago and allegedly remained the only human species until about a quarter of a million years ago. Similarities in stone hand axes among many—but by no means all—of the sites where tools are found provide one reason for this view.

Certainly the humans of eastern Asia were *Homo erectus*, by the strictest nomenclature, because "Java Man" and "Peking Man" were dubbed *erectus* before the southwestern Asian and African discoveries were made; but they comprise two distinguishable subspecies, between which there was a leap in brain size of 20 percent. Moreover, Asian *erectus* did not use hand axes but, like some of his European contemporaries, he preferred to find rounded pebbles and make his choppers from them; in India and parts of Europe these toolmaking traditions overlapped. There seem to have been various northbound waves, with different technologies from different origins. Presumably only one of them gave rise to *Homo sapiens*. In Europe, despite widespread evidence of early human activity, there is no uncontested specimen of *Homo erectus*; on the contrary,

730,000 YEARS AGO

DATED ICE AGES

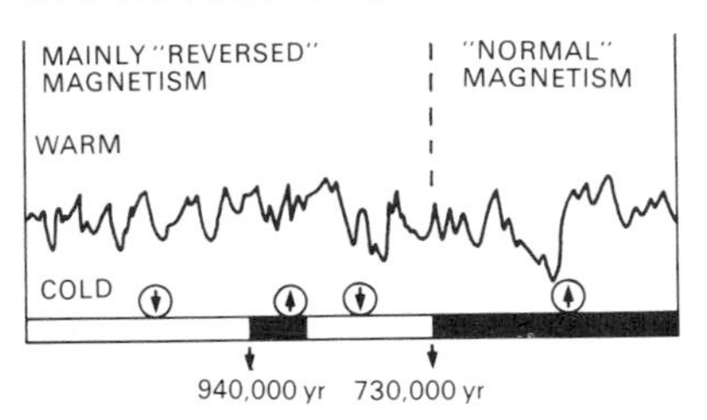

Reversals in the Earth's magnetism provide an accurate timescale for the ice ages. Here, fluctuations in the amount of heavy oxygen (oxygen-18) in small marine fossils measure the amount of ice in the world. Dips in the curve indicate much icing and cold climates. Magnetic reversals, detected in the same ocean-bed deposits as the small fossils, give clear benchmarks. Note the increasing severity around 800,000 years ago.

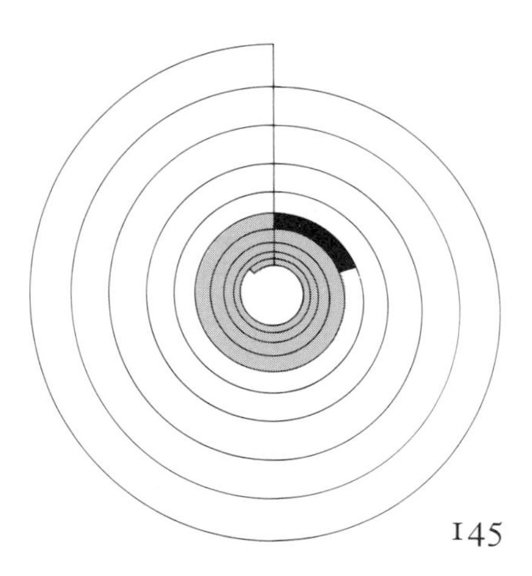

emphatic icing ▼ 800,000 yr

breakout of *Homo erectus* 750,000 yr ▼

Pleistocene disruption ▼ 730,000 yr

Arctic deep freeze; brown bears ▼ 700,000 yr

wolves ▼ 650,000 yr

00,000 | 750,000 | 700,000 | 650,000

600,000 YEARS AGO
THIRD HUMAN SPECIES

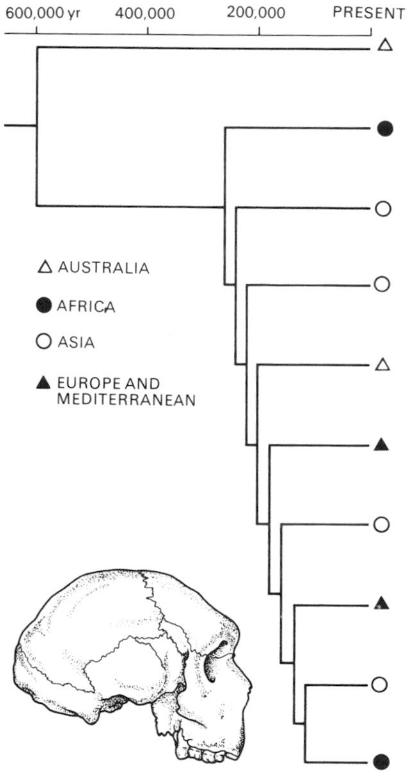

600,000 yr 400,000 200,000 PRESENT

△ AUSTRALIA
● AFRICA
○ ASIA
▲ EUROPE AND MEDITERRANEAN

All human beings alive today are members of the same species, Homo sapiens, *and their shared ancestry traces back 600,000 years, as judged by subtle molecular differences between individuals indigenous to various continents, and exemplified here. A skull from Petralona in Greece is an ancient specimen of* Homo sapiens, *but it may be as young as 200,000 years.*

teeth, jaws, and skull fragments appearing sparsely in the European fossil record all have some resemblances to *Homo sapiens*. More than one human species or subspecies—several kinds of muddlemen—may have existed at this time, and encountered one another anywhere across the Old World.

The supposition that our own species evolved at around this time finds support in genetic evidence. The most recent common grandmother of all living humans lived about 600,000 years ago. Variations in the genetic material in the oxygen-handling inclusions in present-day living cells lead to this estimate, and the female bias in the ancestral information arises because these inclusions, the mitochondria, are inherited only from one's mother. Mitochrondial heredity is remarkably mixed among present-day humans with similar geographical roots, implying that the intervening six thousand centuries have witnessed continual intercontinental movements of women, and presumably their menfolk too, back and forth between Africa, Europe, and Asia. The genetic evidence also indicates that some groups became cut off from the mainstream of human evolution and were reincorporated relatively recently.

A modern species is defined by a common lineage and the possibility of interbreeding, and the evidence of the shared grandmother, dating from long before modern humans burst upon the scene, suggests that the species *Homo sapiens* is about 600,000 years old. Teeth and bones apparently that old, found at various sites in Europe and intermediate in appearance between *Homo erectus* and *sapiens*, may thus belong to the earliest known members of our species. This inception was modest: at first the new species made no distinctive contribution to technology, but persevered with the same tools as *Homo erectus*.

Hunting big animals, as in China, required coordinated action by human teams. In Africa, beside a Kenyan lake, hunters took on baboons of gorilla-like proportions and bagged more than sixty of them. The most striking scenes of early big-game hunting described by archeologists are from Spain 450,000 years ago, during an ice age. Humans lay in wait for big game at a mountain pass, and then set fire to the brush to drive the prey into a swamp. Elephants were taken, along with other animals, including horses, deer, and cattle. The butchery was on a large scale, but not mindless: a human act of symbolism was to arrange lines of the Spanish elephant bones in the form of a T, while the heads of the elephants were evidently taken

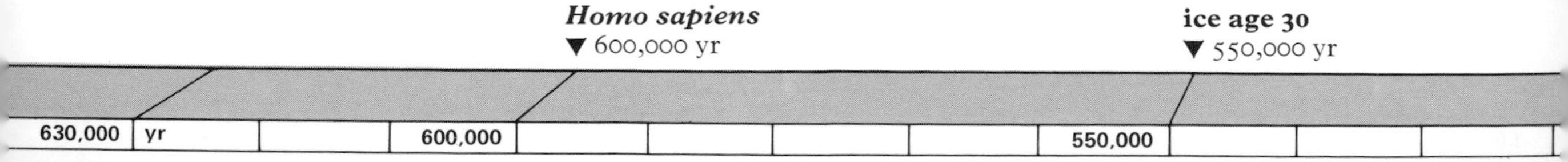

Homo sapiens
▼ 600,000 yr

ice age 30
▼ 550,000 yr

away as trophies. And during the ensuing warm interlude, in France perhaps 420,000 years ago, men and women took springtime vacations on the beach at Nice, where they built large huts, twelve meters long, from branches supported by midline posts and by stones around the base. They also obtained food from the sea: fish, turtles, oysters.

Llamas evolved about 500,000 years ago in the Americas; there too the ice ages kept up their remorseless rhythm. Varying patterns of glaciation, and recent confusions about dating, make it tricky to decide which was the worst of the recent ice ages, but the most extensive was probably the event known variously as the Illinoian in North America and the Elster or Mindel in Europe, which may be identifiable with the major glaciation known to have occurred 430,000 years ago. At that time the edges of the main ice sheet in North America extended south of St. Louis, Missouri; in Europe, the north coast of France was glaciated, and there are indications that the English Channel was filled by glaciers, on this and other occasions.

From the commencement of the present series 3.25 million years ago, down to 550,000 years ago, about twenty-nine episodes of glaciation had already occurred, although the number depends a little on how one distinguishes the wiggles in the climatic record provided by ocean-floor fossils. Since 550,000 years ago, when the thirtieth ice age began, there have been six major cycles of glaciation, each developing to an ice maximum that was followed by a fairly rapid thaw and an erratic warm interlude before the next cycle started. Their increasing clarity makes the logging of the more recent ice ages and warm interludes worthwhile, especially as the dates can now be checked by astronomical calculations.

Ask not why the ice kept coming, but why it sometimes went away. Thick, sprawling ice sheets, accompanied by many glaciers licking the landscape with their icy tongues, represented by this time the normal climate of the Earth. Snow that fell on the northern continents failed to melt completely in the summer, and so the ice sheets thickened. From time to time, though, the summer sunshine became more intense and the ice melted faster than it formed. The explanation lay in the Earth's behavior in its orbit around the sun, as affected by other objects in the solar system.

Gravitational influences of the planets altered the shape of the Earth's orbit, so that sometimes it was almost circular, sometimes

450,000 YEARS AGO

"TUNED" ICE AGES

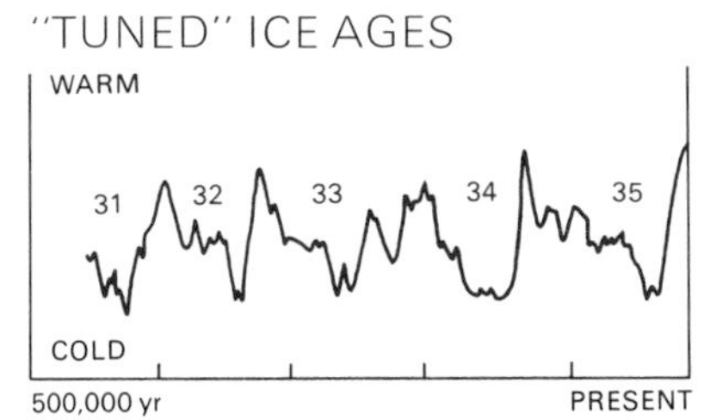

This heavy-oxygen (oxygen-18) record of ice ages has been refined in its dating by knowledge of the various climatic rhythms, which are due to changes in the Earth's attitude and orbit (see overleaf). The numbers show the end of a count of ice ages over the past 3.25 million years.

We were *Homo sapiens*. *The Late Pleistocene epoch and Early Paleolithic industries continued.*

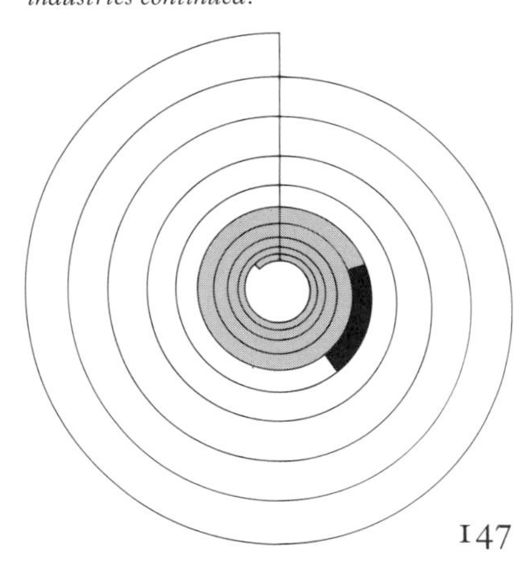

warm interlude; llamas ▼ 500,000 yr

ice age 31 ▼ 475,000 yr

warm interlude; huts; fishing 420,000 yr ▼

human standstill 400,000 yr ▼

500,000 | 450,000 | 400,000

350,000 YEARS AGO
ANTICS IN ORBIT

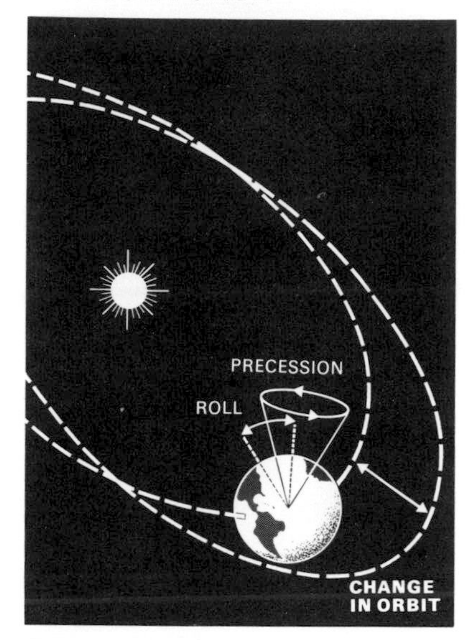

Ever since the origin of the Earth, its orbit and tilt in space have varied, influencing the climate in a cyclical manner. During this relatively recent phase of the planet's history, the cycles were causing exaggerated alternations of icy and mild conditions. For example, feeble summer sunshine in the northern hemisphere provoked a short, sharp plunge to intense cold 350,000 years ago. It ended after 20,000 years, the time taken for the Earth's axis to precess, or swivel, once around.

stretched into an elliptical path, in irregular cycles of roughly a hundred thousand years. The moon and the sun tugging on the Earth's midriff caused the planet to roll slightly, altering the angle of tilt of its axis by a few degrees, and back again, in a period of about forty thousand years. The Earth's axis also swiveled around, like a spinning top off balance, every twenty thousand years or so.

Conspiring together, these factors altered the look of the sun in the sky on midsummer's day in the northern lands: it could appear bigger or smaller, and stand higher or lower in the noonday sky. When the sun was both large and high because of a conjunction in the different cycles, its greater warmth could melt the ice sheets rapidly and usher in a warm interglacial spell. At other times the ice persisted, but it waxed or waned, with the climate warming or cooling, over phases lasting thousands of years, in accordance with a complex interplay of the astronomical rhythms. The simplified annotations on the timescale conceal many lesser fluctuations. Similar cycles acted throughout geological history, but in the ice ages their effects were more obvious.

The ice sheets and glaciers left their imprint in many northern lands, where their former fronts are often delineated by characteristic moraines—ridges bulldozed up by the ice, or heaps of rocks carried by the ice and dropped at its melting snout. Behind the front the surface, including many a long-lost valley, is plastered with till, fine-grained material littered with stones and boulders torn from hills far away. Till and moraines remain substantial enough to dam up the Great Lakes of North America and the Baltic Sea of Europe. Indeed, much of the scenery of the planet today makes sense as direct and indirect work of the ice, which we are allowed to glimpse with the ice temporarily withdrawn. Other ice-age landscapes have disappeared under the sea.

The scoured, furrowed, lake-filled lands of Canada, Finland, and northern Russia testify to the grinding and polishing actions of the major ice sheets, and hard rock that protruded has been fashioned into streamlined hills. Jagged mountain tops tell of cracking by ice, while many broad, rounded valleys are the beds of glaciers; when flooded by the sea, they make fjords and great natural harbors. Far from the ice sheets, fossil sand dunes remain from deserts drier and more extensive than today. The chief planet-wide effect was on the rivers; with the low sea level, the rivers cut deeper and ran farther, to the edge of the continental shelf.

human standstill; yaks
▼ 400,000 yr

ice age 32
350,000 yr ▼

ice maximum
340,000 ▼

warm interlude
330,000 ▼

The grinding of glaciers reduced rock to flour; strong winds carried it as glacial fallout, to be dropped over surrounding areas, thus laying down characteristic deposits of yellow soil called loess. The Yellow River in northern China is colored by the loess it carries to the sea. In China, as in central Europe, the loess was to play a special role in later prehistory, when, at the dawn of agriculture, people would find the fine-grained soil easy to cultivate. The duststorms that made it must have been disagreeable for their ancestors, as were the razor-sharp flakes of rocks, cracked off by the cold, that flew about in some areas like natural arrowheads in the high winds.

Human beings coped with climatic changes for thousands of generations, yet they hardly changed themselves. The hand axes and pebble choppers persisted across wide areas of Africa and Eurasia—a tribute no doubt to the tools' utility, but also making a negative statement about human evolution. Those early members of our species were at a standstill, for all that the tangible relics can show. Give any credit you like for hunting skills adapted to local resources of game and stone; make any allowance for the small sizes of the bands, and for the likelihood that archeologists have failed to perceive the meaning of slight variations in tools; even say, if you will, that these people were sensibly conservative in their habits. The lack of any signs of experiment, or accumulation of skills, suggests that they were incapable of more, and that the stasis would end only with a change in mental performance. Anyone looking for evolutionary news at this time has to content himself with appearances of other animals, such as the yak in Asia. Intangible changes in human beings remain a matter for speculation.

The most polite hypothesis is that this "abysm of time" around 400,000 years ago was occupied with the evolution of good manners. The high rate of technological change in the past few thousand years depended at least as much on cooperation, multiplying brain power, as on the talent of individuals. Human bands remained limited to family groups of twenty or thirty people until less than 30,000 years ago, but to establish the behavior necessary for larger assemblies may have taken a long time. It was a matter of going beyond sexual attraction and family affection, which are the usual bases of social behavior in mammals.

Evolution in other animals had already hit upon a behavioral strategy that moderated fights between unrelated members of the

340,000 YEARS AGO

BURIED LANDSCAPE

During each glacial phase, northern regions of Europe and North America looked like ice caps of today, with valleys and plains lost under thick ice.

We were *Homo sapiens.* *The Late Pleistocene epoch and Early Paleolithic industries continued.*

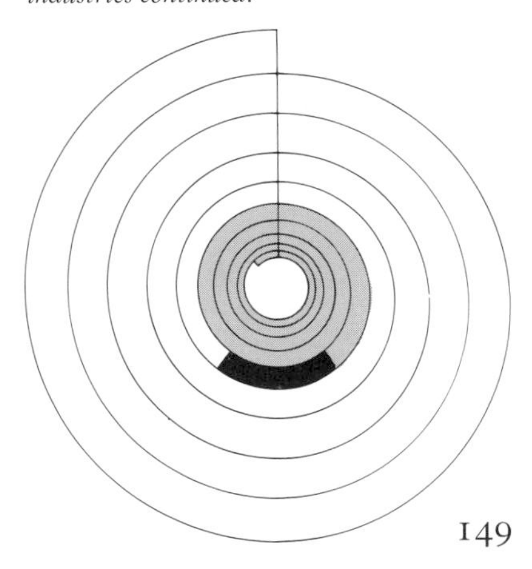

ice age 33
280,000 yr ▼

ice maximum
250,000 yr ▼

300,000

250,000

200,000 YEARS AGO
EMBLAZONED TOOL

The craftsman who made this hand ax during a warm interval in England deliberately shaped it around a fossil embedded in the flint. The fossil itself, a sea urchin about 90 million years old, shows the distinctive five-fold symmetry of the echinoderms.

same species, for example when two males competed for the same female. The golden rule was: Defend yourself if attacked, but don't be the aggressor, and stop fighting as soon as the other relents. Games theorists with computers took thirty years to identify this strategy; evolution needed more time, but found it repeatedly, and long before humans appeared on the scene the golden rule made many conflicts within species less lethal than they might have been. Nevertheless "murder" remains commoner in many species, from gulls to lions, than it is in humans, who are not the atavistic killers they sometimes claim to be.

Human beings took their behavioral ploys much further, to attain cooperation between unrelated people. The powerful human brain could recognize faces, appraise newcomers, and recall previous meetings. Bodily evolution made those tasks easier by creating individualistic and expressive human faces, and supplying voices able at least to utter names. Dealings with nonrelatives could then become more urbane, especially in remembering whether previous encounters were friendly or angry. A variation of the golden strategy emerged as a basis for cooperation and natural justice: offer cooperation to a stranger and, if he defaults, retaliate but bear no grudge. This fostered altruism between nonrelatives, leading eventually to a species able to live in a wary kind of peace in cities of 10 million strangers. The most vehement form of this behavior was the ability of unrelated humans to form instant loyalties to a group, in rivalry with other groups—the basis for the highest human attainments and the worst wars.

Homo sapiens shared with other primates special organs for handling the calculus of social relationships. These were the frontal lobes, massive banks of brain cells lying just behind the forehead and amply coupled to the centers of emotion deep inside the brain. Although scarcely necessary for the routine, workmanlike functions of the brain, in humans these regions became adept at thinking through the consequences of a future course of action, guessing the reactions of other individuals, and blushing with anticipated shame or pride. Friendship and generosity evolved as carefully reckoned behavior, and the frontal lobes were indispensable apparatus for saints and con artists alike. Gestures, glances, smiles, and scowls in the ice-age wilderness, prefigured the plots of a million novels, and the emergent system of social behavior had enough latent contradictions to color politics down to the present. On the whole, though,

ice maximum
▼ 250,000 yr

warm interlude
▼ 240,000 yr

archaic *Homo sapiens*
▼ 230,000 yr

250,000 yr	240,000	230,000	220,000	210,000

it worked and good manners became the norm; just when, no one can say.

An ice age ended 240,000 years ago, and human skulls from Europe in the ensuing warm interlude were sufficiently modern-looking to be securely classified as "archaic" *Homo sapiens*. They included "Steinheim Man," "Swanscombe Man," and well-dated remains from Bilzingsleben, 230,000 years old. Some of these people favored hand axes, others pebble choppers. They were distinctly beetle-browed, with bony eyebrow ridges that would be exaggerated in some of their descendants and much diminished in others. Outwardly these were minor changes compared with the evolutionary inventiveness that conjured cave bears, reindeer, goats, and modern cattle into existence around 200,000 years ago.

An obvious and persistent trend was in cosmetic dentistry: as human brains grew bigger, teeth and jaws became prettier. Between the earliest humans (*Homo habilis*) and the latest, the combined crown area of the teeth diminished by about 40 percent, and in proportion to stature the latest human teeth are only half as big as the earliest. The most rapid phase of the change occurred during the transition from archaic *Homo sapiens* to anatomically modern humans, in the interval from 230,000 to 100,000 years ago. This dental transformation reflects progress in cooking and other methods of making food easier to chew.

A tangible hint of intellectual change in early *Homo sapiens* comes from deposits in southeastern England, dated at about 200,000 years ago. A craftsman selected a piece of flint containing a "shepherd's crown," a fossil sea urchin, and fashioned a hand ax symmetrically around the fossil, so that it appeared as a neat ornament on the implement. In the same deposits were pieces of chert containing fossil corals, treasured objects brought from identifiable fossil beds two hundred kilometers away. The pieces had been artificially stained with red ocher, a material that humans had for long been using to paint their own skins.

The cold returned, and England, like other northern countries, became uninhabitable once again. That next ice age lasted, with occasional relief, from 188,000 to 128,000 years ago. During it, and foreshadowed by "proto-Levallois" tools from Africa up to 700,000 years old, there occurred the first major innovation in stone-tool manufacture for more than a million years. The new tools looked, at first glance, as scrappy and arbitrary as ever, but the Levallois

200,000 YEARS AGO

FREE LUNCH

The aurochs, Bos primigenius, *was the ancestor of modern domesticated cattle. Hunters prized it for its meat, when it appeared in Eurasia 200,000 years ago. The aurochs became extinct, in its wild form, in* AD *1627; this illustration is adapted from a contemporary picture of the last survivor.*

We were *Homo sapiens*. *The Late Pleistocene epoch and Early Paleolithic industries continued.*

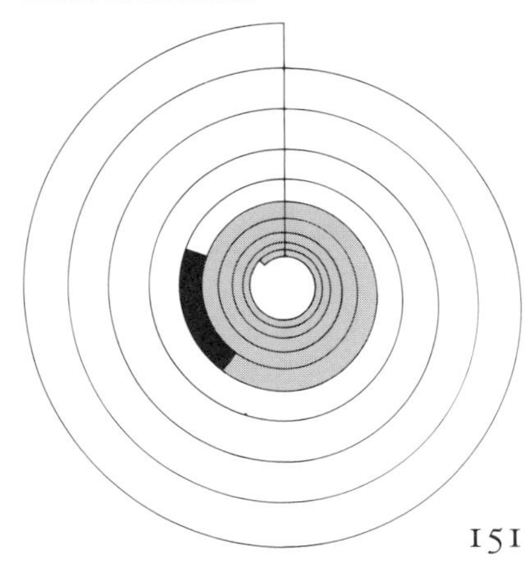

cave bears; goats; modern cattle
▼ 200,000 yr

ice age 34
▼ 188,000 yr

Levallois tools
160,000 yr ▼

200,000 | 190,000 | 180,000 | 170,000 | 160,000

160,000 YEARS AGO
A BETTER TOOL

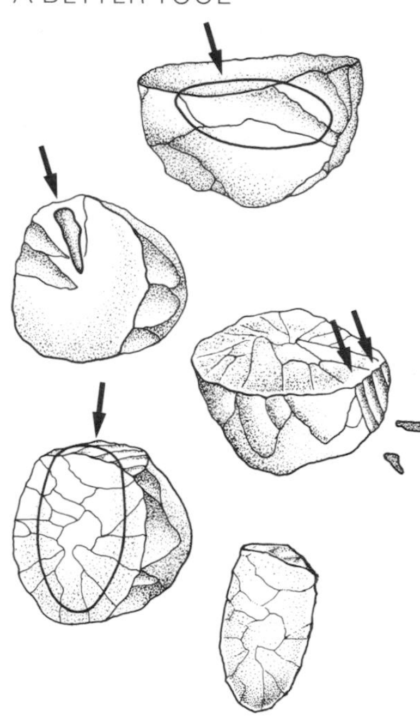

The Levallois technique, anticipated in some early examples of toolmaking like that from Kilombe in Africa, 700,000 years ago, was perfected and widely adopted 160,000 years ago. It was a matter of striking a well-formed, economical tool from a prepared core of stone.

technology, dating from roughly 160,000 years ago, showed the toolmaker planning to extract his implements from inside the material. In essence he whittled away one side of a piece of flint to produce a flattened surface, and then struck out one or two disk-like flakes from that side. The Levallois method greatly increased the length of cutting edge for a given weight of stone, and during the ensuing hundred thousand years it came into use in Africa, parts of Asia, and most conspicuously in Europe, where industrious stone-knappers no doubt licked their lips in anticipation of juicy mammoth steak.

The woolly mammoth, an emblem of life in the most recent ice ages, made its debut about 150,000 years ago. As an elephant with a warm coat, the woolly mammoth evolved from the steppe mammoth and replaced it. The newcomer spread across Eurasia, in sufficient numbers for its giant tusks to make ivory mines for later humans. Eventually the woolly mammoth wandered into North America: to reach the Great Plains, the mammoth found a route over the dried-out Bering Strait and across the realm of the Canadian ice sheets, evidently at a time when there was enough ice on land to lower the sea level but not enough to create an impenetrable wall across the American route.

The ice melted rapidly 128,000 years ago, and from close to that time comes clear evidence of humans eating large quantities of shellfish at the southern tip of Africa. The interlude was exceptionally warm. Hippos and lions lived as far north as London, and the sea rose perhaps eighteen meters above its present level, leaving its mark in raised beaches all around the world. At least two subspecies of *Homo sapiens* shared the Earth between then and now. Those who looked more like modern humans were to turn up in Africa, but despite appearances they were not at first any more talented than the coexistent subspecies, the neanderthalers, who flourished from China to Europe: *Homo sapiens neanderthalensis*. The neanderthalers were stocky and robust, with massive bones, low foreheads, and protruding teeth. Both subspecies had affinities with archaic *Homo sapiens* and represented contrasting evolutionary lines from shared ancestors.

We were *Homo sapiens*. *The Late Pleistocene epoch continued. The warm interlude beginning 128,000 years ago is called the Sangamonian stage in North America, the Eemian in Europe, and core stage 5 in the ocean-floor record. The Middle Paleolithic of the archeologists, and the Mousterian industries (improved stone tools), can both be loosely dated to 120,000 years ago.*

The neanderthalers appeared about 120,000 years ago, at about the same time as wild ancestors of the domestic cat evolved. They were vigorous people with brains that were, if anything, larger than those of modern humans. They bore the brunt of the cold when it

Levallois tools
▼ 160,000 yr

woolly mammoth
▼ 150,000 yr

ice maximum
133,000 yr ▼

160,000 yr | 150,000 | 140,000 | 130,0

returned in Eurasia, and they developed a notable culture of their own that was well in advance of all previous humanity's. Neanderthalers commanded Eurasia for eighty thousand years, as the great have-beens of prehistory. Their rise and fall demonstrates how chancy was the evolution of our own subspecies. One should not, with self-important hindsight, look upon the neanderthalers as apprentices striving unsuccessfully to become like ourselves: their way of life was complete in itself.

The interglacial phase was interrupted 115,000 years ago, and by this time Earth history is sharp enough to show a major climatic change in detail. The cold struck with an observable abruptness that contradicts any comfortable belief about ice ages taking thousands of years to develop. In a bog in northern France, pollen deposits revealed a forest of fir and spruce, spattered with oak, giving way to pine forests typical of central Scandinavia today, near the Arctic Circle. The timing fitted the astronomical account of ice ages, but the pace was faster than that theory would predict. The change was recorded in a layer three millimeters thick, corresponding, in the simplest interpretation, to just twenty years of elapsed time; even if it really took a century or more, the event was disconcertingly rapid. A similar change occurring now, at a vulnerable stage of the present warm interlude, would be called a great disaster. How did it come about?

During different phases of the ice-age cycle, the weather machine of wind, water, and ice ran in distinctive patterns, and it had to change gear to conform to the varying intensity of summer sunshine; evidently it could hesitate and then suddenly go into a new regime. A mechanism for such a switch at the start of glaciation was the snow blitz, in which large areas of northern lands remained covered with unmelted snow all through the summer. Ice sheets then grew, not by the slow creep of glaciers from their mountain lairs, but from the bottom up, as each year's snowfall added to the accumulation. On this occasion 115,000 years ago, the climate approached a full ice age in severity, but it was a false one, with warmer conditions returning quite quickly, by about 108,000 years ago. Quickly by geological standards; in human terms this blip in the climatic curves represents an interval much longer than that which separates us from the pyramid builders of Egypt. Then the same thing happened again. After an interval when the climate was similar to the present, the plunge into another false ice age occurred

115,000 YEARS AGO

THE FALSE ICE AGES

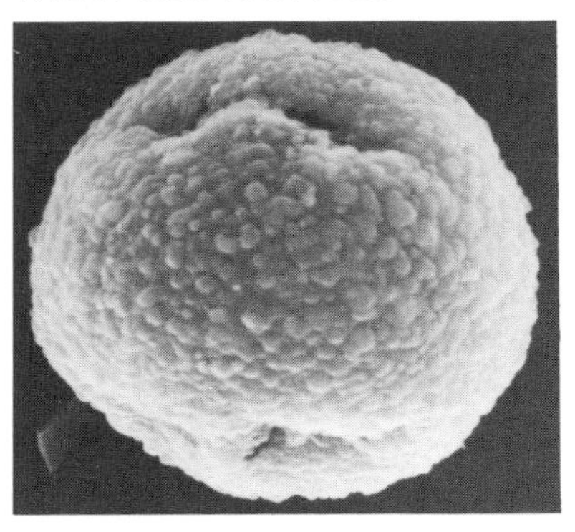

Following the very warm interlude that began 128,000 years ago, two severe cold episodes occurred. Here, a pollen record of trees in northern France is matched to the well-dated record of ice fluctuations in marine fossils. The appearances and disappearances of oak pollen (above) in Europe indicated warmth and cold.

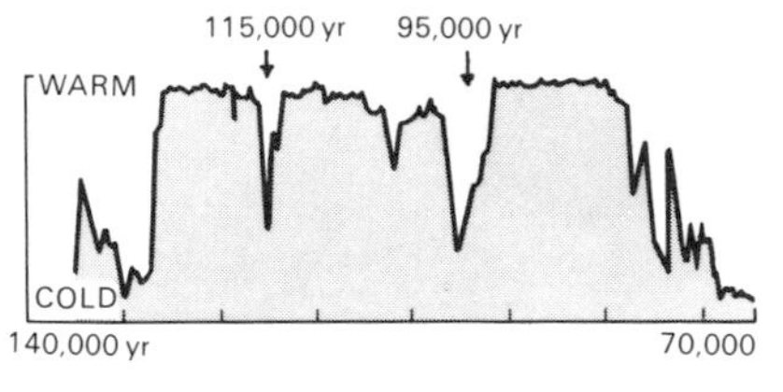

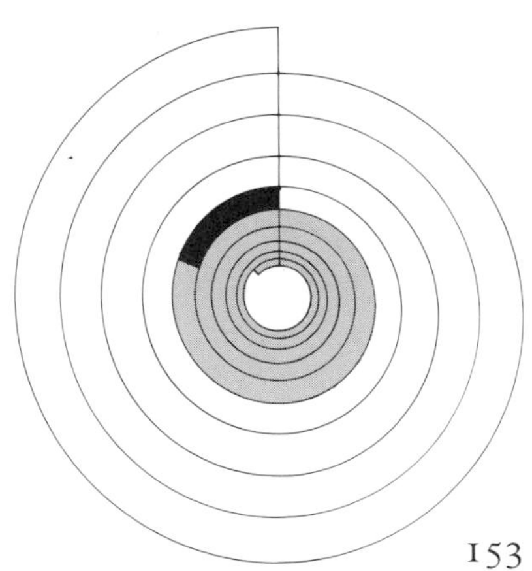

warm interlude; shellfish collecting
▼ 128,000 yr

neanderthalers; wildcats
▼ 120,000 yr

false ice age
▼ 115,000 yr

proto-modern humans
100,000 yr ▼

120,000 110,000 100,000

100,000 YEARS AGO
HUMAN MODERNIZATION

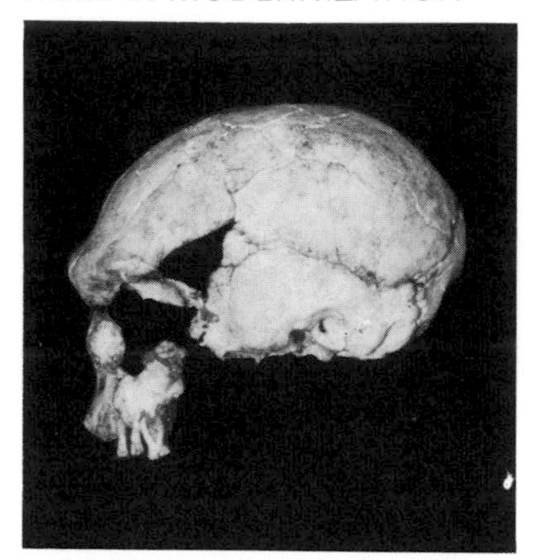

A skull from Tanzania, between 150,000 and 90,000 years old, shows an enlarged, rounded vault like that of modern Homo sapiens, *without the protruding face and jaw of the neanderthaler who lived in colder climates at around the same time. A sloping forehead and moderate bony ridges over the eyes persisted as primitive features.*

We were proto-modern. *The Late Pleistocene epoch and Middle Paleolithic industries continued. The ice age beginning 72,000 years ago is called the Wisconsin glaciation in North America and the Würm or Weichselian in Europe. The first cold phase is core stage 4 in the ocean-floor record.*

about 95,000 years ago, only to have temperatures recovering once more within about seven thousand years. These frequent changes of climate taxed human populations and wildlife, and herds of mammoths perished when frozen ground thawed, letting them sink helplessly into bogs.

In Africa the subspecies to which modern humans belong, *Homo sapiens sapiens*, was already evolving an outward form that looked not at all like that of its neanderthaler cousins. A skull from Tanzania, at least 100,000 years old, was still somewhat beetle-browed, but it possessed a dome-shaped cranium. The face and jaw were much neater, to modern eyes, than those of the neanderthalers, and the construction was less sturdy. The contrasts make good sense in relation to climate; our immediate forebears were tropical creatures. Combined with evidence from other African sites, this find virtually rules out any possibility that the trim build and high foreheads of modern humans evolved from the neanderthalers' very different skeletal forms. On the other hand, we are entitled to ask of those people in Africa, "What kept you?" Their tools were mediocre, and they made no special mark on the world until 40,000 years ago, when they did so quite emphatically. It would be scarcely believable that people with the same endowment as present-day farmers and engineers should do nothing noticeable for sixty thousand years.

The impression that remains is therefore of people who looked like us but did not act like us: graceful, highbrowed figures who hunted and gathered their food in Africa, competently but with the traditional skills. The spark that was missing was probably language, in which case, despite appearances, the repetitive sobriquet of *sapiens sapiens* (wise, wise man) seems even less apt for them than for ourselves. Proto-modern humans is the tag adopted here for these doppelgängers.

The neanderthalers are more conspicuous in the archeological record. One of them shaped and colored a mammoth's tooth, 100,000 years ago, to make a decorative charm, or amulet, in a rare survival from the meager arts of the neanderthalers. In Yugoslavia, perhaps 90,000 years ago, they feasted on sufficient quantities of cooked human meat to indicate a less than reverential habit of cannibalism. Remains of neanderthalers elsewhere show weapon wounds. Neither too little nor too much should be made of such signs of mayhem; scholarly attempts to prove that human beings

proto-modern humans; amulet; ritual burials
▼ 100,000 yr

false ice age
▼ 95,000 yr

100,000 yr | 95,000 | 90,000 | 85,000

are natural pacifists are as futile as the frequent contradictory allegations of superaggressiveness.

The most violent volcanic eruption of recent geological times occurred in Indonesia 73,000 years ago. At Toba in Sumatra, an exploding mountain hurled into the air two thousand cubic kilometers of material—a hundred times more than the famous eruption at Krakatoa in the nineteenth century AD—and created the largest volcanic crater lake in the world, some forty kilometers in diameter. Lesser but still very violent volcanic explosions occurred in the Caribbean and Japan. The dust raised by Toba darkened the sun and chilled the world, but it did not quite start an ice age. Very erratic excursions in the climate occurred, even so, showing up in the European pollen record: a deep chilling followed by rewarming, probably associated with the Toba event.

About 72,000 years ago, and once again swiftly, snow piled upon snow in the northern lands to build ice sheets for the most recent of the true ice ages. It was to persist, with some important fluctuations, for more than sixty thousand years. The polar bear, a relatively young species among the larger animals of the present world, appeared just at the beginning of the ice age. The first known polar bears pursued seals in the river Thames near London, where heat-loving hippos had flourished during the warm interlude.

Signs of the presence of neanderthalers multiplied, from western Europe to central Asia and Afghanistan. Their tools and their habits of hunting and gathering varied from place to place, according to what the surroundings had to offer. On the open plains of eastern Europe, during the early part of the ice age, neanderthalers quarried the bones and tusks of dead mammoths, and used them to build large huts, in which they lighted fires. The principal game was wild horse and reindeer in the north and bison farther south; woolly rhino, deer, and other large animals were hunted less systematically, while smaller animals, notably Arctic fox and wolf, supplied furs for clothing. At the other end of Europe, in northern Spain, the neanderthalers dwelt in limestone caves, and wild cattle were their main prey.

Spirituality and ritual were not lacking among the neanderthalers. They were the first people known to have buried their dead, and they did so with care and ceremony. A circle of goat horns surrounded a boy's grave in Uzbekistan. There is evidence that elderly, sick, and handicapped people were kept alive by the goodwill of

73,000 YEARS AGO
VOLCANIC FALLOUT

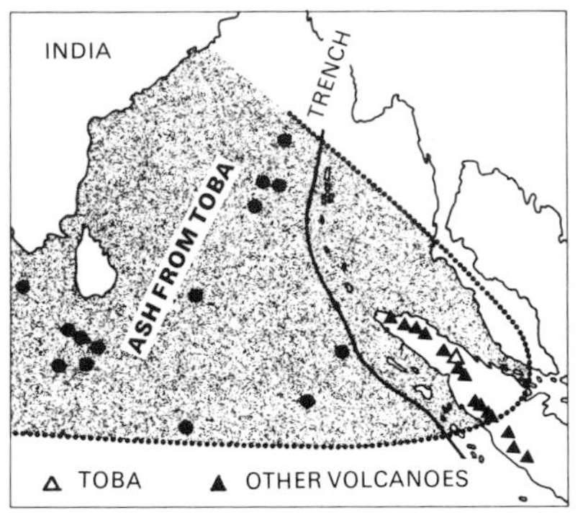

An explosion in the Indonesian island of Sumatra left a lake covering 1300 square kilometers. A layer of ash blown far across the ocean is conspicuous in cores recovered from the ocean bed at the points shown. The wind direction suggests an explosion during the summer.

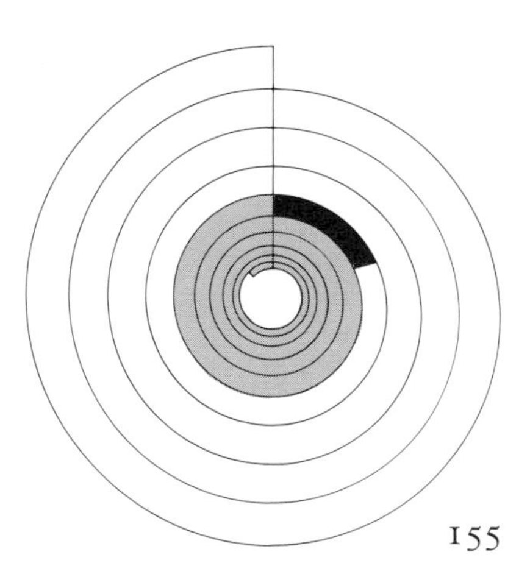

Toba explosion; climatic oscillation 73,000 yr ▼ **ice age 35; polar bear** ▼ 72,000 yr

80,000 | 75,000 | 70,000 | 65,000

45,000 YEARS AGO

CHANGE OF VOICE

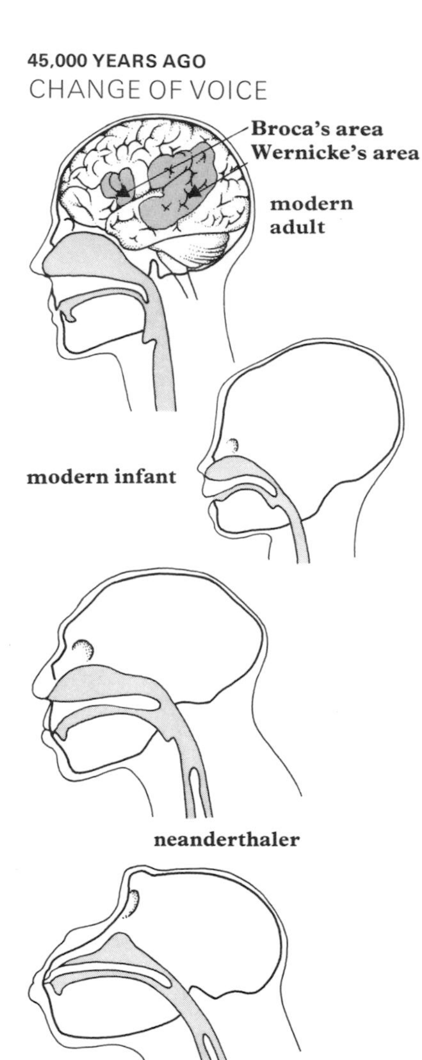

The speech sounds that a neanderthaler could generate were limited by his relatively straight mouth tube (shaded regions), which was like that of a modern baby. But changes in the brain areas concerned with speech were probably at least as important in the mental revolution that created modern human beings.

their associates, and by medical arts as well. When a man living in Iraq perished one springtime about 60,000 years ago, members of his band foraged in the hills for flowers for his grave: yarrow, ragwort, grape hyacinth, hollyhock, and others, to make a bed and wreaths for the body. They were preserved as clusters of pollen. Flowers at a funeral are striking enough; more so is the fact that seven of the eight chosen species of flowers are well known to Iraqis today as traditional folk medicines. Hollyhock, for example, has been called the poorman's aspirin, while the English name "yarrow" means "healer." The occasion seems, therefore, to have been the burial of a medicine man.

Neanderthalers used their big brains to develop their culture along certain narrow tracks. Another sign of that is the first appearance of multiple pigments for body painting. Besides the red ocher (hematite) coveted by human beings through much of prehistory, a cave level in France dating from perhaps 50,000 years ago also contained yellow ocher (geothite), together with many scratched or smoothed blocks and crayons of black manganese dioxide. The European cave bear, standing three meters tall, inspired a cult among neanderthalers, who set cave-bear skulls in niches. In a cave in France, 47,000 years ago, they entombed twenty cave bears under a stone slab weighing almost a ton, as if to bottle up the bears' spirits.

After an ice maximum around 62,000 years ago—not yet the coldest phase of the ice age—the climate relented for a while, and by 58,000 years ago it had turned comparatively mild. Temperatures fluctuated markedly, but the sea level rose, and at times trees spread northward, with summers in the northern lands becoming distinctly warm. Neanderthalers had survived the climatic changes, and they were dourly prospering, standing at an evolutionary pinnacle where skills of hand and eye, high intelligence, and affectionate sociability had paid off. They were the most ingenious predators the world had ever seen, yet their populations were small and scarcely a threat to the survival of the animal species on which they fed. Had they themselves survived, the world of the present warm interlude might have remained a plentiful hunting ground.

Whatever the neanderthalers may have known about justice was not shared by the agencies of evolutionary change. A terrifying race of supermen appeared, created by mutations among the highbrow subspecies living in warm regions. "In the beginning was the

lesser ice maximum 62,000 yr ▼

herbal medicine ▼ 60,000 yr

milder interlude ▼ 58,000 yr

63,000 yr | 60,000 | 55,000

Word,'' a later writer of Scripture declared shrewdly, but neanderthalers had words of a sort, and they must have strained to share their thoughts. The mutations did not quite supply telepathy, but the next best thing: in the beginning was the Sentence. A novelist captures the moment better than any impersonal commentary when he imagines a neanderthaler's astonishment on first overhearing modern speech:

> The sounds made a picture in his head of interlacing shapes, thin, and complex, voluble and silly, not like the long curve of a hawk's cry, but tangled like line weed on the beach after a storm, muddled as water. [William Golding, *The Inheritors*, 1955]

Refinements to the mouth may have been evolving gradually, but the acquisition of fully fledged language, characteristic of modern humans, was probably a matter of rewiring of the parts of the brain responsible for uttering and comprehending speech. The necessary mutations must have occurred by 45,000 years ago, allowing for the time needed for the mutations to establish themselves before their consequences unfolded. Eve sang to Adam, and words strung together in lucid sentences conveyed information, requests, promises, criticism, compassion, recollections, plans, slogans. Sentence making raised confidence and standards of performance, sharpened analytical thought, and above all multiplied brain power by pooling it, not just within the group but from one generation to the next. The mental revolution altered the very nature of evolution. Thenceforward, purposeful technical and social innovations were to set the pace, rather than chancy genetic changes. Thus did a barely tangible novelty in one inconspicuous species become a geological force that destroyed or rearranged plants and animals worldwide, and moved mountains.

Present-day humans are descended from a small population of mutants who, according to genetic mapping, may have lived in western Asia somewhere between the Caspian Sea and the Indian Ocean. Like their forerunners, the proto-modern humans of Africa, they had more slender bones than the classic neanderthalers of Europe. Some of their offspring interbred with neanderthalers and other pre-existing groups in various parts of the world. Yet an account of the origin of modern humans that avoids incredible coincidences must envisage a primary wave of mutants, or at least of their genes, radiating across the world. The talkative hunters could go almost anywhere and survive, and when their breakout occurred

45,000 YEARS AGO

DOOMED CREATURES

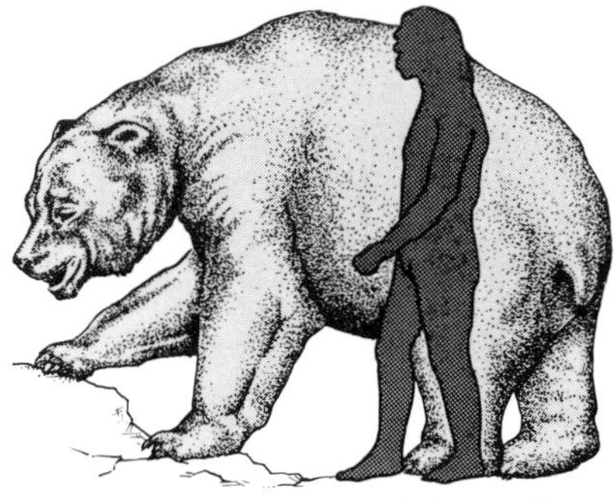

Neither the stocky neanderthaler nor the large cave bear, which he regarded with awe, survived the irruption of fully modern human beings. The last known neanderthaler dates from 34,000 years ago, while the cave bear died out 12,000 to 10,000 years ago.

We learned to speak. *The Late Pleistocene epoch continued. The milder interlude 58,000 to 28,000 years ago corresponds with core stage 3 in the ocean-floor record. Archeologically, the Middle Paleolithic gave way to the Late Paleolithic (much improved stone tools) about 45,000 years ago at some sites, and Late Paleolithic industries were widespread by 40,000 years ago.*

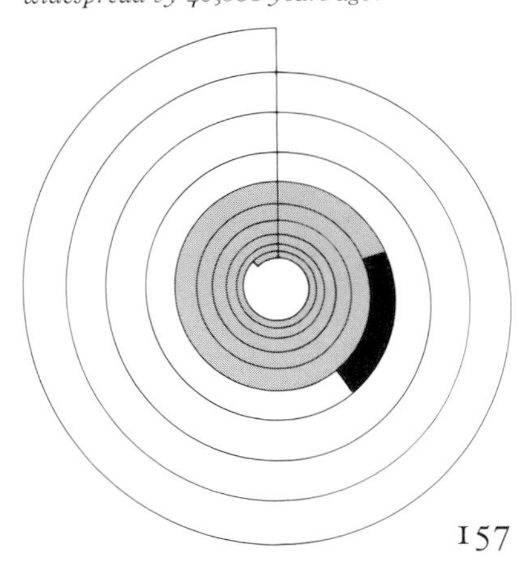

multiple pigments ▼ 50,000 yr

cave-bear cult ▼ 47,000 yr

mental revolution ▼ 45,000 yr

breakout of modern humans 40,000 yr ▼

50,000 | 45,000 | 40,000

40,000 YEARS AGO
STONE RAZORS

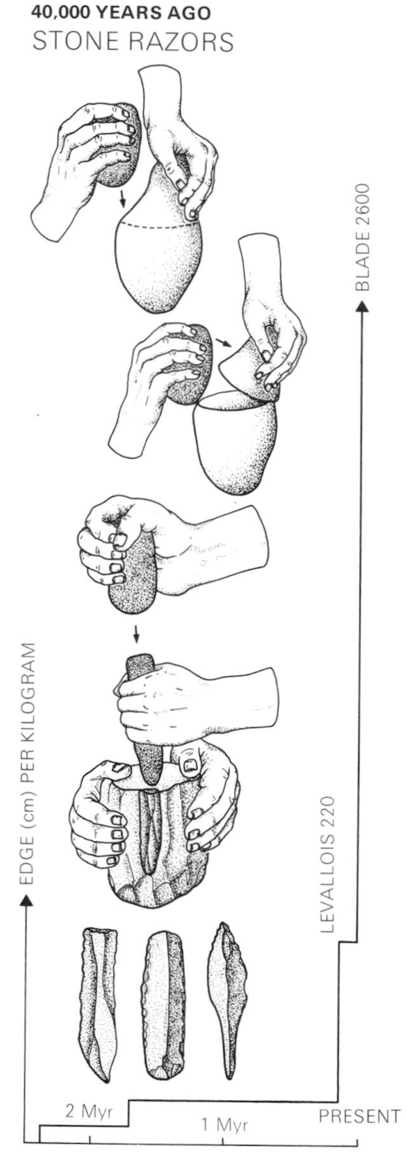

The blade technique, for making stone tools, gave a huge improvement in the length of cutting edge obtainable from a given weight of stone. It was one of the signs of new minds at work.

it was comprehensive. They were in Australia, where no human foot had trodden before, about 40,000 years ago—so quickly that uncertainties in the dating make it look instantaneous. Elsewhere the impression is of an advance into new territory at an average rate of a few kilometers a year, passing right across Europe by about 35,000 years ago, and to central Africa and remote Siberia by about 32,000 years ago. The modern humans may have penetrated to Alaska and the Yukon by 27,000 years ago.

They learned fast, adjusting their ways of life to whatever nature had in store for them, whether coconuts and wild pigs in southeastern Asia, or berries and woolly mammoths in ice-age Europe. Oak might give way to fir, and wild horse to reindeer, either because the climate changed or because groups of hunter-gatherers migrated a few hundred kilometers north. Whatever the reason, the new people coped, but they changed in stature, physiognomy, and pigmentation to suit their environments, while random variations of genes and languages in small bands added to the diversity of mankind.

The talkative hunters put to sea right away, otherwise they could not have reached Australia so promptly: even allowing for the low sea levels of the ice age, they had to cross a minimum of ninety kilometers of open ocean to reach the island continent. Seagoing craft were therefore available from the very earliest times, for fishing, transport, and migrations. Their characteristic blade tools were, from the start, a marked improvement on previous stone technology. But to suppose that the modern humans thought of everything at once would be as foolish as underestimating their resourcefulness. The record as a whole gives an impression of inventions and refinements spread over tens of thousands of years.

Calendar keeping began early. Logging the days between the repetition of events in the twilight sky was quite easy; any diligent person in one lifetime should have been able to establish that the year was about 365.25 days long, and so mapped out the seasons. The rest was obfuscation, in which the solar system connived by displaying a thrilling moon and puzzling planets, out of synchronism with the sun, and by throwing in occasional eclipses to arouse wonder and fear. The phases of the moon were of little practical consequence except for fishermen concerned about the tides, but the lunar cycle divided the year into months too convenient not to use. On a flat piece of bone a person could make notches of different symbolic shapes corresponding to the different phases of the moon. Lunar calendars of that kind were in use in France around 35,000 years ago.

Whether or not the modern humans attacked the neanderthalers or mated with them, they quickly outnumbered them. The neanderthalers were the first of many creatures who found the wide world too small for them and the new people, and the last neanderthaler expired near Cognac in France about 34,000 years ago. The usurpers did not lose their mobility, and groups engaged in seasonal migrations to find the best hunting; they also sought out the best toolmaking and coloring stones, by mining expeditions or trade, over distances of hundreds of kilometers.

The earliest known trade routes in regular use crisscrossed eastern Europe about 30,000 years ago, distributing prefashioned blanks of flint from mines in Poland and Czechoslovakia over a wide area. It was in eastern Europe, too, that the oldest ceramic objects yet discovered were made about 29,000 years ago, from tempered clay fired in a kiln in Czechoslovakia. They included models of animals and also a ceramic "Venus"—a stylized figurine of a woman with no face and an obese torso. And this European site, shared by more than a hundred people, affords the earliest evidence of a quantum jump in the sizes of human groups and the advent of the first communities larger than family bands.

These people, called Gravettians by archeologists, carved figurines from ivory, stone, or coal, and they possessed unusually neat tools, fashioned in some cases to take wooden shafts and handles, in composite implements. In Czechoslovakia they lived amid hoards of mammoth bones and tusks, which they used for propping up roofs, covering graves, and making spades and spear points. Around 28,000 years ago, as the most severe phase of the ice age began, the woolly mammoth returned to lower latitudes of Eurasia, and when the mammoth spread, so did the new cultural package. The characteristic tools and figurines appeared all the way from France to Russia, as a rare marker of an early human response to climatic change.

Artists of the Gravettian school decorated their caves with engraved and painted outlines of animals; abstract works and careful representation coexisted, but anyone who has not stalked a herd of woolly mammoths at the snout of a glacier may be unable to guess the meaning of this art. Nothing was more important than a fur coat

28,000 YEARS AGO

MAMMOTH FANCIER

The well-clad woolly mammoth prospered in a worsening climate, and a culture that exploited the ivory of its tusks extended its range across Europe. Human beings of the time were just as resourceful and imaginative as their present-day descendants.

We were hunters. *The Late Pleistocene epoch and Late Paleolithic industries continued. Cultures in Europe included the Aurignacian starting 40,000 and the Gravettian 29,000 years ago. The cold phase beginning 28,000 years ago corresponds with core stage 2 in the ocean-floor record.*

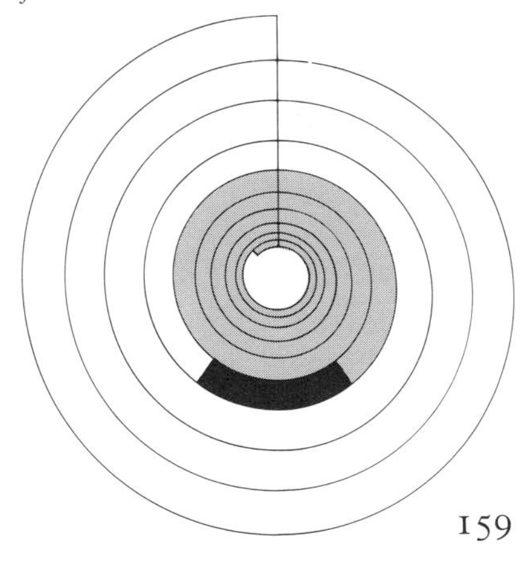

trading 30,000 yr ▼ **communities; ceramics** 29,000 yr ▼ **severe cold; Gravettian heyday** ▼ 28,000 yr

30,000 25,000

25,000 YEARS AGO
FINE CLOTHING

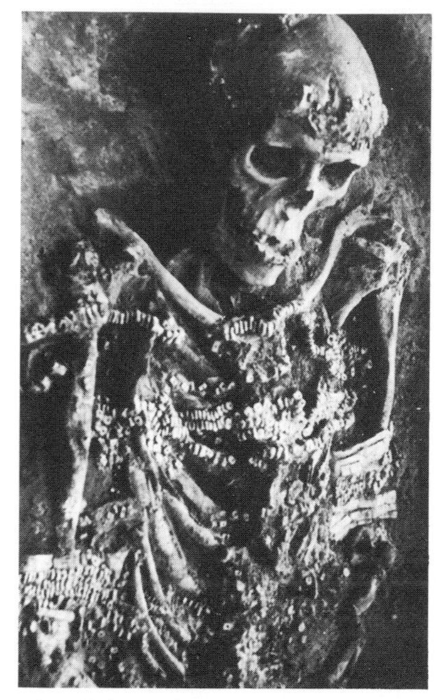

The tailored clothing of a man who died near Moscow 25,000 years ago has rotted away, but an abundance of decorative beads remains.

18,000 YEARS AGO
ICE MAXIMUM

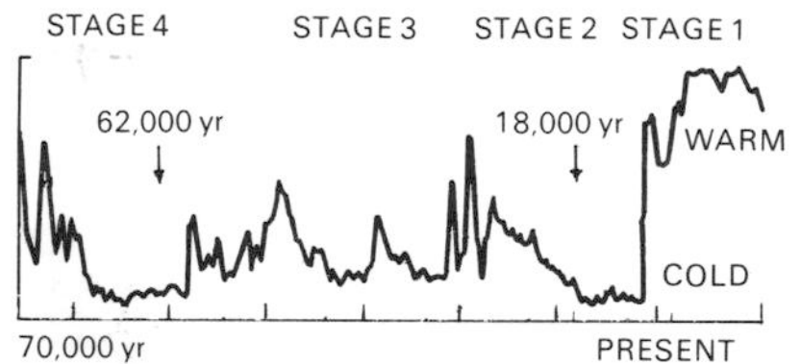

The most recent ice age consisted of two main periods of glaciation (core stages 4 and 2), separated by a relatively mild phase (core stage 3). Stages of alternating warmth and chill were numbered from the top down, in cores from the ocean bed. The vegetation record for northern France is matched to the heavy-oxygen curve.

for survival in the bitter cold of ice-age Europe, and the earliest signs of well-tailored clothing come from the Moscow region, about 25,000 years ago, where marks in the grave of a man and two boys have been interpreted as evidence for jackets, trousers, caps, and moccasins, all possibly sewn with thongs and decorated with remarkable quantities of beads.

The amount of carbon dioxide in the air declined, reinforcing the chill, and the machinery of the world's weather ran very sluggishly. The only redeeming feature of this phase of the ice age was that as it crept toward a glacial maximum of coldness, around 18,000 years ago, there were no very erratic changes of climate. The modern humans could settle to regimes lasting for thousands of years, but innovation and migration did not cease. Objects that look exactly like arrowheads give notification of the invention of the bow and arrow in North Africa or Spain about 20,000 years ago. The first clear signs of human habitation south of the Canadian ice sheets date from about 19,000 years ago, when hunter-gatherers lived in Pennsylvania.

If life was harsh in Africa in the worst of the ice age, it was because of drought rather than cold. But the River Nile and its tributaries still flowed intermittently, and from southern Egypt, 18,000 years ago, comes the earliest evidence for the cultivation of crops—barley and wheat—in riverside gardens. It was a fateful experiment, of the sort that would eventually alter the lives of all mankind. "Why should we plant, when there are so many mongongo nuts in the world?" a hunter-gatherer in the Kalahari Desert responded when a twentieth-century visitor suggested that his people might raise crops. Hunting and gathering gave ample food supplies with far less effort than most systems of cultivation have subsequently demanded, and the switch from the carefree and lazy existence of the hunter-gatherers to the artificial and careworn life of agriculturalists requires a strong explanation. Human beings were seduced into cultivation, because at first it was easy.

As the mud flanking the wadi dried after the floods, it cracked in the sunshine, and feminine laughter no doubt disturbed the water birds when the women came with a bag of seeds to invent crops. Perhaps it was a waste of good food, and nothing to tell the men about—yet it took only moments to poke the seeds into the ready-made cracks in the mud. A crazy experiment, but exactly what was needed, not just to make use of the self-cultivating riverbank, but to

tailored clothing
▼ 25,000 yr

divorce the grasses from their wild relatives so that they would evolve under human hands to suit human purposes. The women knew little of plant genetics, but the grain grew and ripened in their riverside garden before the sun parched the ground entirely, and when they came back with stone sickles, they may have felt a certain goddess-like pride.

The archeological evidence consists of cereal grains associated with milling stones and pestles and mortars. This incipient cultivation, in an area where wild grain would not fare well, persisted for six thousand years in the Nile Valley. Similar experiments were probably in progress elsewhere, perhaps with rice and other crops in southeastern Asia, but the evidence is sketchy. Not far from the Nile, in southwestern Asia, there are clearer signs of another innovation beginning about 18,000 years ago: the herding of docile beasts at the start of a gradual domestication of the animals. Early herdsmen kept gazelles and goats in Palestine and Jordan.

The hunters, meanwhile, were clever at killing but not, in the end, at saving. Mass extinctions of large mammals and flightless birds accompanied the worldwide dispersal of modern humans. Sometimes, no doubt, it was mindless slaughter. Sometimes kill rates sustainable in stable circumstances became overkill when the climate changed sharply and the species in question had troubles enough without being hunted too. Sometimes the extinctions occurred indirectly from competition for the same ecological niches—literally niches in the case of the cave bear of Eurasia, where caves were in demand among humans.

The Pleistocene overkill was first noticeable about 17,000 years ago, at the very time when a new school of hunter-artists, equipped with plant-fiber ropes and tallow lamps, was immortalizing the big game in magnificent cave paintings, as at Lascaux in France. During the next seven thousand years the cave bear would disappear forever, along with the mammoth, the woolly rhino, the musk ox, the steppe bison, the giant Irish elk, and a few other Eurasian species—but many more in other continents where the animals had no previous experience of human hunters. In Australia about 17,000 years ago the early casualties included a number of large kangaroo-like species, together with an entire group of herbivorous marsupials called diprodotons, some as big as hippos, and when they were gone, the Australian hunters were impoverished. The turn of the great beasts of the Americas had yet to come.

17,000 YEARS AGO

THE HUNTER ARTISTS

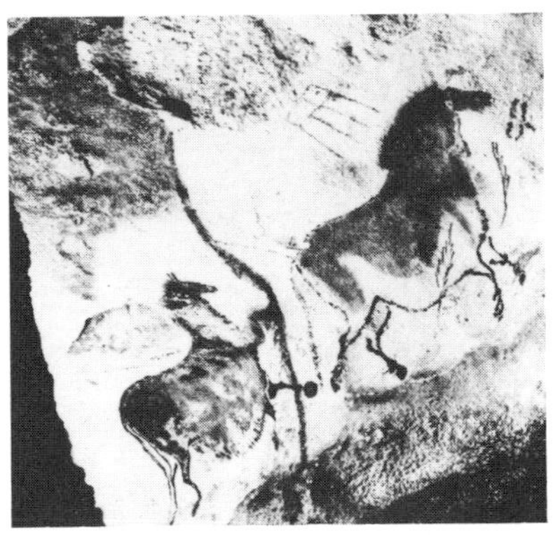

Paintings of game animals on the walls of French caves in France are so fine that, when they were rediscovered in modern times, there was disbelief that they were the work of "primitive" stone-age hunters. Many different pigments were used, some requiring intense heat for their preparation. This example is from Lascaux Cave.

We were hunters. *The Late Pleistocene epoch continued. The Epipaleolithic archeological phase, transitional between the Paleolithic and Neolithic periods, began in southwestern Asia 18,800 years ago, but later elsewhere.*

See map p. 86.

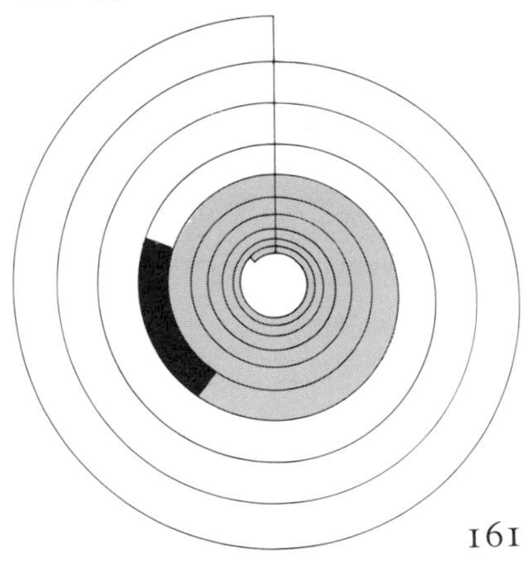

bow and arrow ▼ 20,000

peopling of America 19,000 ▼

latest ice maximum; cultivation; animals herded 18,000 yr ▼

Pleistocene overkill; lamps; ropes ▼ 17,000 yr

20,000 | 19,000 | 18,000 | 17,000 | 16,000

11,000 YEARS AGO
AMERICAN OVERKILL

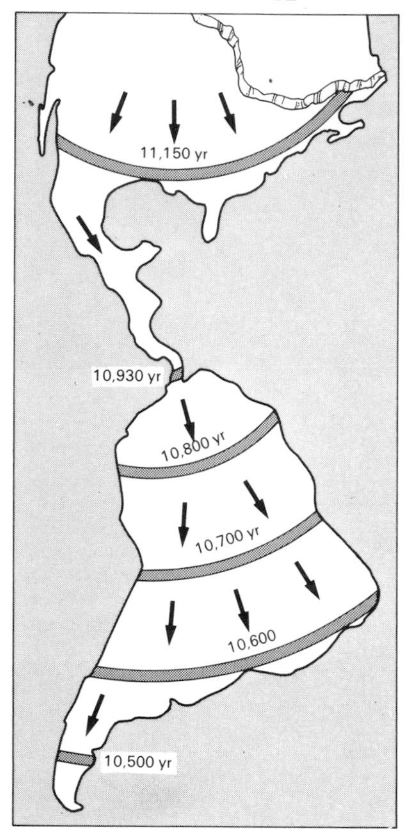

A theoretical map of the wave of advance through the Americas, from north to south, visualizes human beings feeding lavishly but wiping out most of the big game. As a result, later generations of Amerindians were deprived of meat and draft animals.

When the midsummer sun appeared bigger and higher in the sky than it does today, because of the orientation of the Earth's axis and the state of its roll, the northern ice sheets and glaciers melted. The thaw began hesitantly at first, 14,000 years ago; by 12,000 years ago the sea level was rising at a rate of about two meters per century. Meltwater filled the Baltic Sea of Europe and the Great Lakes of North America.

Something more than a change in climate perturbed life in the Americas. The human presence had been muted, while the game grazed in comparative peace. By genetic evidence, the Amerindians originated as hunter-gatherers from Asia, and although the Bering Strait was flooded by 12,000 years ago, assertive newcomers could have walked in from Alaska or arrived by boat. Be that as it may, beginning about 11,300 years ago hunters seem to have passed as an annihilating wave across the landscapes of North America and South America, at an average rate of sixteen kilometers a year. By the time they reached the tip of South America, they had eliminated two thirds of the mammalian species in the twin continents, including nearly all the large animals. The dog was spared, soon to be domesticated as a source of meat. In the high Andes, the llama and its relatives escaped extinction, while the modern bison evolved as a new species, the lonely survivor on the Great Plains.

In a climate setback beginning 10,900 years ago, and lasting for about six hundred years, glaciers advanced again and bulldozed the forests that had sprung up in northern valleys. Young forests retreated in southwestern Asia too, because of drought, and grass replaced them. This relatively short-lived climatic episode gave a most consequential boost to those human groups that were cultivating selected grasses for this grain, or else herding goats and other animals that grazed on the grass.

The earliest animal definitely tamed, and modified as a result of its domestication, was the wolf, becoming the dog around 12,000 years ago in southwestern Asia, where it seems to have been a pet and a hunting companion. Goats and sheep were next: by 10,700 years ago in the hills of Iran, their herdsmen were sufficiently proprietorial about their animals to count them and pledge them, using the oldest known system of accountancy. The information technology that led to writing and numerals began with the invention of small clay tokens in the form of spheres, disks, cones, and cylinders. These appeared in Iran at a site well trodden by goats.

Epipaleolithic phase. *The Neolithic age began 10,600 years ago in southwestern Asia, but apparently later elsewhere. The Pleistocene epoch of the Quaternary period gave way to the Holocene epoch (ocean-floor core stage 1) 10,300 years ago.*

thaw started; fishing gear
14,000 yr ▼

pottery
13,000

16,000 yr — 15,000 — 14,000 — 13,

Twenty different symbols represented commodities such as sheep or measures of grain. The tokens were apparently pledges for offerings of food, to be given ceremonially to a leader for redistribution among other members of the group. If so, they enabled the leader and his clerks to plan communal affairs, in a goatherds' bureaucracy. The tokens turned up two thousand kilometers away in Anatolia at about the same time. Adaptable at will to cover new commodities, this system of tokens flourished over a large area of southwestern Asia, northeastern Africa, and northwestern India, until it evolved into writing five thousand years later.

Gardening first became a distinctive way of life at Jericho, in the rift valley of the Jordan, where domesticated wheat and barley, different from the wild varieties, appeared 10,600 years ago as a product of biotechnology. The world's first town arose at Jericho at that time, and the burghers built a wall around it, four meters high, to protect their stockpile of grain from herdsmen who might sooner steal than trade for it.

The earliest known ceramic pots, dating from about 13,000 years ago, were made by cave-dwelling hunter-gatherers in Japan, so cultivating the soil was certainly not the only road to material success at the end of the ice age 10,300 years ago. On the contrary, subsequent history tells how herders of animals subjugated the gardeners, while fishing in favored places led quickly to assured affluence. As the melting ice dripped into the oceans, and the sea rose and grew warmer, the waters became far more congenial for fish, shellfish, and fishermen, both on the flooded edges of the continental shelf and in the rivers. Harpoons and fishhooks had begun to appear in Europe and southern Africa about 14,000 years ago. The rising sea drowned any sites where earlier coastal fishermen might have been operating during the ice age, but it also flooded large territories, turning hills into islands and valleys into harbors, and obliging many landlubbers to take up fishing to survive.

In Europe, for example, coast dwellers of Denmark may have obtained as much as nine tenths of all their food from the sea, while less specialized people in Greece combined fishing with seaborne trade, fetching from islands the prized volcanic glass called obsidian. Widely scattered mounds of shells on southeastern Asian shores tell of appetites there too, for fish and seafood. Other participants in this worldwide boom of 10,000 years ago were fisherfolk equipped

10,600 YEARS AGO

A PLATE OF CEREALS

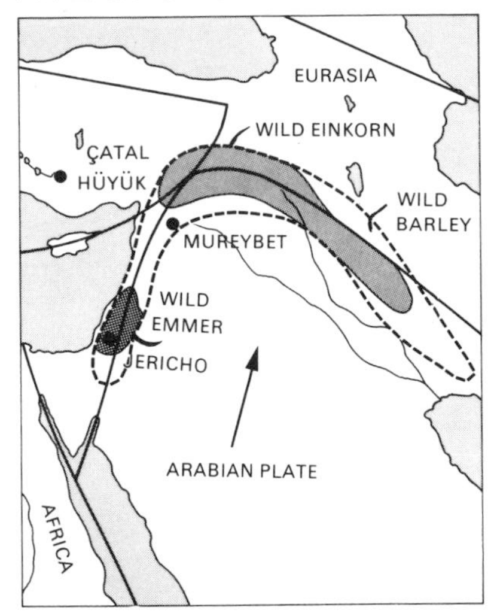

Inhabitants of Jericho domesticated emmer wheat and barley. Einkorn wheat was fully domesticated somewhat later, at Mureybet. The collision between Arabia and Eurasia created mountains watered by rain, where wild grasses flourished. The Fertile Crescent follows the plate boundary and the first city, Çatal Hüyük, arose 8800 years ago, close to it.

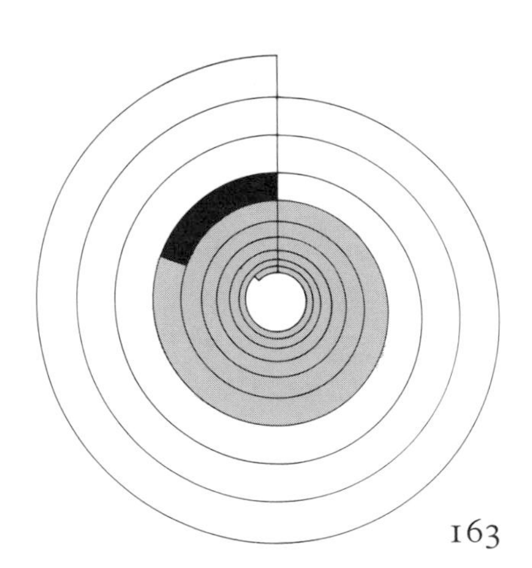

dogs tamed 12,000 yr ▼

sheep and goats tamed; accountancy; bureaucracy 10,700 yr ▼

wheat; town ▼ 10,600

ice age ended ▼ 10,300 yr

12,000 | 11,000 | 10,000

8500 YEARS AGO
CITY OF WEAVERS

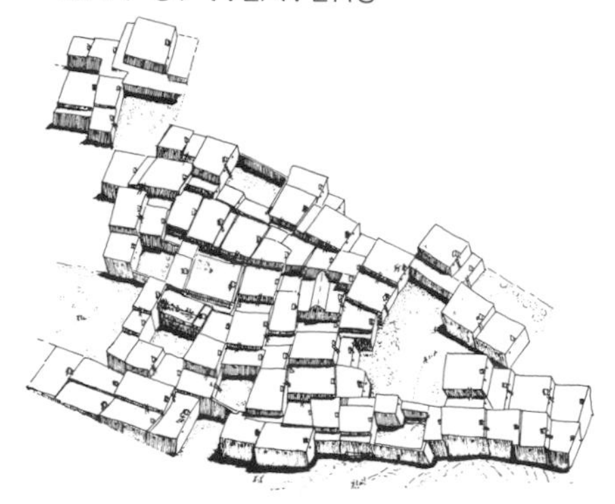

At Çatal Hüyük, houses and shrines were huddled in a mass, without streets. The early specimen of woven animal fiber illustrated above comes from a somewhat later level (VI A) than that depicted in the reconstruction (VI B).

Into the Copper Age. *The Chalcolithic or Copper Age nominally began 8000 years ago in southwestern Asia, but metallic names for archeological ages are not always a clear reflection of the state of technology (see the reference index: archeological timescale; metal industries).*

with canoes, harpoons, and nets, who followed the teeming Kuroshio current and, by 9600 years ago, had settled in Japan. They went out to sea on prolonged fishing trips, supplemented their diet from the forest, cooked their food in fine pots, and lived in permanent homes.

The rising sea level and improved rainfall transformed intermittent torrents into generous rivers that plastered their banks with silt at flood times, thus controlling weeds and preparing the soil for easy riverside cultivation. Humanity's main crops were domesticated in an impressive worldwide roll call: rice in Indochina, sugar cane in New Guinea, millets in Africa and China, beans and maize in the Americas. Altogether about twenty major crop plants came into cultivation in seven to ten distinct regions, between 10,000 and 7500 years ago. In each center of origin women gardeners, no longer as lean and well exercised as their gatherer grandmothers, found themselves cursed with frequent menstruation and babies galore. Local surges of population began a gradual growth in global numbers, and the gardeners spread along the river valleys. In southwestern Asia the breakout of gardeners can be dated at 9500 years ago, when the same biotechnology was to be found all around the Fertile Crescent which curves from Palestine to the Arabian-Persian gulf. Domesticated forms of wheat, barley, sheep, and goats constituted a highly successful package. From earlier Indochinese origins, rice gardeners with water buffalo and pigs turned up in southern China 9000 years ago. By 7500 years ago millet gardeners farther north, beside the Yellow River, were exploiting the wind-blown loess of the ice age, a particularly labor-saving soil.

The gardeners reached southeastern Europe 8700 years ago, taking with them cattle, newly domesticated in Anatolia (modern Turkey). The augmented package vaulted to the Danube Valley, also carpeted yellow with loess, and generations of gardeners followed the loess across Europe until they were on the Rhine 7500 years ago. The same southwestern Asian package reached the Indus Valley about 8000 years ago, and the Nile a thousand years later.

If walled Jericho (population 2000) had been the earliest market town, the first settlement to qualify as a city, by virtue of its bustle and crime rate, was Çatal Hüyük (population 5000), which flourished in Anatolia 8800 to 8000 years ago. Its inhabitants ate beef, traded intensively in obsidian, and were weaving cloth by 8500 years ago. They decorated their shrines with startling multicolored paintings, sculptures, and plaster reliefs. The buildings of the precocious city slickers were huddled together in a continuous mass. Entry was through the roof, and various platforms served for sleeping and working. Skeletons of the dead were buried cozily under the sleeping platforms, after birds had picked them clean.

On the edge of the Mesopotamian plain, some gardeners enlarged their cultivated areas by spreading river water through man-made irrigation ditches. Early cultivation was generally tranquil: women gardened lazily with hoes, while men amused themselves with fishing, hunting, baiting animals, and trading. From about 8000 years ago onward, occasional fights or wars led to the destruction of settlements in southwestern Asia. Some towns were fortified, but warfare was nothing like the habit that it later became.

Hot furnaces developed from bread ovens, and pottery kilns launched a copper industry in Rumania, Bulgaria, and Yugoslavia, where metal smelted from its ores was cast into ax heads 6500 years ago, and within a few centuries the smiths of Europe and southwestern Asia were making hard bronzes by alloying their copper with arsenic. The oldest known stone buildings date from the same time: standing on the island of Guernsey, off the coast of France, a tomb marks an onset of megalithic building in western Europe. But what turned the world upside down was the milking cow.

Certain human beings went pale and started drinking milk. As the gardeners advanced northward into less sunny regions, their children became liable to grow up with rickety bones. Pink skins evolved, which admitted ultraviolet rays and promoted the formation of vitamin D, a factor in bone growth. Milk, too, was good for bones, and a bizarre genetic mutation enabled children to go on drinking milk after weaning. These genetic peculiarities may have taken thousands of years to become normal in a population, so their origin is obscure. Some paleskins lived on the steppes of the Ukraine in eastern Europe, where they began taming horses about 6400 years ago; later, as speakers of Uralic, and especially Indo-European languages, the horse fanciers became as obstreperous as their mounts. But milk-drinking relatives, speakers of proto-Semitic languages living with dairy cattle in the hills of northern Mesopotamia, were the first to impose upon their neighbors.

Animals could evidently supply more than meat and skins, and once the dairymen had made this mental leap, other innovations followed. Someone, clowning perhaps, hitched a hoe to an ox and

7700 YEARS AGO

THE MUTANT CORN

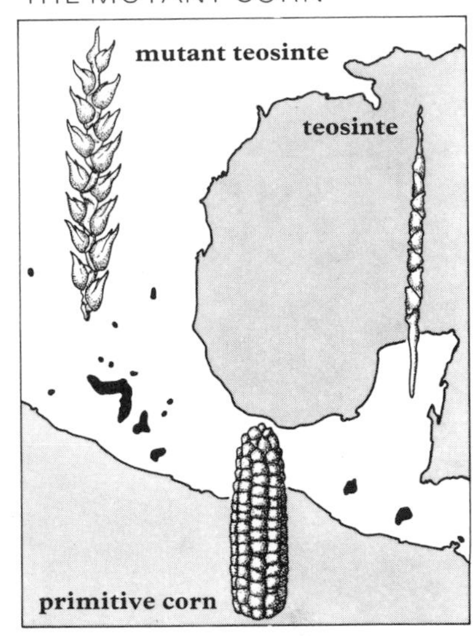

The grass teosinte grows wild in Middle America, in the districts shown. Its name means "God's corn" in the Aztec language, and it makes good popcorn. Modern corn, or maize, derives from a mutation in teosinte that gave rise to soft-cased kernels. Amerindians evidently spotted and nurtured this mutant form, and they were cultivating primitive corn in the Tehuacán region of Mexico, by 7700 years ago.

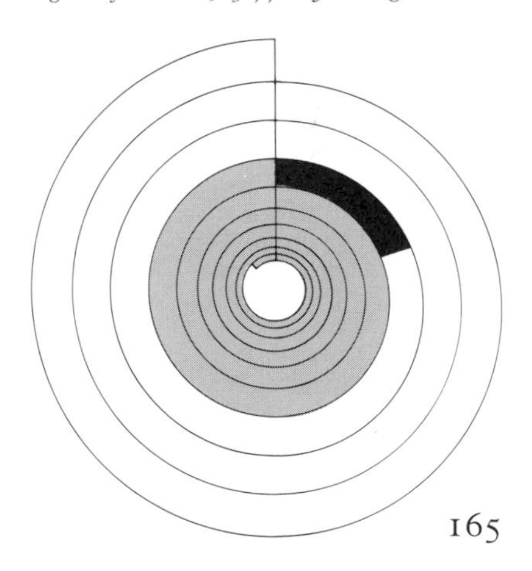

irrigation ▼ 8000 yr

copper smelting; stone buildings 6500 yr ▼

horses tamed ▼ 6400 yr

000 | 7500 | 7000 | 6500

5600 YEARS AGO
TAX DEMANDS

Bound prisoners are depicted on a clay envelope from Susa, in present-day Iran. They showed the consequences of failing to meet a demand for taxes, which took the form of clay tokens, representing various commodities, that were bundled inside envelopes of this type (lower illustration). Writing began in Mesopotamia with simple impressions of the tokens on envelopes; these soon gave way to marks on clay tablets, which at first resembled the envelopes.

invented the plow; thus, animal energy augmented human muscle power. The plow was shared by the Sumerians, inhabitants of southern Mesopotamia who made distinctive ceramics using the potter's wheel, and who began developing the first highly urbanized culture 6000 years ago. The breeding of mutant woolly sheep as a new source of textiles completed the original manifesto of a livestock revolution. These innovations had been gathered together by 6000 years ago, in or near northern Mesopotamia. With the new agriculture of cow and plow came social change. For a start, herding and plowing were tasks for men, while the women were set to spinning and weaving woolen cloth. Not for the last time, people in general found that labor-saving innovations meant harder work.

Semitic-tongued dairymen broke out from their region of origin around 6000 years ago, and climatic disruptions due to a sharp, worldwide cooling 5800 years ago may have accelerated their migrations. Entering Europe, they diffused among the hoe gardeners and hacked down the forests to make new fields for their plows. By 5200 years ago, ox-drawn plows were scratching soil all the way from Britain to the Arabian-Persian gulf. The plow united cultivation and herding in a system of mixed farming very different from the earlier gardening, but well suited to western Asia and Europe, where grazing land often interfingered the best arable land. The Sumerians looked to their defenses while some of the dairymen walked into Africa, where they overbore the inhabitants of the Nile Valley, imposing Semitic grammar on an indigenous vocabulary to create the ancient Egyptian language. Other dairymen led their cattle into what is now the Sahara Desert but was then much better watered. When the Sahara dried out, about 4500 years ago, these milk-drinking nomads were extruded southward into central Africa.

The political lineaments of the modern world began to appear 5600 years ago in Mesopotamia, with the first tax demands issued by parceling up the clay accountancy tokens in clay envelopes marked with official seals. Some bright clerk then thought of dispensing with the tokens and envelopes and making his marks on flat clay tablets. Writing thereby originated, 5500 years ago, in nice time to record the advent of another innovation, the wagon wheel. Pictures on tablets show wheels fitted to a sled, to make a wagon.

Before taxation, human beings gave their personal wealth away, thus winning prestige while maintaining material equality. What bureaucrats demanded ceased to be a moral obligation, and 5600 years ago class distinctions became conspicuous for the first time. Cities of Mesopotamia remained ostensibly democratic, but beside the Nile 5300 years ago a second political invention altered human relationships: chronic warfare. Warriors achieved lasting power over their fellow men by making warfare a perpetual risk instead of a rare calamity. One man ruled most of Egypt by 5200 years ago, and Mesopotamian cities also fell under the sway of warrior kings. Thus were human beings tamed at much the same time as the donkeys, and by 4700 years ago entombing Sumerian royalty at Ur required the slaughter of courtiers, guards, and especially women.

The domestication of donkeys and camels around 5000 years ago made long-haul trading easier, notably carrying supplies of tin for alloying with copper, to make the tin bronze that came into general use in southwestern Asia by 5000 years ago. It was no better than arsenic bronze, but less likely to poison the metalsmiths. Some say the tin came from Afghanistan, others from Malaya in southeastern Asia; at any rate, between Malaya and Mesopotamia bronze-using cities of the Harrapa culture sprang up in Pakistan and northwestern India about 4900 years ago. That was at a time of rapid warming, when a four-meter rise in sea level caused flooding, including the biblical flood in Mesopotamia.

The craft of kingship extended to the subjugation of aliens. To begin with, the aim may have been to prevent interference with essential trade; later, taxes were the chief prize. About 4400 years ago the Akkadian empire, the world's first, covered an important slice of southwestern Asia. The Semitic-tongued emperor from Akkad in Mesopotamia styled himself "king of all the lands of the Earth," and his grandson claimed to be a god. In lands of the Earth not familiar to these emperors, the horsemen were breaking out.

Early speakers of Indo-European languages combined milk drinking with horsemanship. From origins on the steppes near the Volga River in Russia, their first decisive sweep was westward, starting about 4900 years ago, when they entered Anatolia; by 4300 years ago, horses were grazing in Ireland. Greeks and Hittites, who conquered Greece and Anatolia respectively, were among the early identifiable Indo-European speakers. The horses, indifferent beasts by modern standards, were made more formidable by the advent of the lightweight chariot with spoked wheels, showing up in southwestern Asia a little over 4000 years ago. Chariot blitzkrieg became customary around the shores of the eastern Mediterranean. There,

5400 YEARS AGO

DAIRY INDUSTRY

Cattle, calves, and milk jugs stored in huts figure in this Mesopotamian scene – a drawing made from the impression formed by rolling a cylinder seal. Systematic dairying had begun only a few centuries earlier, and the cows were not yet highly bred as milk producers.

Early Bronze Age. *In southwestern Asia, the Early Bronze Age nominally began 6000 years ago, and the Middle Bronze Age 4150 years ago. The first dynasty of Egypt is dated at 5200, and the Early Dynastic period in Sumer at 5100 years ago.*

See maps pp. 88, 90.

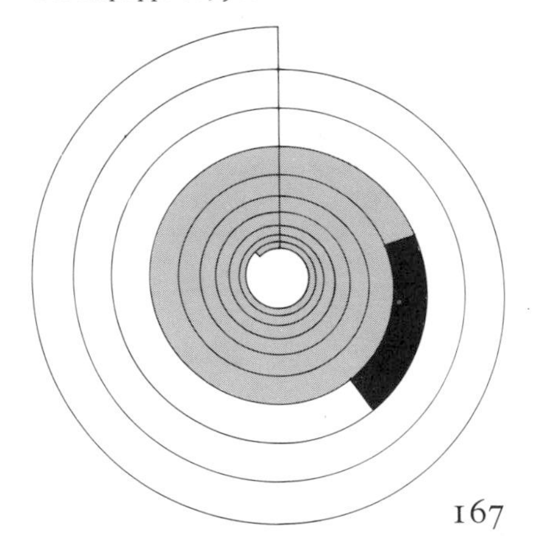

pack animals ▼ 5000
breakout of the horsemen ▼ 4900 yr
North African desiccation 4500 yr ▼
empire ▼ 4400 yr
chariots ▼ 4100 yr

5000 | 4500 | 4000

3500 YEARS AGO
CHINESE CHARIOT

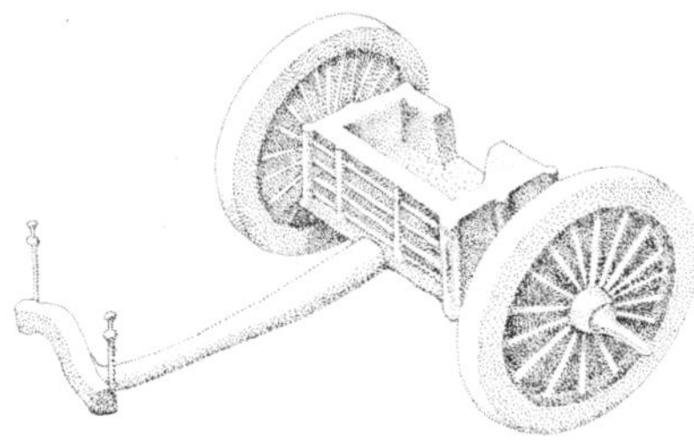

The Shang rulers in China customarily buried chariots, horses, and slaughtered servants, along with their dead kings. This practice preserved a gruesome record of chariot technology.

2550 YEARS AGO
FORGING GREEK CULTURE

Although the classical Greeks remembered Bronze Age heroes of the Trojan wars, they employed newfangled iron for their own tools and weapons. This representation of blacksmiths is from a Greek vase.

Abbreviation: *yr = years before* AD *2000. To convert ages to dates* BC *subtract 2000.*

the alphabet was another salient invention. The arbitrary symbols of earlier scripts had suited the exclusive priesthoods, but were inconvenient for the growing numbers of traders. The alphabet originated among Semitic-speaking tribesmen, perhaps in the Sinai about 3800 years ago; it was widely imitated.

Indo-European charioteers reached the Altai Mountains in the heart of Asia about 3900 years ago, and a century later men of shadowy identity were in China, using chariots to conquer a small parcel of hoe gardeners who grew millet in the valley of the Yellow River. The invaders set themselves up as kings and lords, but adopted the local culture and language. The core area of the Shang Chinese was barely a hundred kilometers wide, but mobilized by the charioteers, they began to expand along the Yellow River about 3650 years ago. Their queries to the gods took the form of scratches on the shoulder blades of cattle (oracle bones) and these gradually evolved into Chinese writing, loosely dated at 3500 years ago.

Other chariots drove to India. The Harrapan city-states in the Indus Valley and Rajasthan had, by 4000 years ago, suffered a climatic disaster; in areas where trees and crops grew by rainfall, the monsoon broke its promise and the culture crumbled. By 3800 years ago, charioteers were harassing the northwest frontier of the Indian microcontinent, and within about two centuries they had occupied the Indus Valley. Descendants of these charioteers retained their Indo-European speech, which is why the mother tongues of 400 million people in Pakistan, India, Bangladesh, and Sri Lanka are related to European languages.

In Middle America, the aristocratic priesthood of the Olmec cult arose about 3250 years ago, in the earliest known break from egalitarian habits in the New World; its hallmarks included jaguar worship and giant "football player" stone heads. Successors of the Olmecs, the Maya, began clearing wide areas of forest for intensive farming, about 2750 years ago. Networks of ditches, serving for drainage, irrigation, and transport, helped gradually to quadruple the Maya population in eight hundred years and then quadruple it again. Other Amerindians in the Valley of Mexico founded their first ceremonial city, Cuicuilo, about 2600 years ago, and the Zapotec script of 2500 years ago made hieroglyphic writing fashionable among priestly elites of Middle America. People living in the complex topography of the Andes of South America began extracting copper from its ores 3100 years ago.

chariots to China and India; alphabet 3800 yr ▼

Chinese writing 3500 yr ▼

sharp cooling rise of the Olmecs 325(

4000 yr | 3500

When the climate cooled sharply again 3250 years ago, severe drought in the eastern Mediterranean seems to have accelerated the collapse of two urban cultures—those of the Mycenaeans in Greece and the Hittites in Anatolia. The exodus of Jews from Egypt at that time marked an enfeeblement of the Nile. With many refugees on the move, and with robbers and pirates cutting the trade links, the metalsmiths were starved of tin for producing bronze. Urgent research and development made a well-known but difficult metal, iron, available for practical use, and blacksmiths on the island of Cyprus produced the first hard steel for knifemaking 3100 years ago. Within half a century iron and steel surpassed bronze in importance around the eastern Mediterranean.

By 2800 years ago, the blacksmiths were on the move against their neighbors. Greeks, Phoenician-Carthaginians, and Etruscans were contending with war galleys and iron weapons for the control of Mediterranean trade. From a base on the Rhine, Celtic men of iron overran much of Europe; from Iran, the Persians tried taking over the world. Knowledge of ironmaking spread to tropical western Africa by 2750 years ago, where Bantu-speaking blacksmiths made high-quality iron in furnaces with tall stacks. Iron users from western Africa turned up far to the east, at Lake Victoria, 2500 years ago, as they began to zigzag and backtrack across the continent for two thousand years, making the Bantu language group the most widespread in Africa.

The Iron Age brought religious ferment to Asia, with Taoism, Confucianism, Jainism, Buddhism, Zoroastrianism, and Judaism all being articulated 2550 to 2500 years ago. Among the Iron-Age voices that ring most modernly are those of social protest, navigation, and science. A Jewish prophet rebuked:

> You cows of Bashan who live on the hill of Samaria, you who oppress the poor and crush the destitute, who say to your Lords, "Bring us drink" . . . [*Book of Amos*, 2750 years ago]

The poem *The Odyssey* was a device to help Greek seafarers remember exact sailing directions for the Mediterranean. And a Greek olive-oil merchant living in Miletus 2600 years ago suggested that the world was made of water; with that guess, not a bad one, he launched the enterprises of physics and philosophy, and put the species to its biggest gamble: the bet that organized knowledge would do more good than harm.

Slaves dug the silver from Athenian mines to pay for the Greek

2500 YEARS AGO

MAGIC FOR BLACKSMITHS

Terracotta sculptures found at early ironworks of the Nok culture in Nigeria may have been religious offerings, made for the sake of quality control in the metal produced.

Babylon and Shang. *Archeology shades into history: the Old Babylonian empire arose 4030 and the Middle Empire 3170 years ago; in China the Shang period began 3800 and the Chou period 3100 years ago. In southwestern Asia the Late Bronze Age dates from 3740 and the nominal Iron Age from 3200 years ago.*

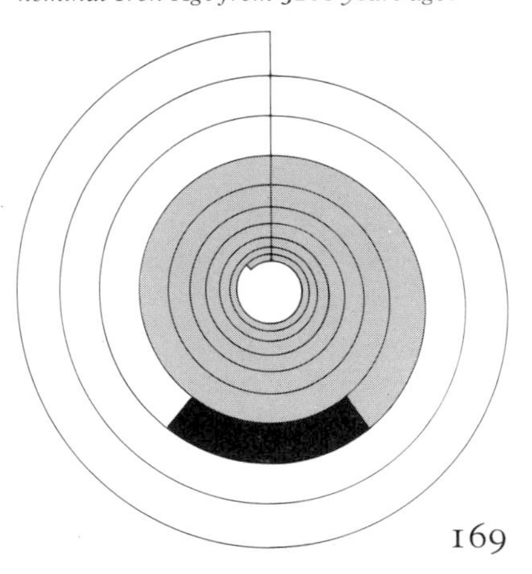

Amerindian copper smelting 3100 yr ▼	iron important ▼ 3050 yr	breakouts of the blacksmiths 2800 yr ▼	rise of the Maya ▼ 2750 yr	Greek science 2600 ▼	religious surge ▼ 2550 yr

3000 — 2500

2000 YEARS AGO
GREAT WALL, SMALL WALL

The Great Wall of China (upper photograph) was made by linking lesser walls 2214 years ago into a military barrier two thousand kilometers long. Although often refurbished, it failed to prevent repeated conquests by mobile herdsmen from the steppes. At the other end of Eurasia, Hadrian's Wall (lower photograph) was built across northern England 1864 years ago (AD 136), to protect a remote Roman colony from troublesome Celts. It was a hundred kilometers long.

ships that smashed the Persian fleet 2480 years ago, and because of its navy, Athens was one of the few prosperous places in Eurasia where soldiers did not rule as if by right. Self-confident and democratic Athenians then lit a beacon of excellence in drama, satire, the visual arts, and reasoning, that two millennia of unreason failed to extinguish. When the democratic experiment ended a century and a half later, conquerors and emigrants scattered Greek arts and sciences through southwestern Asia, across southern Europe, and to Egypt, where geometry flourished 2300 years ago.

Originality passed to the opposite end of Eurasia. Inhabitants of China had cast-iron plowshares at least 2400 years ago, eighteen centuries before Europeans attained the high furnace temperatures needed for casting iron. Ores in China had the advantage of a high phosphorus content, which lowered the melting point of the iron. By 2300 years ago, iron-tax regulations specified that every woman needed a needle and knife, every farmer a hoe and plowshare, and every cartwright an ax, saw, awl, and chisel. Iron tools helped to construct canals in China, for irrigation and transport.

Among Iron Age warriors, two decisive battles occurred 2260 years ago. The people of the Ch'in kingdom, which gave its name to China, beat their arch foes from Ch'i, near the mouth of the Yellow River. Latin-speaking farmers from Rome contrived to defeat old rivals from Carthage at sea, off the coveted island of Sicily. These coincident events launched two empires at opposite ends of Eurasia. The Ch'in snatched control of the rice-growing valley of the Yangtze from its Indochinese inhabitants. The emperor then used forced labor on an unprecedented scale to complete the Great Wall of China 2214 years ago; tracing the natural boundary between northern farming lands and the steppe, the wall was meant to keep the Hun cowboys out and the peasants in. The Han regime took power in 2202, and under its more durable administration the empire's industry boomed. A single iron master in Szechuan province employed a thousand men, while a thriving textile industry supplied hemp garments for the peasants and silk for the well-to-do.

The Chinese also launched an export business in silks to western Eurasia, and the first through caravan arrived in Persia from China via the transasian Silk Road, 2106 years ago. By the birth of Christ (strictly 2007 years ago, or 7 BC) the Chinese superpower encompassed more than 70 million people in what they called Middle Earth. Around another "Middle Earth," the Mediterranean Sea,

Athenian heyday 2480 yr ▼

cast iron in China ▼ 2400 yr

breakouts of the Chinese and Romans ▼ 2260 yr

water energy 2030

2500 yr | 2400 | 2300 | 2200 | 2100

the Romans had imitated Etruscan technology and alphabet, Greek seamanship and arts, and Persian crucifixions to put together an empire only a little less populous than the Chinese, although industrially outclassed by it. The empires shared the first bright idea about energy since draft animals, namely, that running water could turn a wheel. Various sorts of waterwheels appeared, but their origins are obscure. On the other hand, a thousand experts attended the first scientific conference, convened by the Chinese court 1996 years ago (AD 4). Paper was plainly a Chinese invention of 1900 years ago (AD 100).

Adventurous Polynesians occupied the islands of the broad Pacific east from Fiji. Again and again they set off in sailing canoes with fishing gear, stone tools, pigs, and coconuts; steering by the stars, they made landfalls on far-flung islands—first to the Marquesas in mid-Pacific by 1900 years ago (AD 100) and radiating thence to Hawaii in the north, Easter Island in the extreme east, and New Zealand far to the southwest. Whatever dangers they faced in the voids of the ocean, the Polynesians did well to remove themselves far from the disease market of Eurasia.

An epidemic 1838 years ago (AD 162) killed a third of the Chinese army on the northwest frontier; three years later an unfamiliar and deadly disease swept the Roman empire. In both regions the disease was a novel virus, possibly measles, to which the populations had no resistance. It traveled with the textiles by the Silk Road and other trading routes. Epidemic after epidemic ravaged both superpowers. The Chinese empire broke up in AD 220, and seventy years later the Roman empire was divided, after suffering another unaccustomed disease, perhaps smallpox. A chilling of the climate around 1730 years ago (AD 270) reduced the rainfall in the Roman heartland of Italy and also in Arabia and central Asia.

Out of the grasslands of central Asia galloped new raiders, the Huns, driven by drought and lured by a chance to profit from imperial decay. The original Huns spoke the language of Paleosiberian fishermen, but they made a confederacy with other nomads of the steppes who spoke Altaic and Indo-European languages. Horses had improved greatly and the newly invented stirrup made the mounted archers deadly. The Huns invaded China 1696 years ago (AD 304) and took the capital Lo-Yang seven years later. Heading west, they rode into eastern Europe 1630 years ago (AD 370). Retreating Germanic tribesmen, Goths and Vandals, were driven,

1835 YEARS AGO

PLAGUED EMPEROR

Marcus Aurelius was ruler of the Roman empire when the "Antonine plague" struck in AD 165.

Classical empires. *Iran: Achaemenid 2550 years ago, Alexander 2330, Parthian 2141, Sasanomian 1776 (AD 224). India: Mauryan 2332 years ago, Gupta AD 320. China: Ch'in 2221 years ago, Han 2202 (fell AD 221). Rome: first emperor 2207 years ago. Middle America: Classic Maya AD 260.*

Abbreviation: *yr = years before AD 2000. To convert ages greater than 2000 yr to dates BC, subtract 2000.*

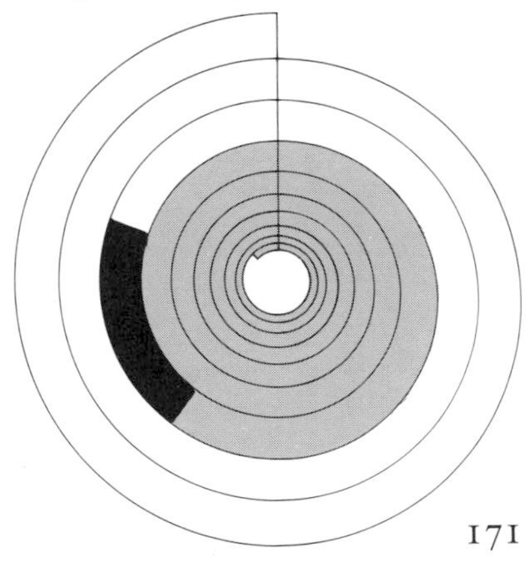

birth of Christ ▼ 2007 yr (7 BC)

paper 1900 yr ▼ (AD 100)

Eurasian epidemics ▼ 1838 yr (AD 162)

breakout of the Huns ▼ 1696 yr (AD 304)

2000 1900 1800 1700 1600

AD 630

HOLY METEORITE

The prophet Mohammed took possession of a peculiar black stone in Mecca and made it the focus of the Islamic religion. The stone is believed to be a carbonaceous meteorite.

AD 770

PRINTED CHARM

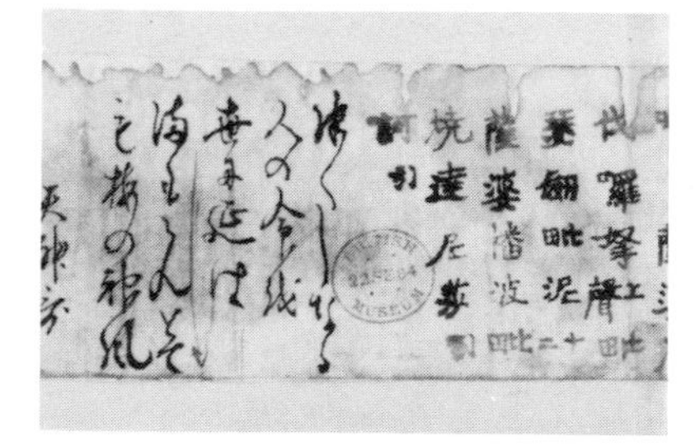

The oldest surviving example of printed words comes from Japan, although the technique of printing evidently originated in China. The charms were long streamers, printed by order of the empress, to be tied to pagodas.

as if by a syringe, into the bloodstream of the Roman empire; they gave vandalism a bad name.

While the Chinese kept returning to empire, the Europeans preferred a quarrelsome plurality of languages and cultures. A device that aided decentralization appeared among the Germans around AD 500: the moldboard plow, which turned over the soil and greatly enlarged the areas that could be cultivated in northern Europe. Germanic Anglo-Saxon tribes expelled the Celts from England and there created a land of scattered villages, where farmers and craftsmen retained unusual control over their own affairs. Germanic Franks in France adopted, around AD 720, the lance-carrying, stirrup-bound knight in armor. This heavy cavalry kept Europe more or less secure by sheer momentum in battle, but it burdened the small continent with bullies on horseback. The knights could not keep diseases at bay, when bubonic plague swept the warmer countries of Eurasia, from AD 542 on.

Successive Chinese emperors forced workers in their millions to build the Grand Canal, the main sections of which were complete in AD 611. Linking the valleys of the Yellow and Yangtze rivers with regions to both north and south, the canals gave the Chinese an advanced transport system. By AD 700 the Chinese were printing words on paper by laborious methods. Soon afterward they made the first mechanical clock with an escapement; besides being an astronomical timekeeper, it was a demonstration of what intricate machinery could accomplish. But as the Chinese reasserted their ingenuity and power, a new force in the world checked them.

A meteorite seeded the breakout of the Arabs, at the startling birth of Islam. The curious black stone had fallen at Mecca in Arabia, and with nomads and traders making pilgrimages to see it, Mecca became a safe and favored marketplace—a rarity fourteen centuries ago in Arabia, where ancient bridges and dams ran across the beds of dried-up rivers and nomadic warriors robbed the trading caravans. A merchant-prophet seized Mecca and adopted the shrine of the black stone as the Islamic holy of holies.

The new religion reconciled the nomads and traders, and directed their military energy to the conquest of the world, starting in AD 632. Camels and zeal carried the Arabs across the deserts like sandstorms, and they fell upon Syria, Palestine, and Egypt. By AD 714 the Arab empire stretched from Spain, via northern Africa and Iran, to northwestern India, and many non-Arabs had adhered to

moldboard plow; abacus AD 500 ▼

bubonic plague AD 542 ▼

Maya heyday AD 600 ▼

breakout of the Arabs ▼ AD 632

1600 yr | 1500 | 1400 | 13

Islam. While the Muslims expanded three thousand kilometers east from Mecca, the Chinese extended their empire three thousand kilometers west from their core domain, and armies of the two superpowers collided in the heart of Asia, at the Talas River. In a battle there in AD 751 the Muslims won decisively. From their Chinese prisoners, Arabs found out how to make paper.

Arab traders arriving in western Africa around AD 800, in search of gold, slaves, and ivory, discovered at Kumbi Saleh the city state called Ghana, which had evidently emerged some centuries earlier. Other complex societies thrived in the tropics at that time. Hindu-Buddhist temples at Prambanan in Indonesia, and Angkor in Indo-china, testified to the influence of Indian traders. In the Valley of Mexico, Teotihuacán was a city possessing a vast market-place, many workshops, and large pyramids. Nearby, the Maya canal systems covered large areas of Middle America, and the population had soared to an estimated 14 million, making the Maya one of the more numerous peoples of the period. Tikal, with its pyramids, had been a major ceremonial center for the elite, since AD 260, but in the heyday beginning in AD 600 there were several prominent cities. By AD 800 the Maya temples and some of the canal networks were being abandoned. Nor were the Eurasian superpowers secure. In AD 800, two cities stood like poles of a magnet, Baghdad of the Islamic Abbasid rulers and Ch'ang-an of the Chinese T'ang, both glittering with power and fine arts. Within little more than a hundred years, both empires were in fragments.

Arithmetic was more durable than empires, and the abacus was invented in Eurasia around AD 500. Modern numerals are called "arabic" because of Islam's role in their dissemination. In fact, the system of numbering traces back to India, and it was completed in southeastern Asia, where a crucial missing element, the symbol for zero, first appeared in AD 683 in Kampuchea and Sumatra. The Maya, who counted in twenties (toes as well as fingers), also invented zero, but their notation was less convenient.

Germanic sea raiders, the Vikings, burst upon their European neighbors, starting with England in AD 793. One venture took them deep into eastern Europe, where they rallied the Slavs and founded the Russian state at Kiev in the Ukraine in AD 882. Another was into France, where, as "Normans," they became a power in the Mediterranean as well as Europe. Two centuries of exceptional northern warmth from AD 930 helped the Vikings to take Iceland from the

AD 800

THE VIKINGS' BREAKOUT

As raiders and traders, fierce seafaring peoples from Scandinavia made all of northern Europe their stamping ground. This is a stem post from a boat found in Belgium.

Early Middle Ages. *The western sector of the Roman empire fell in AD 476. The Frankish empire arose in Europe AD 486, and the Byzantine empire (eastern Mediterranean) dates nominally from AD 610. In Middle America, the Late Classic apogee of the Maya ran from about AD 600 to 800. The Abbasid caliphate dominated Islam from AD 750. The Sui, T'ang, and Sung empires of China date from AD 581, 618, and 960 respectively.*

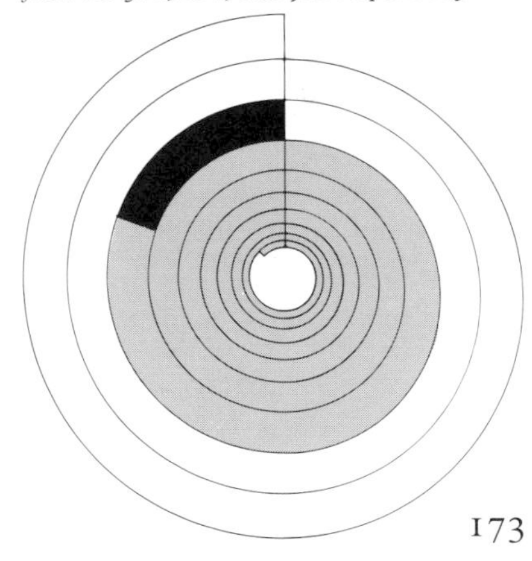

printing
AD 700

warm episode
AD 930 ▼

population surge
AD 1000 ▼

1200 | 1100 | 1000

AD 1100

PRINTED MONEY

The Chinese began using paper bills on a large scale, about 700 years earlier than the Europeans. In this example, the designer took care to depict the equivalence in the iron coinage.

Later Middle Ages. *The Sung empire in China gave way to the Mongol empire, formalized in AD 1260; the post-Mongol Ming empire began in 1368. The Aztecs rose to power in Middle America from 1325, and the Incas in the Andes from about 1400. The Ottoman empire expanded into Europe 1351.*

Irish, settle in Greenland, and explore as far west as Newfoundland in North America. During the period around AD 1000 grain grew in northern Norway and grapes in northern England. High in the mountains of central Europe, abandoned ancient mines were successfully reopened. In New Mexico and adjacent areas generally thought of as arid, crop-growing Amerindians of the Anasazi culture built canyon towns and irrigation works. Rain watered the grasslands of Asia well during the warm centuries, and nomadic horsemen swarmed like locusts.

"Seek learning though it be in China," was an old Islamic saying, and Islam was self-consciously also the custodian of Greek learning. Its treatment of science was often conservative, but mathematics and chemistry flourished, as betokened still by Arabic words like algebra and alchemy. Arabic scholars showed originality in optics, and a textbook written in Cairo around 1038 set out the theory of lenses, on which the invention of the telescope would depend. European science had its origins in Islamic work, and a scholar in Muslim Spain recommended that books should not be sold to Christians, because they plagiarized them.

One of the most fateful of inventions, the navigational compass, appeared in China. The earliest surviving description of a practical compass in a Chinese military manual of 1044 tells of a device in the form of a magnetized iron fish, floating on a bowl of water; before long, compass needles hanging on silken threads were more common. By 1115 the magnetic compass was helping navigators less skilled than the Polynesians to sail safely out of sight of land.

The printers in China progressed in 1064 from carving every item afresh to using movable type, made originally of clay. Administrative reforms starting in 1068 turned the Chinese empire into the first economy managed on modern lines, relying on equitable money taxes rather than forced labor, and on far-reaching interplay between public and private enterprise. Taxes and the supply of paper money were regulated in accordance with sophisticated ideas about avoiding inflation. The economy and the population boomed, in the warm weather, as government loans encouraged farmers to exploit a new variety of rice from Indochina and Chinese seafarers traded widely in southeast Asia.

Around 1200, Middle America was in turmoil, and Aztecs from the north entered the Valley of Mexico. From about 1320 onward they rose to power over their neighbors. It is not clear whether their

population surge; Islamic science
▼ AD 1000

managed economy
AD 1068 ▼

navigational compass
AD 1115 ▼

incursion was related to a downturn in the climate that began about 1190. More definite sufferers of the climatic change were the Anasazi Amerindians, forced by drought to abandon their canyon settlements and concentrate along the Rio Grande. Starting in 1314, repeated famines struck Europe, mountain mines were again abandoned, and by 1342 the Vikings' customary route to Greenland would be blocked by sea ice.

Terrible military events began on the Eurasian steppes. When the previously favorable climate faltered, with rainfall diminishing from 1160 onwards, the numerous horsemen were ready for a warlord to tell them which farmers to attack first. The breakout of the Mongols and their Turkish allies (who outnumbered them) exceeded all previous human explosions from the grasslands in its scope, its generalship, and its success in gutting Eurasia with its arrows. The Chinese wall was breached in 1211. In 1258 Baghdad, amid decaying irrigation works, fell to the Mongols, so ending five thousand years of Mesopotamian prosperity. Iran, Russia, Tibet, and Korea were all conquered, and eventually the whole of China. Only the death of a Mongol king spared western Europe, and a typhoon that smashed the Mongol fleet saved Japan. Later conquests by Turks set up the Ottoman empire of the eastern Mediterranean (1354) and the Mughal empire of India (1525).

A crash in the population of medieval Eurasia, already evident in China by 1290, was made worse by disease conveyed in the trading caravans and supply columns of the Mongol empire. Not for the first time, but now on an unprecedented scale, bubonic plague transferred itself from marmots, the ground squirrels of the Asian steppes, to the rats infesting the cities, ships, and farms of the wider world. Fleas were the unsuspected transmitters. The Black Death first appeared among the Chinese in 1331, killing more than a quarter of them, and in 1346 a Mongol army in southern Russia spread it to Europe. The disease broke the Mongols' power.

When the Chinese revolted against their Mongol overlords in 1356, they used cannons. Gunpowder, a mixture of saltpeter, sulfur, and charcoal, had first been employed five hundred years earlier to scare off demons. Amid the commotion created by the Mongols, experiments in gunmaking began before 1300 in China, India, Islam, and Europe. Guns were a way of ending the dominance of cavalry on the battlefield, and the Chinese expulsion of the Mongols was the first war in which they played an important part. But

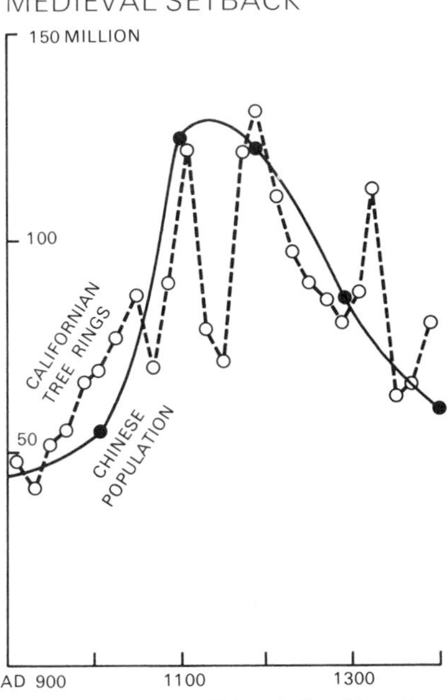

An onset of cool conditions is dated here by a downturn in growing-season temperatures, as revealed in Californian tree rings. The population of China, having soared in previous centuries, crashed thereafter, although any effect of climate on human survival was in this case probably indirect.

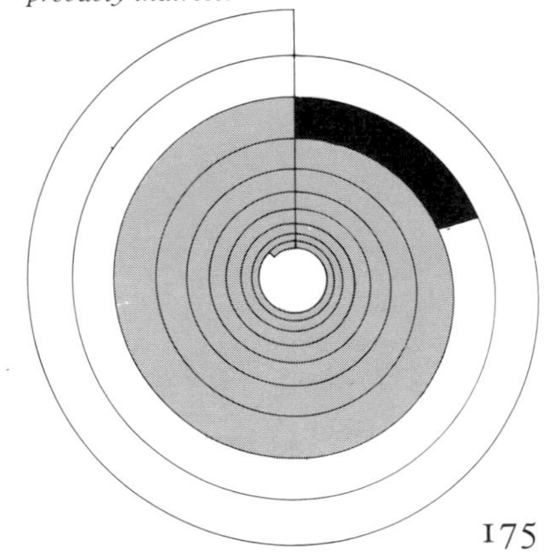

breakout of the Mongols ▼ AD 1211
population crash AD 1290 ▼
rise of the Aztecs AD 1320 ▼
Black Death ▼ AD 1331
gun warfare ▼ AD 1356

800 750 700 650

AD 1500
THE SEAFARERS

Small sailing ships, swarming on the Atlantic shores, provided the barely habitable vehicles in which Europeans conquered the oceans, and thus came to dominate the world. (The scene is Calais, c. 1545.)

early guns were cumbersome, and the ideal vehicle for them was a sailing ship.

Ambitious Chinese saw that the time was ripe for domination of the world by gun-carrying ships. They pulled their high technology together and built dozens of large sailing junks with multiple masts, steered by sternpost rudders, navigated by magnetic compasses, and armed with guns. In AD 1405 a powerful fleet set off to impress the barbarians, and a succession of expeditions overawed half the known world, gathering treasure from as far away as Mecca and Africa. Had that naval policy persisted, this book would be written in Chinese. Officials and accountants persuaded the emperor after less than thirty years to put a stop to it, and eventually they destroyed even the records of the voyages. It was bureaucracy's most breathtaking accomplishment.

The peoples of Islam showed no sign of taking the world by water, so the opportunity went by default to the Europeans—an outcome as surprising as if nowadays the Americans and Russians left space exploration to the Vietnamese. The Europeans had sound agriculture, and the arms race in artillery encouraged a catch-up with Chinese technology: the first European cast iron was made in Flanders in 1380. Every village was using watermills or windmills with increasing ingenuity. Gold from western Africa fueled commercial enterprise, and Italians asserted sea power in the Mediterranean. Mental vigor was manifest in the formulation of the laws of perspective, in Florence around 1415, at the very vanishing point where art, science, mathematics, architecture, and mapmaking all converged. In Germany, printing presses began turning out fine books in 1455. The concept of printing, and the paper too, were Chinese in origin, yet the mechanisms were indigenous, not least in the winepress screw—the Chinese had overlooked the screw.

The products of the presses would include navigational manuals and narratives of voyages, but Bibles had priority, as weapons in religious wars. A European custom was to serve God by burning nonbelievers and heretics at the stake; a similar industry, which nourished the sun by tearing out human hearts, was thriving in the Americas, where the warrior kings of the Aztecs in Mexico and the Incas in Peru were unaware of what Europeans had in store for them. Such practices had little effect on global population numbers, which recovered in the mid-fifteenth century.

The essence of the Chinese maritime technology of ship handling,

Iberian empires. *The Middle Ages nominally ended with the fall of the Byzantine empire AD 1453. Tsarist Russia began 1478. The Portuguese overseas empire started in El Mina in 1482, and the Spanish in Hispaniola in 1493. The Mughal empire in India dates from 1526.*

See map p. 92.

cast iron in Europe AD 1380 ▼
Chinese naval ventures AD 1405 ▼
population recovery AD 1450 ▼
printing press ▼ AD 1455

navigation, and gunnery was known in Europe. The Portuguese flotillas that began groping along the African coast were ludicrously small and ill-found, but as events showed, the world could be snatched by diminutive carracks, without grand fleets of the Chinese sort. Like their horsemen ancestors coming off the steppes, the Europeans made up in daring, avarice, and mutual rivalry what they lacked in imperial wealth and sophistication. The breakout of the European navigators can best be dated from 1492, when a westbound Spanish flotilla stumbled upon the Americas, mistaking them for Asia. Portuguese seamen heading the other way reached India by sea in 1498 and China in 1514. The first circumnavigation of the planet was completed by a Spanish ship, *Vittoria*, in 1522.

The Aztecs and Incas succumbed to Spanish adventurers whose secret weapons were viruses. The troops carried smallpox, measles, and other diseases unknown in the Americas, and from 1520 onward epidemics swept through the empires like forest fires. The invaders, with immunity acquired in childhood, rode about unaffected and that heightened the shock. Thus did some hundreds of Europeans overwhelm millions of warlike people, not only in the two well-known empires, but in many other places including the recently rediscovered Tairona cities of northern Colombia. By coincidence, when the sun was no longer gratified by Amerindian sacrifices, further cooling ushered in the Little Ice Age around AD 1530.

To the invaders the botany of the Americas was unfamiliar, and plants long nurtured as crops by the Amerindians—maize, potato, sweet potato, manioc, chocolate, sunflower, superior cotton, and so on—were taken for growing in the Old World. Crops went the other way, too, and the great worldwide transplantations can be dated from 1560, which was roughly when the Portuguese took sugar cane to Brazil. They shipped West African slaves there, too, to do the work of growing and extracting the sugar. The slaves-and-sugar trade began the process whereby 10 million Africans were kidnapped to convert large tracts of the Americas to agriculture of European conception.

The Russians threw off the Mongols in 1480, enlarged their European territory, and then crossed the Ural Mountains in 1586, at the start of an expansion across Asia. They subdued the peoples of the forests and the steppes, and reached the Pacific Ocean in 1649. During this period, the Russian peasants were reduced to serfdom, just when seafaring powers in northwestern Europe were

AD 1533
DEATH OF THE INCA

Spanish intruders captured the monarch of Peru by skulduggery, and condemned him to death after a mock trial. Cruel treatment was not, though, as deadly to the Amerindians as unfamiliar disease.

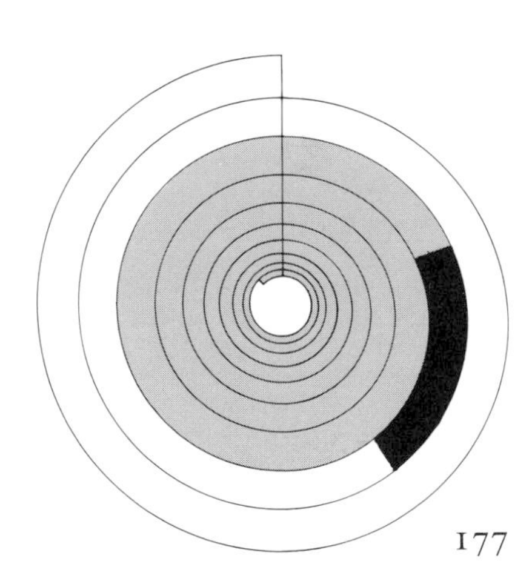

breakout of the navigators ▼ AD 1492

Amerindian epidemics ▼ AD 1520

slaves-and-sugar trade AD 1560 ▼

breakout of the Russians ▼ AD 1586

500 450 400

AD 1650
DENTAL DECAY

Amsterdam became the sugar capital of Europe, and Dutch paintings of the period document, as a price of empire, a catastrophic loss of teeth. (Painting by Jan Steen.)

Northerners' empires. *The French overseas empire dates from AD 1600 (Canada), the English from 1607 (Virginia), and the Dutch from 1619 (Batavia). Louis XIV reigned in France 1643 to 1715. The Manchu (Ch'ing) dynasty took charge in China in 1644. The Swedish Baltic empire was at its zenith in 1658. Unification of Britain occurred 1707.*

mobilizing the human resources of free peoples. A northern European challenge to the Spanish and Portuguese trading empires took shape in the East India Company, founded in London in AD 1600. The Dutch and French followed suit, while other companies in those countries financed the creation of trading settlements in the Americas. The Dutch made deep inroads into the Portuguese empire, securing by force a monopoly of the oriental spice trade, and penetrating Brazil.

Two ideas sprang up that transformed human existence: nationalism and experimental science. Both are so pervasive nowadays that it is hard to imagine a world without nation-states or scientific laboratories. Yet in 1600 anyone could work or fight for an alien regime with no sense of treachery, and traders and scholars moved about freely even in time of war. Equally it was a world inhabited by spirits and demons, where universities taught astrology and imperial courts practiced it, and where if you wanted to understand a natural phenomenon you went, not to a laboratory, but to a library, to see what the Greek or the Arab masters thought about it. Clear expressions, both of the new religion of nationalism and of the social system of modern science, came from the English.

Antecedents for nationalism and the nation-state are to be found in ancient racism, in ethnocentric religions, and in centralized institutions of China, Spain, and France. But by about 1640 nationalism was becoming a matter of public enthusiasm, and an English poet could remark unambiguously: "Let not England forget her precedence of teaching nations how to live" (John Milton 1643). Such a belief was heady stuff for sea captains and traders confronting Polynesian gardeners or Chinese tycoons. Nationalism needed time to take root, even in England, but eventually it did for English what Islam did for Arabic, making it a world language. Liberal ideas in religion and politics were a counterpart to nationalism; developing gradually in the new Europe, liberalism was incorporated in formal philosophy by 1690, and it became part of the explosive that caused European empires to self-destruct.

Science sat more comfortably with pragmatic nationalism and liberalism than with older beliefs. When people engaged in experiments and observations beyond the control of the established order, they invited trouble. The chief progenitor of modern science, in Italy in 1609, improved the telescope until it was worthy of being pointed at the sky. It showed that Jupiter had moons, orbiting

imperial companies ▼ AD 1600
astronomical telescope ▼ AD 1609
Dutch spice monopoly ▼ AD 1621
imperial nationalism AD 1640 ▼
Manchu China ▼ AD 1644
science formalized AD 1662 ▼

around that planet just as, some said, the Earth went around the sun. The astronomer was silenced, but experimentalism was irrepressible, and other discoveries, including the circulation of the blood, the pressure of the air, and the microscope, followed swiftly. The charter of the Royal Society of London in 1662 gave official blessing to the childlike curiosity of the experimental philosophers, and renewed the Greek gamble on rationality: there was nothing human beings should not know. European governments created observatories and botanical gardens, so that navies and planters of the new empires should benefit. One effect of science since then has been to preserve the advantages of the northern nations.

A monumental achievement came swiftly. By 1687 a gravitational theory explained all the motions of the solar system, with the proposition that the laws of nature were the same in the sky as on the Earth; the very same force that steered the moon made the apple fall. The author gave a glimpse of future possibilities when he explained how to put an artificial satellite into orbit around the Earth. An intellectual flowering unseen since Athens in its heyday brought every aspect of nature under scrutiny in Europe, in projects both brilliant and ludicrous. Experiments with air pressure and vacuum were a distinctive European innovation which led to modern meteorology, and also to the steam pump, introduced in England in 1696 and improved in 1712—the antecedent of steam engines. Later dividends of air-pressure science included ballooning in France and, after two centuries, electric lighting and electronics.

The Chinese had been taken over yet again by alien invaders—this time by Manchus from the north, in 1644—but when peace ensued, a great enlargement of the Chinese population occurred, more than doubling it between 1700 and 1800. Epidemics in Eurasia were fewer and less grave, and the Chinese had improved their food supplies by growing maize and sweet potatoes transplanted from the Americas. Their surge in numbers marked the start of the rapid growth in global populations that has continued ever since.

As poachers turned gamekeepers, the Manchu emperors were resolved that never again should the thunder of hooves on the steppes threaten to simplify the complex societies of Eurasia. In 1727 they divided central Asia between themselves and the Russians. But there was more to the fall of the middle Asians than that: the steppes experienced frequent droughts in the Little Ice Age, and oceanic trade routes diminished the value of the central Asian

AD 1700

SURGING POPULATIONS

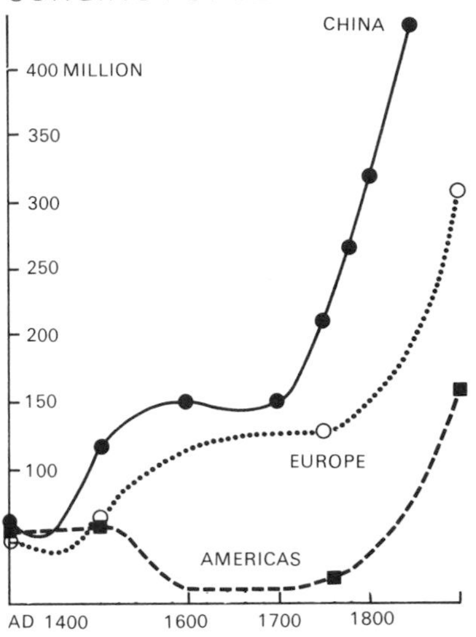

After Eurasian fluctuations and the Amerindian crash, the Chinese led the explosion in human numbers that became a feature of the modern world.

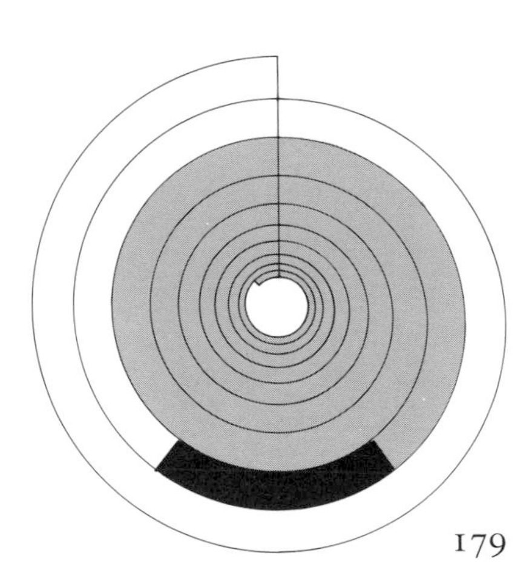

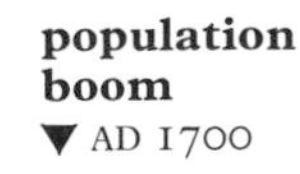

liberalism ▼ AD 1690
population boom ▼ AD 1700
steam energy ▼ AD 1712
carve-up of Asia ▼ AD 1727

300 250

AD 1788
NATIVE AUSTRALIAN

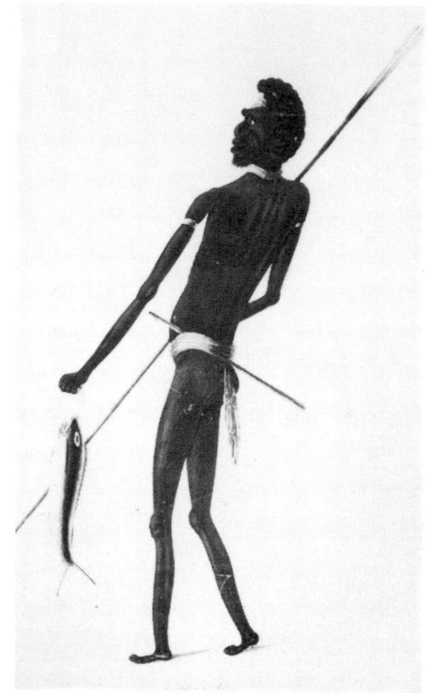

The aboriginal inhabitants of the coasts and dry interior of Australia spoke more than 260 different languages, and their hunter-gatherer traditions, rich in skill and ritual, traced back 40,000 years. The settlement of their best land by intruders, from 1788 onward, was a catastrophe. Although nominally British citizens, they were killed, maltreated, or impoverished.

roads. India was not at first included in the carve-up, but by AD 1765 the British gained control of Bengal, their doorway to the subcontinent.

Scientific chemistry escaped from a quagmire of alchemy, false theory, and empirical industrial practices, when a Swedish chemist discovered oxygen in 1771. Its biological role became evident very quickly, and that helped in the development of scientific farming. Solving the algebra of chemical reactions led eventually to a wealth of dyes, drugs, explosives, and new materials, but bleaches for textiles were the early prize.

The new nationalism allowed you to conjure a nation into existence by uttering its name. American settlers showed the way by revolting against the British and, in 1776, declaring the independence of the United States of America. The British consoled themselves by colonizing Australia in 1788. It was the last continent held exclusively by scattered groups of hunter-gatherers and, like the Amerindians, the Australian and Tasmanian aborigines were vulnerable to diseases imported from Europe, as well as to more deliberate pacification. Even the remote Pacific islands of the Polynesian gardeners became hotbeds of shipborne disease.

The French revolution of 1789 sent shock waves through the courts and clergy of Europe, but nationalism was loud from the outset, and the democratic revolution ended in a military dictatorship dedicated to conquest. Colonists of South America, consciously imitating the Americans and the French, liberated themselves from Spanish and Portuguese rule in the early 1800s, creating a patchwork of nation-states. They were aided by the British, who were keen to sell them cotton cloth.

With ample labor and food to feed it, the British industrialized; by 1800 the agricultural workforce was less than 40 percent of the total. Their cotton industry showed what mechanization and imperial sea power could accomplish in combination. Off the stormy tip of Africa ships carrying raw cotton from India to Britain hailed ships taking finished cotton cloth from Britain to India, while the Indians were disindustrialized. Having taken off during the 1780s, British exports of cotton goods doubled every twelve years until, in the year 1850, they amounted to enough cloth to make a shirt or a skirt for everyone on the planet.

In the global timescale, the industrialization of Europe was at first a matter of catching up with the Chinese in such technologies

British Bengal AD 1765 ▼
oxygen discovered AD 1771 ▼
revolutionary nationalism ▼ AD 1776
French revolution AD 1789 ▼

250 yr | 240 | 230 | 220 | 210

as canals, ceramics, and mechanized spinning and weaving; not until 1856 could Europeans match ancient Chinese methods of making steel cheaply. In a European perspective, mechanization in the textile industry, involving a widespread use of waterwheels, traced back at least five centuries. The steam engine, despite great technical improvements, remained less important than the waterwheel in 1800. The seed of a mechanical revolution was more truly sown in 1803, when compact high-pressure steam engines ran on wheels, along streets in London and rails in Wales.

The introduction of cowpox vaccine to prevent smallpox in 1796 followed a doctor's observation that milkmaids were immune to the disease. The use of smallpox itself as a vaccine was an older art, originating in Asia; with the new vaccine, one of the killer diseases could be more safely contained. But a globe unified by navigators was highly vulnerable to diseases, and in 1817 cholera burst out of Bengal in the first of successive epidemics that spread to much of the world in the ensuing decades. Alarm about cholera prompted demands that cities, sewers, and water supplies be cleaned up. Even while they grew bigger, the cities of Europe and North America grew healthier.

The ascendancy of Europeans at home and abroad, initiated by navigation and supported by science, was confirmed by coal. For more than two centuries, the British and others had replaced dwindling supplies of firewood from living forests by coal from fossil forests 300 million years old. A rapid growth in the use of coal began early in the nineteenth century with coal feeding steam engines and steam engines helping to win coal. Railroads carried coal, needed coal, and also needed iron, which in turn required more coal for its smelting. The puffing smokestack of the first passenger steam train, in Britain in 1825, announced the fossil-energy revolution.

The steam train was the first acceleration in land transport since the horse, and within a hundred years of its invention a million kilometers of railroad would be crisscrossing the continents. It carried the fossil-energy revolution far and wide, and Americans and Germans in particular intensified their mining in the same carboniferous seams of old Euramerica. From their core region east of the Mississippi River, the English-speaking Americans broke out across the North American continent, especially from 1830 onward, into territories of the Amerindians and Spanish-speaking

AD 1809

THINGS TO COME

Decades of experiments and demonstrations, like this one at Euston Square, in London, preceded the worldwide railroad boom.

Age of revolutions. *The Industrial Revolution in Britain is loosely dated from AD 1760. This was also the Age of Enlightenment (notably in France). Political events included the American revolution 1776; the Napoleonic empire 1804 to 1814; and Middle and South American revolutions 1808 to 1826.*

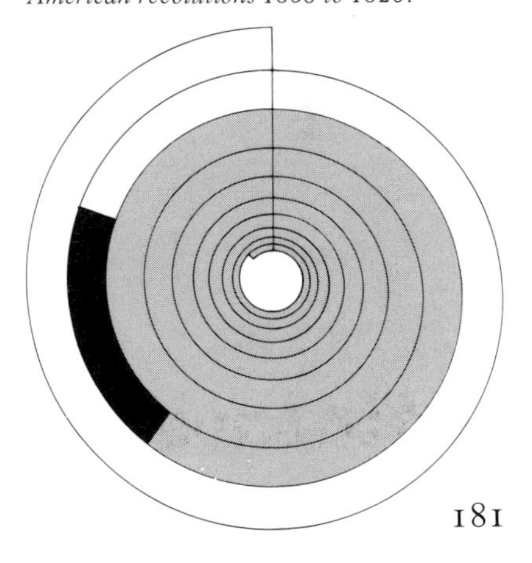

postagricultural British ▼ AD 1800

fossil-energy revolution AD 1825 ▼

breakout of the Americans ▼ AD 1830

200 190 180 170 160

AD 1869

RAILS ACROSS AMERICA

Transcontinental railroads, of which the first is advertised here, took settlers far and wide across North America. In all habitable continents, railroads knifed through terrain that had kept people and cultures apart.

settlers. The first rail link to the Pacific, completed in AD 1869, advertised the conquest of the continent.

The British meanwhile demonstrated their transient mastery of the world by pushing narcotics. Europeans relished tea from China, but the Chinese still had little need for industrial imports, and all that could be offered to them in exchange for tea was opium from India. When the Chinese rulers tried to ban this addictive drug, the British went to war, and by 1842 they had intimidated the Chinese, forcing them to continue taking opium, and to yield Hong Kong. In the mid-1840s the Japanese too were harassed by a succession of naval expeditions, European and American. With their traditional isolation forcibly ended, the Japanese rode the new industrial wave, while their population surged. The Japanese revolution of 1868, the Meiji restoration, signaled a modernization of the social order.

Ships rotted in San Francisco harbor, abandoned by their crews who joined the California gold rush of 1849. Discoveries of gold became a demographic force, luring hopeful people to far-flung places—California, Australia, South America, and Alaska—and accelerating the spread of speakers of European languages across newly acquired territories. Oil prospectors traveled to other unlikely places when, from 1857 onward, North Americans and eastern Europeans industrialized the production, refining, and distribution of petroleum, as the chief alternative to coal in the fossil-energy revolution. Oil encouraged inventors of cars to switch from steam to gasoline, and in 1885 the first automobile powered by an internal-combustion engine heralded another revolution in transport, in which the Amazonian rubber trees played their part. The introduction of pneumatic tires in 1888 perfected the bicycle, one of the most efficient and truly liberating machines in all history.

The Europeans abroad and their ex-colonial cousins, reinforced by 50 million emigrants from Europe, overwhelmed the world. Canneries and refrigeration made possible the shipping of meat across continents and oceans. Ingenious machinery for farmers, and cheap steel plows strong enough to tackle virgin grasslands, enabled American harvests of corn and wheat to increase sevenfold between 1841 and 1900, while the Texan cowboys in 1871, their peak year, drove 600,000 head of cattle two thousand kilometers to market. In benign weather after the Little Ice Age ended around 1850, pioneer agriculture boomed also in South America, South Africa, Australia, New Zealand, and the Russian empire.

Chinese cowed AD 1842 ▼

California gold rush AD 1849 ▼

germ theory of disease AD 1863 ▼

Marxism AD 1867 ▼

160 yr | 150 | 140

Marxism, especially from 1867 onward, commented on how the modes of production interacted with social power. Biologists produced the theories of evolution by natural selection (1858) and heredity by genes (1865). The comprehensive theory of electromagnetism (1864) pointed the way to radio and other marvels, and ranks among the greatest of intellectual achievements. Especially consequential were the French discoveries of microbes causing disease (1863) and of radioactivity (1896). With the identification of "germs," efforts in hygiene and public health redoubled, and disease lost its grip on population growth in the industrialized countries. There were swift geopolitical consequences too. Disease had guarded much of tropical Africa from intrusion, by killing 60 percent of Europeans within a year of their arrival. By 1880 the mortality from tropical diseases had fallen enough to let rival Europeans scramble for colonies in the interior of Africa.

In 1878, at a time when the Americans were surpassing the British in industrial output, the telephone came into practical use, standing at the intersection of two major lines of technology. One was electricity as a novel form of energy: it became useful for general purposes by 1881 and led by 1888 to the commercial production of aluminum, the first important new constructional metal in three thousand years. But the telephone also represented the bustle in information technology, which already included high-speed printing, photography, and transatlantic telegraphs, and was soon extended by the phonograph and motion pictures. Strangely analogous to the movie projector was the machine gun. In 1898 a British army in the Sudan, equipped with machine guns, killed ten thousand fierce warriors who wanted to drive them out of Africa, with a loss of fewer than fifty men.

Why did pale-skinned Europeans become so powerful? All sagas of nations and battles, of explorations and inventions, and even of disease as a factor in conquest, conceal a nutritional key. Whites feed well because they conquered the world, but perhaps they conquered the world because they were well fed. Milk is five times more efficient than meat in converting grass into food energy for humans. The assurance against protein malnutrition conferred by the milk-drinking mutation traces back more than six thousand years to ancestors of speakers of Semitic, Indo-European, and Uralic languages, and to the livestock revolution of cow and plow.

The first sustained challenge since the Mongols was made by

AD 1880

MECHANIZED SLAUGHTER

Machine guns were early products of a technological arms race between the industrialized nations that began in the 1870s and has continued without remission ever since.

Victorian age. *Queen Victoria reigned AD 1837 to 1901 at the apogee of the British empire. Italy was unified in 1861 and Germany in 1871.*

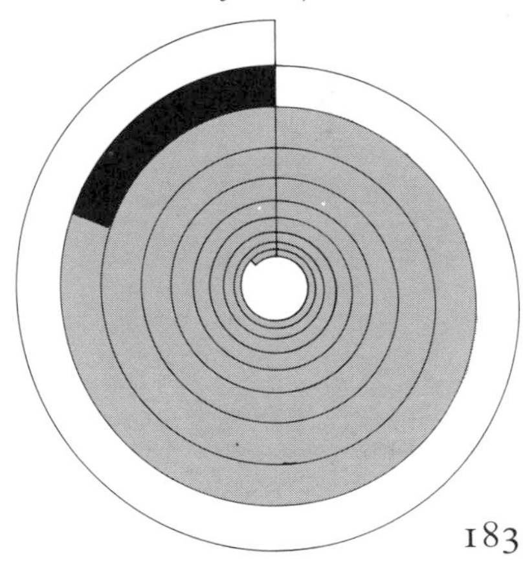

telephone AD 1878 ▼

carve-up of Africa ▼ AD 1880

automobile ▼ AD 1885

radioactivity discovered ▼ AD 1896

120 110 100

AD 1905

TSUSHIMA STRAIT

Out of a fleet of forty Russian warships sent to chastise them, the Japanese sank or captured all but three small vessels, during a two-day battle in the Tsushima Strait between Japan and Korea. Europeans could no longer regard themselves as undisputed masters of the Earth.

people whose protein came from the great fisheries of the Kuroshio Current. When modern ironclad warships met in battle in AD 1905, the Japanese annihilated a Russian fleet that was despatched to put a stop to their territorial expansion in eastern Asia. The Japanese expansion had begun in 1872, with the appropriation of islands and nibbles at territory of the demoralized Chinese, but their naval victory over a major European power is the most apt marker of their breakout, and it gave the Japanese control of Korea and southern Manchuria. The assertiveness of people whom the Europeans and Americans regarded as quaint became thereafter a major theme of twentieth-century history. It was entwined with other scientific and technological developments besides the battleship. Aircraft, for example, the first of which flew in the United States in 1903, became an instrument of war that the Japanese used with particularly devastating effects.

The human mind took wing in 1905, when a patent clerk at Bern in Switzerland established that light consisted of particles and initiated relativity theory. Almost in passing, he explained that matter was frozen energy, which helped to account for the energy found leaking out of radioactive atoms; in due course his equation $E = mc^2$ quantified the heat of the sun and the nuclear energy let loose on a large scale by earthlings. Electronics dawned in 1906 with the invention of the triode vacuum tube. It regulated the flow of electrons in creative ways and transformed the crude radio of the pioneers into an agreeable system, so that radio broadcasting became the fashion by 1922; television, radar, radio astronomy, and electronic computers were not far behind. Again, emphatic expressions of atomic energy and electronics were to involve the Japanese.

War breaking out among the imperial European powers in 1914 gave the Japanese the chance to take over German colonies in mainland China. World War I brought submarines, aircraft, gas, and tanks into battle. It killed 10 million people—30 million if one counts the influenza virus that mutated in the trenches and swept around the world during 1918. Apart from momentarily checking German ambitions and dismantling the empires of Austro-Hungary and Turkey, the consequences of the war were limited. Two involved the chemistry of explosives manufacture. A German method of fixing nitrogen of the air to make ammonia became the basis of the artificial fertilizer industry, which multiplied the output of the world's farms. And the British, grateful for a method of extracting

Dates: AD *for events; scale calibration in years before* AD *2000.*

aircraft 1903 ▼

Japanese breakout 1905 ▼

electronics ▼ 1906

nitrogen fixation 1913 ▼

Russian revolutions 1917 ▼

100 yr | 95 | 90 | 85

acetone from maize, promised its Zionist inventor in 1917 that the Jews could set up a national home in Palestine.

A revolution in Russia toppled the emperor in March 1917; eight months later, another revolution displaced the ensuing liberal-socialist regime and instituted a style of revolutionary one-party government that came to be widely imitated by regimes of the far right, as well as the far left. The Russian revolutionaries' claim to be Marxist communists brought a modern religious dimension into geopolitics. The Japanese were frustrated in an attempt to take over eastern Siberia from revolutionary Russia, but their other territorial gains were confirmed in the peace treaty of 1919.

Around 1930 the population surge that had begun in China in 1700 became a worldwide phenomenon, even in the poorer nations and colonies. Why death rates fell in the tropics is unclear; candidate reasons include public-health measures, the better understanding of nutrition that followed the indentification of vitamins, and science-based farming methods. The weather may have helped too, in many places, as the world went through its warmest spell for nine hundred years and the Indian monsoons became more reliable; on the other hand, the midwest of the United States suffered from drought and duststorms.

In a chronically dry climate, on the island of Bahrain in the Arabian-Persian gulf, oil found in 1932 was the first hint of riches for the clansmen of Arabia who were sitting beside the world's most generous oil fields, created on the Cretaceous shores of the disintegrating Gondwanaland. Discoveries about energy of a different kind came from the physicists, with the atomic nucleus (1911) and quantum mechanics (1926) entirely transforming their ideas about the character and behavior of matter. Antimatter and the neutron followed in 1932, and in 1934 a Japanese physicist turned up at the forefront of nuclear theory-making when he inferred correctly that undiscovered subatomic particles, the mesons, conveyed the strong nuclear force.

The Japanese military breakout had resumed with the seizing of all Manchuria in 1931, and in 1937 their armies invaded China's most productive areas, where Japanese industrialists already had a secure footing. Air raids at Shanghai and elsewhere mimicked attacks on Madrid in 1936, in the course of a Spanish civil war, when European aviators initiated modern bomber warfare. The Japanese soon found more effective uses for their air power than dropping

AD 1913

ASSEMBLY LINE

Mass production of Ford Model T cars, of which 15 million were sold between 1908 and 1927, put the American people on wheels and, by example, helped to transform working and personal lives in all industrialized countries.

Age of dictators. *Chinese republic AD 1912. World War I 1914 to 1918. League of Nations 1920. Fascist era in Italy 1922 to 1943; Stalin era in the U.S.S.R. 1924 to 1953; militaristic rule in Japan 1931 to 1945; Nazi era in Germany 1933 to 1945. The Great Depression began in 1929.*

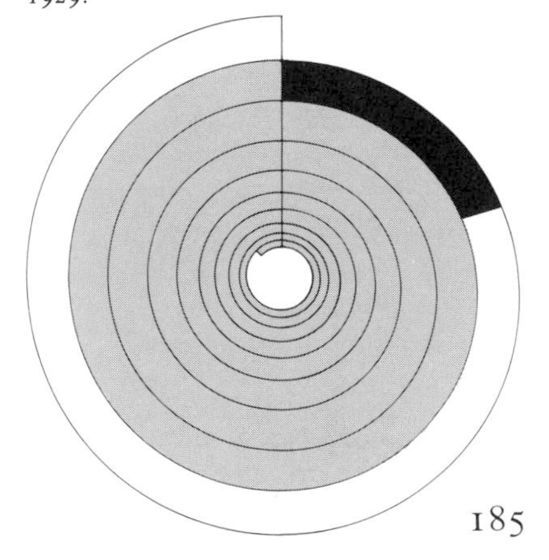

radio broadcasting boom ▼ 1922

global population boom 1930 ▼

Arabian oil ▼ 1932

bomber warfare ▼ 1936

80 | 75 | 70 | 65

AD 1952
ATOMIZED SEA WATER

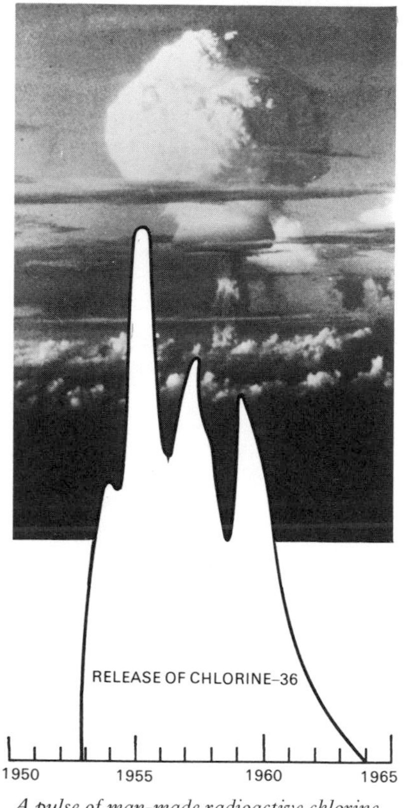

A pulse of man-made radioactive chlorine-36, detectable worldwide, persists as a geological signature of American H-bomb tests in the Pacific in the 1950s. The chlorine comes from the salt in sea water exposed to the thermonuclear explosions, and is not characteristic of parallel Soviet tests. The Mike H-bomb test illustrated was a ten-megaton experimental explosion at Eniwetok in 1952

Modern superpowers. *World War II (conventional dates) AD 1939 to 1945. Japanese Co-Prosperity Sphere 1942 to 1945. United Nations 1945. Soviet-American carve-up of Europe 1945 (Yalta). Cold War esp. from 1947. Korean War 1950 to 1953.*

explosives on civilians. World War II was beginning, and it was to finish as the first nuclear war.

The discovery of nuclear fission in Germany in AD 1938 started the race for the A-bomb, while the Germans irrupted over Europe and North Africa, embarked on a titanic struggle with the Russians, and created an industry devoted to gassing all the Jews on whom they could lay their hands. The defeat of the French in 1940 gave the Japanese the chance to begin occupying Indochina. At the end of 1941, they destroyed by air attacks an American battle fleet at Pearl Harbor in the Pacific and a British battle fleet off Malaya. Before the cherry trees blossomed in 1942, the Japanese had seized southeastern Asia and the islands of the western Pacific, and they were knocking at the doors of Australia and India. The rollback took much longer, but it began when the Americans broke Japanese codes and their dive bombers destroyed three Japanese aircraft carriers in three minutes off Midway one morning in June 1942. By December 1942 the first man-made nuclear chain reaction was sustaining itself in a reactor at Chicago. The war ended with the Americans dropping A-bombs on Japanese cities in August 1945.

Several wartime innovations had consequences far beyond the battlefield. DDT was one; its insecticidal qualities had been quietly noted by a Swiss chemist in 1939, and the Anglo-American armed forces took it up. Then came antibiotics, medicine's most revolutionary innovation: penicillin isolated in Britain in 1940, streptomycin in the United States in 1944. Microwave radar stimulated electronics, and for codebreaking the British assembled the first electronic digital computer in 1943. By 1944 the Germans had developed the liquid-fueled rocket into a ballistic missile; without precise guidance or a nuclear warhead, it was an ineffectual weapon, but it opened the way to a missile race and to space flight. For better or worse, the postwar uses of science and technology promised to determine the fate of the species.

The defeated Japanese, stripped of their territorial gains and forcibly democratized, cast around for a nonviolent strategy. Chinese armies, not for the first time, fought southward from the Yellow River to the Yangtze and the sea, to reunify their country; on this occasion they were communist guerrillas who proclaimed the largest revolutionary one-party state, in October 1949. Otherwise, the upshot of World War II, with its 50 million dead, was that the Russians and Americans divided the world between them, and the

antibiotics 1940 ▼ **nuclear energy** 1942 ▼ **computer** ▼ 1943 **A-bomb** ▼ 1945

empires of the Europeans disintegrated. Japanese wartime victories destroyed any belief in European invincibility. The Indians were among the first to win independence in 1947, and they established the largest parliamentary democracy. By 1960, wholesale decolonization was turning Africa into a nursery of nation-states.

New manufacturing methods reduced the need for labor, and around 1950, the Americans attained the postindustrial condition, when employment in the service and information sectors first exceeded the workforces of industry and agriculture combined. Leisure habits changed drastically during the 1950s in prosperous countries, with the growth of television viewing; between 1949 and 1959 the number of video receivers in American homes increased fiftyfold.

The delicate ingenuity of the genes emerged from X-ray images of DNA molecules, and modern biology began. That was in 1953. In the same year came the H-bomb, capable of destroying or mutating many genes, and a thousand times more powerful than the A-bombs used against the Japanese (1953 is a median date for American and Russian tests of experimental and practical H-bombs). The Russians also adapted an intercontinental ballistic missile, designed for carrying H-bombs, and used it to put the first artificial satellite, *Sputnik 1*, into orbit around the Earth in 1957. Two years later one of their lunar probes made the first escape from the planet's gravity. A feverish competition ensued between the Russians and Americans, in nuclear weaponry and in the technical breakout into space. Scientific and civilian benefits of the space race were dazzling, but those who perceived its strong military flavor were not mistaken; the Americans had photographic spy satellites by 1960.

Another race was between food and babies. Plows were busy as never before, in the 1950s, when the areas of land devoted to growing grain in the poorer regions of the world increased by more than 60 percent in little more than ten years. By this effort the poor countries did better than keep pace with their growing populations. Yet the output of grain rose even faster in rich countries, where territory and manpower devoted to grain production were actually diminishing. As fertilizers and machinery pushed up the yields, the input of inanimate energy began often to exceed the energy value of the crops themselves. Supplies of meat and milk soared, and people in prosperous countries ate so well that dieting became a fad.

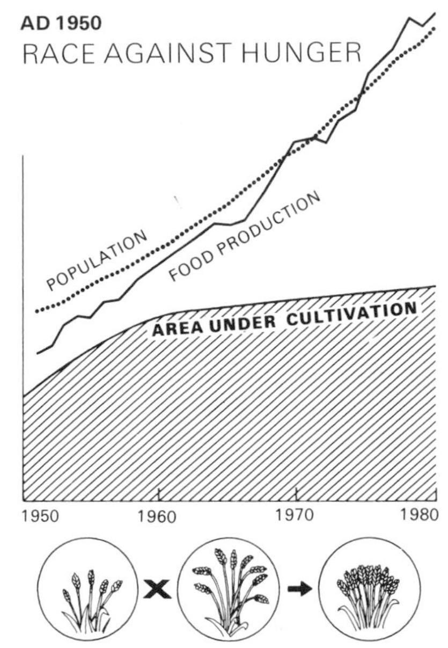

Output of food in the poorer, developing countries increased even faster than their populations. In the 1950s, larger harvests required larger areas but innovations in the later decades included high-yielding, short-stalked cereal varieties.

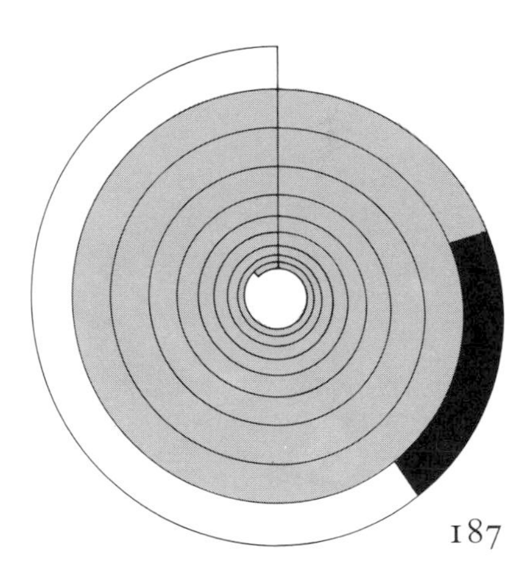

hinese unified 1949 ▼ | television boom ▼ 1950 | gene structure; H-bomb ▼ 1953 | spacecraft ▼ 1957 | African decolonization 1960 ▼

50 45 40

AD 1969
TRIUMPH IN THE SKY

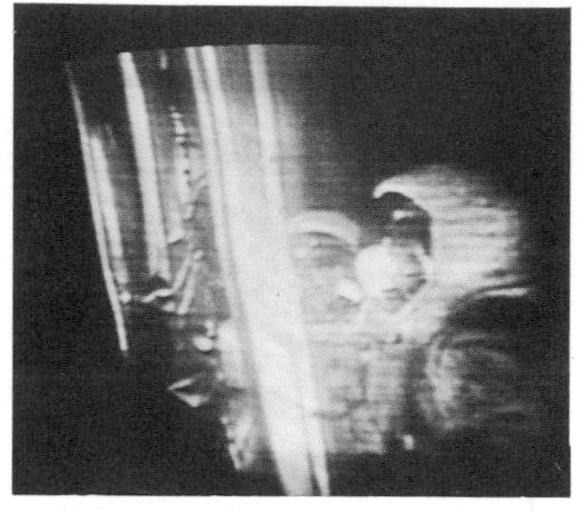

For 100 million people who watched, in real time, the first television show from the surface of the moon, it was the fulfillment of a dream, a victory for American technology in the Cold War, or simply the fascination of seeing Homo sapiens, *in high spirits, defying a wholly strange environment.*

By the 1960s AD, the poor countries were increasing their numbers at a peak rate that could in principle double their populations in twenty-eight years. The Chinese switched large areas of land from sorghum and millet to high-yielding varieties of maize, while in the green revolution, plant geneticists released high-yielding varieties of wheat in Mexico in 1961, followed by rice varieties developed in the Philippines and released in 1966. The first of the new rice varieties were vulnerable to disease, and the grains of the green revolution required expensive inputs, including fertilizers and irrigation, so favoring the well-to-do landowners at the expense of the poorest farmers; nevertheless, their introduction enabled many people to survive who might otherwise have starved.

The Japanese burst back on the scene with supertankers in place of aircraft carriers, at the start of their economic conquests. With the builders of giant ships leading the way, Japanese manufacture increased threefold during the 1960s, and exports fourfold. Much as the British profited from a cotton-led industrial surge around 1800 without growing any cotton of their own, so the islanders at the other end of Eurasia managed with little iron ore, coal, or oil. The Japanese bought ores and oil in large quantities, and they sold steel, cars, electronic equipment, and many other products abroad.

In April 1961 a Russian cosmonaut in the *Vostok-1* capsule circumnavigated the world in 108 minutes. Before the decade was out, as an American president promised, astronauts landed on the moon, in July 1969. Commercial communications satellites having come into service in 1965, many people were able to watch the voyage live on television. Ballyhoo apart, reactions to life's first leap to another cosmic body were curiously muted. The Americans were embroiled in a grim war in Indochina, and growing unease about the thrust of technological innovation provoked an environmentalist movement concerned about pollution, wildlife, and extravagant use of natural resources.

Although the *Apollo* series of missions to the moon sustained only limited enthusiasm, the precision of the guidance systems had implications for the arms race. In 1962 airborne American bombers, loaded with nuclear weapons, menaced the Russians until they withdrew certain missiles newly planted on Cuba, and thereafter the superpowers appeared to achieve mutual nuclear deterrence, especially as seemingly invulnerable missile-launching submarines

green revolution; manned spaceflight
▼ 1961

communications satellite
▼ 1965

came into service. The spread of nuclear weapons compromised the balance. By 1960 both the British and French had the bomb, China followed by 1964, and Israel and India by 1974, although a notable Nobel prize was awarded to a Japanese prime minister in 1974, for not making nuclear weapons. (Proliferation of launching rockets also occurred, with the French putting up a satellite in 1965 and the Japanese and Chinese doing likewise in 1970.) More subtle destabilization began with the American and Russian development of multiple H-bomb warheads for missiles, each independently targetable, as first deployed by the Americans in 1970. Multiple warheads offered an advantage to the side that launched its missiles first, and increasingly accurate guidance, illustrated in the space exploits, aggravated suspicions between the superpowers.

Microelectronics played a part in missile development, but its consequences were much more widespread. The power of computers had been roughly doubling every two years during the 1950s and 1960s, but they were characteristically big main-frame installations giving computing power most conveniently to generals, spies, bureaucrats, policemen, and corporation accountants. The first integrated-circuit microprocessor on a chip of silicon was made in the United States in 1971. It was a liberating invention, like the bicycle, and within a few years calculators were in many pockets and minicomputers in many small businesses.

The species as a whole entered the postagricultural phase around 1970, when the proportion of human employment worldwide devoted to agriculture fell to less than half. Also around 1970, the rate of growth of world population eased for the first time in three centuries. But setbacks in food production began in 1972, when adverse weather hit harvests in many places. In the years that followed, the climate was unhelpfully variable, with weather records being broken repeatedly, and farmers' yields lurching between dearth and glut. Mineral facts of life caught up with the world in 1973, when oil producers began drastically raising their prices, so terminating an era of cheap fossil energy. It was a further misfortune for farmers committed to energy-dependent fertilizers and machinery.

Genes were spliced for the first time in California in 1973, initiating genetic engineering by recombinant-DNA techniques; in Britain in 1975 came the cloning of antibodies by the hybridoma technique, and the first birth of a "test-tube" baby conceived

AD 1972

FAMINE ON THE GROUND

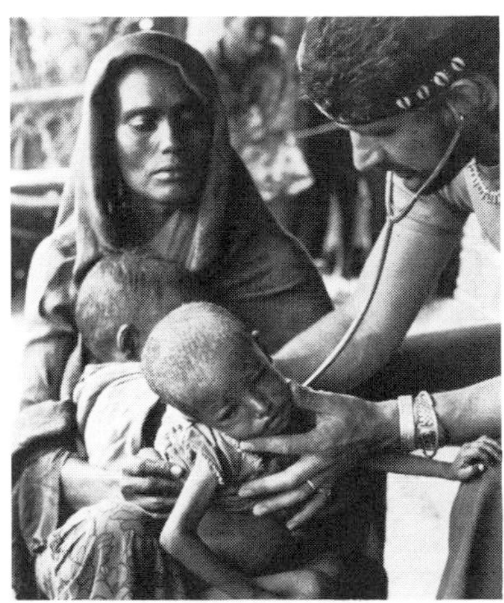

In the early 1970s, people in Ethiopia and the Sahel of Africa perished by drought at a rate of 100,000 a year. The famine was a symptom of a fickle monsoon, of marginal subsistence in a semidesert, and of the seeming inability of the species to distribute the produce of a rich planet.

Indochinese wars. *Americans were involved in the Vietnam War especially AD 1961 to 1973. Cuban missile crisis 1962. Chinese cultural revolution 1966.*

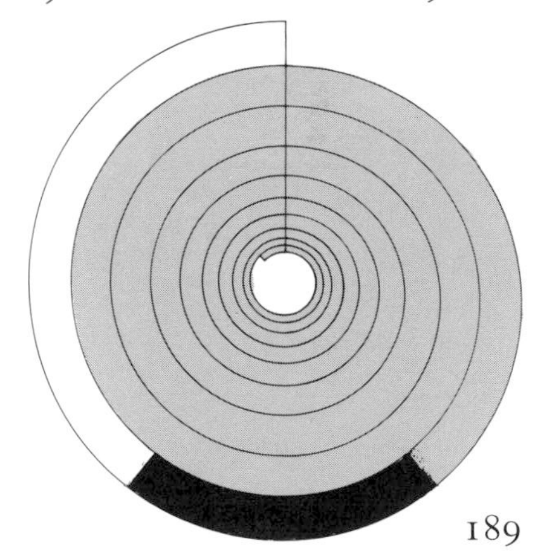

men on the moon ▼ 1969

postagricultural world; multiple warheads ▼ 1970

costly energy; gene splicing ▼ 1973

30 25

AD 1982

MENTAL REACH

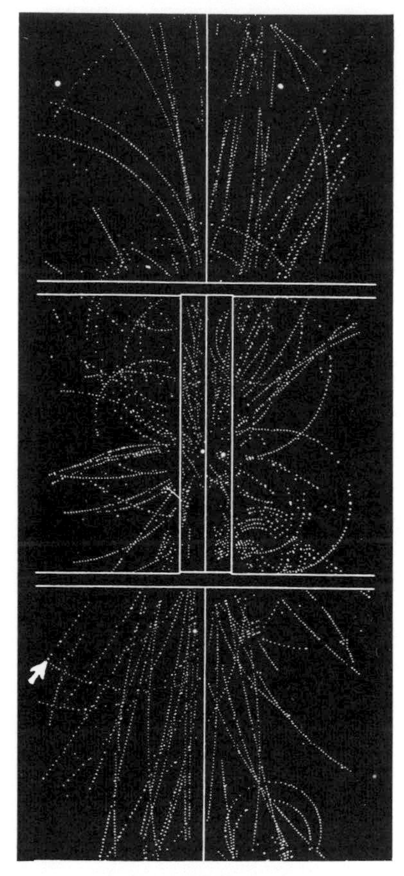

Evidence for the W, a massive force-carrying particle, rewarded a feat of the human intellect. The particle was predicted by a theory that unified two cosmic forces, the electric force and the weak force of radioactivity. The arrowed track, registering among 64 others in an array of particle detectors, shows an extremely energetic electron, a characteristic product of a decaying W particle. It was produced in a giant machine that collided protons and antiprotons at the European Laboratory for Particle Physics (CERN, Geneva).

outside the mother's body. Biotechnology had come a long way since the domestication of animals and plants, around ten thousand years ago; it promised intricate control of living systems, for the purposes of agriculture, industrial production, the manipulation of human reproduction, and medicine. By AD 1982 even the intricate mysteries of cancer began to yield to the reading of genes. Fears about the creation of human castes, or the invention of genetically selective viruses as biological weapons, grew alongside hopes for the conquest of disease and the elimination of hunger and poverty. A program of forcible sterilization in India in 1976 and 1977 showed how easily biomedical practices could become mixed with issues of political power and human rights.

More encouraging was the eradication of smallpox from the planet; by 1977 vaccination had eliminated a disease that ravaged the classical empires, decimated the Aztecs and Incas, and caused untold death and disfigurement even in recent centuries. The most revolutionary technique of cultivation since the plow offered to enhance food production in adverse environments, by dispensing altogether with the soil. As an advanced version of hydroponics developed for greenhouse horticulture in the 1970s, the nutrient-film technique grew crops in gutters by trickling nutrient-rich water over their roots. It offered great savings in water and fertilizer, when growing food or cash crops, provided that the advanced technology could be adapted to suit peasant farmers: that was the aim of trials in Spain in 1982.

American space vehicles landing on Mars, and Russian landers on Venus, advanced the exploration of the solar system, although the Mars landers (1976) dashed expectations of finding life on other planets. A pair of particularly ingenious robot spacecraft set off in 1977 to visit the outer planets. The identification in 1976 of a predicted quality of matter, called charm, spurred the physicists' efforts to comprehend all the particles and forces of the universe in a single scheme. By 1983, European experimenters had found the heavy W particle, agent of the weak force responsible for radioactivity. While the cosmic frame became clearer, fossil hunting and vigorous archeology illuminated the origins of human beings and their institutions. Many events mentioned in this narrative dropped into place, as the inhabitants of the Earth found out where they stood in space and time.

Optical signals charged with large quantities of information began

Mars landers
1976 ▼

smallpox eradicated; outer-planets mission
▼ 1977

carrying commercial telecommunications traffic in 1980, along hair-thick fibers of glass that acted as long-distance cables. With cables, broadcasting satellites, and video recording systems all supplying television, a decentralization of video services followed in the wake of the diffusion of computing power. Even so, central computers still pooled data in nationwide and international communications systems serving large institutions, and supplying the infrastructure for police states.

A statistically minded species could wonder at the transformation of the Japanese that accompanied economic success. Analogous changes may have occurred at Çatal Hüyük or in ancient Athens, but in the twentieth century the data were precise. The Japanese grew taller by more than five centimeters (comparing males born in 1960 with those born in 1940) and they lived longer, adding twenty-four years to male life expectancy between 1950 and 1977, suddenly making the Japanese the longest-lived of the world's major populations. Japanese born after 1946 were at least ten points up in IQ tests, compared with their predecessors, perhaps as the result of a reduction in inbreeding, when half the population moved from the countryside to the cities during half a century.

The transfer of technology to poorer countries, about which the superpowers and their allies had spoken much for thirty years and done little, began in earnest as an accomplishment of Japanese traders. While they monopolized the world market in advanced random-access memory devices using silicon chips and magnetic bubbles, the Japanese found expanding markets in the poor countries for goods of all kinds, and manufacturers of machinery and complete industrial plants sent most of their exports to southeastern Asia, the Pacific, Latin America, Africa, and the Arab world. At the same time, the recently pacified Japanese came under pressure from the Americans to increase their spending on arms.

The first true spaceship, the reusable Space Shuttle that started commercial operations for the Americans in 1982, seemed to some the vehicle for an impending breakout of the human species into the solar system, which would bring life to the deserts of space. To other eyes, the Shuttle sharpened warlike competition in the neighborhood of the Earth. Efforts in self-understanding were racing against the militarism by which states had thrived for five thousand years. The outcome of the great gamble of reasonableness versus dangerous knowledge remained entirely in doubt.

AD 1982

THE SPACEFARERS

Repeated returns to Earth by the same recoverable space vehicle culminated in practical operations for customers and inaugurated a new era in spaceflight. As well as launching satellites, the Space Shuttle Columbia *obtained the radar image shown on page 45.*

Global stagnation. *Economic recession, coupled with high rates of unemployment and monetary inflation, affected many parts of world, especially from AD 1980.*

See map p. 94.

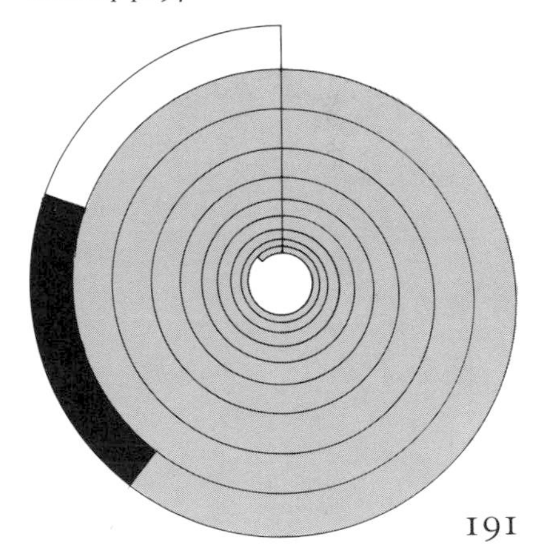

optical-fiber links
▼ 1980

reusable spaceship; cancer mutation pinpointed
▼ 1982

20 19 18 17 16

REFERENCE INDEX

This section is an index of dates, a set of topic-oriented timetables, and a guide to sources of facts and theories that have gone into the narrative and timescale. The more conventional uses of an index are subordinated to these main purposes. Proper names are indexed only in relation to specified events.

Most items in the index have dates associated with them (exceptions include entries concerned with methods of dating and analysis). If the date appears in bold type, the subject is mentioned in the text, illustrations, or chronological notes, on the relevant pages of the narrative. For sources and discussion, the reader is directed to one of the longer entries in the index. For example:

communities (more than 100 people) **29,000 yr**; economic communities esp. AD 1957. *See* government.

This means that the emergence of large human communities (briefly defined in the parentheses) occurred 29,000 years ago, and is mentioned on the relevant narrative pages, in this case 40,000 to 25,000 years ago. The event is put in context, and sources of information indicated, under **government**, where modern regional economic communities (not referred to on the narrative pages) are also mentioned. While *see* in italics refers to other index entries, "see" not in italics, with mention of a page number, refers to the overview or maps section (pp. 1–96). Sources of illustrations are acknowledged on p. 6; sources for the maps appear on p. 79.

Cross references within the index, to items that do not figure in the narrative, are limited to a selection of interesting and distinctive items. It does not attempt to be comprehensive. For example, more than 60 individual mountain-building events are mentioned under **continents**, but these are not cross referenced, except where they are also mentioned on the narrative pages.

Dates are quoted mainly as millions of calendar years ago (Myr) or calendar years (yr) before AD 2000, except in the twentieth century, when they are given as AD dates, to avoid indications that AD 1982, for example, was 18 years ago. Between AD 1 and AD 1899, AD dates are added in parentheses. Where runs of AD dates occur, within an index entry, these may be used in preference to yr to reduce the number of numerals; it will be apparent from the context when this is done.

To convert ages greater than 2000 yr to dates BC subtract 2000 from the number; for example 3250 yr is 1250 BC. To retain this simplicity, no correction is made for the loss of one year occasioned by the lack of year "0" in the BC/AD chronology. BC dates are seldom quoted.

Many dates are approximate, but "c." (circa) is restricted to dates that are vague, or relate to gradual or multiple events. The use of "?" is reserved for a few cases where dating is particularly doubtful.

ABBREVIATIONS

AD calendar years after 1 BC.
a.d. raw radiocarbon date (5568 years half-life).
BC calendar years before AD 1.
b.c. raw radiocarbon date (5568 years half-life).
esp. especially (e.g., becoming notable).
est. estimated, usually on theoretical grounds.
et al. and other authors.
fl. flourished (extant at date mentioned).
Myr millions of years ago.
Myr ditto, mentioned in the narrative pages.
see reference to another index entry.
see reference to another part of the book.
yr calendar years before AD 2000.
yr ditto, mentioned in the narrative pages.

MAJOR TOPICS

The following selection of key words may assist the reader who wishes to look into particular subjects.

On astronomy and cosmology: Big Bang; cosmic impacts; cosmological timescale; elements; galaxies; solar system forming; stars.

On geology and climatology: atmosphere; climate (subdivided by intervals); coal; continents; Earth forming; geological timescale; oceans; ores; petroleum; volcanic events.

On life: animal origins; australopithecines; dinosaurs; forests; invertebrates; life on Earth; mammals; mass extinctions; neanderthalers; plants ashore; primates; vertebrate animals.

On general affairs: archeological timescale; breakouts; cities; disease epidemics; government; historical timescale; human genetics; human origins; language; population history; work.

On technical history: animal domestication; arts and crafts; building; crops; cultivation practice; energy; information technology; livestock revolution; medicine; metal industries; science; stone tools; transport; weaponry.

On dating and analytical methods: dendrochronology; fission-track dating; fossil record; king lists; magnetic stratigraphy; molecular dating; orbital dating; oxygen-18 stratigraphy; paleogeography; radiocarbon dating; radiometric dating; sea-level stratigraphy.

A

abacus c. 1500 yr (AD 500). *See* information technology.

Abbasid (caliphate) **1250 yr (AD 750).** *See* historical timescale.

A-bomb AD 1945. *See* weaponry.

accountancy 10,700 yr. *See* information technology.

Achaemenid empire (Persia) **2550 yr.** *See* archeological timescale.

Acheulian technology 1.9 Myr. *See* stone tools.

Adena culture c. 2750 yr to 2300 yr. *See* climate 10,300 yr.

Afrasian spread (Semitic and related languages) 6000 yr. *See* language.

Africa *ores* 2900 Myr etc., *see* ores; *Limpopo Mountains* **2600 to 2300 Myr**; *Pan-African event* **650 Myr**; *quit North America* **210 Myr**; *rifting of Arabia* **c. 30 Myr**; *Arabia* (still linked) *hit Eurasia* **c. 20 Myr**, *see* continents; *Old World Interchange* **20 Myr**, *see* mammals; *apemen* **4 Myr**, *see* australopithecines; *drought* (eastern) **2.0 Myr**, *see* climate 3.25 Myr; *Homo habilis* **2.0 Myr**; *Homo erectus in Africa* **1.6 Myr**; *proto-modern humans* **c. 100,000 yr**, *see* human origins; *modern humans in Africa* **32,000 yr**; *breakout of Bantu* **2500 yr**; *breakout of Arabs* **1368 yr (AD 632)**; *European colonization* 518 yr (AD 1482), etc.; and other movements, *see* breakouts; language; *agriculture* 18,000 yr etc., *see* animal domestication; crops. *Egypt's First Dynasty* 5200 yr, *see* archeological timescale; *desiccation* (northern) **4500 yr**, *see* climate 10,300 yr; *iron smelting* **2750 yr**, *see* metal industries; *population estimates* 2000 yr, etc., *see* population; *slaves to Americas* **c. 440 yr (AD 1560)**, *see* slaves-and-sugar trade; *colonial carve-up* **120 yr (AD 1880)**, *see* carve-up of continents; *decolonization* **esp. AD 1960**, *see* government. See also maps pp. 81–95.

Agade period 4470 yr. *See* archeological timescale.

agriculture: cultivation experiments and herding c. 18,000 yr etc., *see* animal domestication; breakouts; crops; cultivation practice: *cow-and-plow revolution* **est. 6000 yr**; *see* livestock revolution; plow.

aircraft AD 1903, *see* transport; (military) AD 1908; (air forces) AD 1913, *see* weaponry.

air pressure (discovered) **357 yr (AD 1643).** *See* science.

Akkad period 4470 yr, *see* archeological timescale; **empire 4400 to 4330 yr**, *see* government.

Alaska gold rush 104 yr (AD 1896). *See* gold rushes.

alchemy esp. 1000 yr (AD 1000). *See* science.

alcohol esp. 2500 yr. Beer is hard to date; it was drunk in Mesopotamia 8000 yr ago, but may well be much older (A. L. Demain and N. A. Solomon, *Scientific American*, Vol. 245, No. 3, 1981, p. 42). Grape wine may be about as old as domesticated grapes, est. 6000 yr. Liquor was being distilled in Asia 2500 yr ago, as judged by literary references to "elephant trunks," identifiable with ceramic stills known from 2150 yr at Charsada, Pakistan (F. R. Allchine, *Man*, Vol. 14, 1979, p. 55).

Alexander's empire 2334 to 2323 yr. *See* breakouts (blacksmiths); historical timescale.

algae esp. 1500 Myr. *See* life on Earth.

alpacas domesticated c. 7000 yr. *See* camelids.

alphabet c. 3800 yr. *See* information technology.

Alpine Mountains 100, **esp. 10 Myr** to present. *See* continents; Swiss Alps.

Altaic (languages, spread westward) 760 yr (AD 1240). *See* languages.

aluminum (ore) c. 130 Myr and contemporary, *see* ores; *smelted* **112 yr (AD 1888)**. *See* metal industries.

Amazonia (supercontinent) **1500 Myr.** *See* continents.

America. *See* American Interchange (mammals); American intrusive peoples; Amerindians; North America (geology); South America (geology).

American Interchange 3.0 Myr. South American animals lived in isolation for most of the period since at least 85 Myr ago, except for a mysterious influx of rodents and monkeys from Africa 35 Myr ago. The last sea gap between North and South America, the Bolivar Trough between Colombia and Panama, closed 3.0 Myr ago. Some waif animals had passed in both directions during a few million years before the closure, but at about 3.0 Myr many animals walked between the continents. The chief overall effect was the success of the North American mammals in South America, where they came to account for half of all the mammalian genera on that continent. (M. C. McKenna, in R. L. Ciochon and A. B. Chiarelli, eds., *Evolutionary Biology of the New World Monkeys and Continental Drift*, Plenum,

1981; L. G. Marshall, et al., *Science*, Vol. 204, 1979, p. 272; L. G. Marshall in M. H. Nitecki, ed., *Biotic Crises in Ecological and Evolutionary Time*, Academic Press, 1981). Illus. p. 139.

American intrusive peoples 508 yr (AD 1492) etc. *Spanish* in Hispaniola AD 1493 and Mexico 1519; *Portuguese* in Brazil 1549; *Spanish* in Florida 1565; *French* in Canada 1600; *English* in Virginia 1607; *Dutch* in New York 1625; *Russians* in California 1812, *see* breakouts (navigators; Russians); *carve-up of South America* AD 1494; *North America* 1818, *see* carve-up of continents; *population estimates* AD 1500 etc., *see* population; *African slaves* to Brazil etc., **440 yr (AD 1560)**, *see* slaves-and-sugar trade; *American revolution* (United States) **225 yr (AD 1775)**; *Latin American revolutions* **192 yr (AD 1808)**, *see* historical timescale; *revolutionary nationalism* **224 yr (AD 1776)**, *see* government; *breakout of Americans* (United States) **(AD 1830)**, *see* breakouts; *postagricultural Americans* **c. 110 yr (AD 1890)**; *postindustrial Americans* **c. AD 1960**, *see* work. See also maps pp. 92–95.

Amerindians *peopling of America* **19,000, esp. 11,300 yr** (overkill), *see* breakouts; *language split* before 11,000 yr, *see* language; *in Maya region* 11,000 yr; *Olmecs* **3250 yr**; *Maya* **esp. 2750 yr**; *Maya heyday* **1400 yr (AD 600)**; *Aztecs* **esp. 680 yr (AD 1320)**, *see* archeological timescale; *agriculture* c. 10,000 yr etc., *see* animal domestication; crops; *copper smelting* **3100 yr** and other metalworking, *see* metal industries; *population estimates* 2000 yr etc., *see* population; *epidemics* **480 yr (AD 1520)**, *see* disease epidemics. See maps pp. 86–95. *See also* Adena; Anasazi; Hopewell; Incas; Independence; Mill Creek; Old Copper Culture; Tairona.

ammonites (coiled mollusks) **230 to 67 Myr.** *See* invertebrates.

amniotes (reptiles, followed by mammals and birds) **313 Myr.** *See* vertebrate animals.

amphibians 370 Myr; modern c. 240 Myr. *See* vertebrate animals.

amulet (neanderthaler's) **100,000 yr.** *See* arts and crafts.

Anasazi (at Chaco Canyon) **c. 1000 to 700 yr (AD 1000 to 1300).** *See* climate 10,300 yr.

Anatolia (modern Turkey; innovative region) **esp. 8800 yr** (city). *See* animal domestication; cities; crops; metal industries.

Andes esp. 100 and **4.5 Myr**. The Andean mountains of western South America are a jumble, rich in minerals, of active and extinct volcanoes, uplifted plateaus, and island arcs that have piled up during long exposure to the effects of oceanic plates being driven to destruction under the margin of the continent. The building of the Andes may have been in progress since 2000 Myr ago. Consolidations of new continental material occurred around 230, 215, 150, 100, 60, 35, and 6 Myr ago; of these, the 100 Myr event was the main one. (Emplacements of granitic rocks in Chile, as tabulated by R. W. R. Rutland, *American Journal of Science*, Vol. 273, 1973, p. 811; dates simplified and adjusted to the revised geological timescale.) Relatively recent surface movements include a major event that began 12 Myr ago and then, 4.5 to 2.5 Myr, doubled the altitude of the tropical Andes, from two thousand to four thousand meters (B. S. Vailleumier, *Paleobiology*, Vol. 1, 1975, p. 273). As the ocean floor of the Nazca plate dives under the continental margin, activity continues, marked by volcanoes and destructive earthquakes.

angiosperms 123, esp. 114 Myr. *See* plants ashore.

animal domestication 18,000, esp. 10,700 yr. From persistent shadowing of a particular herd by hunters to selective killing, castration, selective breeding, and eventual separation of the stock from its wild relatives, was probably a gradual transition, and E. S. Higgs suggested that reindeer, gazelle, and North African sheep were in practical terms domesticated by 20,000 yr ago (*La Recherche*, Vol. 7, 1976, p. 308). A zoologist examining bones cannot certify that an animal was domesticated until it is demonstrably different from wild counterparts of the time. Nevertheless, a heavy reliance on particular species of meek animals began in the Kebaran culture of western Asia, best known from Palestine and Jordan. For example, goats provided 82 percent of the meat at a site near Petra and gazelles supplied 74 percent on Mt. Carmel, indicating that the animals were at least closely herded (Mellaart, reference below). Dates for the Kebaran culture cover a long interval and range far back (e.g., 16,300 b.c., Mt. Carmel, adjusting to 18,800 yr). To avoid any spurious contrast with cultivation experiments (see crops), the same date is adopted for these two ill-defined transitions: i.e., cultivation experiments and animal herding, 18,000 yr.

Wolves were probably the first animals to show the effects of full domestication, as dogs. A dog-like jaw, with crowded teeth suggestive of domestication and about 12,000 years old, turned up in Palegawra cave,

Iraq (P. F. Turnbull and C. A. Reed, *Fieldiana Anthropology*, Vol. 63, 1974, p. 81). Dogs were independently domesticated in northwestern Europe, North America (Idaho and Illinois), and Japan, possibly all around 10,000 yr ago. The next domesticated animals were goats and sheep, by skeletal evidence from the site of Aq Köprück in Afghanistan, 10,560 yr (8260 b.c.). As there is other evidence that goats and sheep were items of wealth in Iran by 10,700 yr (*see* information technology) it would be pedantic to insist on skeletal proof of domestication in these cases.

The first bulky animal to be tamed was the water buffalo of eastern Asia. It was in use, together with the pig, in southern China by c. 9000 yr (Hemudu, Lower Yangtze; K. C. Chang, *American Scientist*, Vol. 69, 1981, p. 148). If it was domesticated in Indochina, the event probably occurred several centuries earlier. The water buffalo apart, western Asia and Europe were better favored in respect of domestic animals than any other part of the world, and that gave their peoples great advantages, especially after the advent of animal traction and milking cows (*see* livestock revolution). Mounted nomadic pastoralism emerged as a characteristic mode of life on the Asian steppes by 4000 yr ago.

The timetable draws upon J. Mellaart, *The Neolithic of the Near East*, Thames & Hudson, 1975; J. R. Harlan, *Scientific American*, Vol. 235, No. 3, 1976, p. 88; A. G. Sherratt, ed., *Cambridge Encyclopedia of Archaeology*, Cambridge University Press, 1980; and A. G. Sherratt, in I. Hodder, et al., eds., *Pattern of the Past*, Cambridge University Press, 1981. These data are supplemented and amended from various sources, including those noted in this introduction and under horses; and for the camelids, J. W. Pires-Ferreira, et al., *Science*, Vol. 194, 1976, p. 483.

DOMESTICATION: TIMETABLE

herding of goat and gazelle: **18,000 yr.** Palestine (e.g., Mt. Carmel).
dog: **12,000 yr.** Iraq/Palestine.
goat: **10,700 yr.** Iran/Afghanistan.
sheep (for meat): **c. 10,700 yr.** Iran/Afghanistan. Secondary uses (wool, ?milk) c. 7300 yr, Iran.
water buffalo: **est. 9500 yr.** ?Indochina; in southern China by c. 9000 yr.
pig: **9500 yr.** Eastern Asia; in southern China by 9000 yr; independently in Anatolia, 9400 yr.
cattle: **8800 yr.** Southeastern Anatolia. Secondary uses (traction, ?milk), 7800 yr; in northern Mesopotamia.
chicken: ?7000 yr. Southern Asia.
camelids (llama, alpaca): c. 7000 yr. Peru.
horse: **6400 yr.** Ukraine.
silkworm: 6000 yr. China.
zebu cattle: 5500 yr. Thailand.
onager (half donkey): est. 5200 yr. ?Mesopotamia.
donkey and mule: **c. 5000 yr.** Palestine.
camel (two-humped) **c. 5000 yr.** Iran.
camel (one-humped): **c. 5000 yr.** Egypt or Arabia.
elephant: c. 5000 yr. India.
yak: 4500 yr. Tibet.
cat: 4500 yr. Egypt.
guinea pig: 4000 yr. Peru.
reindeer: 2500 yr. Central Asia (Sayan Mountains).
turkey: 2300 yr. Mexico.

animal energy **esp. 6000 yr.** *See* energy.

animal origins **670** to 425 Myr. Organizing the many-celled bodies of animals (metazoa) was one of the most difficult steps in the whole evolutionary story, yet it seems to have happened more than once. Sponges are descended from one group of single-celled proto-animals (protozoa), more complex animals from another. The oldest definite traces of animal life are from Newfoundland, dating from c. 670 Myr (early Vendian). All were soft-bodied animals, principally jellyfish and sea pens. Although the period 670 to 570 Myr ago might be called an age of jellyfish, the most striking evolution involved the coelomates, modest-looking burrowing worms with hollow bodies.

Because soft-bodied worms are not well preserved, the fossil record does not necessarily reflect the early course of evolution; studies of present survivors give a clearer picture of relationships (J. W. Valentine, *American Zoology*, Vol. 15, 1975, p. 391, and in A. Hallam, ed., *Patterns of Evolution*, Elsevier, 1977, and other writings). The earthworm illustrates the initial advantages of the hollow body of the coelomates; pressurized with fluids it becomes quite rigid, and it can move by stretching and contracting (peristalsis).

Mollusks (nowadays snails, clams, octopuses, etc.) resemble flatworms in using simple repetitive organs for muscles, gills, etc., and they possess only simple hollow spaces, chiefly around the heart. Sipunculids (nowadays peanut worms) represent another very simple hollow-body design that was successful enough in its own terms but led to no more ambitious forms. Annelids (nowadays earthworms, fan worms) used a segmented body plan making repetitive use of hollow modules, and its evolutionary possibilities quickly manifested themselves in the arthropods (nowadays crabs, shrimps, insects, spiders, etc.) in which the development of jointed legs on particular segments was the key to success.

Phoronids (horseshoe worms) seem to be the prototype of a body plan with a characteristic crown of tentacles attached to a hollow trunk that has only two or three separate regions; the same body plan turns up in brachiopods (e.g., lampshells) and in moss-like bryozoa. Something very like a horseshoe worm would also be the best candidate for the forerunner of a wide range of other hollow-bodied animals including echinoderms (nowadays starfishes, sea urchins, sand dollars, etc.), pogonophora (deep-sea worms), hemichordates (acorn worms), urochordates (sea squirts), and chordates (animals with spinal columns). These last, the chordates, include the vertebrates, the group that in turn includes human beings.

Data in the timetable follow S. M. Stanley, in *American Journal of Science*, Vol. 276, 1976, p. 276, and *Macroevolution*, Freeman, 1979; H. A. Lowenstein *Science* Vol. 211, 1981, p. 1126; and S. Conway Morris, in D. E. Smith, ed., *Cambridge Encyclopedia of Earth Sciences*, Cambridge University Press, 1982. The dating is vague; a basis for refinements and revisions has been laid by global correlations of P. Cloud and M. F. Glaessner, *Science*, Vol. 218, 1982, p. 783. See also the map on p. 82 and illus. pp. 96, 114, 116. For later events *see* invertebrates; vertebrate animals.

EARLY ANIMAL EVOLUTION: TIMETABLE

jellyfish, sea pens, and relatives (coelenterates): **670 Myr**. Early Ediacaran fauna, Newfoundland.

flatworms, annelid worms, and arthropods: **620 Myr**. Late Ediacaran fauna, Australia and U.S.S.R.

early hard animals: **600 Myr**. Experimental hard-part mineralization, 66 percent phosphate.

beardworms (pogonophora): c. 590 Myr.

conodontophorida: 580 Myr. Makers of enigmatic tooth-like phosphate fossils, extinct c. 200 Myr.

sponges: c. 580 Myr.

many hard animals: **570 Myr**. Archaeocyathid reef builders, mollusks, and hyolithids. Tommotian shelly fauna; 50 percent calcareous. The archaeocyathids were impoverished 550 Myr, and especially by 530 Myr.

trilobites (arthropods), *brachiopods*, and *echinoderms:* **560 Myr**.

chordates (including ancestors of vertebrates): **550 Myr**. Priapulid worms appeared at about the same time.

vertebrates (initially fishes): **c. 510 Myr**.

bryozoa (moss-like animals): c. 500 Myr.

animals ashore (invertebrates): **est. 425 Myr**.

Antarctica *part of Gondwanaland* **500 Myr**; *India quit* **100 Myr**, *New Zealand quit* 90 Myr, *Australia quit* **50 Myr**, *see* continents; Gondwanaland; *marsupials* **c. 100 Myr**, *see* mammals; *isolation* **30 Myr**; *East Antarctic deep freeze* **14 Myr**; *West Antarctic deep freeze* **6.6 Myr**, *see* climate 67 Myr. Maps pp. 81–87; illus. p. 14.

antelopes **19 Myr.** *See* bovids.

antibiotics **AD 1940.** *See* medicine.

antimatter (discovered) **AD 1932.** *See* science.

apemen split from apes **est. 5 Myr,** *see* primates; *fossil apemen* **4.0 to 1.4 Myr**; *robust* **2.2 Myr**, *see* australopithecines.

apes (dryopithecines) **20 Myr**; (orangutans) **10 Myr**; (chimpanzees and gorillas) **est. 5 Myr**. *See* primates.

Apollo (moon flights) **AD 1969.** *See* transport.

Appalachian Mountains c. 380, and **350 to 280 Myr**. *See* continents.

Aptian turnover 117 Myr. The disappearance of the mighty brachiosaurs in the Aptian stage, and their subsequent replacement by other herbivorous dinosaurs, may be related to a major fall in sea level that occurred 117 Myr ago, suggesting global environmental changes affecting life. Flowering plants diversified at this time. *See* dinosaurs; floral revolution; mass extinctions; sea-level stratigraphy.

Arabia (microcontinent) *oil forming* **esp. 93 Myr**, *see* petroleum; *quit Africa* **30 Myr** onwards; hit Eurasia **20 Myr**, *see* continents; *oil discovered* **AD 1932**, *see* energy; *Arabic* (language origins) est. 3800 yr, *see* language; *droughts* **c. 1730 yr (AD 270)**, *see* climate 10,300 yr; *breakout of Arabs* **1368 yr (AD 632)**, *see* breakouts. *See also* religious surge.

arachnids 395 Myr. *See* invertebrates.

Aramaic (language) est. 5200 Myr. *See* language.

arch 4700 yr. *See* building.

archaeobacteria (metabacteria) est. 2500 Myr. *See* life on Earth.

archaeocyathids (hard animals) **570 Myr**. *See* animal origins.

Archaeopteryx (early bird) **150 Myr**. *See* birds.

Archean era **3800 to 2500 Myr.** *See* geological timescale.

archeological timescale Anticipating what is to come in the principal archeological timescale for southwestern Asia, the first table gives a conspectus of Old World archeology to clarify what archeologists may have in mind when they use terms such as "Neolithic" or "Bronze Age." *See also* human origins; metal industries; stone tools. To convert ages from yr to dates BC, subtract 2000.

OLD WORLD CONSPECTUS

(All dates are for the starts of the phases.)

Paleolithic (Old Stone Age)	
Early (or Lower) Paleolithic	2.4 Myr
Middle Paleolithic	120,000 yr
Late (or Upper) Paleolithic	40,000 yr
Epipaleolithic (or Mesolithic)	18,800 yr
Neolithic (New Stone Age)	
Aceramic (potless) Neolithic	10,600 yr
Ceramic (with pots) Neolithic	8500 yr
Chalcolithic (or Copper Age)	8000 yr
Bronze Age	
Early Bronze Age	6000 yr
Middle Bronze Age	4150 yr
Late Bronze Age	3740 yr
Iron Age	
Early Iron Age	3200 yr
Full Iron Age	3000 yr

The Late Paleolithic is best documented in Europe. A brief summary of European archeological phases is given here; for more details, *see* arts and crafts; breakouts; Gravettian heyday; metal industries; neanderthalers; stone tools. (The main source for the early cultures is P. Phillips, *The Prehistory of Europe*, Allen Lane, 1980.)

EUROPE: TIMESCALE

Aurignacian (widespread)	40,000 yr to 20,000 yr
Châtelperronian/Perigordian	36,000 to 34,000 yr
Gravettian	29,000 to 10,000 yr
Solutrean	21,000 to 16,000 yr
Magdalenian	20,000 yr to 10,000 yr
Epipaleolithic/Mesolithic	10,300 yr
Neolithic began (Greece)	8700 yr
copper industries (Balkans)	6500 yr
megalithic structures (western Europe)	6500 yr
plow farming (across Europe)	5200 yr
Indo-Europeans (Greece)	4350 yr
Beaker culture (esp. western Europe)	4300 yr
Minoan culture (Crete)	4000 to 3450 yr
Mycenaean culture (Greece)	3600 to 3250 yr
Celtic expansion (Hallstatt culture)	2800 yr
Etruscan expansion	2700 yr
Athenian heyday	2480 yr

(For general European events, 2500 yr to present, *see* breakouts; historical timescale.)

Although, or perhaps because, southwestern Asia is a heartland of archeology, the only comprehensive attempt to tune the timescale to radiocarbon dates is due to J. Mellaart (*Antiquity*, Vol. 53, 1979, p. 6). His admittedly controversial scheme is here adopted in its entirety for the interval 6000 to 3750 yr ago (4000 to 1750 BC), except that a key related date for Egypt, the start of the First Dynasty, is taken to be 5200 yr (J. Mellaart, *Antiquity*, Vol. 54, 1980, p. 225). The timescale is extended backward and forward in time. The earliest phase follows another Mellaart table (in his book *The Neolithic of the Near East*, Thames & Hudson, 1975). That was presented in uncalibrated radiocarbon terms, which are here adjusted or calibrated by specified rules; *see* radiocarbon dating. The latest phases of the timescale follow C. Burney (*From Village to Empire*, Phaidon, 1977), for the interval 3170 to 2550 yr, although he is not to blame for the hiatus after the Kassite period.

SOUTHWESTERN ASIA: TIMESCALE

Epipaleolithic Age (Palestine)	
Kebaran (Mt. Carmel)	18,800 yr
Natufian (Mt. Carmel, end of Kebaran)	12,300 yr
Proto-Neolithic (Jericho)	11,500 yr
Aceramic Neolithic Age (Palestine)	
Early Prepottery Neolithic A (Jericho)	10,600 yr
Late Prepottery Neolithic A (Jericho)	9900 yr
Early Prepottery Neolithic B (Beidha IX)	9500 yr
Late Prepottery Neolithic B (Beidha IX)	9100 yr
Ceramic Neolithic Age (Mesopotamia)	
Umm Dabaghiya (U.D. IV)	8500 yr
Early Halaf (Tell Halaf)	8200 yr
Chalcolithic Age	
Middle Halaf (Arpachiya X)	8000 yr
Late Halaf (Arpachiya VI)	7600 yr
Early Ubaid (Eridu XII)	7300 yr
Late Ubaid (Eridu VII)	7100 yr
Early Bronze Age	
Early Uruk (Warka XIV = EBA IA)	6000 yr
Middle Uruk (Warka XII)	5900 yr
Late Uruk (Warka VII = EBA IB)	5750 yr
Jemdet Nasar (Warka III = EBA IIA)	5400 yr
Early Dynastic I (Warka II = EBA IIB)	5100 yr
Early Dynastic II (Warka I = EBA III)	4900 yr
Early Dynastic IIIA (Sumer)	4800 yr
Early Dynastic IIIB (Sumer)	4650 yr
Agade (Akkad)	4470 yr
Lagash II	4330 yr
Ur III (Third Dynasty)	4260 yr
Middle Bronze Age (Mesopotamia)	
Isin-Larsa	4150 yr
Old Babylonian empire	4030 yr

Late Bronze Age (Mesopotamia)	
Kassite empire	3740 yr
Iron Age (Mesopotamia)	
Middle Babylonian empire	3170 yr
Neo-Babylonian	2626 yr
Achaemenid empire (Persian)	2550 yr

For archeology in China, the following brief summary follows K. C. Chang (*Shang Civilization*, Yale University Press, 1980, and *American Scientist*, Vol. 69, 1981, p. 148); also B. Hook, ed., *Cambridge Encyclopedia of China*, Cambridge University Press, 1982. A standard date for the start of the Yang-shao millet, cultivating culture, possibly ancestral to the core "Han" Chinese, is 7000 yr. Here the cultivation of millet is set somewhat earlier, at 7500 yr, in view of a clustering of dates for a widespread proto-Yang-shao culture, in the late sixth millennium BC, in the Wei and Yellow River regions. The date of arrival of the intrusive Lung-shan culture in northern China is from an uncalibrated radiocarbon date of 2174 BC given for Lungshanoid ware at Miao-ti-kou by J. G. D. Clark, *World Prehistory*, Cambridge University Press, 1977. As Chang observes, radiocarbon dating has made a succession of early dynasties (Hsia, Shang, Chou) an obsolete concept; these cultures overlapped.

CHINA: TIMESCALE

proto-Yang-shao culture	7500 yr
Yang-shao culture	7000 yr
Miao-ti-kou culture	5200 yr
Lung-shan culture	4700 yr
Hsia culture	4200 to 3400 yr
Shang culture	3800 to 3100 yr
Chou culture	3300 to 2221 yr

(For general events, 2260 yr to present, *see* breakouts; historical timescale; population.)

In Middle America, the central location of the Maya among once-prosperous Amerindians makes their culture a barometer of events in the Americas before the arrival of Europeans. Formerly regarded as passive victims of invaders, the Maya are now seen as warlike and subject to continual tension between the cities and the provinces; in Late Preclassic times a single powerful lineage of rulers was in charge (G. R. Willey, *Man*, Vol. 15, 1980, p. 249). The Maya lowlands, the chief area of cultural development, lay mainly in Yucatán (Mexico), Belize, and northern Guatemala; the Maya highlands were in southern Guatemala. In the Classic period, Tikal (60,000 people) was the dominant city; during the Late Classic period there were several prominent cities. In the Post-classic period, there was Toltec (or Putun Maya) domination at Chichén Itzá.

The timetable follows primarily G. R. Willey, *Science*, Vol. 215, 1982, p. 260. An eruption of the volcano Ilopango in El Salvador apparently triggered the rise of the Classic Maya by disrupting old trade routes (P. D. Sheets, *Natural History*, Vol. 90, No. 8, 1981, p. 32) and Willey's Classic Maya start-date is postponed by ten years to conform with it; his date of AD 250 was, he said, arbitrary, and set a few years before the appearance of the *stelae* (carved stone slabs). Aztec and Spanish dates are from G. Barraclough, ed., *The Times Atlas of World History*, Times Books, 1978. Suggestions of a calamitous Maya population crash after AD 800 become doubtful when attention is directed away from the elite (R. Sidrys and R. Berger, *Nature*, Vol. 277, 1979, p. 269), but disaster came with the Spanish invasion of Mexico (*see* disease epidemics; population).

MAYA REGION: TIMESCALE

hunter-gatherer occupation	11,000 yr
maize farming and pottery	4000 yr
Olmecs, hierarchical neighbors	3250 yr
Maya farming boom	2750 yr
Late Preclassic Maya	2400 yr
hieroglyphic writing	?2000 yr
Classic Maya	1740 yr (AD 260)
hiatus	1450 yr (AD 550)
Late Classic Maya	1400 yr (AD 600)
Terminal Classic	1200 yr (AD 800)
Postclassic (Toltec)	1000 yr (AD 1000)
Late Postclassic (Mayapán)	800 yr (AD 1200)
Aztecs, militant neighbors	680 yr (AD 1326)
destruction of Mayapán	559 yr (AD 1441)
Spanish invasion of Mexico	481 yr (AD 1519)

architecture (brick houses) 10,600 yr etc. *See* building.

Arctic deep freeze 700,000 yr. *See* climate 3.25 Myr.

argon-argon dating *See* potassium-argon dating.

aristocracy 4700 yr. *See* government.

arms race (current) 129 yr (AD 1871). *See* government.

army (professional warriors) 5300 yr. *See* weaponry.

arthropods 620 Myr. *See* invertebrates.

arts and crafts 1.6 Myr to present. Red is the favorite color of apes and humans, and *Homo erectus* 1.6 Myr ago (Olduvai, Tanzania) collected red minerals, presumably to use for cosmetic or ritual body painting. About **200,000 yr** ago (Swanscombe, England), an archaic *Homo sapiens* shaped a hand ax around a fossil sea urchin

(K. P. Oakley, *Philosophical Transactions of the Royal Society*, Vol. B292, 1981, p. 205). About **100,000 yr** ago (Tata, Hungary), a neanderthaler worked on a piece of a mammoth's tooth, eleven centimeters long, to make a decorative amulet (H. P. Schwarcz and I. Skoflek, *Nature*, Vol. 295, 1982, p. 590). Otherwise, arts and crafts are scanty before the emergence of modern humans est. 45,000 yr ago, when music and oral literature may have begun as plausible accompaniments of the development of language.

Systematic use of animal skins must have been one of the earliest crafts (?*Homo erectus*, 1.9 Myr) but these are perishable items, and no transition to curing and tanning processes can be usefully dated. Basketry too is perishable, and must be very ancient; the oldest surviving examples seem to be the plaited birch-bark work from Pennsylvania **19,000 yr** ago (T. E. Stile, Minneapolis meeting, Society for American Archeology, 1982). As for pottery, the earliest known containers come from Japan **13,000 yr** ago (Fukui Cave, 10,450 or 10,750 b.c.; J. G. D. Clark, *World Prehistory*, Cambridge University Press, 1977). The pots were round-bottomed and ornamented with linear relief. Large fired-clay vessel-like objects, not really portable, appeared in southwestern Asia c. 9300 yr ago (Ganjdareh, Iran, c. 7000 b.c.). In the Americas, known pottery dates from perhaps 7000 yr ago (Pre-Valdivia, Ecuador). Natural volcanic glass (obsidian) was used from the earliest human times; man-made glass appeared simultaneously in Egypt and Mesopotamia c. 3400 yr ago, and may have derived from glazes used in Egyptian pottery (T. A. Wertime, in T. A. Wertime and J. D. Muhly, eds., *The Coming of the Age of Iron*, Yale University Press, 1980).

Virtually all arts except for those using science-based techniques were known in the Stone Age and the dates probably understate their antiquity. The principal source for early visual-arts elements in the timetable is D. Schmandt-Besserat (in Wertime and Muhly, reference above) but with dates adjusted (*see* archeological timescale). Other entries derive from A. Leroi-Gourhan, *Scientific American*, Vol. 246, No. 6, 1982, p. 80; D. Stordeur, *La Recherche*, Vol. 12, 1981, p. 452; A. G. Sherratt, ed., *Cambridge Encyclopedia of Archaeology*, Cambridge University Press, 1980; and general sources including *Encyclopaedia Britannica* (1981), and *Guinness Book of Answers*, Guinness Superlatives, 1980. *See also* information technology, metal industries, stone tools.

ARTS AND CRAFTS: TIMETABLE

"leather" (use of skins): ?1.9 Myr. Western Asia.
pigments: 1.6 Myr. Eastern Africa.
artistic hand ax: **c. 200,000 yr**. England.
amulet: **100,000 yr**. Hungary.
multiple pigments: **c. 50,000 yr**. France.
music and oral literature: est. 45,000 yr. Asia.
pestles and burnt pigments: c. 34,000 yr. France.
bone carving and graffiti: c. 34,000 yr. France.
ceramics and sculpture: **29,000 yr**. Czechoslovakia.
stencils of hands: ?25,000 yr. France.
tailored clothing: 25,000 yr. U.S.S.R.
paintings: 20,000 yr. Spain.
basketry: 19,000 yr. Pennsylvania.
action scenes: **17,000 yr**. France.
ropes: **17,000 yr**. France.
pottery: **13,000 yr**. Japan.
metalworking: 9900 yr. Anatolia.
life-size plaster figures: c. 9000 yr. Palestine.
complementary colors: 8600 yr. Anatolia.
painted pottery: 8600 yr. Iran
weaving and printed textiles: **8500 yr**. Anatolia.
wheel-thrown pottery: **6000 yr**. Mesopotamia.
written literature: 5000 yr. Mesopotamia.
man-made glass: 3400 yr. Egypt and Mesopotamia.
musical scale theory: 2600 yr. Babylon.
formal drama: 2500 yr. Greece.
musical notation: 2100 yr. China.
glassblowing: c. 2050 yr. Syria.
opera: c. 1350 yr (AD 650). China.
clockwork (escapement): **1275 yr (AD 725)**. China.
porcelain: c. 1150 yr (AD 850). China.
painting in oils: c. 600 yr (AD 1400). Europe.
formal perspective: c. 585 yr (AD 1415). Italy.
Italian opera: 393 yr (AD 1607).
photography: **161 yr (AD 1839)**. France.
synthetic dyes: **135 yr (AD 1865)**. Britain.
motion pictures: **105 yr (AD 1895)**. France.
fully synthetic plastics: AD 1910. United States.
talking pictures: AD 1927. United States.
fully synthetic fibers: esp. AD 1935. United States.
holograms: esp. AD 1963. United States.

Asia *ores* 2500 Myr, etc., *see* ores; *Baikalian* 1000 to 800 Myr and other mountains; *Kazakhstan, Tarim, North Tibet, and North China hit Siberia* 350 to 280 Myr; *Siberia hit Euramerica* **c. 300 Myr**; *South China hit* **230 Myr**; *Korea hit* c. 140 Myr; *Kolyma hit* c. 100 Myr; *India hit* **c. 45 Myr**; *Japan quit* **c. 30 Myr**; *Arabia hit* **c. 20 Myr**; *see* continents; Himalayan uplift; *coal* **esp. c. 260 Myr**, *see* coal; *Turgai Sea* 160 to 29 Myr, *see* continents flooded; *Old World Interchange* **20 Myr**, *see* mammals; *Homo erectus* **1.9 Myr**; *Peking Man* ?650,000 yr; *neanderthalers* **120,000 yr**, *see* human origins; neanderthalers; *Asian origin of early modern humans* **est. 45,000 yr** *see* human genetics; *modern humans in Siberia* **32,000 yr** and many subsequent movements of Asian peoples; *breakout of Russians* **414 yr (AD 1586)**, *see* breakouts; language; *archeology* (southwestern Asia and China), *see* archeological timescale; *agriculture* **18,000 yr**, etc., *see* animal domestication; crops; *metalworking* 9900 yr, etc., *see* metal industries; *cow-and-plow revolution* **6000 yr**, *see* livestock revolution. *Religions* esp. **2550 yr**, *see* religious surge; *population estimates* 2000 yr, etc., *see* population history; *epidemics* (Eurasian) **esp. 1838 yr (AD 162)**, etc., *see* disease epidemics; *droughts* **esp. 1730 yr (AD 270)**, *see* climate 10,300 yr; *carve-up* (Russians and Chinese) **esp. 273 yr (AD 1727)**, *see* carve-up of continents. See maps pp. 81–95. *See also* Anatolia; China; historical timescale; India; Indochina; Islam; Japan; Mesopotamia; Palestine.

assembly line **esp. AD 1913**. *See* work.

asteroids **4550 Myr.** *See* cosmic impacts; solar system forming.

astronomical telescope **391 yr (AD 1609).** *See* science.

Athenian heyday **2480 yr.** In a sea battle off Salamis the Greek navy smashed the fleet of Persian invaders. During the interval 2480 to 2338 yr (Macedonian conquest), people in Athens were astoundingly creative: in the theater (e.g., Aeschylus, Sophocles, Euripides, Aristophanes); in sculpture and architecture (e.g., Pheidias, Ictinus, Callicrates); in history (Thucydides); in philosophy and science (Socrates, Plato, and latterly Aristotle). The population of Athens, city and country, c. 2431 yr ago, is estimated at a quarter of a million, of whom about one third were slaves (M. I. Finley, *The Ancient Greeks*, Pelican, 1977).

atmosphere est. 4500 Myr to present. Ideas about the composition of the atmosphere on the young Earth changed markedly in the 1970s with the realization that the atmospheres of neighboring planets (Venus and Mars) are mixtures of carbon dioxide and water vapor, and that the early atmosphere of the Earth was probably of the same kind. Cirrus clouds could have acted like greenhouse windows, and exerted a warming effect in early periods when the sun was considerably feebler than today (A. Henderson-Sellers, personal communication, 1981). Of all the oxygen produced by bacteria and plants during the Earth's history, perhaps only one twentieth remains free today in the atmosphere or dissolved in the oceans; about 95 percent has been bound chemically by sulfur and iron. Banded iron formations were a characteristic early product—these evidently required a little oxygen but not too much. About 1800 Myr ago free oxygen appeared in the atmosphere (*see* oxygen revolution). The timetable follows a framework due to A. Henderson-Sellers (in C. Ponnamperuma, ed., *Cosmochemistry and the Origin of Life*, Reidel, 1982), with the interpolation of "living environment" into her "evolutionary environment" for the circumstances of 4000 Myr ago.

ATMOSPHERE: TIMETABLE

astrophysical environment: est. 4500 Myr (principal constituents, carbon dioxide and water vapor). During the final phase of the planet's assembly, objects resembling comets and carbonaceous meteorites, rich in frozen or chemically bound carbon dioxide and water, fell on the Earth and created the initial atmosphere, partly by direct vaporization and partly by subsequent volcanic out-gassing from below the surface.

evolutionary environment: est. 4200 Myr (principal constituent, carbon dioxide). As the rate of impacts eased somewhat, the atmosphere consisted of carbon dioxide, plus some water vapor, nitrogen, carbon monoxide, sulfur dioxide, and a very little hydrogen. The atmospheric pressure was perhaps a little less than at present. Most of the water had already condensed to form oceans.

living environment: est. 4000 Myr (principal constituents, carbon dioxide and nitrogen). Subtleties in the proportions of carbon atoms of different atomic weights imply that bacteria were involved in making rocks 3800 Myr old (banded iron formation, Isua, Greenland; M. Schidlowski, 1979, cited by Henderson-Sellers). Biochemical mechanisms available to early photosynthetic bacteria did not necessarily involve the production of oxygen, but by 2500 Myr ago, when banded iron formations appeared in abundance, oxygen release was probably intensive. The amount of carbon dioxide diminished as a result of the formation of carbonate rocks. Nitrogen gradually became the principal component of the atmosphere, because of its relative inertness.

modern environment: 1800 Myr (principal constituents, nitrogen and some oxygen). The "rusting" of the planet in the worldwide appearance of red beds signals the appearance of free oxygen at first perhaps less than a millionth of present atmospheric abundance. Effects on living things were far-reaching.

buildup of oxygen: 1800 Myr to present (principal constituents, nitrogen, and oxygen increasing to 21 percent). If the oxygen increased steadily over the past 1800 million years, it would have been at about 60 percent of the present level 670 Myr ago, when animals first appeared; actual levels may have been higher or lower.

atomic nucleus (discovered) AD 1911. *See* science; *see also* energy; weaponry.

atoms forming 300,000 years after origin. *See* Big Bang.

Aurignacian culture 40,000 yr to 20,000 yr. *See* archeological timescale.

Australia *living reefs* **3500 Myr**, *see* stromatolites; *ores* 2700 Myr, etc., *see* ores; mountain-building c. 2600 Myr, etc., *quit Antarctica* **c. 50 Myr**, *see* continents; also Circumantarctic Current; *marsupials arrived* after **125 Myr**; *overkill* c. 17,000 yr, *see* mammals; *humans arrived* **c. 40,000 yr**, *second wave* before 13,000 yr; *colonized* **212 yr (AD 1788)**, *see* breakouts; *gold rush* 149 yr (AD 1851), *see* gold rushes; *female emancipation* 139 yr (AD 1861), *see* government. See maps pp. 80–96. *See also* human genetics; language.

australopithecines 4.0 to 1.4 Myr. The earliest apemen that walked on two legs are probable forerunners of human beings, but the later ones are certainly not, because they coexisted in Africa with early *Homo* species. They had ape-sized brains, and more difference in stature between the sexes than in humans. The chief distinction is between "gracile," or lightly built, and "robust," or heavily built, forms (D. C. Johanson and T. D. White, *Science*, Vol. 203, 1979, p. 321; also, J. E. Cronin, et al., *Nature*, Vol. 292, 1981, p. 113).

Apart from the merest traces of teeth and bones, the earliest remains are skull fragments indicating a brain about the size of a great ape's, and a thigh bone quite different from an ape's and adapted to upright walking (J. D. Clark, T. D. White, et al., reports in press, 1982). The owner of this femur stood perhaps 1.3 meters (4.5 feet) tall. The fossils come from the Middle Awash Valley in Ethiopia; the skull is from twelve meters below, and the femur from six meters above, a layer of volcanic ash dated at 4.0 Myr (R. Walter). Australopithecines were probably about as intelligent as present-day chimpanzees. The most radical hypothesis, not adopted here, is that australopithecines carried fire (J. D. Clark, quoted in *New Scientist*, Vol. 95, 1982, p. 20).

Together with well-preserved fossils of early gracile australopithecines from Tanzania, c. 3.7 Myr, the Ethiopian remains have been dubbed *Australopithecus afarensis*. The skeleton called "Lucy" from Afar (c. 3.0 Myr) is said to be one of the early australopithecines, although the lapse of time is considerable. Footprints **3.7 Myr** old discovered at Laetoli in Tanzania show remarkably "human" feet and a two-legged gait (M. D. Leakey, *Antiquity*, Vol. 52, 1978, p. 133), as verified by photogrammetric comparisons with modern footprints (M. H. Day and E. H. Wickens, *Nature*, Vol. 282, 1980, p. 385). The stride was short and inefficient compared with that of human beings (W. L. Jungers, *Nature*, Vol. 297, 1982, p. 676).

Descendants of *Australopithecus afarensis* gave rise to early humans, *see* human origins, and also to new species of apemen, first as *Australopithecus africanus*, which may have evolved in South Africa 3.0 Myr ago; there is tenuous evidence for this species appearing in eastern Africa 2.8 Myr. *A. africanus* survived until c. 2.0 Myr, alongside more robust australopithecines known from **c. 2.2 Myr** (*Australopithecus boisei* in eastern Africa, and *A. robustus* in South Africa). These robust australopithecines ate mainly fruit and nuts protected by tough skins (A. C. Walker, et al., *Science*, Vol. 201, 1978, p. 908) and may have themselves been eaten by early human beings (J. A. J. Gowlett, et al., *Nature*, Vol. 294, 1981, p. 120). They became extinct **c. 1.4 Myr** ago.

Austronesians' breakout before 6000 yr. *See* breakouts (gardeners); language.

automobile 115 yr (AD 1885). *See* transport.

Aztecs esp. 680 yr (AD 1320). *See* archeological timescale.

B

baboons 4 Myr. *See* primates.

Babylonian empires 4030 yr etc. *See* archeological timescale.

bacteria est. 4000 Myr. *See* life on Earth.

Baikalia (supercontinent) **800 Myr**. *See* continents.

Baikalian Mountains 1000 to 800 Myr. *See* continents.

ballistic missiles AD 1944. *See* weaponry; spacecraft.

Bantu (languages) **esp. 2500 yr**. *See* breakouts (blacksmiths).

barley domesticated 10,600 yr. *See* crops.

basketry 19,000 yr. *See* arts and crafts.

bats 55 Myr. *See* mammals.

beans domesticated c. 9,500 yr. *See* crops.

bears 30 Myr. Fossil *Cephalogale* judged to be a member of the bear family, or ursids, is known from the Late Oligocene. The modern genus *Ursus* made its debut **c. 4.5 Myr** (5 or 4 Myr) in the Early Pliocene (adapted from A. W. Gentry personal communication, 1982, citing Q. B. Hendy, 1980, and R. Ballesio, et al., 1973).

Among recent species, brown bears (*U. arctos*) evolved **c. 700,000 yr** (early Late Pleistocene); cave bears (*U. spelaeus*) **c. 200,000 yr** (e.g. Swanscombe); and polar bears (*U. maritimus*) **72,000 yr** first known from London, England, at the start of the last ice age (dates adjusted from data in B. Kurtén, *Pleistocene Mammals of Europe*, Weidenfeld & Nicolson, 1968). Cave bears became extinct between 12,000 and 10,000 yr ago (B. Kurtén, personal communication, 1972). Illus. p. 157.

beer before 8000 yr. *See* alcohol.

beetles 330 Myr. *See* invertebrates.

bennettitales (primitive flowering plants) **235 to 90 Myr**. *See* plants ashore.

bicycle esp. 112 yr (AD 1888). *See* transport.

Big Bang 13,500 Myr. For the dating of the frenzied explosion at the origin of the universe, and the dating model adopted for subsequent events, *see* cosmological timescale. The reddening of the wavelengths of light of distant galaxies and quasars indicated to astronomers that the contents of the universe were flying apart as if from an explosion. Radiation pervading the universe (microwave background) showed them that empty space was warm, as they would expect after such an event. In any case, the law of gravity ruled out the possibility of a static situation: virtually any universe must either explode or collapse.

Powerful particle accelerators re-create, in effect, conditions prevailing early in the history of the universe, and details of the Hot Big Bang have changed even since S. Weinberg's masterly introduction *The First Three Minutes* (Basic Books, 1977), and J. Silk's *The Big Bang* (Freeman, 1980). The account in the narrative draws upon theories that seek to unify the fundamental forces and particles in a single grand scheme (H. Georgi, *Scientific American*, Vol. 244, No. 4, 1979, p. 40; S. Weinberg, *Scientific American*, Vol. 244, No. 6, 1979, p. 64; A. Guth, *Physical Review*, Vol. D23, 1981, p. 347; M. J. Rees, personal communication, 1982).

The main cosmic forces are four in number: familiar *gravity*, *electromagnetism*, the *weak force* which transmutes particles, as in radioactivity, and the *nuclear force*, used here as a shorthand both for the color force that binds quarks together, in protons, and for its by-product, the so-called strong nuclear force that binds protons and neutrons and powers the sun. The electromagnetic, weak, and nuclear forces are different manifestations of a single electronuclear force, or X force. A mechanism for creating radiation and matter out of empty space was discovered, theoretically, by S. W. Hawking (*Nature*, Vol. 248, 1974, p. 30), who realized that very strong tidal forces, for example near a black hole, could create particles in abundance. Events thereafter accord with the theory of the Hot Big Bang, originated by G. Gamow (1946) and subsequently much refined. The slight imbalance in favor of "normal" particles may have arisen at around 10^{-35} second, because of a fleeting survival of X particles, carriers of the primeval X force. The survival of matter is in any case associated with the instability of the proton, as adumbrated by theories of A. D. Sakharov and S. Weinberg.

The temperature of the microwave background radiation above absolute zero is 2.7 kelvin; in the past it was inversely proportional to its diameter, back to a time when the temperature was so high that the radiation was more massive than the matter. If the density of matter is approximately 20,000 times the density of the background radiation, they were equal 4773 years after *zero*, rounded (logarithmically) to 10,000 years. The temperature of the universe was then 54,000 kelvin. In earlier phases, when radiation predominated, the age, temperature, and size of the universe were related in more complex ways. Size is a subtle concept in curved space, and in any case, there is no telling whether or not the observable universe is just part of a much larger entity. With these caveats, a rough impression of scales is possible. Assuming the present diameter of the observable universe to be 27,000 million light-years, and taking the unrounded figures for the start of the matter area, one can say that the observable universe was then 1.3 million light-years wide, or 1.2×10^{24} centimeters. Before then, the diameter was proportional to the square root of the age, so that at 10^{-35} second it was 10 centimeters, somewhat larger than a tennis ball in size.

Contemporary accounts of the origin of the universe hinge on the concept of "inflation," developed by American and Soviet theorists (A. H. Guth, *Physical Review*, Vol. D23, 1981, p. 347, and later authors). In this theory, the universe is a loaves-and-fishes event, in which much is created out of almost nothing, in defiance of the conventional law of conservation of energy. A very small volume of spacetime, in some pre-existing timeless foam, expanded a little, and cooled. So-called quantum gravity gave way to normal gravity. Soon afterward, a cosmological force, acting like repulsive gravity, inflated the universe very rapidly and filled it with latent energy.

In the transformation to the Hot Big Bang, this energy was converted into matter and radiation, and the universe was committed to particular tendencies: expansion, atom making, and clumping by gravity. These trends set the direction of time's progress (P. C. W. Davies, *Nature*, Vol. 301, 1983, p. 398). Note that some accounts have the inflation occurring between 10^{-35} and 10^{-32} second, rather than 10^{-38} to 10^{-35} second, as given here.

The sequence of events is given in the timescale and narrative, beginning on p. 98, but here some more formal names for the eras are noted.

normal gravity: **10^{-43} second**. Planck time.
grand repulsion: **10^{-38} second**. Cosmological "constant."
Hot Big Bang: **10^{-35} second**. A technical term.
proton era: **10^{-6} second**. Hadronic era.
electron era: **10^{-4} second**. Leptonic era.
helium-making: **3 minutes**. Nucleosynthesis, or fireball.
atoms forming: **300,000 yr**. Decoupling era.

For the dating of the last item, and subsequent cosmic events, *see* cosmological timescale; elements; galaxies; stars.

big-game hunting **c. 750,000 yr.** *See* hunting.

big-man system c. 5000 yr. *See* government.

biology (modern) AD 1953. *See* gene structure.

birds **150 Myr.** Birds evolved from small dinosaurs, the coelurids. The famous feathery fossil *Archaeopteryx* of 150 Myr ago (Kimmeridgian, Late Jurassic; Europe) was a poor flier and may not be close to the ancestral line of any modern bird, but it was highly developed in its own curious direction. It had reptile-like teeth and tail, and the status of *Archaeopteryx*, its precise antecedents, and its relationship to modern birds, are matters of perennial debate among paleontologists. Feathers may have evolved initially for the sake of warmth: eiderdown is one of the best heat insulators known. Flamingo-like birds are known from c. 133 Myr ago, earliest Cretaceous, and by latest Cretaceous, c. 67 Myr ago, a wide variety of water birds had evolved (W. M. McFarland, et al., *Vertebrate Life*, Collier Macmillan, 1979). Fossils of birds are scarce, and molecular comparisons may be the best guide to the overall evolution of the birds. The following data are from E. M. Prager and A. C. Wilson, *17th Congressus Internationalis Ornithologici*, Berlin, 1980.

BIRDS: TIMETABLE

paleognathous (flightless) and *neognathous* (typical) birds split c. 80 Myr.

paleognathous birds: ancestors of kiwis and ostriches split c. 70 Myr; ancestors of kiwis and emus split c. 60 Myr.

neognathous birds: ancestors of gallo-anseriformes (e.g., chickens) and other orders split c. 70 Myr; chickens and ducks split c. 60 Myr.

other orders: ancestors of pigeons and water birds split c. 60 Myr; pigeons, blackbirds, and other orders split at the same time.

water birds and allies: ancestors of herons, grebes, gulls, and owls split c. 50 Myr.

water birds: ancestors of herons, penguins, loons, and albatrosses split c. 40 Myr.

bison (genus *Bison*) **1.8 Myr**. *See* bovids.

Black Death **669 yr (AD 1331).** *See* disease epidemics.

black holes ?13,500, **esp. 12,200 Myr**. *See* galaxies; stars.

blacksmiths' breakout **2800 yr.** *See* breakouts.

blade tools **esp. 40,000 yr.** *See* stone tools.

blood circulation (discovered) **372 yr (AD 1628).** *See* science.

blue-green algae (bacteria) **est. 1600 Myr**. *See* life on Earth; stromatolites.

boats 40,000 yr. *See* transport.

bomber warfare **esp. AD 1936.** *See* weaponry.

bone **510 Myr.** *See* vertebrate animals.

bone carving c. 34,000 yr. *See* arts and crafts.

bovids (antelopes, cattle, etc.) **19 Myr**. Ranging from graceful gazelles to heavy cattle, bovids are the dominant hoofed mammals of the present world, and they have always interested human beings, whether for hunting, herding, or modern mixed farming. Bovids are essentially antelopes that arose from pronghorn-like, even-toed, hoofed animals. The earliest bovids date from c. 19 Myr (Miocene; A. W. Gentry personal communication, 1982, citing L. Ginsburg). They diversified into many kinds of antelopes, and these gave rise to other forms. Early cattle (genus *Leptobos*) date from **c. 3.5 Myr**, Early Villafranchian. Diversification of cattle and their relatives then occurred, with the modern cattle genus (*Bos*), the water buffalo (*Bubalus*), and the bison genus (*Bison*) showing up together in India **c. 1.8 Myr** (some time between 2.9 and 1.5 Myr) (Pinjor formation; various authors, in *Palaeogeography, Palaeoclimatology, Palaeoecology*, Vol. 37, 1980, p. 1, cited by Gentry). The modern water buffalo (*Bubalus bubalus*) was domesticated **est. 9500 yr** ago, in ?Indochina. The yak (*Bos grunniens*) evolved **c. 400,000 yr** (C. C. Flerov, 1980, cited by Gentry) and was domesticated 4500 yr ago in Tibet.

Bison bison, the surviving American species, is a newcomer, having evolved 5000 or 4000 yr ago (Holocene; J. N. McDonald, 1981, cited by Gentry).

The modern cattle species *Bos primigenius* is dated at **c. 200,000 yr** (Gentry, citing H. de Lumley, 1976). It was domesticated in Anatolia **8800 yr** ago. Goats and ibexes of the modern genus *Capra* appeared in Europe **c. 200,000 yr** ago (Gentry citing H. de Lumley). Fossils of true sheep (*Ovis*) are known in Europe from **c. 1.8 Myr** (Late Villafranchian; B. Kurtén, *Pleistocene Mammals of Europe*, Weidenfeld & Nicolson, 1968). Sheep and goats were first domesticated **10,700 yr** ago, and woolly sheep were a later mutation. (*See* animal domestication; livestock revolution.)

bow and arrow **c. 20,000 yr.** *See* weaponry.

brachiopods 560 Myr. *See* animal origins.

brachiosaurs (gigantic herbivores) **175 to 117 Myr**. *See* dinosaurs.

brains **c. 620 Myr**, *see* life on Earth; *human brain evolving* **2.0 Myr to 45,000 yr**, *see* human origins.

breakout of *Homo erectus* **c. 750,000 yr.** *See* human origins.

breakouts 40,000 yr to present. These are defined by their genetic, linguistic, and political impacts; intention is implied. Traders may or may not be sufficiently assertive and numerous to constitute a breakout. Military expansions that are quickly repressed (e.g., Napoleon's esp. AD 1804 to 1814, and Hitler's AD 1936 to 1945) may have little lasting consequence. Although it is too early to assess current Japanese activities, they are included among the breakouts, as the chief twentieth-century analog of ancient adventures.

Among the following modes of genetic and/or linguistic expansion, one contrast offered is between "tribal expansion," where the aim is acquisition of territory for living, and "imperial expansion" where the aims are political, especially the enlargement of tax revenues.

1. *Into vacant regions:* genetic and linguistic differentiation, by dispersal (e.g., early hunter-gatherers and Polynesians).
2. *Into vacant niches in inhabited regions:* genetic mixing (or castes?), and eventual linguistic dominance or submersion (e.g., gardeners, proto-Semitic, proto-Indo-European, proto-Uralic, and Bantu).
3. *Into trading colonies:* genetic dilution, prominent languages of commerce (e.g., Phoenicians and Greeks; also English as a world language).
4. *Imperial expansion:* limited genetic effect, prominent languages of administration and religion (e.g. Chinese and Romans; Mughals and Manchus adopted local languages).
5. *Tribal expansion:* major genetic and linguistic replacement, sometimes temporary (e.g. Celts, Iranians, Germans, Huns, Khmer, Thais, Mongols, and Turks; Vikings adopted local languages).
6. *Mixed modes (2, 3, 4, 5)*: (e.g. Arabs, European navigators, Russians, Americans, and Japanese).

The most striking lacuna in early prehistory concerns the peopling of the Americas, with opinions divided between "early entry" and "late entry." Here, an uneasy compromise offers Meadowcroft, Pennsylvania, **19,000 yr** ago as the earliest firm place and date for any humans south of the Canadian ice sheet (albeit with anachronistic fauna and flora), but puts the main dispersal southwards much later, **c. 11,300 yr** ago. In the earlier phase, the evidence is extremely thin, because human populations at the time were thin (J. M. Adovasio, personal communication, 1982). More than one wave of hunter-gatherers may have made their way from Asia to the Americas, either overland or by water.

Special sources used in this compilation include L. Freedman and M. Lofgren, *Nature*, Vol. 282, 1979, p. 298, for the second wave to Australia; R. C. Carlisle and J. M. Adovasio, eds., Meadowcroft papers for Society for American Archaeology, Minneapolis, 1982, and J. L. Bischoff and R. J. Rosenbauer, *Science*, Vol. 213, 1981, p. 1003, for the early Amerindians, with supporting arguments about game kills from P. S. Martin and A. Long, *Science*, Vol. 186, 1974, p. 638; A. G. Sherratt, in I. Hodder, et al., eds., *Pattern of the Past*, Cambridge University Press, 1981, for the dairymen; and D. W. Phillipson, *Scientific American*, Vol. 236, No. 4, 1977, p. 106, for the Bantu blacksmiths. General sources are A. G. Sherratt, ed., *Cambridge Encyclopedia of Archaeology*, Cambridge University Press, 1980; P. Phillips, *The Prehistory of Europe*, Allen Lane, 1980; C. Burney, *From Village to Empire*, Phaidon, 1977; J. Mellaart, *The Neolithic of the Near East*, Thames & Hudson, 1975; G. Barraclough, ed., *The Times Atlas of World History*, Times Books, 1978; and R. I. Moore, ed., *The Hamlyn Historical Atlas*, Hamlyn, 1981.

BREAKOUTS: TIMETABLE

modern humans (hunter-gatherers): **c. 40,000 yr**. From ?west-central Asia, to Borneo, Australia, and eastern Europe; later to western Europe, c. 35,000 yr; to eastern Siberia (Lena basin) and Africa (Zaire) c. 32,000 yr; to Namibia and perhaps to northwestern America by 27,000 yr; and to central North America (Pennsylvania) by 19,000 yr. See map p. 86. Among secondary breakouts of hunter-gatherers were these:

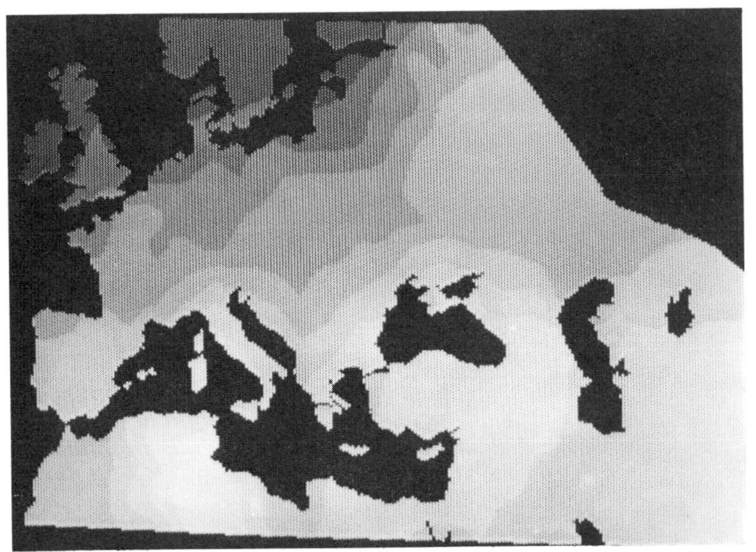

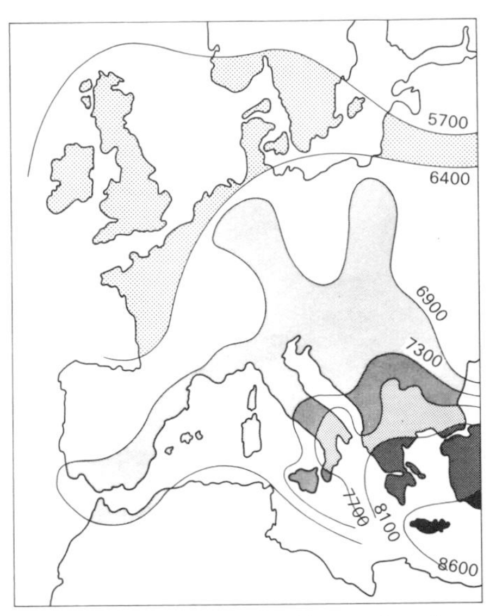

Genetic analysis can disentangle broad features of genetic histories even in a region as complicated as Europe (P. Menozzi and others, Science, *Vol. 201, 1978, p. 786). The technique consists of finding clumps of preferred genetic variants that seem to fingerprint important ancestral populations: three such clumps ("principal components") account between them for more than half of all the genetic variation among native Europeans in Europe. One clump, in which the marker genes include HLA-B5 and HLA-B7, is mapped here. It seems to correspond with events of 8700 to 5000 yr ago when gardeners spread across Europe from western Asia, as shown on the accompanying map adapted from* Man Before Metals, *British Museum, 1979.*

1. A second wave to Australia with more robust skeletons, perhaps via western Australia, before 13,000 yr (Kow Swamp).
2. Amerindians from eastern Asia or North America: through southern North America **c. 11,300 yr** (11,000 yr at Del Mar, California); thence to farthest South America by c. 10,500 yr. See map p. 88.
3. Eskimos from Alaska: spreading c. 4500 yr across Arctic North America to northern Greenland by 4000 yr.

gardeners (horticulturalists without plows): **9500 yr**. (See map p. 89.) From several independent centers, including the following:

1. Wheat and barley gardeners through southwestern Asia 9500 yr, thence slowly to Anatolia and Europe, the Indus Valley, and the Nile Valley.
2. Rice gardeners from Indochina, est. 9500 yr, spreading to southern China by 9000 yr, and to Burma.
3. Austronesian seafarers with yams, coconuts, pigs, etc., from Indonesia, date uncertain, to New Guinea, the Philippines, and Taiwan; also east to the Lapita core area on islands beyond New Guinea by 6000 yr; west to Malagasy (Madagascar) c. 1500 yr (AD 500); and north to Malaya c. 800 yr (AD 1200). *See also* below: Polynesians.
4. Amerindian maize gardeners from Middle America, ?6300 yr, south to Peru, and onward to Chile.

dairymen (milk-drinking, proto-Semitic speakers with milking cows and plows): **6000 yr**. (See map p. 89.) From the steppe fringe of the Fertile Crescent, into Egypt and across the Sahara Desert, arriving in western Africa 4500 yr; also into southern Mesopotamia where the Semitic-language empire of Akkad emerged 4400 yr. People of unknown linguistic affinity were plowing in Britain by 5200 yr. For later movements of Semitic speakers, *see* below: Arabs.

horsemen (milk-drinking, proto-Indo-European speakers with milking cows and tamed horses): **4900 yr**. (See map p. 89.) From the lower Volga River (Yamnaya culture 5000 yr), into semiarid niches in southwestern Asia and across the Eurasian steppes; into Anatolia as the Luvians 4900 yr, and the Hittites 4000 yr; into southern Greece 4350 yr, and technologies into western Europe (e.g., France, Ireland) by 4300 yr; also east into the region beyond the Caspian Sea c. 4000 yr, and on into central Asia; south into southwestern Asia (in Syria as Mitanni, fl. 3600 yr), and permanently into Iran as Persians and Medes by c. 3350 yr. *See also* below: charioteers, blacksmiths, Romans, Vikings, navigators, Russians, and Americans.

northern dairymen (milk-drinking speakers of proto-Uralic languages): **est. 4900 yr**. From the middle Volga River into the northern Eurasian forests; to the Baltic Sea as proto-Finnish and proto-Lappish speakers; also west as Magyars into Hungary 1105 yr (AD 895).

charioteers (Indo-European speakers and others): **c. 4000 yr**. (See map p. 89.) From western Asia east to the Altai Mountains of Mongolia by c. 3900 yr. Unidentified charioteers in China founded the Shang kingdom 3800 yr, which expanded 3650 to 3400 yr (Middle Shang). Aryans (Iranian branch of the Indo-Europeans) began assaults on India c. 4000 yr and took the Indus Valley ?3600 yr; Aryans migrated to the Ganges Valley by 3000 yr and Sri Lanka by 2500 yr; they founded the Mauryan empire, c. 2325 yr, which by c. 2250 yr (Asoka) encompassed most of the Indian subcontinent except the far south. Illus. (China) p. 168.

blacksmiths (Iron Age peoples): **c. 2800 yr**. (See map p. 89.)

1. Phoenician traders becoming Carthaginians (Semitic speakers) from the Levant (Phoenicia) c. 2880 yr along the Mediterranean, esp. to Carthage c. 2800 yr; thence to Spain, Sardinia, Sicily, etc.
2. Greeks (Indo-Europeans) from Greece and western Anatolia c. 2800 yr to France, Italy, Sicily, Tripoli, and beside the Black Sea; starting 2334 yr the Greeks under Alexander attacked the Persian empire. Illus. p. 168.
3. Celts (Indo-Europeans) c. 2800 yr from the Rhine Valley east and west across Europe; by 2200 yr they thrived from the Atlantic to the Lower Danube.
4. Etruscans (?Ugric speakers) c. 2700 yr from Italy to Corsica, Sardinia, and Spain.
5. Iranians (Indo-European Persians and Medes) from Iran 2550 yr into Anatolia, Egypt, and India (Sind).
6. Bantu speakers c. 2500 yr from western Africa east to Lake Victoria 2500 yr; from that secondary center west to Angola 1900 yr (AD 100) and then east into Zaire and Zambia AD 500; also south to Transvaal and Swaziland by AD 400; and from a tertiary center in southwestern Zaire, throughout eastern and southern Africa c. AD 1000 (Great Zimbabwe founded c. AD 1200). Illus. p. 169.
7. Ch'u (non-Chinese, iron-pioneering inhabitants of China's Yangtze Valley) were expanding forcibly c. 2400 yr.

Germans (Indo-European speakers): c. 2500 yr. From northeast Germany and Denmark into Poland, Scandinavia, the Rhineland, and the Netherlands; then east to the northern Balkans and Ukraine c. 1850 yr (AD 150), whence the Germanic tribes (Goths, Vandals, etc.) were dislodged by the Huns (see map p. 89); they invaded the Roman empire from AD 378 onward, proceeding around it from Greece, through Spain, to Africa (AD 429). A second breakout from the core area of Germany and Denmark began c. 1572 yr (AD 428), with Franks (AD 428) and Burgundians (AD 443) into France; Anglo-Saxons into England (c. AD 440); and Lombards into Italy (AD 558). *See also* below: Vikings, navigators, and Americans.

Chinese (Sino-Tibetan speakers): **2260 yr**. The Ch'in kingdom of northwestern China (from which the name "China" derives) beat their archrivals, the Ch'i, in a decisive battle 2260 yr, and then 2230 yr ago they struck decisively east along the Yellow River, taking the Han region 2230 yr and Wei 2225 yr; and the Yangtze Valley, largely inhabited by peoples of Indochinese affinity, was seized by 2223 yr. For later expansion, see map p. 89. Trading settlements from c. 2350 yr (Malaya, Indochina, and Indonesia) to 651 yr (AD 1349, Singapore) scattered Chinese peoples widely in eastern Asia. Speakers of Mandarin Chinese have for long been the most numerous single-language entity in the world.

Romans (Indo-European speakers): **2260 yr**. After revolting against the Etruscans 2510 yr, Latin-speaking peoples from Rome slowly built their power in mainland Italy. The first large Roman navy beat their archrivals the Carthaginians at the sea battle of Mylae off Sicily 2260 yr; and the Romans took Sicily by 2241 yr. For later expansion, see map p. 89.

Polynesians (gardening and fishing Austronesians): **1900 yr (AD 100)**. From the Lapita core area east of New Guinea, across the Pacific to Marquesas by AD 100 (a.d. 40) or possibly a century or so earlier; thence to Hawaii (c. AD 300), Easter Island (c. AD 400), and New Zealand (by AD 1000).

Huns (Paleosiberian speakers, with mixed allies): **1696 yr (AD 304)**. (See map p. 89.) From east-central Asia into China in AD 304 (Lo-Yang fell, AD 311); also west into Europe, AD 370, reaching France by AD 451, where they were beaten; and south into India (AD 480) and Persia (AD 484). Few linguistic or genetic traces of this irruption remain.

Slavs (Indo-European speakers): c. 1450 yr (AD 550). From eastern Europe into central Europe and Balkans, spreading the Slavic languages.

Arabs (Semitic speakers): **1368 yr (AD 632)**. (See map p. 89.) From Arabia into western Asia, northern Africa, and southwestern Europe. Religious and commercial connotations of the Arabic language encouraged its wide adoption.

Indochinese peoples (various linguistic groups): c. 1150 yr (AD 850). Khmer from Angkor c. AD 850, achieving their greatest extent AD 1180; Thais from Siam c. AD 1150 south into Malaya, then east into Angkor empire 1394; Vietnamese from Tonking south into Champa 1471.

Vikings (seafaring Germanic speakers): **1207 yr (AD 793)**. From Scandinavia they raided and settled in the British Isles and the Frankish empire, and traded in Kiev and Novgorod, founding the Russian state at Kiev, AD 882; they took Iceland and settled in Greenland, and explored west to Canada c. AD 1000. From a secondary base in the Seine Valley in France, other Vikings (Normans) reached the Mediterranean and invaded Sicily in 1061; they also assimilated Anglo-Danish England, 1066. Illus. p. 173.

Mongols and Turks (Altaic languages): **789 yr (AD 1211)**. (See map p. 89.) From Karakorum, Mongolia, south into China, AD 1211, and west into southwestern Asia and eastern Europe. Turkish in Turkey is the chief linguistic trace. Turks carried out secondary conquests as follows:

1. Ottoman Turks from Anatolia, west into Balkans, 1351; later south into Egypt and Mesopotamia.
2. Timur from central Asia, west to Turkey 1370, and south to India.
3. Babur from central Asia, into India, 1526, founding the Mughal dynasty.

navigators (seafaring Romance and Germanic speakers): **508 yr (AD 1492)**. From Europe worldwide. After preliminary

Portuguese forays along the western African coast, c. AD 1415, Spanish and Portuguese flotillas began exploring the oceans in earnest in AD 1492 (see map p. 90). Seeding colonizations proceeded as follows: El Mina (Ghana), 1482; Hispaniola, 1493; Mozambique, 1507; Goa, 1510; Malacca, 1511; Mexico, 1519; Peru, 1533; Brazil and Paraguay, 1549; Florida (Spanish), 1565; Philippines, 1571; Angola, 1576; Canada (French), 1600; Virginia, 1607; Batavia, 1619; New York (Dutch New Amsterdam), 1625; Cape of Good Hope (Dutch), 1652; Bombay, 1660; Mississippi Valley (French "Louisiana"), 1684; Calcutta, 1690; Australia, 1788; New Zealand, 1840; and Indochina, 1859. The chief modern languages spread by this breakout are Romance (esp. Spanish, from 1492) and Germanic (esp. English, from 1607). Illus. pp. 61, 176.

Russians (Slavic speakers): **414 yr (AD 1586)**. From Moscow into Asia, reaching the Pacific (AD 1649), Alaska (1784), and California (1812); Russians also, esp. from 1772, expanded south and west into the Caucasus, the Ukraine, and Poland; by 1948 they dominated eastern Europe. Russian is now a mother tongue for people distributed from Vladivostok in the east to Leningrad in the west.

Manchus (Altaic speakers): **364 yr (AD 1636)**. From Manchuria into China, where they seized Peking in AD 1644 and the rest of China and Taiwan by 1685; then west into Outer Mongolia (1696), Tibet (1720), and Tarim (1759); thereafter the tide receded, but in China the Manchu empire (Ch'ing dynasty) lasted until 1912.

Americans (Germanic, i.e. English speakers): **170 yr (AD 1830)**. From eastern North America westward, dated by the Indian Removal Act, AD 1830; into Texas ("liberated," 1836) and California (invaded, 1846); northwest to Oregon (nominally 1848, by treaty) and Alaska (1867, by purchase); then overseas, taking Hawaii and the Philippines 1898, and the Panama Canal Zone, 1903.

Japanese (independent language): **AD 1905**. Radially from Japan, dated for the Tsushima Strait battle. Japanese had already taken the Ryuku Islands in AD 1872 and Formosa in 1895, but they gained control of Manchuria and Korea in 1905 and of former German territories in China in 1915; in 1941 and 1942 they occupied southeast Asia and the western Pacific; defeated and pacified in 1945, they lost nearly all of their territorial gains; from c. 1960 onward the Japanese resumed their challenge as nonviolent traders. Illus. p. 184.

brick houses 10,600 yr. *See* building.

bridges (arched) c. 2850 yr. *See* building.

British (initially English) *East India Company* **400 yr (AD 1600)**, *see* imperial companies; *settlements overseas* (Virginia) **193 yr (AD 1607)**, *see* breakouts (navigators); *unified* 293 yr (AD 1707); *Victoria* **163 yr (AD 1837)**, *see* historical timescale; *postagricultural economy* **c. 200 yr (AD 1800)**, *see* work.

British Bengal 235 yr (AD 1765). Following the dislodgement of their French rivals, by the defeat of an Indian ruler at the battle of Plassey in AD 1757, militant traders of the East India Company had by 1765 gained territorial control over much of eastern India, including Bengal, Bihar, and Orissa. It was the start of the British conquest of India. Calcutta was the headquarters for the trade with Bengal, the products of which included cotton and silk textiles, indigo, and opium. (A. Calder, *Revolutionary Empire*, Cape, 1981.)

Bronze Age 6000 yr. *See* archeological timescale.

Brunhes geomagnetic epoch 730,000 yr. *See* geological timescale.

bryozoa 500 Myr. *See* invertebrates.

bubonic plague 1458 yr (AD 542); **Black Death 669 yr (AD 1331)**. *See* disease epidemics.

Buddhism esp. 2500 yr. *See* religious surge.

buffalo (bison and water buffalo) **1.8 Myr**. *See* bovids.

building ?420,000, esp. 10,600 yr. Traces of large and well-constructed huts made from tree branches are known from Terra Amata in southern France (H. de Lumley, Musée de l'Homme, Paris). The find is assigned to the "Holsteinian" interglacial, possibly corresponding to core stage 11, 421,000 to 347,000 yr ago; hence the adoption of ?420,000 yr as a possible date. In the timetable, dates for brick buildings follow J. Mellaart, *The Neolithic of the Near East*, Thames & Hudson, 1975; those for early stone buildings are from I. Kinnes, personal communication, 1982, and C. Renfrew, *Before Civilisation*, Cape, 1973. The other items, cursory headlines for a complex subject, are from diverse sources, including J. Hawkes, *The Atlas of Early Man*, Macmillan, 1976 (dates adjusted). *See also* cities.

BUILDING: TIMETABLE

huts: **?420,000 yr**. France (Terra Amata).
brick houses: 10,600 yr. Palestine (Jericho).
upper story: 8100 yr. Anatolia (Hacilar).
brick temples: 7800 yr. Mesopotamia.
stone buildings: **6500 yr**. Guernsey.
stone temples: 5000 yr. Malta.
arches and domes: 4700 yr. Mesopotamia.
Egyptian pyramids: 4600 yr. Dahshur.
arched bridges: 2850 yr. Smyrna.
Amerindian pyramids: 2800 yr. La Venta.
Greek architecture: 2650 yr. Smyrna.
concrete: ?2200 yr. Italy (e.g., Praeneste).
large domes: esp. 1876 yr (AD 124). Rome.
Islamic architecture: 1310 yr (AD 690). Jerusalem.
iron buildings (pagodas): c. 1000 yr (AD 1000). China.

Gothic architecture: 840 yr (AD 1160). Chartres.
reinforced concrete: 133 yr (AD 1867). France.

bureaucracy 10,700 yr. *See* government; information technology.

burials, ritual ?100,000 yr. *See* neanderthalers.

Byzantine empire 1390 to 547 yr (AD 610 to 1453). *See* historical timescale.

C

calendars c. 35,000 yr. *See* information technology. *See also* king lists; religious surge.

California gold rush 151 yr (AD 1849). *See gold rushes.*

Cambrian disruptions 550 to 520 Myr. Several events, three to five in number, disrupted life in the sea during the latter part of the Cambrian period (A. R. Palmer, Conference on Large Body Impacts, Snowbird, 1981). *See also* cosmic impacts.

Cambrian period 570 to 520 Myr. *See* geological timescale.

camelids (camels and llama-like animals) 40 Myr. Of the modern genera, *Camelus* appeared **c. 4.5 Myr** (5 to 4 Myr) and *Lama*, **?500,000 yr**, dating uncertain (A. W. Gentry, personal communication, 1982, citing D. J. Golz, 1976, and S. D. Webb, 1974). Llamas and their alpaca relatives were domesticated c. 7000 yr, and camels, c. 5000 yr. *See* animal domestication.

canals esp. 4200 yr; *Grand Canal of China* **1389 yr (AD 611)**. *See* transport.

cancer mutation pinpointed AD 1982. *See* medicine.

canids 35 Myr. *See* dogs.

cannon esp. 644 yr (AD 1356). *See* weaponry.

carbon dioxide (in atmosphere) est. 4500 Myr; (declining) 2500 Myr. *See* atmosphere.

carbon-14 dating *See* radiocarbon dating.

Carboniferous era 360 to 290 Myr. *See* geological timescale.

carnivores (mammalian order) **est. 90, esp. 60 Myr**. *See* mammals.

carnosaurs (lizard-hipped meat-eaters) **c. 225 to 67 Myr**. *See* dinosaurs.

carve-up of continents 506 yr (AD 1494), **esp. 273 yr (AD 1727), 120 yr (AD 1880), and AD 1945**. From time to time, powerful nations have sought to divide large tracts of the world between themselves, to minimize superpower conflict, without regard for the wishes of indigenous peoples. The most ambitious agreement of that kind was the formal carve-up of the undiscovered world between the Portuguese and Spanish empires, by the Treaty of Tordesillas in AD 1494; one lasting effect is the linguistic division of South America, with Portuguese spoken in Brazil, and Spanish elsewhere. Similar events include the Treaty of Nerchinsk in 1689 which defined boundaries between the Russian and Chinese empires, and a no-man's-land between them; the carve-up of Asia reached a climax in **AD 1727**, when the Treaty of Kiakhta permitted the Chinese to advance across much of the Eurasian steppe. Western North America was divided between Americans (United States) and British (in Canada) along latitude 49°N in 1818; this line was extended to the Pacific in 1848. In a carve-up of Africa beginning **c. AD 1880** (dated for Belgian activity in the Congo in 1879 and the French occupation of Tunisia in 1881), a scramble of Europeans divided most of the continent between preexisting colonialists (Portuguese, Dutch, British, and French) and newcomers (Belgian, German, Spanish, and Italian). Another event of this sort was the division of Europe into Western (esp. American) and Soviet spheres of influence in **AD 1945**, toward the end of World War II. (Sources include W. H. McNeill, *A World History*, Oxford University Press, 1979; G. Barraclough, ed., *The Times Atlas of World History*, Times Books, 1978; and S. E. Morison, *Oxford History of the American People*, Oxford University Press, 1965.)

cast iron (China) **c. 2400 yr**; (Europe) **620 yr (AD 1380)**. *See* metal industries.

catastrophes 670 Myr, etc. *See* mass extinctions.

cats 35 Myr. The cat family (felids) appeared as a branch of the mammalian carnivores in the Early Oligocene (*see* mammals). The cheetah (*Acinonyx*) and the sabertooth "tiger" (*Megantereon*) became widespread **3.2 Myr** ago, in the Late Pliocene; the latter disappeared from Africa, Europe, and western Asia 1.8 Myr ago, but the sabertooth *Smilodon* survived elsewhere, before becoming extinct in North America c. 10,000 yr ago, early in the Holocene. (Data primarily from C. A. Repenning and O. Fejfar, *Nature*, Vol. 299, 1982, p. 344.) Modern cats (genus *Felis*) evolved **c. 8 Myr** ago (9 to 8 Myr, Late Miocene). Modern species include leopards (*F. pardus*) and lions (*F. leo*) **1.8 Myr**, end of

Pliocene; jaguars (*F. onca*) 1.7 Myr, Early Pleistocene; tigers (*F. tigris*) 600,000 yr, Late Pleistocene; and wildcats (*F. sylvestris*) **c. 120,000 yr**, in the last interglacial. (Data from A. W. Gentry, personal communication, 1982, citing R. H. Tedford, G. de Beaumont, H. de Lumley, A. J. Stuart, B. Kurtén and E. Anderson, and E. H. Colbert and D. A. Hooijer, in various publications, 1963 to 1980.) The wildcat was domesticated in Egypt 4500 yr ago.

cattle 3.5 Myr, esp. 200,000 yr, *see* bovids; *tamed* **8800 yr**, *see* animal domestication; *milking* ?7800 yr, *see* livestock revolution.

cave-bear cult 47,000 yr. *See* neanderthalers. For cave-bear origins, *see* bears.

cells (bacteria) **est. 4000 Myr**; (modern, eukaryotic cells) **est. 1700 Myr**. *See* life on Earth.

Celtic breakout c. 2800 yr, *see* breakouts (blacksmiths); language c. 2500 yr, *see* language.

Cenomanian turnover 95 Myr. In the Late Cretaceous, changes in the dinosaur populations were accompanied by some extinctions and innovations among marine animals, and a major fall in sea level. *See* dinosaurs; mass extinctions; sea-level stratigraphy.

Cenozoic ("new life" era) **67 Myr** to present. *See* geological timescale.

ceramics 29,000 yr. *See* arts and crafts.

cereals *See* crops; grass.

Chalcolithic period (Copper Age) **8000 yr.** *See* archeological timescale.

chalk deposits esp. 93 to 67 Myr. *See* continents flooded.

chariots 4100 yr, *see* transport; weaponry; *breakout of charioteers* c. 4000 yr; *chariots to China and India* **c. 3800 yr**, *see* breakouts (horsemen).

cheetahs 3.2 Myr. *See* cats.

chelicerates (pincered arthropods) **560 Myr**. *See* invertebrates.

chemical fertilizers c. 180 yr (AD 1820), **esp. AD 1913**. *See* cultivation practice.

chemistry esp. 229 yr (AD 1771). *See* oxygen discovered.

chimpanzee est. 5 Myr. *See* primates.

Ch'in (kingdom) **esp. 2260 yr**, *see* breakouts; (empire) **2221 yr**, *see* historical timescale.

China *coal* **esp. 260 Myr**, *see* coal; *assembled* (south China hit North China) **esp. 230 Myr**; *coastal provinces added* c. 140 Myr, *see* continents; Himalayan uplift; *tungsten* c. 140 Myr, *see* ores; *Peking Man* **?650,000 yr**, *see* human origins; *agriculture* **9000 yr**, etc., *see* animal domestication; crops; *Chinese language* ?5000 yr, *see* language; *chariots to China* **3800 yr**, *see* breakouts (charioteers); *Chinese writing* **3500 yr**, *see* information technology; *cast iron* **2400 yr**, *see* metal industries; *breakout* **2260 yr**, *see* breakouts (Chinese); *Great Wall* **2214 yr**, *see* weaponry; *population censuses* **1998 yr (AD 2)** to AD 1982, *see* population history; *science* **esp. 1996 yr (AD 4)**, *see* science; *Grand Canal* **1389 yr (AD 611)**, *see* transport; *naval ventures* **595 yr (AD 1405)**, *see* Chinese naval ventures; *Manchu China* **356 yr (AD 1644)**, *see* breakouts (Manchu); *Chinese cowed (opium wars)* **158 yr (AD 1842)**; *reunified* **AD 1949**, *see* historical timescale. See also maps pp. 81–95.

Chinese naval ventures 595 yr (AD 1405). The Chinese had established a trading colony in Singapore in AD 1349. This was followed (1405) by a brief burst of large-scale Chinese maritime expeditions, in which large imperial fleets roamed from the Philippines to Africa and the Arabian-Persian gulf; it ceased abruptly and inconsequentially in 1433, leaving mainland China officially passive in matters of foreign trade, although illegal trade continued at a considerable rate. (A. Calder, *Revolutionary Empire*, Cape, 1981; P. Curtin, personal communication, 1981.)

Ch'ing empire 356 yr (AD 1644). *See* breakouts (Manchus); historical timescale.

Chinling Mountains c. 230 Myr. *See* continents.

chloroplasts (photosynthetic units) **est. 1500 Myr.** *See* life on Earth.

cholera epidemics 183 yr (AD 1817). *See* disease epidemics.

chopper tools c. 2.0 Myr, **esp. 750,000 yr**. *See* stone tools.

chordates 550 Myr. *See* animal origins.

Chou period 3300 to 2221 yr. *See* archeological timescale.

Christ born 2007 yr (7 BC); *Christianity legalized* 1387 yr (AD 313). *See* religious surge.

chromosomes est. 1700 Myr. *See* life on Earth.

cinematography 105 yr (AD 1895). *See* arts and crafts.

Circumantarctic Current c. 30 Myr. *See* climate 67 Myr.

circumnavigation (by sailing ship) **478 yr (AD 1522)**; (by manned spacecraft) **AD 1961**. *See* transport.

cities 8800 yr. Here, Jericho (Palestine, modern West Bank) is called the first town, founded **10,600 yr** ago (8350 b.c.), and Çatal Hüyük (Anatolia, modern Turkey) is offered as the first city, founded **8800 yr** ago (6250 b.c.). Their populations have been estimated at 2000 and 5000 respectively. By the criterion that a city generates economic growth from its own local economy, J. Mellaart (*The Neolithic of the Near East*, Thames & Hudson, 1975) suggested that even Jericho was a city state. Illus. pp. 27, 163, 164.

The world's first predominantly urbanized society arose in Mesopotamia, where settlement of the alluvium at the head of the Arabian-Persian gulf began c. 7300 yr ago (Ubaid period). In the Uruk period beginning 6000 yr ago, Uruk itself (7000 people), and Nippur and Adab (5000 people), emerged as notable cities, with about 46 percent of the inhabitants of the region living in settlements of more than 1000 people. This proportion increased sharply to 70 percent in the Nippur-Adab region **5800 yr** ago, the date adopted here for urbanization (rounded from 5750 yr, Middle/Late Uruk transition). In Early Dynastic times (5100 yr) Uruk attained a population of more than 40,000. (R. M. Adams, *Heartland of Cities*, University of Chicago Press, 1981.)

City dwellers were population surplus to the needs of agriculture, who had to generate their own economy or starve; as a result they developed the city-slicker skills called civilization. Urban regions waxed and waned in various parts of the world; and some notable cities of 500 yr ago (AD 1500) appear on the map p. 92. Villages and towns became larger than the earliest cities, yet in AD 1800 only 2.5 percent of the world's population lived in settlements of more than 20,000 people. A surge in urbanization that began in Britain c. 1825 brought the proportion up to 16 percent by 1930. At that time, urbanization became a global phenomenon, and the urban population more than doubled by 1960, with the fastest growth in the largest cities. North America became the first urbanized continent, c. 1950, with half its population in settlements of 20,000 or more (G. Breeze, ed., *The City in Newly Developing Countries*, Prentice-Hall, 1969). By the 1970s, some reversal of the flow of population, turning outward from the largest cities, was becoming apparent in prosperous countries.

Cities take in food, water, fuel, and materials, and they excrete sewage, solid wastes, and pollutants of air and water. They are also hotbeds of disease, and until c. AD 1900 cities killed more people than they bred: they not only mopped up surplus agrarian population but buried much of it. The cleanup can be dated from AD 1866 (*see* medicine).

class distinctions esp. 5600 yr. *See* government.

climate 3800 to 67 Myr Generalizations about climatic conditions in past eras should be treated with reserve, because there may have been erratic oscillations (*see* sea-level stratigraphy). Hints of a very warm and humid world 3000 Myr ago include an estimate of 70° Centigrade for the mean annual ground temperature in South Africa (L. P. Knauth and S. Epstein, *Geochimica and Cosmochimica Acta*, Vol. 40, 1976, p. 1195). The first ice ages, occurring **c. 2300 Myr** ago, may be associated with a reduction of carbon dioxide in the atmosphere by vigorous growth of photosynthetic bacteria (J. E. Lovelock, personal communication, 1982).

During the past 600 million years there have been two phases of generally high sea level, Silurian to Devonian 430 to 360 Myr, and Jurassic and Cretaceous 170 to 70 Myr ago. Before, between, and after these peaks, there were ice ages. A. G. Fischer called the alternations "greenhouse" and "icehouse," and suggested they followed a supercycle of 300 million years' duration (in M. H. Nitecki, ed., *Biotic Crises in Ecological and Evolutionary Time*, Academic Press, 1981). The dates for icehouse/greenhouse transitions are here adjusted to general sea-level changes. In the following timetable, data on ice ages are based on M. J. Hambrey and W. B. Harland, in L. R. M. Cocks, ed., *The Evolving Earth*, Cambridge University Press, 1981, and L. A. Frakes *Climates throughout Geologic Time*, Elsevier, 1979, with dates adjusted; other, very broad and provisional descriptions of climate in the timetable follow mainly Frakes. The Late Proterozoic starting date is here linked to climatic events, rather than the formal geological timescale.

CLIMATIC TIMETABLE 3800 TO 67 MYR

Archean: 3800 Myr. Hot and humid.

Early Proterozoic: 2500 Myr. Warm and less humid, but with ice ages **c. 2300 Myr** ("Huronian" etc., 2400 to 2100 Myr) in three episodes of glaciation, with evidence from North America, Africa, and Australia.

Middle Proterozoic: 1600 Myr. Warm and moderately humid.

Late Proterozoic: 950 Myr. Cold, with repeated series of ice ages beginning **950 Myr** (evidence from North America, Africa, and Australia), **770 Myr** (evidence from Africa, Australia, etc.), and **670 Myr** ("Varangerian"; evidence from most continents); the 670 Myr series may have been the coldest episode in the history of the Earth.

Paleozoic: 570 Myr. Warm and moderately humid during the Fischer "greenhouse" phase **430 to 360 Myr**, but with two ice-age series: "Ordovician" **440 Myr** (Late Ordovician) affecting Gondwanaland, esp. Africa, then at the South Pole; and "Permo-Carboniferous" **290 Myr** (earliest Permian, Sakmarian times) with glaciation in Gondwanaland, esp. Antarctica, Australia, India, South Africa, and South America.

Mesozoic: 245 to 67 Myr. Inclining to dryness, c. 240 Myr (Middle Triassic), but warm and moderately humid **170 to 70 Myr**, in the second Fischer "greenhouse."

climate 67 to 3.25 Myr Overviews of climatic events come from oxygen-isotope analyses of surface-living and bottom-living forams at high latitudes in the Pacific (N. J. Shackleton and J. P. Kennett, in *Initial Reports of the Deep Sea Drilling Project*, Vol. 29, 1975; J. P. Kennett, *Marine Geology*, Prentice-Hall, 1982). The origin of the vigorous Circumantarctic Current, or West Wind Drift, has been pinpointed (J. P. Kennett et al., *Nature*, Vol. 239, 1972, p. 51). These data, reinforced by Kennett's interpretations, are the basis of the timetable, which also takes into account major sea-level falls that may be indicators of ice amassing on land (*see* sea-level stratigraphy). The sea-level curves show marked and rapid fluctuations around a relatively high sea level until 29 Myr ago, and fluctuations around the present sea level thereafter. Geologists working on land in eastern Antarctica suspect that this continent was covered with a large ice sheet 20 Myr ago, earlier than deep-sea drilling data have indicated (R. A. Kerr, *Science*, Vol. 216, 1982, p. 972). Illus. p. 128.

CLIMATIC TIMETABLE 67 TO 3.25 MYR

Paleocene: 67 to 55 Myr. Overall warming occurred, but a major fall in sea level 61 Myr ago suggests an interruption.

Eocene: 55 to 37 Myr. A sharp oceanic cooling of several degrees Centigrade, c. 50 Myr ago, coincided with a major fall in sea level. Thereafter the temperature declined more gradually, with another marked fall in sea level c. 41 Myr ago.

Oligocene: 37 to 25 Myr. The first major lurch toward colder conditions, by several degrees Centigrade, occurred **37 Myr** ago (*see* Eocene terminal turnover), with wintry conditions beginning on land at high latitudes. After a "blip" of somewhat warmer temperatures, the Circumantarctic Current formed **30 Myr** ago, and a second major cooling (again of several degrees), c. 29 Myr, was associated with a very large fall in sea level, possibly implying the temporary accumulation of Antarctic ice sheets. Temperatures recovered gradually afterward.

Miocene: 25 to 6.3 Myr. By c. 15 Myr ago sea-surface temperatures had regained the level of 40 Myr ago, although there was a major fall in sea level 20 Myr ago. Following disturbances 15 Myr ago (*see* Miocene disruption), East Antarctica was in deep freeze with a permanent ice sheet, **14 Myr** ago. A major sea-level fall occurred 10 Myr ago, and the earliest known northern glaciers of recent times appeared in Alaska **9 Myr** ago. West Antarctica acquired a permanent ice sheet **6.6 Myr** ago, the sea level fell again, and glaciers appeared in South America at about the same time. Sea-bottom temperatures were by this time in a continuous decline, but surface temperatures recovered a little by 6 Myr ago.

Early Pliocene: 5 to 3.25 Myr. Shoaling water between North and South America, 3.9 Myr ago, stimulated the Gulf Stream. Temperatures were falling again, and **3.25 Myr** ago the current series of ice ages began.

climate 3.25 Myr to 128,000 yr After the onset of the current series of ice ages, temperatures were often lower than they are during the warm, interglacial interlude in which we live. The onset can be dated fairly precisely at **3.25 Myr**, by magnetic stratigraphy and oxygen-isotope data (N. J. Shackleton and J. P. Kennett, *Nature*, Vol. 270, 1977, p. 216; date adjusted to take account of new dates for magnetic reversals). Then, **2.4 Myr** ago, the glaciations became deeper. Glaciation in high latitudes is often associated with drought in the tropics, and a dramatic decrease in rainfall occurred in eastern Africa **2.0 Myr** ago (oxygen-isotope data, J. E. Cerling et al., *Nature*, Vol. 267, 1977, p. 237; P. I. Abell, *Nature*, Vol. 297, 1980, p. 321).

The ice ages were inherently cyclical, alternating with warm interludes (N. J. Shackleton and N. D. Opdyke, *Quaternary Research*, Vol. 3, 1973, p. 39). About 34 main glaciations can be counted in the interval 3.25 Myr to 128,000 yr ago (*see* orbital dating). The rhythm became more emphatic **800,000 yr** ago, and the Arctic pack ice began to persist even in summer from **700,000 yr** ago (Y. Herman and D. M. Hopkins, *Science*, Vol. 209, 1980, p. 557). For divisions of this Late Pliocene-Early Pleistocene period into Villafranchian and Biharian phases, *see* geological timescale. Illus. pp. 35, 145, 147, 149.

climate 128,000 to 10,300 yr For the last major glacial cycle, a clear picture comes from oxygen-isotope data from ocean cores, adjusted by orbital dating and reconciled with radiocarbon dating, together with oxygen-isotope data from the Greenland ice sheet and pollen data from various sites, notably from Grande Pile in northern France (trees versus herbs). It started with a very warm interlude, the Eemian or Sangamonian (European and North American nomenclature). The timetable follows mainly G. M. Woillard and W. G. Mook (*Science*, Vol. 215, 1982, p. 159), with some trivial adjustments to dates, and "Dryas" data cited in H. H. Lamb, *Climate: Present, Past and Future*, Vol. 2, Methuen, 1977. The presence in northwestern Europe of the flower mountain avens (*Dryas octopetala*) first signaled an improvement in the climate, but the Older Dryas and the Younger Dryas marked sharp reversals to colder conditions. *See also* ice ages, false; volcanoes (Toba). Illus. pp. 153, 160.

CLIMATIC TIMETABLE 128,000 TO 10,300 YR

Eemian (warm interlude): **128,000 yr**.
warm maximum: 120,000 yr.
false ice age: **115,000 yr**.
St. Germain 1 (warm spell): 108,000 yr.
false ice age: **95,000 yr**.
St. Germain 2 (warm spell): 88,000 yr.
pre-Brørup (very cold oscillation): **c. 73,000 yr**.
start of ice age: **72,000 yr**.
lesser ice maximum: **62,000 yr**.
milder interlude (with oscillations): **58,000 yr**.
Moershoofd (mainly mild spell): 52,000 to 47,000 yr.
Hengelo (mild spell): c. 40,000 yr.
Denekamp (inc. mildest spells): c. 30,000 yr.
severe cold: **28,000 yr**.
ice maximum (cold and dry): **18,000 yr**.
Oldest Dryas (less frigid): 15,000 yr.
Bølling (warm spell; thaw started): **14,000 yr**.
Older Dryas (recooling): 13,000 yr.
Alleröd (warm spell): 12,000 yr.
Younger Dryas (severe recooling): **10,900 yr**.
ice age ended: **10,300 yr**.

climate 10,300 yr to present Fluctuations since the end of the ice age have been complex, but they are simpler to analyze on the assumption that changes in the Earth's attitude in orbit and variability in the sun's output have been responsible. Changes are then fundamentally a matter of warming or cooling, and they are assumed to be synchronous worldwide, with variations in sea level providing an important source of data. Globally (not necessarily locally) warmth goes with generous rainfall, and cooling with drought.

The following timetable draws especially on botanical changes (W. M. Wendland and R. A. Bryson, *Quaternary Research*, Vol. 4, 1974, p. 9); on growing-season temperatures in California shown by a 5421-year tree-ring series (V. C. LaMarche, *Science*, Vol. 183, 1973, p. 1043); on emergences and transgressions showing sea-level fluctuations (R. W. Fairbridge, *Science*, Vol. 191, 1976, p. 353); on stream erosion and glaciation (G. R. Brakenridge, *Nature*, Vol. 283, 1980, p. 655); on mid-twentieth-century interpretations (Climatic Research Unit, Norwich); and on diverse other data (H. H. Lamb, *Climate: Present, Past and Future*, Vol. 2, Methuen, 1977). Adjusted radiocarbon dates are used for the phase 10,300 to 5400 yr ago. Thereafter, maxima and minima in the Californian tree-ring data calibrate the onset of trends; because they relate to growing-season temperatures, the tree-ring data do not necessarily indicate the severity of trends. Illus. pp. 175, 189.

CLIMATIC TIMETABLE 10,300 YR TO PRESENT
(Dates of starts of episodes, all approximate)

Preboreal: 10,300 yr. Ice age ended; cool.
Boreal: 9800 yr. Warmer; monsoon in India; warmest phase in southern hemisphere.
Atlantic: 9000 yr. Very warm (Europe); heavy rain (Africa) until 8000 yr; Older Peron transgression 7000 yr; Ur floods 6500 yr.
Early Subboreal: 5800 yr. Sharp cooling; Bahama emergence 5360 yr (further cooling); Younger Peron transgression 4920 yr (sharp warming); Kish (biblical) flood 4900 yr.
Late Subboreal: 4120 yr. Cooling; Crane Key emergence 3280 yr (sharp cooling).
Subatlantic: 2960 yr. Sharp warming, Abrolhos transgression; cooling 2840 yr; warming 2480 yr; Florida emergence 2120 yr (cooling); Rottnest transgression 1850 yr (AD 150) (warming).
Scandic: 1730 yr (AD 270). Cooling; warming AD 630; cooling AD 670.
Neoatlantic: 1070 yr (AD 930). Very warm.
Pacific: 810 yr (AD 1190). Cooling; warming AD 1470.
Little Ice Age: 470 yr (AD 1530). Cooling; warming AD 1700; cooling AD 1810.
Modern: 150 yr (AD 1850). Warming; cooling AD 1940; variable AD 1972).

Evidence for general and particular relationships between climatic and cultural changes comes from the work of R. A. Bryson and H. H. Lamb (references above), although archeologists and historians often remain skeptical. The following climatic events that bore on human affairs are singled out for special notice. At the very least they say something interesting about the environment within which events occurred; but causal connections with the events mentioned here are judged to be plausible.

CLIMATES AND PEOPLES: TIMETABLE

ice age ended: **10,300 yr**. Fisheries boom, c. 10,000 yr; breakouts of gardeners, 9500 yr.

sharp cooling: **5800 yr**. Subboreal episode onset; in Mesopotamia, the Middle/Late Uruk transition, 5750 yr, and taxes, 5600 yr.

sharp warming: **4900 yr**. Younger Peron transgression; the Indus Valley, the rise of Harrapan culture; in the Americas, the rise of Independence culture; in Eurasia, the breakout of horsemen, and the biblical flood.

North African desiccation: **4500 yr**. After oscillations the level of Lake Chad fell decisively, 4500 yr (J. Maley, *Nature*, Vol. 269, 1977, p. 573); nomadic peoples were displaced southwards from the Sahara.

Harrapan collapse: **4000 yr**. In the Indus Valley, cultural withering, following cooling c. 4120 yr (Early/Late Subboreal transition).

sharp cooling: **c. 3250 yr**. Crane Key emergence; in Middle America, the rise of the Olmecs, c. 3250 yr; in eastern Mediterranean region, Trojan War, c. 3250 yr, poor Nile floods, and the fall of Hittites and Mycenaeans, c. 3200 yr.

cooling: c. 2840 yr. In Middle America, the Maya farming boom in presently swampy terrain; in eastern North America, the rise of the Adena culture; in Japan, the intrusion of the Yayoi culture, all c. 2750 yr.

droughts: **c. 1730 yr (AD 270)**. Scandic episode onset; in North America, retreat of the Hopewell people (successors of Adena) from the Great Lakes region, c. AD 300; in central Eurasia, a fall in the level of the Caspian Sea, and the breakout of the Huns, AD 304; in northern Africa, a fall in level of the Nile floods; in Arabia, recurrent droughts before the breakout of the Arabs, AD 632.

warm episode: **c. 1070 yr (AD 930)**. Neoatlantic onset; in Europe, crops and Vikings prospered, and a spate of church building occurred; in North America, the Anasazi moved into dry Chaco Canyon territory.

climatic adversity: **810 yr (AD 1190)**. Pacific episode onset; in North America, the Mill Creek culture of Iowa declined, from AD 1200, and the Anasazi retreated by 1300; in mid-latitude central Eurasia, previously high Crimean rainfall showed a decline from 1160 onward, before the breakout of the Mongols in 1211.

Little Ice Age: **c. 470 to 150 yr (AD 1530 to 1850)**. Glaciers advanced in North America and Europe; breakouts of the Russians, AD 1586, and the Manchus, 1636.

climatic adversity **c. 810 yr (AD 1190).** *See* climate 10,300 yr.

climatic oscillation **73,000 yr.** *See* climate 128,000 yr.

climatic variability **c. AD 1972.** *See* climate 10,300 yr.

clockwork (escapement) **1275 yr (AD 725)**. *See* arts and crafts.

clothing, tailored **25,000 yr.** *See* arts and crafts.

club mosses **395 Myr.** *See* plants ashore; also forests.

clumping of matter **est. 10^{-35} second** after origin. *See* galaxies.

clusters of galaxies **?12,500 Myr.** *See* galaxies.

coal **esp. 320, 260, and 70 Myr.** Nature makes the best coal by growing trees in warm coastal swamps like those in Florida today, where the crust is subsiding, and then burying the resulting peat deep underground, where heat and pressure gradually transform it into coal of increasing quality. Coal traces formed from primitive marine organisms occur in old rocks (for example in Greenland, 2000 Myr old), but land plants had to evolve before generous deposits of coal could accumulate. This process began 370 Myr ago, in the Late Devonian period, and continued in the Carboniferous, the name of which means "coal-bearing." Trees with fern-like foliage were outstanding contributors to the coal formed during the Carboniferous. Coal is arguably the chief industrial mineral, and the industrial ascendancy of northern Europe and the northeastern United States during the nineteenth century AD depended on the fact that, in a favorable geological phase around 300 Myr ago, these regions lay in the best location, astride the equator, where swampy forests produced excellent coal. Coal making climaxed in these regions 320 to 290 Myr (Late Carboniferous). For the world as a whole, the period of most rapid formation of surviving coal occurred 260 to 245 Myr (Late Permian), when 30 percent of the world's reserves were created, especially in China and the U.S.S.R. (B. Tissot, *Nature*, Vol. 277, 1979, p. 469; dates adjusted). Another important coal-forming episode occurred in western North America 70 to 37 Myr ago. The United States, Britain, and Germany share between them 40 percent of the world's reserves of coal; China and the U.S.S.R. have another 40 percent. *See also* energy.

coelenterates 670 Myr. *See* animal origins.

coelomate worms **est. 670 Myr.** *See* animal origins.

coelurids (small dinosaurs) **235 Myr**. *See* dinosaurs.

coins esp. 2600 yr. *See* information technology.

cold, severe **28,000 yr.** *See* climate 128,000 yr.

Cold War **esp. AD 1947.** *See* historical timescale.

colonization (global) esp. 507 yr (AD 1493) etc. *See* breakouts (navigators); government.

comets **4550 Myr.** *See* cosmic impacts; solar system forming.

communications satellite **esp. AD 1965.** *See* information technology.

communities (more than 100 people) **29,000 yr**; economic communities esp. AD 1957. *See* government.

compass (navigational) c. 885 yr (AD 1115). *See* transport.

composite tools c. 29,000 yr. *See* stone tools.

computer AD 1943. *See* information technology.

concrete ?2200 yr; (reinforced) 133 yr (AD 1867). *See* building.

Confucianism esp. 2500 yr. *See* religious surge.

conifers 350 Myr. *See* plants ashore.

continents 3800 Myr to present. The outer shell of the Earth manufactures basaltic ocean floor at rifts, and destroys it at ocean trenches, where the more buoyant granitic material of continents is manufactured. Existing continents and island arcs are pushed around like flotsam that collides and breaks up (*see* paleogeography). These processes are the key to continental drift, and the geometric theory of plate tectonics initiated modern geology by describing the motions of rigid oceanic plates on a sphere (D. P. McKenzie and R. Parker, 1967, and W. J. Morgan, 1968). Unlike oceanic plates, continents are not rigid. They can stretch by nearly vertical faulting between blocks, creating sedimentary basins in which thick deposits can accumulate (D. P. McKenzie, *Earth and Planetary Science Letters*, Vol. 40, 1978, p. 25).

Evidence of past events includes relics of oceans along sutures that mark continental collisions; radiometric ages of granite masses, built by plate action, which constitute the characteristic buoyant basement of continents; torn edges of continents, and associated rift valleys and volcanic dikes, which mark episodes of breakup; and ancient "transform" fault lines where one continental block has slid past another.

The oldest rocks known on the Earth are in western Greenland, **3800 Myr** old (S. Moorbath, personal communication, 1980). The main history of the continents concerns the origin and movements of the underlying basement rocks. Radiometric dating shows when segments of the continents were last melted in ancient mountain-building (orogenic) events. Because of the time taken for deep-lying rocks to cool and solidify, the radiometric dates are typically about 50 million years younger than the mountain-building events. Mountain building does not necessarily require the collision of continents, but continental collisions have been a means not only of building mountains but of welding small continents together. From time to time, collisions have created supercontinents in which virtually all of the world's land masses were gathered in a single unit.

A predominant opinion among geologists is that continents have grown with the passage of time (J. F. Dewey and B. F. Windley, *Philosophical Transactions of the Royal Society*, Vol. A301, 1981, p. 189; for a contrary view see R. L. Armstrong, same volume, p. 443). Estimates of the proportion of present continental material formed in different phases of the Earth's history are as follows:

3800 to 3000 Myr	15 percent
3000 to 2500 Myr	70 percent
2500 Myr to present	15 percent

A peak in continent making beginning **2800 Myr** ago may have coincided with the onset of the modern tectonic regime, although the previous regime was not very different (N. H. Sleep and B. F. Windley, *Journal of Geology*, Vol. 90, 1982, p. 363).

High rates of mountain building have alternated with very low rates. The most recent "low" corresponded with the completion and initial breakup of the supercontinent of Pangaea 220 Myr ago. From earlier "lows" in the radiometric data, W. R. Dickinson (*Philosophical Transactions of the Royal Society*, Vol. A301, 1981, p. 207) visualized three earlier supercontinents, at 2300, 1300, and 800 Myr ago. Of these, he regarded the 1300 Myr supercontinent as "possible" only; here it is regarded as firmer, but by Dickinson's own data, and information from other sources, it is backdated to 1500 Myr ago.

Names for the earlier supercontinents employed in this book are unauthorized but not arbitrary. *Kenora* **(2300 Myr)** is named for the Kenoran mountain-building event in North America. *Amazonia* **(1500 Myr)** refers to the Amazonian shield (Guayana and Guapore cratons) that consolidated c. 1700 Myr ago. *Baikalia* **(800 Myr)** owes its name to the Baikalian mountain-building event extending for three thousand kilometers across eastern Asia, as a major joint in the third supercontinent. The name *Pangaea* **(220 Myr)** was coined by A. Wegener, early advocate of continental drift.

Almost every piece of the basement of the Earth's land surface has been involved in one or more mountain-building events, in the recent or remote past. Such events are therefore beyond enumeration, and only examples can be given. The following timetable of some major mountain-building events (orogenies) draws upon H. H. Read and J. Watson, *Introduction to Geology*, Vol. 2, Macmillan, 1975; B. F. Windley, *The Evolving*

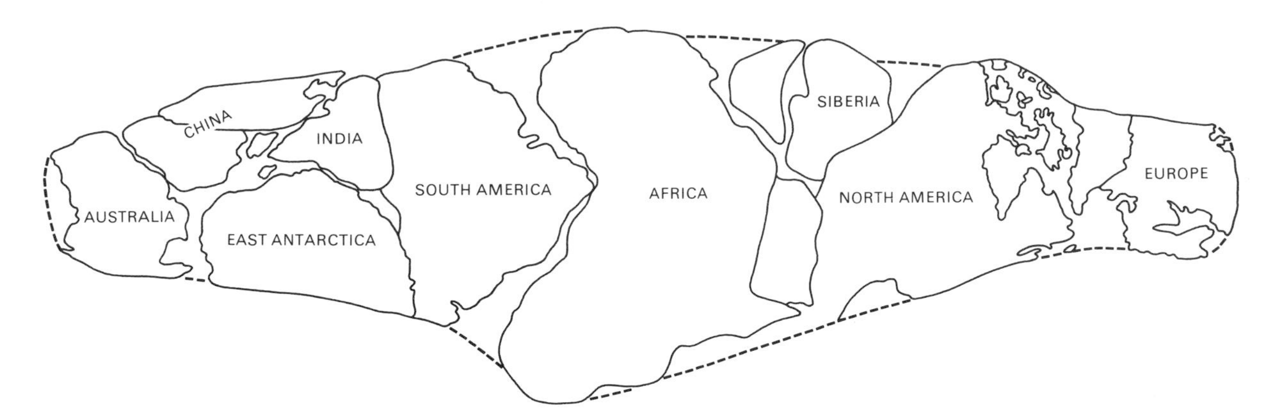

Continents, Wiley, 1977, and personal communication, 1982; D. R. Derry, *A Concise World Atlas of Geology and Mineral Deposits*, Mining Journal Books, 1980; J. F. Dewey, personal communication, 1981; and S. A. Drury, in D. G. Smith, ed., *Cambridge Encyclopedia of Earth Sciences*, Cambridge University Press, 1982. The names of some orogenies (e.g., "Variscan," "Alpine") are sometimes used globally to refer to mountains built anywhere during a certain interval; here they are applied mainly in a regional sense. Dates are smeared because the events can take tens of millions of years to complete. (For more details on some selected events, *see* American interchange; Andes; Gondwanaland; Himalayan uplift; North American uplifts; Pangaea; Swiss Alps.) See also maps pp. 81–85, and illus. pp. 109, 110, 113, 134, 135.

MOUNTAIN-BUILDING PULSES: TIMETABLE
(All dates approximate)

pulse: **2800** to 2300 Myr. Including Kenoran, Scourian, and Saamide, 2700 to 2300 Myr, in North America and Europe; Dnieprovian, 2700 to 2300 Myr, in Ukraine; Singbhun, 2700 Myr, in India; Limpopo and Dharwar, 2600 to 2300 Myr, in Africa and India; Yilgarn (reworking), c. 2600 Myr, in Australia.

lull: 2300 Myr. First supercontinent, "Kenora."

pulse: **2200 Myr.** Including Penokean and Hudsonian, 2200 to 1950 Myr, in North America; Eburnean, 2200 to 1800 Myr, in Africa; Aravalli, 2100 to 1900 Myr, in India; Coronation, 2100 to 1750 Myr, in North America; Transamazonian, c. 2000 Myr, in South America; Usagaran, c. 2000 Myr, in Africa; Ubendian, Ruwizi, and Ruwenzorian, 2000 to 1800 Myr, in Africa; Arunta and Gawler, 2000 to 1700 Myr, in Australia; Arequipa, c. 1950 Myr, in South America; Svecofennian, 1900 to 1700 Myr, in Europe; Laxfordian, c. 1750 Myr, in Europe; Circumungava, 1750 to 1600 Myr, in North America.

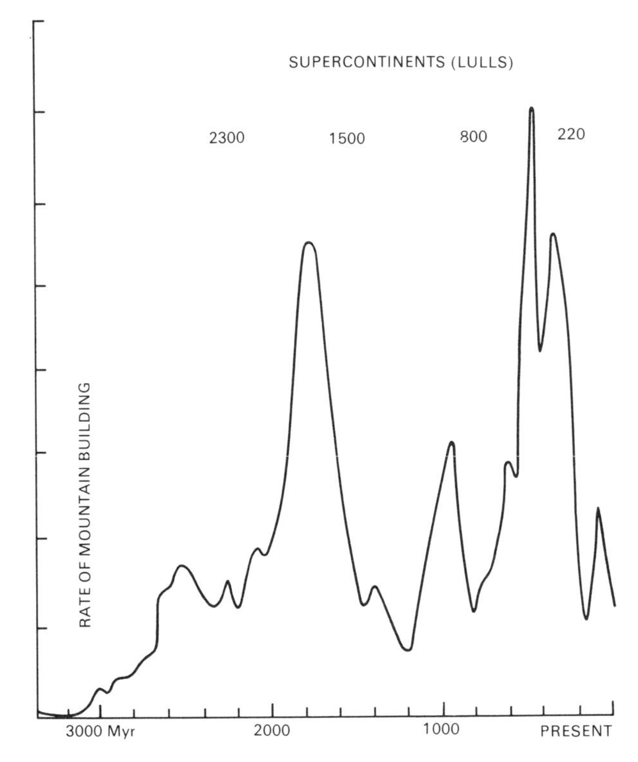

*The supercontinent of 800 million years ago has been reconstructed by J. D. A. Piper (*Earth and Planetary Science Letters, *Vol. 59, 1982, p. 61). Contrary to the view adopted here, Piper argues that this supercontinent persisted, essentially intact, from 2600 to 570 Myr ago. The major blocks drifted in company, by the evidence of fossil magnetism, and features of many ages seem to be shared by adjacent blocks. On the other hand, the emphatic pulses of mountain-building, evident in the accompanying graph, are most readily interpreted as the phases of assembly of four successive supercontinents.*

lull: 1500 Myr. Second supercontinent, "Amazonia."

pulse: **1400 Myr.** Including Karagwe-Ankolean, 1500 to 800 Myr, in Africa; Riphean, 1200 to 900 Myr, in Europe; Grenville and Moinian, 1150 to 850 Myr, in North America and present Europe (Scotland); Baikalian, 1000 to 800 Myr, across Asia.

lull: 800 Myr. Third supercontinent, "Baikalia."

pulse: 650 Myr. Including Pan-African, 650 to 450 Myr, in southern continents and northwestern France; Adelaidean, 650 to 500 Myr, in Australia; Tasman, 600 to 250 Myr, in Australia; Transantarctic, c. 500 Myr, in Antarctica; Caledonian, 460 to 390 Myr, in Europe, Greenland, and North America; Acadian (early Appalachian), c. 380 Myr, in North America; Tienshan, 350 to 280 Myr, at Tarim-Kazakhstan boundary; Hercynian (Variscan) and Alleghenian (later Appalachian), 350 to 280 Myr, across Europe, North America, and South America; Altai, c. 300 Myr, at Kazakhstan-Siberia boundary; Uralian, 300 to 260 Myr, at Europe-Asia boundary; Pamir-Kunlun-Altyn Tagh, c. 300 Myr, at Tibet-Tarim boundary; Yablonovy-Stanovoy, c. 300 Myr, at Siberia-Mongolia boundary; Chinling c. 230 Myr, across China; Tanglha-Red River, c. 220 Myr, from Tibet to Vietnam; in southeastern Asia, c. 220 Myr, at Malaya-Indonesia collision.

lull: 220 Myr. Fourth supercontinent, Pangaea.

pulse: 200 Myr. Including Andean and Cordilleran, esp. 200 Myr to present, on western margins of South America, Antarctica, and North America; Yenshanian, 200 to 50 Myr, around China, Korea, and Japan; Sikhote Alin, c. 100 Myr, east of Manchuria; Verkoyansk, c. 100 Myr, Kolyma-Siberia boundary; Alps, 100 Myr, esp. 10 Myr, in Europe; Atlas, 100 to 50 Myr, in Africa; early Rockies, c. 60 Myr, North America; Caucasus, c. 60 Myr, in Georgia, U.S.S.R.; Pyrenees, c. 40 Myr, in Europe; Himalayan, 45 Myr, esp. 5 Myr, at India-Tibet boundary; Zagros-Bitlis, c. 20 Myr, in Iran and Turkey; Basin and Range, and Rockies rejuvenation, c. 8 Myr, in North America; Andean uplift, c. 4.5 Myr, in South America.

Around 500 Myr ago in the Ordovician period, the main continent was Gondwanaland (present-day Africa, Arabia, South America, Antarctica, India, Australia, and southern Europe). The chief fragments were Siberia, Laurentia (mainly eastern North America), Baltica (northern Europe), and various pieces of China and southeastern Asia. The following timetable for what happened thereafter is based primarily upon information from C. R. Scotese and J. F. Dewey, personal communications, 1981, and B. F. Windley, personal communication, 1982, supplemented from other sources, including E. Irving, *Nature*, Vol. 270, 1977, p. 304, and A. Hallam in G. Nelson and D. E. Rosen, eds., *Vicariance Biogeography*, Columbia University Press, 1981. All dates are approximate and are adapted to the timescale used here; *see* geological timescale. The date of the North American/Africa split conforms with botanical data (H. L. Cousimer), paleogeography (A. G. Smith), and the age of the Palisades basalt (New York, **195 Myr**). See also maps pp. 81–85.

MAJOR COLLISIONS AND RIFTS: TIMETABLE
(All dates approximate)

Europe hit North America: **460** to 390 Myr.
Euramerica hit Gondwanaland: **350** to 280 Myr.
Kazakhstan, Tarim, North Tibet, and North China piled up on Siberia: 350 to 280 Myr.
Siberia hit Euramerica: **300** to 260 Myr.
Pangaean slide: 260 to 220 Myr. *See* Pangaea.
South China hit North China: **230 Myr**.
Indochina and Central Tibet hit Asia: 220 Myr.
Pangaea essentially complete: **220 Myr**.
North America quit Africa: **210 Myr**.
India quit Africa: 140 Myr.
Eurasia and Africa eased apart: 140 Myr.
Korea and coastal parts of China joined Asia: 140 Myr.
South America eased from Africa: 130 Myr. An ocean fully open between Africa and South America did not develop until **85 Myr**.
Gondwanaland disintegrating: **130 Myr**. Dated from the Africa/South America rift.
Italy (etc.) quit Africa: **110 Myr**.
Kolyma and Sikhote Alin hit Eurasia: 100 Myr.
India quit Antarctica: **100 Myr**.
New Zealand quit Antarctica: 90 Myr.
Spain swung away from France: 90 Myr.
Greenland slid from North America: 90 Myr.
North America quit Europe: **60 Myr**. Greenland/Eurasia split.
Italy (etc.) hit Europe: 60 Myr.
Australia quit Antarctica: **50 Myr**.
India hit Eurasia: **45 Myr**.
South America quit Antarctica: 35 Myr.
Baja California quit Mexico: 35 Myr.
Japan quit Eurasia: **30 Myr**.
Arabia quit Africa: **30 Myr**. Oceanic crust in the Red Sea dates from **3.5 Myr**.
Arabia hit Eurasia: **20 Myr**.
Corsica and Sardinia quit Europe: 14 Myr.
Italy drove again into Europe: **10 Myr**.
North America and South America reconnected: **3 Myr**.

continents flooded 93 Myr. Exceptionally high sea level in the Late Cretaceous flooded about one third of the total areas of the continents. The onset is here dated from the late Cenomanian stage, after a major sea-level fall, 95 Myr ago. It was a time of great production of chalk and oil (*see* petroleum). The flooding also influenced the distribution of land animals, which were trapped on continental islands as much by the shallow seas as by the opening of new oceans due to the breakup of Pangaea (M. C. McKenna, *Annals of the Missouri Botanical Garden*, Vol. 62, 1975, p. 335, and A. Hallam in G. Nelson and D. E. Rosen, ed., *Vicariance Biogeography*, Columbia University Press, 1981). Flooding between South America, Antarctica, and Australia cut the routes by which placental mammals might have reached Australia. A North American seaway ran north-south across the continent from Alaska to the Gulf of Mexico c. 93 to 62 Myr (end Danian), and its effect was to separate dinosaurs and mammals of western North America and eastern Asia from those of eastern North America and Europe. A north-south seaway, the Turgai Sea (160 to 29 Myr, Callovian to Oligocene times), already divided Eurasia, with its narrowest part at the Turgai Strait northeast of the Caspian Sea.

cooking ?1.4 Myr. *See* firemaking.

cooling events (global) **10,900 yr**, *see* climate 128,000 yr; **5800 and 3250 yr**, *see* climate 10,300 yr.

Copper Age (Chalcolithic) **8000 yr**. *See* archeological timescale.

copper boom (Katanga) **850 Myr**. *See* ores.

copper smelting (Europe) **6500 yr**; (Amerindian) **3100 yr**. *See* metal industries.

corals 490, esp. 235 Myr. *See* reefs.

cosmetics (red ocher) 1.6 Myr. *See* arts and crafts.

cosmic forces $\mathbf{10^{-35}}$ **second** after origin. *See* Big Bang.

cosmic impacts **4450 Myr** to present. The Earth was assembled from small planetisimals akin to the comets, asteroids, and meteorites of the present-day solar system (*see* Earth forming) and a storm of cosmic impacts was still in progress at the completion of the Earth **4450 Myr** ago. Impacts were abating **3900 Myr** ago, as judged by a characteristic age of 4000 or 3900 Myr for many rocks of the lunar highlands. Direct evidence for more recent events on the Earth comes from known craters and also tektites—small droplets of molten rock scattered by the impacts. Most traces of cosmic impacts will have been destroyed or hidden by the renewal of the ocean floor and by geological erosion or burial on land. The detection of a thin worldwide layer of material enriched with exotic elements such as iridium, known in meteorites but very scarce on the Earth's surface, gave another kind of indication of a cosmic impact 67 Myr ago. For that discovery and possible mechanisms of slaughter *see* Cretaceous terminal catastrophe. See also illus. pp. 23, 107, 125.

The statistics of small, potentially menacing objects in the solar system suggest impacts by asteroids or "dead" comets one kilometer wide occurring every 250,000 years on average, impacts by five-kilometer objects about once in 20 million years, and impacts by fifteen-kilometer objects every 100 million years or so (G. W. Weatherill, *Icarus*, Vol. 37, 1979, p. 96; also E. M. Shoemaker, Conference on Large Body Impacts, Snowbird, 1981). A relatively small, one-kilometer object would release energy equivalent to about 100,000 megatons of TNT and dig a crater about twenty kilometers wide. A fifteen-kilometer object would be some thousands of times more energetic than that. Live comets must contribute to the most severe impacts because they can hit the Earth head-on.

Disruptions due to cosmic impacts operated like a ragged drumbeat all through the history of life on Earth, and were an important influence on the course of evolution. The following list of direct signs of cosmic impacts may include only a few percent of events; most objects would have fallen in the oceans, and the preponderance of Canadian and Soviet craters implies that the discovery rate is far from uniform across the land masses. Some large objects may have broken up before impact, producing several smaller craters instead of one large one: examples of known and possible multiple cratering appear in the list. Plausible connections with turnovers in life are noted in cross references (*see also* mass extinctions). The timetable includes major probable impact craters more than twenty kilometers in diameter (after R. A. F. Grieve and P. B. Robertson, *Icarus*, Vol. 38, 1979, p. 212; R. A. F. Grieve, personal communication, 1981; and K. J. Hsü, et al., *Science*, Vol. 216, 1982, p. 249), and tektite strewnfields (after B. P. Glass, Conference on Large Body Impacts, Snowbird, 1981; and P. J. Smith, *Nature*, Vol. 300, 1982, p. 217).

COSMIC IMPACTS: TIMETABLE

metals from space: **c. 1900 Myr**. Vredefort, South Africa, crater (140 kilometers) c. 1970 Myr, associated with chromium deposits; Sudbury, Ontario, crater (140 kilometers) and iron-nickel mass c. 1840 Myr, associated with nickel deposits.

Carswell, Saskatchewan, crater (37 kilometers): c. 485 Myr.

Siljan, Sweden, crater (52 kilometers): c. 365 Myr. The Charlevoix, Quebec, crater (45 kilometers) c. 360 Myr could be the same event. *See* Frasnian catastrophe (370 Myr).

Clearwater Lakes, Quebec, double craters (32 and 22 kilometers): c. 290 Myr.

Araguainha Dome, Brazil, crater (40 kilometers): less than 250 Myr.

Manicouagan, Quebec, crater (70 kilometers): c. 210 Myr. *See* Norian turnover (216 Myr).

Puchezh-Katunki, U.S.S.R., crater (80 kilometers): c. 183 Myr. *See* Toarcian turnover (182 Myr).

Rochechouart, France, crater (23 kilometers): c. 160 Myr.

Gosses Bluff, Australia, crater (22 kilometers): c. 130 Myr. Strangways, Australia, crater (24 kilometers), more vaguely dated, could be the same event.

Boltysh, Ukraine, crater (25 kilometers): c. 100 Myr.

Steen River, Alberta, crater (25 kilometers): c. 95 Myr.

worldwide iridium-rich layer: **67 Myr**. *See* Cretaceous terminal turnover (67 Myr).

possible multiple cratering: 65 Myr. Manson, Iowa (32 kilometers), Kamensk, U.S.S.R. (25 kilometers), ?Karsk, U.S.S.R. (60 kilometers), and smaller craters dating to around 65 Myr.

Kara, U.S.S.R., crater (50 kilometers): c. 57 Myr.

Mistastin, Labrador, crater (28 kilometers): c. 38 Myr. The much larger Popigai, U.S.S.R., crater (100 kilometers) also dates nominally to c. 38 Myr, although with estimates ranging from 70 to 20 Myr. *See* Eocene terminal turnover (37 Myr).

multiple events: 38 to 30 Myr. Evidence includes iridium-rich deposits in Caribbean, and North America, tektite strewnfield c. 35 Myr. *See* Eocene terminal turnover (37 Myr).

Libyan tektite strewnfield: 29 Myr.

Ries, Germany, crater (24 kilometers): **15 Myr**. Also the related European tektite strewnfield (*see* Miocene disruption, 15 Myr; radiometric dating). Haughton Dome, Northwest Territories (Canada), crater (20 kilometers), c. 15 Myr, may be the same event.

Bosumtwi, Africa, crater: 1.3 Myr. Also the well-related Ivory Coast tektite strewnfield 1.3 Myr.

Australasian tektite strewnfield: 730,000 yr. Possible oceanic impact. *See* Pleistocene disruption 730,000 yr.

Barringer, Arizona, crater (1.2 kilometers): c. 25,000 yr. The largest of recent craters.

Tunguska explosion, Siberia (about 12 megatons of TNT equivalent): AD 1908. The impact of a fragment of Comet Encke.

cosmological timescale Recent estimates of the age of the universe have ranged from more than 20,000 Myr to perhaps as little as 8000 Myr, but the date adopted here for the origin is **13,500 Myr** ago. A first analysis arrives at an age of 15,500 to 13,500 Myr (D. Schramm, personal communication, 1981). Ages of the oldest observable stars lie between 18,000 and 13,000 Myr. The average age of the element rhenium in meteorites is about 8700 Myr, and other elements seem a little older (*see* elements), giving a probable age of the universe between 14,000 and 10,000 Myr. Big Bang computations, involving the possible density of the universe and the cosmic abundances of helium and heavy hydrogen (deuterium), then narrow down the age of the universe to between 15,500 and 13,500 Myr.

A second approach leads to an age between 13,300 and 6700 Myr, from considerations of the history and fate of the universe. Present motions of galaxies backtrack to a notional origin (Hubble time) roughly 20,000 to 10,000 Myr ago, not allowing for slowdown due to gravity. A recent value (D. A. Hanes, 1982) gives a Hubble time of 13,000 Myr. If the density and gravity of the universe are sufficient, the galaxies may halt in their tracks and fall together again in a Big Crunch. Galaxies are now known to be heavier than they look, and swarms of novel particles, such as heavy neutrinos, gravitinos, or photinos, envisaged in theories of particle physics, may pervade the universe, adding to its mass (J. Silk, *Nature*, Vol. 297, 1982, p. 102; M. K. Gaillard, *Nature*, Vol. 298, 1982, p. 420). Powerful modern ideas about the initiation of the universe by inflation (*see* Big Bang) require a density not very different from the minimum

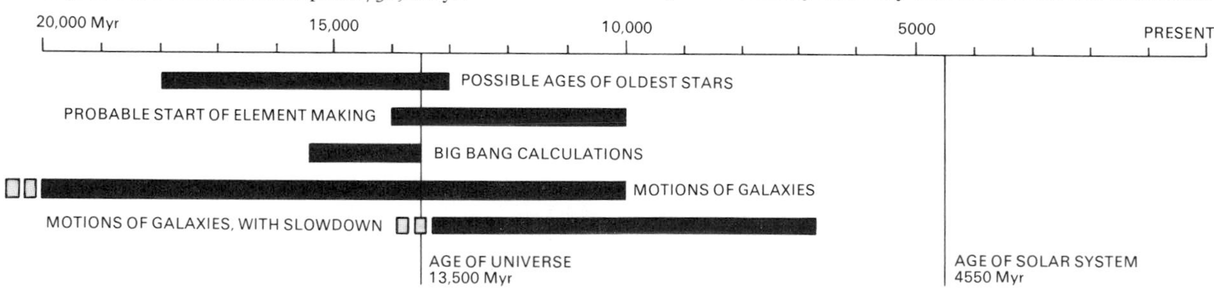

The age of the universe is pinpointed for the purposes of this book by reconciling lines of evidence and argument that define loose but not easily violable limits to the possible age. The age that comes closest to satisfying all of the stipulated conditions is 13,500 Myr.

necessary to bring about a collapse (M. J. Rees, personal communication, 1982).

That gives a good reason for adopting a mathematical model of the universe equipoised between expansion forever and eventual collapse. The Einstein–de Sitter model meets that requirement, and makes the age of the universe two thirds of the Hubble time, i.e., 13,300 to 6700 Myr. But by the considerations given earlier, the age should lie between 15,500 and 13,500 Myr. To adopt an age of 14,000 Myr would be taking liberties with the Hubble time, while 13,000 Myr would raise problems about the interplay of physical factors in the early universe. An age of 13,500 Myr minimizes special pleading.

For an Einstein–de Sitter universe 13,500 Myr old, the Hubble time is 20,250 Myr. The model specifies the age of an object in relation to its red shift. The light of a distant galaxy or quasar is shifted toward the red end of the spectrum, compared with emissions from nearby objects. A red shift $z = 3$, for example, means a 300 percent addition to the wavelength of light, or a fourfold $(z+1)$ multiplication of the wavelength. The universe is now $(z+1)$ times wider than it was when the object emitted the light and it is $(z+1)^{3/2}$ times older. Arithmetic relates the red shift z to the "look-back" time, the age on the cosmological timescale.

RED SHIFT AND AGE

z	*age from start*	*Myr ago (rounded)*
infinite	0	13,500
5	900	12,500
4	1210	12,300
3	1690	11,800
2	2600	10,900
1	4770	8700
0.5	7350	6200
0.4	8150	5400
0.3	9110	4400
0.2	10,270	3200
0.1	11,700	1800
0.05	12,550	1000
0	13,500	present

The volume of the universe was a quarter of its present value 6750 Myr ago (distances 63 percent, at $z = 0.59$), and a half at 3950 Myr (distances 78 percent, at $z = 0.26$). These ages are rounded to **6700 and 4000 Myr** for the quarter-size and half-size universe. More technically, if the present 2.7 kelvin microwave background radiation was at a temperature of 4000 kelvin when atoms first formed, that corresponds to $z = 1500$, meaning that the event began 230,000 years after the start, rounded here to **300,000 years**. Atom making was completed **one million years** after the origin, creating a transparent universe (J. Silk, 1980, reference above).

cotton domesticated c. 6300 yr, *see* crops; *British cotton industry* **esp. 200 yr (AD 1800)**, *see* work.

cow-and-plow revolution **c. 6000 yr.** *See* livestock revolution.

Crater Lake 6400 yr. *See* volcanic events.

craters (cosmic impacts) 1970 Myr to present. *See* cosmic impacts.

Cretaceous period **138 to 67 Myr.** *See* geological timescale.

Cretaceous terminal catastrophe **67 Myr.** The simultaneous extermination of dinosaurs, sea monsters, and pterosaurs ended reptilian dominance in the Mesozoic ("middle life") era. All land animals greater than twenty-five kilograms by weight were extinguished, and major groups of marine animals such as the ammonites were also obliterated. The event was not nearly as severe as the Permian terminal event of 245 Myr ago, but it was wide ranging and sudden. Life disappeared from surface waters of the oceans (J. Smit, *Proceedings of the Cretaceous-Tertiary Boundary Events Symposium*, Vol. 2, 1979, p. 156).

In a layer of clay at the very top of Cretaceous strata in Italy and Denmark, peculiar metals such as iridium were detected, elements scarcely known at the Earth's surface but occurring in meteorites (L. Alvarez, et al., *Science*, Vol. 208, 1980, p. 1095). The layer was dated at 66.7 Myr (W. Lowrie and A. Alvarez, *Geology*, Vol. 9, 1981, p. 392). The iridium anomaly was subsequently traced all around the world. The most direct interpretation is that a giant meteorite—an asteroid or comet—collided with the Earth and killed most living things. Possible mechanisms of destruction include blacking out the sunlight with dust, blast waves, poisoning, destruction of the ozone layer, and disruption of food chains. Changes in temperature were recorded in material recovered by deep-sea drilling (K. J. Hsü, et al., *Science*, Vol. 216, 1982, p. 249). A marked fluctuation in the proportions of carbon-13 and oxygen-18 atoms, close to the iridium-rich clay layer, indicates a brief, sharp cooling followed by a more prolonged heating, perhaps due to the release of large quantities of carbon dioxide from the almost dead oceans.

An impact of the magnitude envisaged would have created a crater one hundred kilometers wide, or larger.

The volcanic "hot spot" under the Hawaiian Islands chain has been offered as a possible product of the impact (H. J. Smith and R. Smoluchowski, Conference on Large Body Impacts, Snowbird, 1981). But half of the ocean floor of 67 Myr ago, representing about one third of the Earth's surface, has been recycled since then, so all evidence of cratering may have been erased. Potassium-rich beads at the bottom of the iridium-rich layer are reminiscent of those found in Siberia after a small piece of a comet hit the Earth in AD 1908, and give some support to the idea that the impacting object at the end of the Cretaceous was a large comet (J. Smit, personal communication, 1981). For comments, see p. 22.

crocodiles c. 230 Myr. *See* vertebrate animals.

crop rotation (no fallowing) esp. 300 yr (AD 1700). *See* cultivation practice.

crops 18,000 yr, esp. 9500 yr. Indications of what may be awaiting discovery come from hints of cultivation in Thailand perhaps 14,000 yr ago, and of forest clearances in Taiwan 11,000 yr, and Sumatra 7500 yr ago (B. K. Maloney, *Nature*, Vol. 287, 1980, p. 324). In a case where geological chance has preserved the traces, cultivation of wild barley and einkorn wheat turns out to have been in progress from 18,000 to 12,000 yr ago in the Nile Valley (F. Wendorf, et al., *Science*, Vol. 205, 1979, p. 1341). Systematic harvesting of wild grain began in Syria and Palestine at least 15,800 yr ago (grinding stones, Ain Gev), and more emphatically by 13,600 yr ago (Natufian). Domesticated varieties of wheat and barley appeared at Jericho 10,600 yr ago (rounded from 10,650 yr, 8350 b.c.). Harvesting of wheat by cutting the stalks selected for nonshattering heads, and preferences for cereals of different grain qualities and different growing seasons, produced a wide repertoire of cereal varieties in southwestern Asia and southeastern Europe within a few thousand years, 10,600 to 8000 yr ago.

Peoples in other parts of the world achieved similar horticultural feats at much the same time. There are invidious difficulties due to the spottiness of research, the definition of cultivation, criteria for identifying domesticated crops, and dating. Only in southwestern Asia is the story more or less clear. The following timetables should therefore be taken as indicative only, and no strong conclusions should be drawn from them, least of all about the cleverness or backwardness of the peoples involved. Data are mainly from various authors in A. G. Sherratt, ed., *Cambridge Encyclopedia of Archaeology*, Cambridge University Press, 1980; various authors in N. W. Simmonds, ed., *Evolution of Crop Plants*, Longman, 1976; and J. Mellaart, *The Neolithic of the Near East*, Thames & Hudson, 1975, supplemented from diverse sources. The origin of rice crops in Indochina is estimated from the presumably later appearance of cultivated rice outside the wild-rice region in southern China c. 9000 yr (K. C. Chang, *American Scientist*, Vol. 69, 1981, p. 148). There may be a case for regarding Indochina, Indonesia, and New Guinea as a single realm of creative gardeners. Illus. pp. 163, 165.

DOMESTICATION OF MAJOR CROP PLANTS: TIMETABLE

Western Asia and Europe

emmer wheat: **c. 10,600 yr**. Palestine.
barley: **c. 10,600 yr**. Palestine.
rye: c. 9500 yr. Syria.
einkorn wheat: c. 9000 yr. Syria.
durum (macaroni) wheat: c. 9000 yr. Anatolia.
flax: ?8500 yr. Southwestern Asia.
aestivum (bread) wheat: c. 8000 yr. Southwestern Asia.
lentils: c. 8000 yr. Southwestern Asia.
grapes: c. 6000 yr. Turkestan.
olives: c. 5500 yr. Crete.
alfalfa (lucerne): c. 4000 yr. Iran.
oats: c. 3000 yr. Central Europe.
apples: undated. ?Caucasus.
coffee: c. 600 yr (AD 1400). Arabia (ex Ethiopia).
sugar beet: 199 yr (AD 1801). Silesia.

Eastern Asia, Indonesia, and New Guinea

rice: **est. 9500 yr**. Indochina.
sugar cane: **?9000 yr**. New Guinea.
yams: ?9000 yr. ?Indonesia.
bananas: ?9000 yr. Indonesia.
coconut: ?9000 yr. Indonesia.
citrus fruit: ?8000 yr. ?Indochina.
foxtail millet: **c. 7500 yr**. Central China.
peach: ?7500 yr. Central China.
date palms: ?6500 yr. India.
cotton: c. 5000 yr. India.
black pepper: undated. India.
soybean: ?3500 yr. ?Manchuria.
cloves: ?2500 yr. Indonesia.
tea: ?2500 yr. ?Tibet.
jute: c. 200 yr (AD 1800). India.

Africa

bulrush millet: **c. 8000 yr**. Southern Algeria.
finger millet: **?8000 yr**. Ethiopia.
oil palm: c. 6000 yr. Sudan.
sorghum: c. 6000 yr. Sudan.
African yam: est. 4500 yr. Western Africa.
African rice: ?3500 yr. Western Africa.

Americas
potatoes: ?10,000 yr. Peru.
sweet potatoes: ?10,000 yr. Peru.
pumpkins: ?10,000 yr. Middle America.
beans (diverse): **c. 9500 yr**. Peru.
squashes (diverse): ?8000 yr. Mexico.
maize: **c. 7700 yr**. Mexico.
avocado: c. 7000 yr. Mexico.
cotton: c. 6300 yr. Mexico.
peanuts: c. 4500 yr. Peru.
cacao: undated. South America.
manioc (cassava): c. 3300 yr. South America.
sunflowers: c. 3300 yr. North America.
tobacco: c. 1600 yr (AD 400). Bolivia.
rubber: c. 128 yr (AD 1872). Brazil.

EXAMPLES OF EARLY TRANSPLANTATIONS

rice: from Indochina to southern China, c. 9000 yr.
wheat and barley: from Asia to Europe, c. 8500 yr.
maize: from Middle to South America, c. 6300 yr.
sorghum: from Africa to India, by 3000 yr.
grapes: from Asia to Europe, c. 3000 yr.
barley and rice: from Korea to Japan, c. 2300 yr.
Champa rice: from Vietnam to China, c. 1000 yr (AD 1000).

WORLDWIDE TRANSPLANTATIONS SINCE 440 yr (AD 1560)

sugar cane and coffee: to Brazil.
wheat, etc.: from Europe to the Americas.
maize and potatoes: from the Americas to Europe.
maize and sweet potatoes: from the Americas to China.
sunflowers: from North America to Russia.
manioc: from South America to Africa and India.
American cotton: displaced Old World cotton.
yams: redistributed around the tropics.
cacao: from South America to western Africa.
rubber: from Brazil to Asia (via Britain).
sugar beet: from Europe to California.
soybean: from Asia to Illinois.

cultivation practice **18,000 yr** to present. Crop farming is by far the largest human assault on the natural environment, and by AD 1980 the best 6 percent of the world's land was cultivated. Cultivation practice is a matter of land clearance, weed control, water supply, and sensible use of the available packages of crops and tools. Prehistorians have often supposed that a primeval method of cultivation was slash-and-burn (also called swiddening), but A. G. Sherratt (*World Archaeology*, Vol. 11, 1980, p. 313) suggested that it was a late development in tropical rain forests and cool northern forests. The earlier part of the following timetable is based on Sherratt's framework; it draws also on archeological sources noted under crops, and on J. Needham and C. A. Ronan, in B. Hook, ed., *Cambridge Encyclopedia of China*, Cambridge University Press, 1982, for some Chinese data. It is completed from diverse sources, especially for the later phases. *See also* animal domestication; crops; livestock revolution; plow. Illus. pp. 47, 187.

CULTIVATION PRACTICE: TIMETABLE

floodwater gardening: **c. 18,000 yr, esp. 9500 yr**. Nile Valley and southwestern Asia.
mixed gardening and herding: c. 9500 yr. Eastern and western Asia.
irrigation: **c. 8000 yr**. Mesopotamia (Choga Mami).
rainwatered cultivation: c. 7500 yr. Northern Mesopotamia.
cistern irrigation: 6300 yr. Jordan.
animal energy (cow-and-plow revolution): **esp. 6000 yr**. Northern Mesopotamia.
paddy: c. 4000 yr. Southeastern Asia.
multiple cropping: c. 3400 yr. China.
rain-forest drainage: c. 3000 yr. Middle America.
domesticated fodder (alfalfa): c. 3000 yr. Iran.
slash-and-burn: ?2500 yr. ?Southeastern Asia.
moldboard plow: **est. 1500 yr (AD 500)**. Germanic people.
continuous rotation (no fallowing): esp. 300 yr (AD 1700). Netherlands.
chemical fertilizers (guano): esp. 180 yr (AD 1820). Europe.
combine harvester: 164 yr (AD 1836). United States.
selective weed killers: 104 yr (AD 1896). France.
scientific plant breeding: AD 1900. Canada.
nitrogen fixation: **AD 1913**. Germany.
tractors: esp. AD 1915. United States (World War I).
chemical pesticides (DDT): **esp. AD 1939**. Switzerland.
green revolution (high-yielding cereals): **AD 1961**. Mexico.
nutrient-film technique: **esp. AD 1982**. Britain.

cycads **280 Myr.** *See* plants ashore.

cynodonts (mammal-like reptiles) **esp. 240 Myr**. *See* vertebrate animals.

D

dairymen's breakout (proto-Afrasian speakers) **6000 yr**; *northern* (proto-Uralic) **est. 4900 yr**. *See* breakouts; livestock revolution.

decolonization (India) **AD 1947**; (Africa) **esp. AD 1960**. *See* government.

deer (cervids) **20 Myr**. Deer evolved in the Early Miocene. The modern genus *Cervus* evolved c. 4.5 Myr (5 or 4 Myr) in the Early Pliocene. Among modern species, the reindeer (*Rangifer tarandus*) appeared **est. 200,000 yr**, Late Pleistocene. (A. W. Gentry, personal communication, 1982, citing L. Ginsburg, R. Ballesio, et al.) The reindeer has been domesticated for c. 2500 yr (*see* animal domestication).

dendrochronology The dating of tree rings relies on direct counts of the years registered in rings of annual growth, in trees in temperate forests. Trees within a small region often share the same experiences of good and bad years, with resulting patterns of wide and narrow rings that can be matched from tree to tree, from living trees to dead trees, and even to timber, charcoal, etc., used by human beings. By that procedure, long sequences can be built up, and use of the bristlecone pine (*Pinus longaeva*) has achieved a chronology of nearly 9000 years for western North America. One important application of dendrochronology has been in calibrating the radiocarbon timescale. Another is in studies of past climates. Rainfall, temperature, sunshine, humidity, and wind, in the different seasons, and the precise setting of each tree, can all affect the widths of rings, and elaborate statistical analyses are needed to reconstruct climatic circumstances in the past. These procedures have been followed most thoroughly in the United States (H. C. Fritts, in A. Berger, ed., *Climatic Variations and Variability*, Reidel, 1981). The use of a single series of tree rings from one place to calibrate global climatic changes (as done here, *see* climate 10,300 yr) is somewhat naive by modern standards, but against the difficulties of interpretation has to be set the advantage of accurate dating of emphatic changes. Illus. p. 29.

Devonian period 410 to 360 Myr. *See* geological timescale.

diamonds 1700 Myr, esp. 300 Myr. *See* ores.

dinosaurs 235, esp. 225, to 67 Myr. One recent study that sheds light on the origin of dinosaurs offers a revised chronology of extinctions and appearances, based on correlations of fossils and strata in Europe, North America, and southern Africa (P. E. Olsen and P. M. Galton, *Science*, Vol. 197, 1977, p. 983). Another concerns the displacement of the mammal-like reptiles (A. J. Charig, personal communication, 1981). A crucial phase of the transition appears in the Ischigualasto formation in Argentina (Late Triassic, Karnian stage, 230 to 225 Myr ago), where the first herbivorous dinosaur made its debut.

The first true dinosaurs were the relatively slender coelurids, which appeared about **235 Myr** ago. The age of the dinosaurs began in earnest **225 Myr** ago, with the emergence of giant carnosaurs and sauropods. All dinosaurs and many of their reptilian relatives vanished from the scene **67 Myr** ago (*see* Cretaceous terminal catastrophe). Intermediate turnovers of large plant-eating dinosaurs, noted below, follow data of R. T. Bakker (in A. Hallam, ed., *Patterns of Evolution*, Elsevier, 1977), with dates adjusted. Illus. pp. 84, 124.

DINOSAURS: TIMETABLE

coelurids (small "lizard-hipped" meat eaters): **235 Myr** (Middle Triassic). Typically less than two meters long, and including the ancestors of the birds.

ornithischians ("bird-hipped" plant eaters): **228 Myr** (mid-Karnian; *Pisanosaurus*). Small at first, but more spectacular in the back-plated stegosaurs 164 Myr; camptosaurs and the small hypsilophodons appeared 154 Myr.

carnosaurs (large "lizard-hipped" meat eaters): **225 Myr** (mid-Middle Keuper = early Norian; *Megalosaurus*). Early *Megalosaurus* was a six-meter, two-ton animal with ten-centimeter teeth; the later twelve-meter tyrannosaurs and twenty-meter *Deinocheirus* were in the same group.

prosauropods and sauropods (gigantic "lizard-hipped" plant eaters): **225 Myr** (*Plateosaurus*, *Anchisaurus*, *Melanorosaurus*). The last named was already twelve meters long; the enormous brachiosaurs evolved **175 Myr** ago (mid-Bajocian), together with cetiosaurs, and diplodocids arose 159 Myr.

Kimmeridgian turnover: **145 Myr**. Several families of large plant eaters (including stegosaurs, cetiosaurs, and camptosaurs) were extinguished and the diplodocids were much depleted; new families rose to prominence 133 Myr, notably the ornithischian iguanodons, panoplosaurs, and pachycephalosaurs, and the sickle-clawed deinonychids.

Aptian turnover: **117 Myr**. The brachiosaurs died out, and the small hypsilophodons were much depleted.

Cenomanian turnover: **95 Myr**. The iguanodons died out in western North America; after the event, the ornithischian hadrosaurs and protoceratopsids appeared, 89 Myr, and ceratopsids 84 Myr (inc. *Triceratops*, esp. c. 70 Myr); tyrannosaurs were conspicuous as top predators 84 to 67 Myr.

Cretaceous terminal catastrophe: **67 Myr**. Every remaining dinosaur became extinct.

disease epidemics est. 5000 yr to present. When the Pilgrim Fathers arrived at Plymouth, Massachusetts, in AD 1620, they found the district conveniently empty of Amerindians, owing to a disease that had swept the region in 1616. Far from being a coincidence, this was entirely typical of events in the Americas, where European diseases brought catastrophe. Illus. pp. 55, 171.

By skeletal evidence, tuberculosis existed prehistorically in Egypt and in the Americas, while yaws was prevalent on Pacific islands; possible evidence for poliomyelitis occurs in ancient Egypt (D. R. Brothwell, *Digging up Bones*, British Museum, 1981). Malaria was endemic in the early farming city of Çatal Hüyük c. 9000 yr ago (J. Mellaart). By epidemiological theory, the major infectious diseases of mankind became important est. 5000 yr ago, when dense population, especially in cities, created conditions favorable for the transmission of microparasitic diseases (R. M. Anderson and R. M. May, *Nature*, Vol. 280, 1979, p. 361). In Europe, life expectancy increased from twenty-five or thirty years 300 yr ago (AD 1700) to about seventy or seventy-five years in AD 1970. Better nutrition and hygiene were in part responsible, but there may have been evolutionary changes in the pathogens, and also among the human populations (R. M. Anderson and R. M. May, *Science*, Vol. 215, 1982, p. 1053). *See also* population history. The timetable (based on W. H. McNeill, *Plagues and People*, Blackwell, 1977) covers events of the past 2000 yr.

EPIDEMICS: TIMETABLE

Eurasian epidemics: **1838 yr (AD 162)**. A "new" disease (?measles) in China, and then to the Roman empire in AD 165 ("Antonine plague"); an equally lethal novelty (?smallpox) appeared in the Roman empire in AD 251, and in China about seventy years later (?AD 317).

bubonic plague: **1458 yr (AD 542)**. In the Mediterranean region ("plague of Justinian"), then to China in AD 610.

Black Death: **669 yr (AD 1331)**. Bubonic plague to China; then to western Eurasia in AD 1346, penetrating to northern Europe.

syphilis: 506 yr (AD 1494). From Naples, globally.

Amerindian epidemics: **480 yr (AD 1520)**. Smallpox from Hispaniola to Mexico, then to Peru in 1525; measles and other introduced diseases followed.

influenza: **434 yr (AD 1556)**. In Europe and the Americas, with high mortality.

yellow fever: 352 yr (AD 1648). From South America, to the Caribbean and Middle America.

cholera: **183 yr (AD 1817)**, also 174 yr (AD 1826). From Bengal, globally.

influenza: **AD 1918**. From northern France, globally, with 20 million dead.

smallpox eradicated: **AD 1977**. Worldwide.

distillation 2500 yr. *See* alcohol.

DNA (deoxyribonucleic acid, structure discovered) AD 1953. *See* gene structure; human genetics; life on Earth; molecular dating.

dogs 35 Myr. The family of dogs (canids) evolved from earlier carnivorous mammals in the Early Oligocene (*see* mammals); modern dogs and jackals of the genus *Canis* appeared **c. 6 Myr** ago at the end of the Miocene (data from A. W. Gentry, personal communication, 1982, citing R. H. Tedford). The modern species of true wolves (*Canis lupus*) dates from **c. 650,000 yr** (early Late Pleistocene, Europe, and China; B. Kurtén, *Pleistocene Mammals of Europe*, Weidenfeld & Nicolson, 1968). The wolf was tamed, as *the* dog, **12,000 yr** ago (*see* animal domestication), and showed extreme genetic plasticity, producing varieties from Great Danes to toy poodles.

domes 4700 yr, esp. 1876 yr (AD 124). *See* building.

domestication (of animals and plants). *See* animal domestication; crops; cultivation practice.

donkeys tamed c. 5000 yr. *See* animal domestication.

dragonflies 330 Myr. *See* invertebrates.

drama (formal) 2500 yr. *See* arts and crafts.

Dravidian (languages shrank southward) 3000 yr. *See* language.

droughts esp. 1730 yr (AD 270). *See* climate 10,300 yr.

Dutch: *language* after 1000 yr (AD 1000), *see* language; *drift nets* 584 yr (AD 1416), *see* fishing; *East India Company* **398 yr (AD 1602)**, *see* imperial companies; *overseas empire* (Batavia) **381 yr (AD 1619)**, *see* breakouts (navigators); *continuous rotation* esp. 300 yr (AD 1700), *see* cultivation practice.

Dutch spice monopoly 379 yr (AD 1621). In Indonesia, the Dutch East India Company conquered the Banda Islands in AD 1621 and thereafter systematically choked off all non-Dutch access to the spices of the Moluccas (Spice Islands), the most highly valued product of the region. Crops were destroyed to push up the prices. (A. Calder, *Revolutionary Empire*, Cape, 1981.)

E

Early Dynastic period (Sumer) **5100 yr.** *See* archeological timescale.

Earth (forming) **4550 Myr**; (complete) **est. 4450 Myr**. According to a leading theory the Earth grew by the coalescence of smaller bodies, planetisimals, orbiting the sun (*see* solar system forming). The process began as indicated by the ages of meteorites, 4550 Myr (rounded from 4555 Myr by uranium-lead dating; J.-F. Minster, et al., *Nature*, Vol. 300, 1982, p. 414). After a calculated 20 million years, half the Earth's mass had accumulated, after 50 million years the planet was five sixths grown, and it was essentially complete after 100 million years, **est. 4450 Myr ago**. The earlier, inner part may have been thoroughly melted, during the first 50 million years. Molten iron, mixed with nickel, sulfur, and some other elements, sank to the middle, forming the Earth's core, and the rocky material now comprising the Earth's lower mantle floated on the top. In the later stage, 4500 to 4450 Myr ago, planetisimals were still impacting very violently and molten iron from them sank through the mantle to augment the core, but the outer rocks were only partially melted.

The latecomers building the outer mantle may also have had a somewhat different chemical composition, if they came from farther away in the solar system. The present flow of heat from the Earth still includes heat generated in its violent birth, as well as heat due to the continual decay of radioactive atoms; at the time of the Earth's completion 4450 Myr ago, the heat flow through the surface was perhaps three to fifteen times the present rate, although some planetary scientists would argue for much more. In any event, volcanic activity would have been substantially more intense than it is now. (This account is based on J. V. Smith, *Philosophical Transactions of the Royal Society*, Vol. A301, 1981, p. 401.)

earthquakes These expressions of plate actions, present and past, reshape the world continuously, and kill people, but they have curiously little effect on the currents of human history and are not treated here.

East Antarctic deep freeze 14 Myr. *See* climate 67 Myr.

Easter Island peopled 1600 yr (AD 400). *See* breakouts (Polynesians).

East India Companies 400 yr (AD 1600). *See* imperial companies.

echinoderms 560 Myr. *See* animal origins; invertebrates.

economic management (by government) **932 yr (AD 1068)**; (by regional economic community) esp. AD 1957. *See* government. *See also* work.

egalitarian communities 9500 yr. *See* government.

eggs (amniotic) **313 Myr.** *See* vertebrate animals.

Egypt *chronic warfare and warrior kings* **5300 yr**, *see* government; weaponry; First Dynasty **5200 yr**, *see* archeological timescale; pyramids 4600 yr, *see* building.

electric force 10^{-35} second after origin. *See* Big Bang.

electrical energy esp. 119 yr (AD 1881). *See* energy.

electromagnetism (theory) **136 yr (AD 1864).** *See* science.

electron (Leptonic) era 10^{-4} to 1 second after origin. *See* Big Bang.

elements esp. 11,000 Myr. Interest focuses on the dates of formation of the elements heavier than hydrogen and helium composing the solar system and the Earth, the solar elements. The manufacture of elements has continued elsewhere in the universe since the solar system formed 4550 Myr ago, but any subsequent contributions to the Earth from cosmic rays or interstellar dust clouds have been negligible. The nuclei of hydrogen and most helium atoms were formed in the Big Bang, one microsecond and three minutes respectively after the start (*see* Big Bang), but the heavier elements (carbon, oxygen, iron, etc.) were made later, by nuclear reactions in exploding stars, in accordance with a well-developed theory due principally to E. M. Burbidge, G. Burbidge, W. A. Fowler, and F. Hoyle (1957).

The average age of the elements heavier than helium, in meteorites and the Earth, can be determined by radioactive dating using long-lived atoms. The best round figure for the average age of the solar elements is **10,000 Myr** (D. M. Schramm, personal communication, 1981) although there were significant late injections from other stars that exploded 4650 Myr and 4550 Myr ago, shortly before the formation of the solar system (*see* presolar supernovas). If the solar elements were made entirely or mainly within the Milky Way Galaxy, and if the Galaxy formed about 12,500 Myr ago (*see* galaxies), then the early rate of element making must have peaked sharply **est. 11,000 Myr** ago. Galaxies of 6000 Myr ago (redshift $z = 0.5$) exhibit more blue, element-making stars than nearer galaxies do. The average age of elements in the

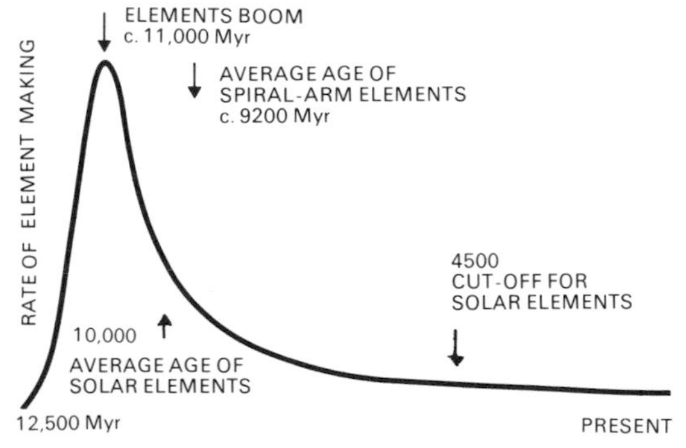

A freehand graph of the making of heavy elements by exploding stars in the outer regions of the Milky Way Galaxy. Given an average age of elements in the solar system of 10,000 Myr, a peak rate aroud 11,000 Myr ago seems probable, almost irrespective of the precise shape of the curve.

spiral arms of the Galaxy, where element making has continued during the past 4550 Myr, is probably about 800 million years younger than that of the solar elements; in other words, 9200 Myr, rounded here to **est. 9000 Myr**. In the center of the Galaxy, and in globular clusters, stars and elements may be older. Illus. pp. 104, 105.

elephants c. 7 Myr. Trunked animals go back 50 Myr, and the mastodons (now extinct) 20 Myr, but the true elephant family, elephantids, evolved c. 7 Myr ago (*see* mammals). *Archidiskodon*, the "great southern elephant," flourished 2.4 Myr ago. The recent genera *Elephas* and *Mammuthus* appeared c. 4 Myr. *Mammuthus primigenius*, the classic mammoth of the late ice ages, evolved from the steppe mammoth *M. trogontherii*, in the penultimate glaciation in Europe, taken here as **150,000 yr**; it is now extinct. The modern Asian species (*E. maximus*) evolved comparatively recently, perhaps in the Late Pleistocene. (A. W. Gentry, personal communication, 1982, citing V. J. Maglio, R. Ballesio, et al.) *E. maximus* was tamed c. 5000 yr (*see* animal domestication). Illus. p. 159.

empire 4400 yr *See* government. For various empires by name, *see* archeological timescale; breakouts; historical timescale.

energy 2.0 Myr to present. An athletic human being can exert 3 horsepower for a few seconds and 0.1 horsepower for several hours, using chemical energy from the oxidation of food. Once they had mastered fire, *Homo erectus* and early *Homo sapiens* could use firewood for heat, light, cooking, and splitting rocks; volcanic heat and hot springs would have been known to them, and coal, petroleum, and natural gas may have had more than curiosity value where they leaked from the crust. Also available to them were animal dung and tallow, and they could feel the force of running water and the wind. Systematic exploitation of these forms of energy came slowly.

The following account is based on diverse sources including those cited in the cross references, and also A. Leroi-Gourhan, *La Recherche*, Vol. 11, 1980, p. 142; C. Susskind, *Understanding Technology*, Johns Hopkins University Press, 1973; C. M. Cipolla,, *The Economic History of World Population*, 7th ed. Penguin, 1978; J. Needham, *Science and Civilisation in China*, esp. Vol. 4, Part 2, Cambridge University Press, 1966; N. Smith, *Scientific American*, Vol. 242, No. 1, 1980, p. 114; S. Nikitin, *Social Science*, Vol. 13, No. 3, 1982, p. 100; D. Fishlock, ed., *A Guide to the Laser*, Macdonald, 1967; and D. R. Hill, personal communication, 1982. Some key dates have been checked against biographies of inventors in *Encyclopaedia Britannica*, 1981. (For further aspects of energy use, *see* cultivation practice; transport; weaponry..)

The first clear reference to a waterwheel was by a Greek poet, Antipater, **2030 yr** ago and more technical descriptions came from the Roman engineer Vitruvius. The Romans used vertically mounted wheels, turning on a horizontal axle; the Chinese, on the other hand, mounted their wheels horizontally. There are vague references to Chinese waterwheels in the first century BC, and by AD 31 they found advanced applications, to drive trip hammers and blow air for ironmaking. Waterwheels were the chief source of industrial energy across Eurasia for nearly two thousand years. In AD 1800 there were more than half a million waterwheels in Europe, while steam engines were counted in hundreds. When improved designs emerged in France (esp. B. Fourneyron, near Dijon, c. 1827), waterwheels were called water turbines for publicity purposes. Even in the late twentieth century, a quarter of the world's electricity was generated by water turbines.

The tags "steam energy" and "fossil-energy revolution" fix dates in the complex story of coal and steam. Flooding was a chronic problem in coal mines, and nonrotary steam pumps were developed by T. Savery in AD 1698, and more especially by T. Newcomen, using a piston, **AD 1712**. By c. 1800, Britain was producing 90 percent of the world's coal, low-pressure steam engines of J. Watt (1765 and 1780) were well known, and high-pressure steam engines were under development, e.g., R. Trevithick's demonstration steam locomotives, 1803. The breakthrough came with the synergism between high-pressure steam engines, the coal industry, and the iron industry; the adopted date is **AD 1825**, a time of rapid growth in steam power and also the year in which the first passenger steam train ran in Britain.

There was no British monopoly of activity or ideas. A steam car ran in France in 1769 (N.-J. Cugnot), a patent for a high-pressure steam engine was obtained by O. Evans in the United States in 1787, a passenger railroad was planned in Austria as early as 1807, and the first fully practical steamboat was the American *Clermont* (J. Fulton, 1807). The internal combustion engine also has a long history, but the key date is that of the first gasoline-driven automobile leading to further development (C. Benz, Germany, 1885), as opposed to earlier French and Austrian experiments, 1862 to 1864.

Even after the discovery of radioactivity, in 1896, nuclear energy was a vague dream until the unexpected detection of uranium fission in 1938 (O. Hahn, Germany). The self-sustaining nuclear chain reaction achieved at Chicago in December **1942** (E. Fermi) is the natural milestone, although it was part of a bomb-making program.

ENERGY SOURCES: TIMETABLE

human muscle: c. 2.0 Myr. Africa.

firewood: **1.4 Myr**. Africa (*see* firemaking).

lamps: **17,000 yr**. France (Lascaux). Hollow stones accommodated tallow for fuel, and twigs for wicks.

wind energy (for boats): 6900 yr. Sumer (or earlier, *see* transport).

animal muscle: **esp. 6000 yr**. Northern Mesopotamia (*see* livestock revolution).

natural gas: c. 2900 yr. China. Wells were drilled, and bamboo pipes were distributing the gas, by c. 1800 yr ago (AD 200). The American natural-gas industry began AD 1821 (New York State), with big growth from c. 1890.

water energy: **c. 2030 yr**. Eurasia (*see* above).

coal: c. 2000 yr. China (systematic use in Han times). A shortage of firewood encouraged a surge in coal-mining in England c. AD 1600.

windmill: est. 1400 yr (AD 600). Persia (vertical axle); a virtual reinvention was the more effective horizontal-axle windmill (France ?AD 1162).

explosive: by 1100 yr (AD 900). China (gunpowder).

steam energy: **esp. 288 yr (AD 1712)**. Britain (*see* above).

coal gas (town gas): 208 yr (AD 1792). Britain (for lighting).

fossil-energy revolution: **c. 175 yr (AD 1825)**. Britain (*see* above).

petroleum industry: **143 yr (AD 1857)**. Canada (commercial production in Ontario). World petroleum production surpassed coal in energy value by AD 1965. The largest oil fields, in Arabia, were first encountered at Bahrain in **AD 1932**.

electrical energy: **119 yr (AD 1881)**. United States, Italy, and Britain (domestic supplies). Electricity was smelting aluminum commercially in **AD 1888**, and the first large-scale production of hydroelectricity was at Niagara, 1895.

internal-combustion engine: **115 yr (AD 1885)**. Germany (dated for the automobile). Diesel engines (Germany, AD 1895) and jet propulsion (Germany, 1939) followed.

rocket motors: esp. AD 1926. United States (liquid-fuel engine).

nuclear energy: **AD 1942**. United States (*see* above). Nuclear energy by 1982 generated 9 percent of the world's electricity, or 2 percent of total commercial energy supplies.

laser (coherent light): AD 1960. United States (*see* information technology).

costly energy: **AD 1973**. The price of oil trebled by 1975, and trebled again between 1978 and 1980. This favored the retrieval of oil from difficult offshore sources, and made other forms of energy more competitive.

English *people* (Anglo-Saxons) c. 1560 yr (AD 440); (Vikings and Normans) 1207 to 934 yr (AD 793 to 1066), *see* breakouts (Germans; Vikings); *language* after 1000 yr (AD 1000), *see* language. *See also* British.

Enlightenment (European) **esp. 240 yr (AD 1760)**. *See* historical timescale.

Eocene epoch 55 to 37 Myr. *See* geological timescale.

Eocene terminal turnover 37 Myr. First noted as a "big break" (*la grande coupure*) in the character of European mammals by H. G. Stehlin in 1909, the transition from the Eocene to the Oligocene epoch is also marked by major changes in vegetation. In northwestern North America, for example, broad-leafed evergreen forests were replaced by deciduous forests, representing a fall in mean annual temperature of about 12 degrees Centigrade at a latitude of 60°N, and an onset of cool winters. The annual temperature range was 3 to 5 degrees in the Eocene, becoming 21 to 25 degrees in the Oligocene. In the oceans, extinctions among forams, mollusks, and other marine animals were accompanied by a cooling of the sea surface by 3 degrees or more (South Pacific) and by a distinctive deposition of a layer of carbonates in the deep sea (J. Van Couvering, et al., *Palaeogeography, Palaeoclimatology, Palaeoecology*, Vol. 36, 1981, p. 321; J. P. Kennett, *Marine Geology*, Prentice-Hall, 1982). The early primates and many old-fashioned mammals suffered badly, and soon afterward several modern families of mammals put in their first appearances (*see* mammals). Water birds diversified est. 40 Myr ago (*see* birds).

Ages given for turnovers in flora and fauna on land are typically 38 or 37 Myr, but sometimes 35 or 34 Myr. On the other hand, corresponding changes in marine life have been set as recently as 32.5 Myr ago. A swarm of roughly contemporary events included many magnetic reversals, two impact craters (Popigai and Mistastin) nominally dating from 38 Myr, and various scatterings of glassy tektites (*see* cosmic impacts). Two distinct coolings of bottom-water temperatures (oxygen-18 data) occurred, one at the Eocene-Oligocene boundary, the other 5 to 10 million years later. A bid to set the Eocene-Oligocene boundary at 34.4 Myr ago relied in part on tektites (R. Ganapathy, *Science*, Vol. 216, 1982, p. 885; W. Alvarez, et al., *Science*, Vol. 216, 1982, p. 886), but there was more than one tektite event. The age adopted here therefore concurs with refined dating of the more conventional Eocene-Oligocene boundary (36.8 Myr, somewhat above magnetic anomaly number 15; J. A. Van Couvering, citing W. A. Berggren and others, personal communication, 1982).

epidemics *See* disease epidemics.

Epipaleolithic period **18,800 yr.** *See* archeological timescale.

Eskimos (spreading) c. 4500 yr. *See* breakouts (modern humans).

Etruscan (expansion) **c. 2700 yr**. *See* breakouts (blacksmiths).

eukaryotes (modern cells) **est. 1700 Myr**. *See* life on Earth.

Euramerica (union of North America and Europe) **c. 460 Myr**; (hit Gondwanaland) **c. 350 Myr**; (hit Siberia) **c. 300 Myr**; (split) **c. 60 Myr**. *See* continents. See also map pp. 84–85.

Eurasia (union of Europe, Siberia, etc.) **c. 300 Myr**, *see* continents; *divided* (by Turgai Sea) 160 to 29 Myr, *see* continents flooded; *Eurasian epidemics* **1838 yr (AD 162)**, *see* disease epidemics. *See also* Asia, Europe.

Europe: *Scourian* 2700 Myr and later mountains; *Baltica hit North America* **c. 360 Myr**; *hit Gondwanaland* **c. 350 Myr**; *hit Siberia* **c. 300 Myr**; *quit North America* **c. 60 Myr**; *Italy hit Europe* c. 60 Myr, **esp. 10 Myr**; and other continental movements, *see* continents; Swiss Alps; *ores* 2500 Myr, etc., *see* ores; *also* coal; petroleum; *Old World Interchange* **20 Myr**, *see* mammals; *early humans* **c. 750,000 yr**; *archaic* Homo sapiens **c. 230,000 yr**; neanderthalers **c. 120,000 yr**, *see* human origins, neanderthalers; *archeology* 40,000 yr, etc., *see* archeological timescale; *modern humans* 40,000 yr; and many subsequent movements of peoples esp. breakouts of European *navigators* **508 yr (AD 1492)**; *Russians* **414 yr (AD 1586)**, *see* breakouts; *copper smelting* **c. 6500 yr**, etc., *see* metal industries; *Greek science* **c. 2600 yr**; *modern science formalized* **338 yr (AD 1662)**, *see* science; *population estimates* 2000 yr to present, *see* population; *epidemics* (Eurasian) **esp. 1835 yr (AD 165)**, *see* disease epidemics; *imperial nationalism* **c. 360 yr (AD 1640)**, *see* government; *carve-up* **AD 1945** *see* carve-up of continents. See maps pp. 81–95. *See also* animal domestication; crops; historical timescale.

explosives (gunpowder) c. 1100 yr (AD 900); (TNT) AD 1904; (nuclear fission) AD 1945; (nuclear fusion) c. AD 1953. *See* weaponry.

extinctions (of animals) **550 Myr**, etc. *See* mass extinctions.

extraterrestrial impacts 4450 Myr, etc. *See* cosmic impacts.

eyes in vertebrates est. 510 Myr, *see* vertebrate animals; *forward-looking* (in primates) c. 55 Myr, *see* primates.

F

fabrics (woven) 8500 yr. *See* arts and crafts (weaving).

false ice ages **115,000 and 95,000 yr.** *See* ice ages, false.

farming (cultivation experiments) c. 18,000 yr; (plow) est. 6000 yr. *See* animal domestication; crops; cultivation practice; livestock revolution. Here "farming" is used preferentially for plow farming, as distinct from hoe cultivation, or "gardening."

feathers **150 Myr.** *See* birds.

female prestige **9500 yr**; *voting* (Australia) 139 yr (AD 1861). *See* government.

ferns 380 Myr. *See* plants ashore.

Fertile Crescent (Palestine, Syria, and Mesopotamia): *forming* **20 Myr**. *See* continents (Arabia hit Eurasia); *also* Mesopotamia; Palestine. Illus. p. 163.

fertilizers, chemical 180 yr (AD 1820), **esp. AD 1913**. *See* cultivation practice.

fibers (wholly synthetic) esp. AD 1936. *See* arts and crafts.

Finno-Ugric (Uralic languages origin) est. 4900 yr. *See* breakouts (northern dairymen); language.

fireball, cosmic **3 minutes** after origin. *See* Big Bang.

firemaking **1.4 Myr.** At Chesowanja, Kenya, from a site dated to more than 1.42 Myr, pieces of burned clay in the presence of man-made tools and animal bones, including bones of eaten apemen, suggest a meal eaten around a campfire by *Homo erectus* (J. A. J. Gowlett, et al., *Nature*, Vol. 294, 1981, p. 125). Magnetic tests showed that the clay had been heated to about 400° Centigrade, perhaps too hot for subsurface clay under a wild brush fire, and too cool for the baking around a burning tree stump. This interpretation has been questioned

(G. L. Isaac, personal communication, 1981). Later traces of firemaking include hearths at Přezletice in Czechoslovakia, perhaps 750,000 yr old (P. Phillips, *The Prehistory of Europe*, Allen Lane, 1980).

fishes (jawless) **510 Myr**; (jawed) **425 Myr**; (ray-finned) **415 Myr**; (fleshy-finned) **400 Myr**. *See* vertebrate animals.

fishing esp. ?420,000, 128,000, and 10,000 yr. People at Terra Amata, France, **?420,000 yr** ago caught fish and turtles and collected oysters and limpets (P. Phillips, *Prehistory of Europe*, Allen Lane, 1980; for dating, *see* building). More intensive shellfish collecting **c. 128,000 yr** ago is attested by mounds of shells at Klasies River Mouth, South Africa; other candidate sites are in Libya and Gibraltar, and later evidence comes from Saldhana Bay, South Africa (T. P. Volman, *Science*, Vol. 201, 1978, p. 912). Fishhooks, harpoons, and other specialized fishing gear appeared in Europe and South Africa c. 14,000 yr ago (R. Klein in A. G. Sherratt, ed., *Cambridge Encyclopedia of Archaeology*, Cambridge University Press, 1980).

A surge in shellfish collecting c. 10,500 yr has been observed in a fivefold increase in the ratio of shell to animal bone at a site in Spain (La Riera); a marked reduction in the sizes of shells may be evidence for overfishing (L. G. Straus, et al., *Scientific American*, Vol. 242, No. 6, 1980, p. 120). Off Japan, deep-sea fishing with nets was in progress c. 9600 yr ago (Natsushima, 7290 b.c.; J. G. D. Clark, *World Prehistory*, Cambridge University Press, 1977). Early settlers in southern Mesopotamia c. 8000 yr ago were fishermen, as were inhabitants of desert regions of South America, c. 5300 yr ago.

Danish coast dwellers of 7500 to 6000 yr ago may have obtained more than 70 percent of their food from the sea, judging by the proportion of carbon-13 in their bones; careful sieving revealed vast numbers of fish bones at the sites. (H. Tauber, *Nature*, Vol. 292, 1981, p. 332; for a criticism of the carbon-13 method, see M. J. Schoeninger and M. J. DeNiro, *Nature*, Vol. 297, 1981, p. 577.)

An important medieval invention was of drift nets, in use in the Netherlands c. 584 yr (AD 1416). In recent times the world fish catch trebled, from 20 to more than 60 million tons a year, between AD 1948 and 1968, but then tended to level off as a result of overfishing, and also climatic factors, especially off South America (*Agriculture: Towards 2000*, FAO, 1981).

fission-track dating Glassy materials and certain kinds of once molten rocks (including volcanic ash, basalt, and granite) preserve submicroscopic damage in the form of tunnel-like tracks made by fragments of uranium-238 atoms that have spontaneously undergone fission. Unlike radioactive decay, in which an atomic nucleus throws out particles of little mass, fission disrupts the nucleus into two heavy lumps. Etching with hydrofluoric acid makes the fission tracks easier to see and to count. The number of tracks depends on how old the example is, and on how much uranium it contains. The amount of uranium can be measured by exposing the sample to neutrons in a reactor and seeing how many new tracks are produced by the fission (in this case) of uranium-235. Applications include the dating of layers of volcanic rocks and of ancient tools made from obsidian (volcanic glass).

flatworms by 620 Myr. *See* animal origins.

flies 330 Myr. *See* invertebrates.

flint trade c. 30,000 yr. *See* trading.

flood (biblical) **4900 yr.** *See* climate 10,300 yr.

floral revolution 114 Myr. Modern flowering plants, the angiosperms, are not known before 123 Myr ago (Early Cretaceous, Barremian). They began to diversify very rapidly 117 Myr ago (early-middle Aptian), with primitive leaves and pollen. Patterns of diversity, and differences in the times of appearance of major pollen types, suggests that the novel plants originated in Africa and South America (then northern Gondwanaland). Sea barriers were quickly crossed to give fossil appearances in England c. 123 Myr ago. The floral revolution is here dated, a little arbitrarily, at the Aptian/Albian boundary, 114 Myr ago, when a "pulse" of novel immigrants from Africa entered North America. By mid-Albian, 105 Myr ago, striking advances in leaves and pollen had occurred. Angiosperms dominated most landscapes by 92 Myr ago (Cenomanian/Turonian), displacing conifers from many environments, and eclipsing cycads, ginkgoes, and the flowery pioneers, the bennettitales. Adaptation to cool latitudes seems to have taken some time: the first modern flowering plants appeared in northern North America and northern Siberia 95 Myr ago (Cenomanian). (J. A. Doyle in A. Hallam, ed., *Patterns of Evolution*, Elsevier, 1977, and personal communication, 1982.) *See also* plants ashore.

flowers (bennettitales) **235 Myr**; (angiosperms) **123 Myr**. *See* floral revolution; plants ashore.

food sharing 1.7 Myr. *See* human origins.

footprints (prehuman) **3.7 Myr**. *see* australopithecines.

foraminifera 550 Myr. *See* invertebrates.

forests 370 Myr. *See* plants ashore.

fortifications c. 8000 yr. *See* weaponry.

fossil-energy revolution c. 175 yr (AD 1825). *See* energy.

fossil record Comparisons between living organisms and their molecules can tell a great deal about ancestral relationships—that humans, for example, are more like bats than lizards. The fossils found in rocks are nevertheless indispensable for establishing what ancient animals looked like, for revealing lineages now extinct, for studying the modes of evolution, and for calibrating a timescale. Scavenging animals, bacteria, and physical stresses destroy the remains of most organisms after death. Hard parts have more chance of being preserved, although even these may dissolve in deep water. Soft tissues (of trees, for example), may be preserved by "petrification," turning to stone as a result of chemical changes, but pollen generally survives better than whole plants. Traps of amber or ice may preserve soft organisms, or these may leave impressions of their bodies stamped in fine sand. Fossilized footprints, burrows, and droppings (coprolites) often give valuable information. Some prime sites where body fossils occur in abundance owe their existence to accidents such as landslides or underwater slumping that suddenly buried large numbers of organisms. (S. Conway Morris, in D. G. Smith, ed., *Cambridge Encyclopedia of Earth Sciences*, Cambridge University Press, 1981.)

Chemical fossils are carbon compounds, found in tarry deposits in old rocks, the chemical structures of which carry undoubted signs of their origin in living processes. For example, petroleum affects polarized light in a manner characteristic of chemicals in organisms. In the 1970s, chemical analysis by gas chromatography, high-speed mass spectrometry, and computer interpretation greatly increased the number of known chemical fossils. Most useful so far are the lipids, which are easily extracted from rocks with chemical solvents, and derive from subtle steroid compounds, such as chlorophyll. Although they are altered from their original composition, skeletal signatures persist, and different classes of organisms (diatoms, higher plants, sponges, etc.) leave characteristic chemical fossils (A. S. Mackenzie, et al., *Science*, Vol. 217, 1982, p. 491).

Certain abundant and widespread organisms provide key markers for biostratigraphy—distinguishing rocks of different ages, and recognizing rocks of the same ages in different places. "Clock" fossils in different geological eras include stromatolites in the Proterozoic, graptolites in the Paleozoic, ammonites in the Mesozoic, and foraminifera in the Cenozoic.

Frasnian catastrophe 380 Myr. Widespread development of reefs and organic limestones in the Late Devonian, at a time of high sea level, was brought to a sudden end with the destruction of most corals and stromatoporoids (*see* reefs). Bottom-dwelling marine animals, including brachiopods and trilobites, also took severe punishment, while surface-dwelling organisms were largely unaffected. The pattern of extinctions suggests the action of huge tidal waves (D. J. McLaren, *Journal of Paleontology*, Vol. 44, 1970, p. 801). This occurrence has been called "the third largest extinction event of the past 500 million years" (G. R. McGhee, et al., Conference on Large Body Impacts, Snowbird, 1981). *See also* cosmic impacts; mass extinctions.

French *people* (Franks) 1572 yr (AD 428), *see* breakouts (Germans); *Frankish empire* **1514 to 1157 yr (AD 486 to 843)**; *overseas empire* (Canada) **400 yr (AD 1600)**; *Louis XIV* **357 yr (AD 1643)**; *revolution* **211 yr (AD 1789)**, *Napoleonic empire* **196 to 186 yr (AD 1804 to 1814)**, *see* historical timescale; *language* after 1000 yr (AD 1000), *see* language; *East India Company* **346 yr (AD 1664)**, *see* imperial companies. *See also* breakouts (Vikings; navigators).

frogs c. 240 Myr. *See* vertebrate animals.

fungi est. 1700 Myr. *See* life on Earth.

G

galaxies est. 12,500 Myr. Galaxies, the most obvious units of which the wider universe is composed, are congregations of billions of visible stars, bound together by gravity so that they do not expand while the universe expands, but move farther apart like ships dispersing in an ocean. The disk-shaped Milky Way Galaxy, to which the sun belongs, is seen edge on from the Earth. The Galaxy is also surrounded by a halo of largely invisible material. The simplest way of arriving at an age for the galaxies is to note that their haloes would have been touching when the universe was younger and smaller. With a separation between galaxies one sixth of the present average, the haloes would have been smeared together. That would have been the situation at a time corresponding to a redshift of $z=5$, or about 12,500 Myr ago (*see* cosmological timescale).

The chief problem about galaxies is to explain why they happened at all, given that the Big Bang produced a remarkably uniform universe that might well have expanded as a featureless gas. The origin of clumpiness lay with pressure waves, analogous to sound waves, passing through the material of the Big Bang (B. J. T. Jones, *Philosophical Transactions of the Royal Society*, Vol. A296, 1980, p. 289, and personal communication, 1981). A certain disorderliness attending the creation of matter 10^{-35} second after the origin of the universe may have initiated the pressure waves. The resulting clumps attained the mass of galaxies after about ten years (Y. B. Zel'dovich, 1980, cited by J. Silk, *Nature*, Vol. 292, 1981, p. 410). This is the representative time adopted here for "galaxies seeded."

Pressure waves that "twanged" the radiation in the universe as well as the subatomic particles would have tended to die out, except at long wavelengths, when the universe became transparent (*see* Big Bang). Soviet theorists showed how galaxies could form by the breakup of very large clumps of matter produced by long waves, each equivalent in mass to about a thousand galaxies. A contrary theory, called clustering, has been more widely favored by Western cosmologists; here the pressure waves involved matter only, and short waves survived in the early universe. Typical clumps were much less massive, by this theory, and had to cluster to make galaxies. It is not foolish to suppose that both processes, breakup and clustering, were at work in the young universe.

However the trick was done, major galaxies certainly existed by the time quasars appeared. These are extremely bright, small, and mainly very distant objects. A quasar is here taken to be a massive black hole sitting at the center of a galaxy and having a mass billions of times greater than the sun's, so that it can swallow stars and gas amid a great uproar of electromagnetic energy in all frequency bands. The oldest quasar known is **12,200 Myr** (object designated PKS 2000–330, redshift $z=3.78$; B. Peterson, et al., press release, 1982), and quasars became abundant later ($z=2$, rounded to **11,000 Myr**) when many of the major galaxies had become fully organized.

Quasars became scarce after 3200 Myr ($z=0.2$), although the most conspicuous is 2700 Myr old (object 3C 273, $z=0.16$). An example of a "young" quasar, 1100 Myr old, is object MR 2251–178 ($z=0.06$), and lesser explosions in galaxies continue virtually to the present. An inference is that central black holes in galaxies have generally become starved of nourishment. The most distant visible normal galaxies are about 6200 Myr old ($z=0.5$).

Giant galaxies evolved by swallowing smaller galaxies. From the outset, a collision between a pair of galaxies, most probable in large clusters, could strip both galaxies of their gas, which was left as a stray cloud within the cluster. The galaxies were then sterilized: without gas they were unable to breed new stars. Eventually the gas in some clusters built up to a density such that stripping of residual gas from many galaxies began. A runaway process ensued because, as more gas was stripped, the greater its stripping power became. Catastrophic sterilization of many galaxies in large clusters began at an estimated redshift of $z=0.5$ (J. Silk, *The Big Bang*, Freeman, 1980), corresponding on the timescale adopted here to est. 6200 Myr ago, rounded to **6000 Myr**. X-ray emissions from the gas in clusters confirm the presence of heavy elements, probably stolen from the galaxies. Illus. p. 77.

gamma-ray era **1 second** after the origin. *See* Big Bang.

gardeners' breakout (hoe farmers) **9500 yr**. *See* breakouts; population history.

gas (natural) c. 2900 and 179 yr (AD 1821); (coal gas) 208 yr (AD 1792). *See* energy.

gazelles herded **18,000 yr.** *See* animal domestication.

gene structure **AD 1953**; *splicing* **AD 1973**. The existence of units of heredity, the genes, was correctly inferred by G. Mendel of Brünn, **135 yr (AD 1865)**. Modern biology was launched with the interpretation of X-ray photographs of genetic material, the nucleic acid DNA. These revealed a twisted double strand (double helix) and suggested how genes replicated, by one strand acting as a template for the other. (F. H. C. Crick, J. D. Watson, and M. H. F. Wilkins shared the Nobel prize for this work.) Other discoveries about molecular structures and the genetic code followed rapidly. Splicing of genes, for genetic engineering by the recombinant-DNA technique, was done first at Stanford, California (P. Berg, et al., personal communication, 1973). While the manipulative technology developed, the reading of genes became almost routine. This led, in 1982, to the pinpointing of a genetic mutation in cancer (*see* medicine). (For applications of gene studies in chronography, *see* human genetics; molecular dating.)

geological timescale For early phases of the Earth's history, radiometric dating is the chief way of setting events on a timescale. Before 1000 Myr ago, dates may be in error by as much as 100 million years. For the phase starting with the appearance of abundant fossils 570 Myr ago, the sequence is best defined by fossils in successive periods, epochs, and stages; developing the timescale is then a matter of assigning radiometric dates to this stratigraphic sequence. Then, by interpolation, a date can be offered for a geological event, or for the bed in which a fossil has been found.

Many recent publications used a timescale promulgated by the Geological Society of London in 1964; by 1982 the Geological Society was at work on a new version. The US Geological Survey issued, in 1980, a revised timescale which is the basis for the timescale used in this book. Modifications have been made in the light of studies up to 1982, with advice from W. Alvarez, R. L. Armstrong, W. A. Berggren, A. Hallam, D. V. Kent, W. S. McKerrow, N. D. Opdyke, R. Pankhurst, J. A. Van Couvering, J. V. Watson, and D. A. Wright.

The Late Proterozoic start conforms with recommendations of the International Union of Geological Sciences. For the start of certain other periods, or epochs, the following dates are adopted: Ordovician (Tremadocian) 520 Myr, Late Carboniferous 335 Myr and Triassic 245 Myr (McKerrow), Jurassic 208 Myr (Armstrong), Tertiary 67 Myr (Alvarez, 66.7 Myr), and Oligocene 37 Myr (Berggren, 36.8 Myr). Interpolated stage dates follow the relative durations of the stages illustrated by Armstrong (in G. V. Cokee, et al., eds., *Contributions to the Geologic Timescale*, American Association of Petroleum Geologists, 1978), except for the Late Triassic and Early Jurassic, where stages conform with Armstrong in G. S. Odin, *Numerical Dating of the Stratigraphic Column*, Wiley, 1982.

For the Late Cretaceous and Tertiary periods, where magnetic stratigraphy rules, the timescale mainly follows W. Lowrie and W. Alvarez (*Geology*, Vol. 9, 1981, p. 392), who incorporated and updated work by Berggren, Van Couvering, and other geochronologists. A few late dates in the Lowrie-Alvarez list are varied, for example to set the most recent magnetic reversal at the start of the Late Pleistocene at 730,000 yr, *see* magnetic stratigraphy.

At a Geological Society meeting in London in 1982, several groups of geologists proposed chronologies for various segments of the timescale (*Geochronology and the Geological Record*, Geological Society, in press, 1983). The chief disagreement concerned the start of the present fossil-rich Phanerozoic era (Cambrian period), with some workers preferring a rather late date. Otherwise the discrepancies between the timescale adopted here and other dates on offer in London were less than 2 percent; similarly, R. L. Armstrong (personal communication, 1981) noted that most key dates in four studies, 1978 to 1981, varied by about 2 percent, and that was also typical of the intrinsic accuracy quoted by the investigators. Armstrong suggested that car speedometers were less accurate.

A 2 percent uncertainty in 300 million years amounts to 6 million years either way, hardly a negligible moment of time. Why then offer a date such as "313 Myr" for the origin of reptiles? Figures carrying a false air of precision are necessary to preserve relative information, when using dates in preference to statements puzzling to non-geologists, of the form "Late Carboniferous, one third of the way above the base of the Westphalian." The "true" date might be anything from 320 to 306 Myr, but what matters more is how the event stands in relation to other events only a few million years before or after it. Because systematic errors are often greater than statistical errors, uncertainties (e.g., standard errors) are not stated in this book, although they are often taken into account in rounding dates and in correlating events.

On terminology: an alternative Soviet scheme for the Proterozoic starts a Riphean period at 1700 Myr (Middle 1350 Myr, and Late 950 Myr), and a Vendian period at 680 Myr. The Tremadocian stage is regarded by some experts as belonging wholly or in part to the Cambrian, rather than the Ordovician period. In North America, the Early and Late Carboniferous are often called the Mississippian and Pennsylvanian respectively, but the transition between them occurs at about 325 Myr (mid-Namurian). Other terms in common use are Precambrian and Phanerozoic, which mean, respectively, before and after the start of the Cambrian period, 570 Myr ago, when abundant fossils first appeared. Obsolete terms are Primary and Secondary for the Paleozoic and Mesozoic eras. Paleogene for Early Tertiary, and Neogene for Late Tertiary, are better avoided because of possible confusion between Paleogene and Paleocene. Many other names of stages are used in various parts of the world;

DATES OF STARTS OF PHASES IN THE GEOLOGICAL TIMESCALE

	Myr
Hadean era	est. 4450
Archean era	3800
Proterozoic era	
Early Proterozoic	2500
Middle Proterozoic	1600
Late Proterozoic	900
Paleozoic era	
Early Cambrian	
Georgian	570
Middle Cambrian	
Acadian	550
Late Cambrian	
Potsdamian	530
Early Ordovician	
Tremadocian	520
Arenigian	507
Llanvirnian	493
Llandeilian	476
Late Ordovician	
Caradocian	458
Ashgillian	447
Silurian	
Llandoverian	435
Wenlockian	430
Ludlovian	423
Downtonian	417
Early Devonian	
Gedinnian	410
Siegenian	399
Emsian	394
Middle Devonian	
Eifelian/Couvinian	389
Givetian	383
Late Devonian	
Frasnian	378
Famennian	370
Early Carboniferous	
Tournaisian	360
Visean	348
Late Carboniferous	
Namurian	335
Westphalian	316
Stephanian	306
Early Permian	
Sakmarian	290
Artinskian	278
Kungurian	268
Late Permian	
Kazanian	256
Tatarian	249

Mesozoic era	*Myr*
Early Triassic	
Scythian	245
Middle Triassic	
Anisian	240
Ladinian	235
Late Triassic	
Karnian	230
Norian	225
Rhaetian	216
Early Jurassic	
Hettangian	208
Sinemurian	205
Pliensbachian	197
Toarcian	188
Aalenian	182
Middle Jurassic	
Bajocian	177
Bathonian	170
Late Jurassic	
Callovian	164
Oxfordian	159
Kimmeridgian	154
Tithonian	145
Early Cretaceous	
Berriasian	138
Valanginian	133
Hauterivian	126
Barremian	123
Aptian	120
Albian	114
Late Cretaceous	
Cenomanian	96
Turonian	92
Coniacian	89
Santonian	88
Campanian	84
Maastrichtian	72

Cenozoic era	*Myr*
Tertiary period	
Early Paleocene	
Danian	67 (66.7)
Late Paleocene	
Thanetian	61 (61.5)
Early Eocene	
Ypresian	55 (54.9)
Middle Eocene	
Lutetian	50 (50.3)
Bartonian	45
Late Eocene	
Priabonian	41 (41.0)
Early Oligocene	
Rupelian	37 (36.8)
Late Oligocene	
Chattian	33
Early Miocene	
Aquitanian	25 (24.6)
Burdigalian	19.5
Middle Miocene	
Langhian	14.7
Serravallian	13.3
Late Miocene	
Tortonian	11.5
Messinian	6.7
Early Pliocene	
Zanclean, etc.	5.3
Late Pliocene	
Piacenzian	3.25
Quaternary period	
Early Pleistocene	1.80 Myr
Late Pleistocene	730,000 yr
Holocene	10,300 yr

LATE MAMMAL STAGES (EUROPE)

Late Pliocene	
Early Villafranchian	3.25 Myr
Late Villafranchian	2.60 Myr
Pleistocene	
Early Biharian	1.90 Myr (?1.80 Myr)
Middle Biharian	c. 1.5 Myr
Late Biharian	730,000 yr
Toringian	480,000 yr

when they are cited elsewhere in this index, the equivalents from the geological timescale are given, except when they are dated by more direct means. The groupings of stages within periods and epochs into "Early," "Middle," and "Late" follow the 1964 Geological Society timescale; "early" rather than "Lower," because readers may not all believe as readily as geologists do that time runs uphill.

The names of the late mammal stages in Europe are often used in discussions of the evolution of mammals (including human beings) in the Old World. The table opposite sets the start of Villafranchian at the onset of the ice ages, and avoids the term Villanyian, but otherwise follows C. A. Repenning and O. Fejfar, *Nature*, Vol. 299, 1982, p. 334. If the Villafranchian/Biharian transition is equated with the Pliocene/Pleistocene boundary, then 1.80 Myr would be more consistent with the main geological timescale.

geology (modern) AD 1967. *See* continents.

geomagnetic reversals *See* magnetic stratigraphy.

geometry esp. 2300 yr. *See* science.

germ theory of disease 137 yr (AD 1863). *See* medicine.

Germans *breakouts* 2500 and 1572 yr (AD 428), *see* breakouts (Germans); *Germanic languages* by 2000 yr; *German* after 1000 yr (AD 1000), *see* language; *Hanseatic League* **c. 720 yr (AD 1280)**; *unified* **129 yr (AD 187)**; *Nazis* **AD 1933 to 1945**; *split* (carve-up of Europe) **AD 1945**, *see* historical timescale; *postagricultural* by c. 110 yr (AD 1890), *see* work.

gingkoes 280 Myr. *See* plants ashore.

giraffes 5.5 Myr. *See* mammals.

glaciations *See* climate; ice ages; orbital dating.

glaciers (Antarctica) **37 Myr**; (northern) **9 Myr**; (South American) **c. 6.6 Myr**. *See* climate 67 Myr.

glass (man-made) c. 3400 yr; *glass blowing* c. 2050 yr. *See* arts and crafts.

global: *colonization* 507 yr (AD 1493); *institutions* 125 yr (AD 1875), *see* government; *population boom* **c. AD 1930**, *see* population history; *postagricultural condition* **c. AD 1970**; *economic recession* **c. AD 1980**, *see* work.

globular clusters ?13,500 Myr. *See* stars.

goats c. 200,000 yr, *see* bovids; *herded* **18,000 yr**; *tamed* **10,700 yr**, *see* animal domestication.

gold boom 2900 Myr. *See* ores.

gold rushes 151 yr (AD 1849). The first major gold rush of modern times was prompted by the discovery of gold in California in AD 1848; other gold rushes followed discoveries in Australia in 1851, South Africa in 1886, and the Klondike in 1896. The chief effect was to lure people, mainly Europeans, to remote parts of the world (V. Buranelli, *Gold*, Red Dembner, 1979).

Gondwanaland 650 to 130 Myr. Australia, Antarctica, India, Africa, and South America were formerly anonymous provinces in the large continent of Gondwanaland. Although magnetic evidence suggests that pieces of Gondwanaland cohered 1000 Myr ago or earlier, widespread mountain building occurred within Gondwanaland c. 650 to 450 Myr ago, in the so-called Pan-African orogeny, which also involved eastern South America, Antarctica, Australia, and India. This episode represented at least a consolidation of Gondwanaland, if not its assembly. As one of the major geological events of the past billion years, it remains a puzzle to geologists (B. F. Windley, *The Evolving Continents*, Wiley, 1977). Gondwanaland was united with Euramerica c. 350 Myr ago in the supercontinent of Pangaea (complete 220 Myr). Gondwanaland began to break up 160 Myr ago and its disintegration was well under way by 130 Myr ago; *see* continents; Pangaea. Illus. p. 115.

gorillas est. 5 Myr. *See* primates.

Gothic architecture esp. 840 yr (AD 1160). *See* building.

government 29,000 yr to present. Redistribution of wealth, and gathering resources for public projects, were probably achieved in the earliest communities by gifts, exacted if need be by social pressures on the prosperous and healthy. For D. Schmandt-Besserat (*American Anthropologist*, Vol. 84, No. 4, 1982), the first accountancy tokens of 10,700 yr ago were instruments of power, for the leader of an "incipient ranked redistributive economy." They represented pledges of offerings by others, and the leader could plan their redistribution, in a bureaucratic fashion. Schmandt-Besserat also suggested (*Technology and Culture*, Vol. 21, 1980, p. 357), that the appearance of clay envelopes with official seals and markings signals the time when donations to the Mesopotamian temples became compulsory taxes (*see* information technology). Many of the seals portray bound prisoners—an undisguised threat of punishment. They may mark a turning point in human social relationships; when gifts became compulsory, tax avoidance was legitimized and the moral obligation to share was compromised.

A long-standing view that government evolved in a more-or-less linear, continuous fashion has been challenged by studies of early agricultural communities in central Europe (A. G. Sherratt, in A. C. Renfrew and S. J. Sherman, eds., *Ranking, Resource and Exchange*, Cambridge University Press, 1982). Sherratt describes instead a cyclical process, in eastern Europe 8000 to 5500 yr ago, where the character of communal organization depended on technology, settlement patterns, and trading relationships. The chief purposes of social organization are to allocate land, facilitate trade, and sustain alliances and defenses. As a result, organization will be relatively tight, but not necessarily authoritarian, in stable, concentrated settlements. These are preceded and followed in the cycle by more dispersed patterns of settlement. Dispersals not associated with much trading may tend to have a more relaxed social order, while those that are trade oriented may become more authoritarian. As new technologies outmode the economic systems of the traditional centers of economic power, wealth and political initiative will move to the periphery.

Ancient cultures showed variety in their social systems: transient, casual leadership versus nontransient, formal leadership; dispersal versus concentration of populations; egalitarian sharing of wealth versus class distinctions; rare versus frequent warfare; a wide spectrum of sexual politics; public versus private religious activity; and low versus high value attaching to trade. Modern nation-states show very little variability among this great range of options. Most are characterized by nontransient leadership, concentration in cities, class distinctions, war or anticipation of war, male domination, public religious or quasi-religious rhetoric, and love of trade.

After states with these characteristics evolved in Mesopotamia c. 5100 yr ago, they persisted and spread with only modest variations, despite the continuing operation of the Sherratt cycle of dispersal and reconcentration. The frequent examples, back to Uruk times in Mesopotamia, of busy cities ruled with a light hand argue against complexity or population density as being sufficient causes of the rise of the state. Stress is a strong promoter of central authority and nontransient leadership, and it may be that one need look no further than the onset of chronic warfare, coinciding with the rise of kings, as a stress under human management. The state as an entity maintained by the stress of war and threats of war endured through the metamorphosis of the state from king-oriented, city-oriented, and imperial forms into the modern nation-state of the seventeenth century AD. Stress is nowadays sustained by a continuous technological arms race, which dates back to the aftermath of the Franco-Prussian war, AD 1871. (P. Noel-Baker, in N. Calder, ed., *Unless Peace Comes*, Viking, 1968.)

The following summary is influenced by Schmandt-Besserat's and Sherratt's archeological concepts, and it draws on diverse sources for the historical phase.

GOVERNMENT: TIMETABLE

communities: **29,000 yr**. Dated for Dolní Věstonice, Czechoslovakia, site of the earliest groups of more than one hundred people (B. Klima's estimate).

communal accountancy and bureaucracy: **10,700 yr**. Tokens at Ganjdareh, Iran (*see* above).

egalitarian communities with female prestige: **c. 9500 yr**. Dated for the breakout of the gardeners in southwest Asia. Gardening without the plow is typically women's work, and female goddesses were conspicuous in gardening communities.

temple-market communities: 7800 yr. Dated for Eridu (level XVIII), southern Mesopotamia.

male dominance: **esp. 6000 yr**. Dated for the breakout of the dairymen from northern Mesopotamia. Male dominance characterized some militarily assertive groups of the Semitic and Indo-European language families.

taxes: **5600 yr**. Envelopes at Uruk, southern Mesopotamia (*see* above).

class distinctions: **esp. 5600 yr**. A minority of blatantly rich graves at Gawra (level X), northern Mesopotamia.

warrior kings: **5300 yr**. Dated for chronic warfare in Egypt, leading to a single kingdom 5200 yr.

state: 5100 yr. Onset of Early Dynastic period in Sumeria.

big-man system: c. 5000 yr. In Europe: more easygoing than kingship, and surviving until c. 2000 yr ago.

aristocracy: 4700 yr. In Mesopotamia, dated for the Royal Cemetery at Ur.

empire: **4400 yr**. Akkadian empire (Sargon's) took possession of Sumer and Ebla, and penetrated Anatolia.

legalistic empire: 3900 yr. Amorite emperor Hammurabi promulgated a law code.

urban democracy: esp. 2800 yr. Early cities in Mesopotamia c. 6000 yr may have been quasi-democratic, but Greek city states experimented with assemblies of all free men (a minority of the population) c. 2800 yr onward.

managed economy: **932 yr (AD 1068)**. In Sung China; reforms of Wang An-shih, using paper money and money taxes.

global colonization: 507 yr (AD 1493). Dated for the Spanish colony in Hispaniola; by AD 1939 most of the world was under the control of Europeans.

nation-state evolving: c. 400 yr (AD 1600). In Europe, e.g., Dutch republic AD 1609.

imperial nationalism: **360 yr (AD 1640).** Nationalism was articulated in England, and served the cause of empire building and national unification (e.g., Britain 1707, Italy 1861, Germany 1870).

parliamentary monarchy: 312 yr (AD 1688). In England, separating titular and actual powers at a time when parliament represented landowners and merchants.

liberalism: **310 yr (AD 1690).** Dated for the second *Treatise* of J. Locke; incorporating a belief in individual freedom and the possibility of improving the world without too much inconvenience.

revolutionary nationalism: **224 yr (AD 1776).** At the inception of the United States; evident soon afterward in the revolutions in France and South American colonies.

female emancipation: 139 yr (AD 1861). Women won the vote in Australia; slowly elsewhere.

Marxism: **133 yr (AD 1867).** Dated for the first volume of K. Marx's *Capital*; a belief that modes of production determine political relationships, and make the triumph of communism inevitable.

current arms race: 129 yr (AD 1871). (*See* above.)

global institutions: 125 yr (AD 1875). First, the Universal Postal Union; later, the League of Nations 1919, and United Nations Organization 1945.

revolutionary one-party government: **AD 1917.** Following the Bolshevik revolution in Russia; widely imitated by right-wing, left-wing, and national-liberation movements.

decolonization: **AD 1947 and esp. AD 1960.** Dated for Indian independence in 1947, and for rapid decolonization in Africa in 1960, where 17 new nation-states came into existence that year, and a further 10 in the next three years.

economic communities: esp. AD 1957. The European Economic Community, keeping the peace among quarrelsome Europeans by enmeshing them in regulations; similar institutions emerged in eastern Europe (revamped Comecon 1959), South America, and elsewhere.

Grand Canal of China 1389 yr (AD 611). *See* transport.

Grand Canyon c. 2 Myr. *See* North American uplift.

grand repulsion 10^{-38} second after origin. *See* Big Bang.

grass 24 Myr. The earliest accepted fossils of grasses are from the Harrison formation, Nebraska (Early Miocene), and finds of similar age in Europe also seem secure. Many reports of older ages involved misidentifications. Undoubted grass fossils, apparently older than the Elias finds, are under study (J. R. Thomasson, personal communication, 1982). Thomasson notes that by the early-mid-Miocene (c. 15 Myr ago) savanna grasslands were widespread in North America, and perhaps elsewhere, and a wide variety of fossil herbivores (horses, rhinos, etc.) are found associated with the fossil grasses. A worldwide surge in grazing mammals occurred c. 11 Myr ago (*see* mammals). Treeless grasslands, as on the high plains of North America today, are probably a development of the past 3 million years. As for those special grasses, the cereals, molecular studies in a number of Old World species (repeated DNA) have revealed their evolution from a common cereal ancestor in the sequence: oats, barley, rye, and then various species of wheat (R. B. Flavell, in G. A. Dover and R. B. Flavell, eds., *Genome Evolution*, Academic Press, 1982). Maize (corn) is a mutant of teosinte, an unpromising-looking ancestor, and the early Amerindian cultivators probably collected and fostered the rare mutant (G. Beadle, personal communication, 1972). For comments on the importance of grass see p. 43. Illus. p. 130.

Gravettian culture 29,000 yr; *heyday* **28,000 yr.** The Gravettian hunter-gatherer culture of Europe persisted for nearly 20,000 years through the coldest phase of the most recent ice age. The Gravettians may have originated in the Ukraine, but the early phase is especially well exhibited at Dolní Věstonice in Czechoslovakia, 29,000 yr (26,390 b.c.). Innovations included the first community with more than one hundred people (*see* government), composite tools with handles and shafts (*see* stone tools), and notable artistry including the first known use of ceramics (*see* arts and crafts). Distinctive features of the Gravettian culture, also found at Dolní Věstonice, are obese "Venus" figurines and extensive reliance on mammoth hunting. From 28,000 yr ago, when the climate deteriorated, Gravettians spread to Poland, Austria, Germany, Italy, and France; by 20,000 yr ago, other peoples asserted themselves in France and Spain (Solutrean and Magdalenian cultures), but the Gravettians continued to thrive elsewhere until c. 10,000 yr ago. (Data mainly from P. Phillips, *The Prehistory of Europe*, Allen Lane, 1980.)

gravity, law of 313 yr (AD 1687). I. Newton's *Principia*, published in London, consolidated classical physics. The idea of gravity acting in the cosmos as on Earth was prevalent in the late seventeenth century AD, but it was Newton who mastered the complexities of it. Although replaced by A. Einstein's general relativity (1915), Newton's theory rates as a model of excellence in science.

gravity (quantum) at time *zero*; (normal) **10^{-43} second** after origin. *See* Big Bang.

grazing animals (surge in) **11 Myr**. *See* mammals.

Great Barrier Reef 20 Myr. *See* reefs.

Great Wall of China **2214 yr.** *See* weaponry.

Greeks: *Indo-Europeans into Greece* **c. 4350 yr**; *breakouts* **2800 yr and 2334 yr**, *see* breakouts (Indo-Europeans; blacksmiths); *language* by 3000 yr, *see* language; *architecture* c. 2650 yr, and other arts, *see* arts and crafts; building; *science* **2600 yr**, *see* science; *Athens* **esp. 2480 yr**, *see* Athenian heyday.

green algae **1500 Myr.** *See* life on Earth.

"greenhouse" phases **430 and 170 Myr.** *See* climate 3800 Myr.

Greenland (microcontinent): *oldest rocks* **3800 yr**; *quit Europe* **c. 60 Myr**, *see* continents; *Eskimos* by 4000 yr; *Vikings* by 1000 yr (AD 1000), *see* breakouts.

green revolution **AD 1961.** *See* cultivation practice.

Grenville Mountains **1150 to 850 Myr.** *See* continents.

Gulf Stream 3.9 Myr. *See* climate 67 Myr; for variations, ice age to warm phases, see maps pp. 86, 90.

gunpower 1100 yr (AD 900); *gun warfare* **644 yr (AD 1356)**. *See* weaponry.

gymnosperms 370 Myr. *See* plants ashore.

H

Hadean era **4450 to 3800 Myr.** *See* geological timescale.

hand axes **1.9 Myr**, *see* tools; *artistic* **c. 200,000 yr**, *see* arts and crafts.

Han empires **2202 yr to 1779 yr (AD 221).** *See* historical timescale.

handedness **?1.9 Myr.** *See* human origins.

handles (for tools) **esp. 29,000 yr**. *See* stone tools.

hang-nose primates **est. 36 Myr**; (fossils) **29 Myr**. *See* primates.

Hanseatic League **720 yr (AD 1280).** *See* historical timescale.

harbors (man-made) esp. 2800 yr. *See* transport.

hard animals (mineral skeletons) **600, esp. 570 Myr**. *See* animal origins.

harnesses (for horses) esp. 2250 yr. *See* transport.

harpoons **14,000 yr.** *See* fishing.

Harrapa culture **4900 to 4000 yr.** Cultivation, using western Asian crops, flourished in northwestern India (including modern Pakistan) from c. 8000 yr onward, and after 4900 yr ago an urban, bronze-using culture arose in the Indus Valley and the adjacent present desert area of Rajasthan. At the time (4120 yr ago) of a sharp change in world climate (*see* climate 10,300 yr) Rajasthan became desiccated 4100 yr ago, with former freshwater lakes turning to salt; the economy collapsed at about the same time (G. Singh, et al., *Philosophical Transactions of the Royal Society*, Vol. B267, 1974, p. 467; also R. A. Bryson, *Monsoon 1981*, University of Wisconsin, 1981). Indo-European invaders harassed the surviving inhabitants and eventually, with their chariots, took possession of the Indus Valley ?3600 yr ago. Controversy surrounds suspected linkages between climatic change, economic collapse, and the arrival of the Indo-Europeans.

Hawaii: *volcanic activity in chain* **14 Myr**, *see* volcanoes; *populated* **1700 yr (AD 300)**, *see* breakouts (Polynesians).

H-bomb **c. AD 1953.** *See* weaponry.

Hebrew (language) est. 3800 yr. *See* language.

Hegira (Mohammed) 1378 yr (AD 622). *See* religious surge.

helium making **three minutes** after origin. *See* Big Bang; elements; stars.

hemoglobin (active molecule in blood) est. 480 Myr. *See* vertebrate animals.

herbal medicine **60,000 yr.** *See* medicine.

Hercynian Mountains **350 to 280 Myr.** *See* continents.

herding **18,000 yr.** *See* animal domestication.

hieroglyphics (Egyptian) 5000 yr, *see* information technology; (Middle American) ?2000 Myr, *see* archeological timescale.

highways 5500 yr. *See* transport.

Himalayan uplift **c. 5 Myr.** The world's highest mountains are the wreckage of the collision between India and Tibet, which began 45 Myr ago when Indian promontories first kissed Asia, as dated by animal traffic. An island arc c. 50 Myr old, hemmed between the continents stands on edge in the Karakoram range. Tibet has been lifted as a high plateau, by the margin of India driving about 500 kilometers under it, and large blocks have been extruded sideways (P. Molnar and P. Tapponnier, *Science*, Vol. 189, 1975, p. 419, and M. Cowan, et al., *Nature*, Vol. 284, 1980, p. 218). The main central thrust began c. 20 Myr ago (Miocene), and deformation along the main frontal zone of the Himalayan range began c. 5 Myr ago (Pliocene). The collision continues vigorously, with India moving north at five centimeters a year and the mountains still rising, at rates of up to nine millimeters a year in some places (B. F. Windley personal communication, 1982, citing P. K. Zeitler).

Hinduism est. 3600 yr. *See* religious surge.

hippopotamuses 5.5 Myr. *See* mammals.

historical timescale This compilation of certain political events and periods is intended only as an analog for the geological and archeological timescales. It provides links with conventional history, and furnishes chronological notes for the narrative pages during the historical period, and a few entries for the timescale. It has no analytical purpose. Dates are from sources cited elsewhere (*see* archeological timescale; breakouts; population), supplemented from G. Barraclough, ed., *The Times Atlas of World History*, Times Books, 1978.

HISTORICAL TIMESCALE

Babylon and Shang	
Old Babylonian empire	4030 yr
Shang period, China	3800 yr
Middle Babylonian empire	3170 yr
Chou period, China	esp. 3100 yr
Neo-Babylonian period	2626 yr
Classical empires	
Persian (Achaemenid) empire	2550 yr
Macedonian (Alexander's) empire	2334 yr
Mauryan empire, India	2332 yr
Ch'in empire, China	2221 yr
Han empire, China	2202 yr
Parthian empire	esp. 2141 yr
Rome's first emperor	2027 yr
Han empire fell	1779 yr (AD 221)
Sasanian empire, Iran	1776 yr (AD 224)
Classic Maya period	1740 yr (AD 260)
Gupta empire, India	1680 yr (AD 320)
western Roman empire fell	1524 yr (AD 476)
Early Middle Ages	
Frankish empire, Europe	1514 yr (AD 486)
Sui empire, China	1419 yr (AD 581)
Late Classic Maya period	c. 1400 yr (AD 600)
Byzantine empire	1390 yr (AD 610)
T'ang empire, China	1382 yr (AD 618)
Abassid caliphate, Islam	1250 yr (AD 750)
end of Late Classic Maya	1200 yr (AD 800)
Sung empire, China	1040 yr (AD 960)
Later Middle Ages	
Hanseatic League	c. 760 yr (AD 1280)
Mongol (Yuan) empire	740 yr (AD 1260)
Aztecs to power	c. 675 yr (AD 1325)
Ming empire, China	632 yr (AD 1368)
Incas to power	c. 600 yr (AD 1400)
Byzantine empire fell	547 yr (AD 1453)
European empires	
Tsarist Russia	522 yr (AD 1478)
Portuguese empire	518 yr (AD 1482)
Spanish empire	507 yr (AD 1493)
Italian Renaissance, zenith	500 yr (AD 1500)
Mughal empire, India	474 yr (AD 1526)
French empire	400 yr (AD 1600)
English empire	393 yr (AD 1607)
Dutch empire	381 yr (AD 1619)
Louis XIV in France	357 yr (AD 1643)
Manchu (Ch'ing) empire	356 yr (AD 1644)
Swedish empire, zenith	342 yr (AD 1658)
Britain unified	293 yr (AD 1707)
Age of revolutions	
Industrial Revolution (Britain)	c. 240 yr (AD 1760)
Age of Enlightenment, zenith	240 yr (AD 1760)
American revolution	225 yr (AD 1776)
French revolution	211 yr (AD 1789)
Napoleonic empire	196 yr (AD 1804)
Latin American revolutions	192 yr (AD 1808)
Napoleonic empire fell	186 yr (AD 1814)
Victorian Age	
Victoria in Britain	163 yr (AD 1837)
Chinese cowed (opium war)	158 yr (AD 1844)
Italy unified	139 yr (AD 1861)
Japanese revolution	132 yr (AD 1868)
Germany unified	129 yr (AD 1871)
Age of dictators	
Chinese republic	AD 1912
World War I	AD 1914
Russian revolutions	AD 1917
League of Nations	AD 1920
Fascist Italy	AD 1922
Stalin in U.S.S.R.	AD 1924
Great Depression	AD 1929
militarism in Japan	AD 1931
Nazi Germany	AD 1933
World War II	AD 1939
Japanese Co-Prosperity Sphere	AD 1942
World War II ended	AD 1945
Modern superpowers	
United Nations	AD 1945
carve-up of Europe (Yalta)	AD 1945
Cold War	AD 1947
Chinese People's Republic	AD 1949

Korean War	AD 1950
Death of Stalin	AD 1953
Americans in Vietnam	esp. AD 1961
Cuban missile crisis	AD 1962
Chinese cultural revolution	AD 1966
Americans quit Vietnam	AD 1973
global recession	AD 1980

Hitler's empire AD 1936 to 1945. *See* breakouts.

Hittites *identifiable* **4000 yr**, *see* breakouts (horsemen); *language* before **3000 yr**, *see* language; *collapse* **3250 yr**, *see* climate 10,300 yr.

hogs **c. 1.8 Myr.** *See* pigs.

Holocene epoch **10,300 yr** to present. *See* geological timescale; climate 10,300 yr.

holograms esp. AD 1963. *See* information technology.

***Homo* species** **2.0 Myr**, etc. *See* human origins.

Hopewell culture 2300 yr; *in retreat* **1700 yr (AD 300)**. *See* climate 10,300 yr.

horsemen's breakout (proto-Indo European speakers) **4900 yr**. *See* breakouts; livestock revolution.

horses **55 Myr, esp. 3.7 Myr.** The family of horses (equids) emerged **55 Myr** ago (Eocene); the genus *Equus* of true horses and zebras evolved from *Pliohippus* **3.7 Myr** ago in North America and was prominent in that continent by 3.3 Myr ago (E. H. Lindsay, et al., *Nature*, Vol. 287, 1980, p. 135). Horses spread to Eurasia 2.5 Myr, and Africa 1.9 Myr ago, becoming zebras. The American horse became extinct more than 8000 yr ago, evidently as a result of overkill by hunters. Horses were tamed **c. 6400 yr** ago, when a subspecies of *Equus caballus*, the tarpan, was domesticated in the southern Ukraine. Large numbers of horse bones occur at settlements of cattle herdsmen c. 6400 yr, and a site at Dereivka yielded a skull of a domesticated horse 6250 yr old (calibrated from data in S. Bökönyi, *La Recherche*, Vol. 11, 1980, p. 919). Some archeologists place the domestication of the horse farther south, and later, e.g., in Turkey or Iran c. 5500 yr, but the steppe was the natural habitat of the wild horse, so the earlier location is reasonable. Horses served for meat, for traction, and eventually for riding, which may have started among Indo-European (Yamnaya culture) of the Lower Volga Valley c. 5000 yr ago (M. Zvelebil, in A. G. Sherratt, ed., *Cambridge Encyclopedia of Archaeology*, Cambridge University Press, 1980). *See also* breakouts; language; livestock revolution; transport.

horsetails **395, esp. 360** to 290 Myr. *See* plants ashore.

host cells (proto-eukaryotic) **est. 2100 Myr**. *See* life on Earth.

Hot Big Bang **10^{-35} second** after origin. *See* Big Bang.

houses (brick) 10,600 yr. *See* building.

human genetics The key molecules of life, genes and proteins, are a surer guide to the history of human populations than supposedly obvious markers such as skin color, stature, physiognomy, and skull shapes, which are often mere adaptations to the local climate. Particular variants of genes occur more frequently in some populations than in others, and modern high-speed methods of testing molecular variations among proteins, products of the genetic instructions, led to new studies of the evolution of human populations, pioneered by L. L. Cavalli-Sforza and his colleagues (e.g., A. Piazza, et al., *Tissue Antigens*, Vol. 5, 1975, p. 445).

Studies of thirty-nine independent genes created a thought-provoking global picture, see map p. 88 (A. Piazza, et al., *Proceedings of the National Academy of Sciences*, Vol. 78, 1981, p. 2638). Genetic gradations outward from a presumed Asian center are modulated by apparent radiations from subcenters. An inference from the genetic analysis of proteins is that all human beings alive today had common ancestors in the fairly recent past—say, 50,000 to 40,000 yr ago—rather than evolving independently in different regions. Interbreeding of those ancestors with other groups is not ruled out.

Direct studies of the DNA composing genes of the mitochondria, the oxygen-handling units of cells, gave further insight (R. L. Cann, et al., 6th International Congress of Human Genetics, in press; R. L. Cann and A. C. Wilson, personal communication, 1981). Variations at 346 locations in the mitochondrial genes of one hundred living individuals show divergences tracing back to an evident origin of the species *Homo sapiens* **600,000 yr** ago (*see also* human origins). Interactions between populations clearly persisted until the supposed dispersal of the populations after 50,000 years ago. Anomalies suggesting a long separation show up in some Australian aborigines.

The San Bushmen in southern Africa were the unusual ones, in another study (M. J. Johnson, et al., *Journal of Molecular Evolution*, in press, 1982). This reconstructed a family tree of types of mitochondrial DNA molecules. The investigators accounted for a relatively large genetic distance between the Bushmen and other peoples by invoking a higher rate of molecular evolution in the Bushmen. Constant rates of evolution (*see* molecular dating) would imply a mitochondrial split from ancestors of other humans, some hundreds of thousands of years ago. For general comments on "races" and history, see p. 54. For relevant archeological and historical data, *see* breakouts; for related studies of molecules, *see* molecular dating; for the milk-drinking mutation, *see* livestock revolution. Illus. pp. 57, 146.

human origins 2.0 Myr to 40,000 yr. For precursors, *see* australopithecines. "Handyman," in Latin *Homo habilis*, is the name assigned to the first of three human species that fossil hunters generally recognize, the others being *Homo erectus* and *Homo sapiens*. D. C. Johanson and T. D. White (*Science*, Vol. 203, 1979, p. 321), supposed *Homo habilis* to be directly descended from apemen, *Australopithecus afarensis*. If small stone tools in the Omo of Ethiopia, **c. 2.4 Myr** were the work of *Homo habilis*, then the origin of this species goes back farther than the oldest fossil remains, which include the skull 1470 from Koobi Fora, Kenya. After much controversy it was dated, by interlocking potassium-argon, argon-argon, and fission-track methods, at "more than 1.9 Myr" (G. H. Curtis, *Philosophical Transactions of the Royal Society*, Vol. B292, 1981, p. 7; for comments see p. 24). Similar creatures lived **c. 2.0 Myr** ago in the Omo Valley of Ethiopia, and c. 1.8 Myr ago at Olduvai Gorge, Tanzania (Bed I). They had hands adapted to precise gripping and they ate meat; remains of a meal in Kenya indicates that they scavenged the corpses of large animals and shared their food in a systematic, human fashion 1.7 Myr ago (G. L. Isaac, same volume, p. 177). *Homo habilis* may have survived until c. 1.45 Myr (Omo), but the evidence is scanty; at the prime sites in Kenya and Tanzania, the picture is of *Homo habilis* being summarily replaced by the larger and brainier *Homo erectus* 1.6 Myr ago. Illus. pp. 140, 141, 142, 144.

A *Homo erectus* child from Sangiran, Java, has a possible date of 1.9 Myr, due for remeasurement at the time of writing. Other evidence for early *Homo erectus* in Asia comes from 'Ubeidiya in the Jordan Valley, with stone tools, including hand axes, typical of *Homo erectus*. Associated fossil mammals indicate a probable age between 2.6 and 1.9 Myr (C. A. Repenning and O. Fejfar, *Nature*, Vol. 299, 1982, p. 344). The date adopted here for *Homo erectus* is **1.9 Myr**. In Africa, the first clear signs of *Homo erectus*, a skull and characteristic hand axes, are dated at **1.6 Myr**. *Homo erectus* probably also mastered fire by **1.4 Myr** (*see* firemaking). Humans using pebble-chopper tools broke out of warm latitudes into Europe **c. 750,000 yr** ago, somewhat earlier than the start of the Brunhes magnetic epoch 730,000 yr ago (Isernia site; M. Coltorti, et al., *Nature*, Vol. 300, 1982, p. 173).

The brain of *Homo erectus* increased in size. In Indonesia, early Java Man (*Homo erectus erectus*) had a mean cranial capacity of about 875 cubic centimeters; the later Peking Man (*Homo erectus pekinensis*; ?650,000 yr, Late Biharian glacial episode) had 1050 cubic centimeters. For comparison, modern humans have about 1300 cubic centimeters. Handedness, the preferred use of one hand for skilled tasks, may have begun with *Homo erectus*. Otherwise there is so little sign of cultural change in *Homo erectus* between 1.4 Myr and 750,000 yr that it seems necessary to declare a "standstill" at **1.4 Myr**. (For a contrary view, see J. E. Cronin, et al., *Nature*, Vol. 292, 1981, p. 113.)

The modern species, *Homo sapiens*, evolved soon after the migration from the tropics. By molecular dating, the most recent common grandmother of all humans alive today lived **est. 600,000 yr** ago (*see* human genetics). Fossil evidence is meager: it includes a jawbone from Mauer in Germany, and skull bones and teeth from Vértesszöllös in Hungary, said to be early *Homo sapiens*, and loosely dated between 800,000 and 500,000 yr (L. G. Freeman, personal communication, 1981; K. Valoch, 1976, cited by P. Phillips, *The Prehistory of Europe*, Allen Lane, 1980). Beginning **400,000 yr** ago another apparent standstill occurred. Piagetian tests applied to stone tools from Isimila, Tanzania, which may be as much as 330,000 yr old (by uranium-series dating), are said to indicate that the makers were as intelligent as modern humans (T. Wynn, *Man*, Vol. 14, 1979, p. 371). If so, they consciously resisted change. Speculations about evolution of social behavior, **c. 400,000 yr**, follow R. Axelrod and W. D. Hamilton, *Science*, Vol. 211, 1981, p. 1390. Illus. p. 146.

Various skulls from Europe **c. 230,000 yr** ago (Steinheim Man, Swanscombe Man, etc.) are identified as "archaic" *Homo sapiens*; uranium-series dating puts a relevant site, Bilzingsleben in eastern Germany, into a warm interlude 228,000 yr ago (R. S. Harmon, et al., *Nature*, Vol. 284, 1980, p. 132). Neanderthalers (*Homo sapiens neanderthalensis*), with brains larger than those of modern humans, appeared during the last interglacial phase, which began 128,000 yr ago. Phillips offers 125,000 or 100,000 yr as possible starting dates for the neanderthalers; the date given here, **c. 120,000 yr**, takes into account disruption of the interglacial spell, 115,000 yr ago. The neanderthalers disappeared **c. 34,000 yr** ago after the breakout of another subspecies (*see* neanderthalers).

A fossil skull with a mixture of modern and archaic features is known from Laetoli, Tanzania, c. 120,000 yr ago (M. H. Day, et al., *Nature*, Vol. 284, 1980, p. 55), and other "intermediate" skulls of about the same date have been found in Zambia, Kenya, and Ethiopia (G. Kennedy, *Nature*, Vol. 284, 1980, p. 11). In Border Cave in South Africa a piece of skull and a jaw, lighter in build than the neanderthalers and anatomically indistinguishable from modern humans (*Homo sapiens sapiens*), are said to be 105,000 to 90,000 yr in age (P. Beaumont, et al., 1978, cited by L. G. Freeman, in A. G. Sherratt, ed., *Cambridge Encyclopedia of Archaeology*, Cambridge University Press, 1980). These fossils give a datum for "proto-modern humans," **c. 100,000 yr**. Illus. p. 154.

Modern humans spread over most of the world **c. 40,000 yr** ago (*see* breakouts). Fully-fledged language is the most plausible evolutionary innovation responsible for the mental revolution that let loose the ingenuity and exuberance that have continued to the present (P. Lieberman, *La Recherche*, Vol. 6, 1975, p. 751). One must allow a little time for a mutation to become commonplace in a population, and take account of tools, said to be transitional to those of modern humans, which date from c. 46,500 yr by uranium-series dating (Nahal Zin, Negev Highlands; H. P. Schwarcz, et al., *Nature*, Vol. 277, 1979, p. 560). Accordingly, the supposed language mutation and the origin of modern humans can be set at **est. 45,000 yr** ago, at the latest (*see* mental revolution).

HUMAN ORIGINS: TIMETABLE

stone tools: **c. 2.4 Myr**.
Homo habilis: **c. 2.0 to 1.6 Myr**.
Homo erectus (in Asia): **c. 1.9 Myr**, in Africa, **1.6 Myr**, standstill, **c. 1.4 Myr**, breakout, **c. 750,000 yr**.
Homo sapiens: **c. 600,000 yr**.
human standstill: **c. 400,000 yr**.
archaic Homo sapiens: **c. 230,000 yr**.
neanderthalers: **c. 120,000 yr**, extinct, **c. 34,000 yr**.
proto-modern humans: **?100,000 yr**.
modern humans: **est. 45,000 yr**, conspicuous, **c. 40,000 yr**.

Huns' breakout 1696 yr (AD 304). *See* breakouts.

hunter-gatherers' breakout c. 40,000 yr. *See* breakouts.

hunting c. 1.9 Myr. *Homo erectus* appeared c. 1.9 Myr ago (*see* human origins), and evidently obtained meat by active hunting, as opposed to scavenging by *Homo habilis*. Big-game hunting is attested at Isernia, Italy, **c. 750,000 yr** ago (M. Coltorti, et al., *Nature*, Vol. 300, 1982, p. 173), at Choukoutien, China **?650,000 yr** ago, and at Olorgesailie in Kenya, **?500,000 yr** ago, where giant baboons were taken. Large-scale hunting of elephants and other game took place in Spain at the sites of Terralba and Ambrona; it was before the cold climax of an ice age, and "Late Elsterian, pre-Holsteinian" (L. G. Freeman, personal communication, 1981). A correlation with the phase before the glacial maximum of core stage 12 (429,000 yr ago) gives a date for this large-scale hunting of **?450,000 yr** (for core-stage 10, the date would be c. 345,000 yr; *see* orbital dating). For consequences of later large-scale hunting, *see* mammals.

Huronian ice ages c. 2300 Myr. *See* climate 3800 Myr.

huts ?420,000 yr. *See* building.

hydrogen nuclei (protons) **1 microsecond** after the origin. *See* Big Bang.

I

ice age, most recent 72,000 yr; *lesser ice maximum* **62,000 yr**; *milder interlude* **58,000 yr**; *severe cold* **28,000 yr**; *latest ice maximum* **18,000 yr**; *thaw started* **14,000 yr**; *recooling* **10,900 yr**; *ended* **10,300 yr**. *See* climate 128,000 yr.

ice ages, ancient c. 2300, 950, 770, 670, 440, and 290 Myr, *see* climate 3800 Myr; *current series* **3.25 Myr**; *deeper* **2.4 Myr**; *emphatic* **800,000 yr**, *see* climate 3.25 Myr; *individual recent* **550,000 yr**, etc; *worst* **430,000 yr**, *see* orbital dating.

ice ages, false 115,000 and 95,000 yr. These events, predictable by orbital dating, correspond with stages 5d and 5b in the ocean-bed core record of climatic fluctuations, and show up clearly in the European pollen (Grande Pile, France, herbs replacing trees; G. M. Woillard and W. G. Mook, *Science*, Vol. 215, 1982, p. 159). In both cases they lasted five to ten thousand years, and the onset of cold conditions was extremely rapid. First indications from low-resolution ocean-bed cores were of a switch from mild to glacial conditions in a few centuries (J. P. Kennett and P. Huddleston, *Quaternary Research*, Vol. 2, 1972, p. 384). Oxygen isotope studies of the Camp Century, Greenland, ice core (W. Dansgaard, et al.) indicated a switch in less than a century. The shortest interval, a couple of decades, was suggested by the pollen work at Grande Pile (G. M. Woillard).

"icehouse" phases 360 and 70 Myr. *See* climate 3800 Myr.

Illinoian glaciation ?430,000 yr. *See* orbital dating.

imperial companies 400 yr (AD 1600). English, Dutch, and French trading-post empires developed in Asia, initially at the expense of the Portuguese; these ventures were signaled by the creation of the East India Companies: English AD 1600, Dutch 1602, French 1664. (A. Calder, *Revolutionary Empire*, Cape, 1981.)

imperial nationalism 360 yr (AD 1640). *See* government.

Incas' rise c. 600 yr (AD 1400). *See* historical timescale.

Independence culture (Amerindian) **c. 4900 yr**. *See* climate 10,300 yr.

India (microcontinent; modern India, Pakistan, and Bangladesh) *quit Antarctica* **100 Myr**; *hit Eurasia* **45 Myr**, *see* continents; Himalayan uplift; *monsoon* c. 9800 yr, *see* climate 10,300 yr; *cultivation* **c. 8000 yr**, *see* breakouts (gardeners); *Harrapa period* **4900 to 4000 yr**, *see* Harrapa culture; **chariots to India c. 3800 yr**; Mauryan empire **c. 2250 yr**, *see* breakouts (charioteers); *Hinduism* ?3600 yr; *Buddhism* **c. 2500 yr**, *see* religious surge; *Vedic Sanskrit* by 3000 yr, *later Indic languages, and retreat of Dravidian, see* language. *"Arabic" numerals evolving* c. 2250 yr, *see* information technology; *Gupta empire* **1680 yr (AD 320)**, *see* historical timescale; *Mughal empire* **632 yr (AD 1576)**, *see* breakouts (Mongols and Turks); *settled by Europeans* (Goa) 490 yr (AD 1510), *see* breakouts (navigators); *British control growing* **by 235 yr (AD 1765)**, *see* British Bengal; *decolonization* **AD 1947**, *see* government. See also maps pp. 81–95.

Indian Ocean c. 160 Myr. *See* oceans.

Indochina (microcontinent) *hit Asia* **c. 220 Myr**, *see* continents; *domesticated rice, water buffalo, and pigs* **est. 9500 yr**, *see* animal domestication; crops; *languages, relationships obscure, see* language; *bronze* ?5400 yr, *see* metal industries; *expansions by Chinese* 2223 yr; *Malays* c. 800 yr (AD 1200); *Khmer* (Angkor) esp. 1150 yr (AD 850); *Thais* esp. 850 yr (AD 1150); *Vietnamese* esp. 529 yr (AD 1471); *Europeans* 141 yr (AD 1859); *Japanese* **AD 1941**; *see* breakouts. *Americans in Indochina* esp. **AD 1961 to 1973**, *see* historical timescale. See also maps pp. 81, 84–95.

Indo-European (languages origin) **c. 4900 yr.** *See* breakouts (horsemen, etc.); language.

Indonesia (island arc): *volcanism noted* 32 Myr to present; *Toba explosion* c. 73,000 yr, *see* volcanic events; Homo erectus *in Java* ?1.9 Myr, *see* human origins; *modern humans* (Borneo) c. 40,000 yr, *see* breakouts (hunter-gatherers); *domesticated yams, bananas, coconuts, etc.* ?9000 yr, *see* crops; *Austronesian migrations* 6000 yr, etc., *see* breakouts (gardeners). *Dutch colonization* (Batavia) 381 yr (AD 1619), *see* breakouts (navigators); Dutch spice monopoly. See also maps pp. 81, 84–95.

industrial revolution (China) c. 2400 yr, *see* metal industries (cast iron); *Industrial Revolution* (conventionally, Britain) **240 yr (AD 1760)**, *see* historical timescale.

Indus Valley esp. 4900 yr, *see* Harrapa culture.

inflation era (early growth of universe) **10^{-38} second** after origin. *See* Big Bang.

influenza esp. 434 yr (AD 1564), and AD 1918. *See* disease epidemics.

information technology est. 45,000 yr to present. One of the gods of the ancient Egyptians was the Plan Maker, and a reasonable guess is that sketch maps were one of the first means by which modern humans of est. 45,000 yr ago supplemented speech (*see* mental revolution).

The forerunner of writing in southwestern Asia was a technique of accounting that used clay tokens of distinctive geometric shapes to represent commodities such as sheep or measures of grain. In Iran **10,700 yr** ago (Tepe Asiab and Ganjdareh, c. 8450 b.c.), goatherds and shepherds had twenty different symbols: spheres, cones, disks (some with incisions), cylinders, tetrahedrons, and so on. This system of reckoning spread throughout southwestern Asia and to the Indus Valley and the upper Nile, persisting with little change for five thousand years. In Mesopotamia **5600 yr** ago (Uruk V/IVB) the practice grew up of putting tokens into a clay envelope, or bulla, and marking the outside of the envelope with an official seal and a note of the contents. The step to writing was a matter of using flat pieces of clay (tablets) instead of envelopes, and drawing pictures of the tokens instead of enclosing them. The transitional step was simply to press the tokens on the clay, leaving indentations. The elaborate written tablets that appeared in Mesopotamian cities **5500 yr** ago (Uruk, level IVA; *not* IVB as often stated), made thorough use of drawings of tokens as ideographs and pictographs. (D. Schmandt-

Besserat, *Visible Language*, Vol. 15, 1981, p. 321, other publications, and personal communication 1982). *See also* government.

The shapes of modern "Arabic" numerals can be traced in India from Asoka times onward (c. 2250 yr). "Place value" by which, for example, the numerals in the expression "555" have quite different meanings to their position (five hundred, fifty, and five) is first known from Kampuchea, AD 604. A culminating invention was the zero, appearing simultaneously in AD 683 in Kampuchea and Sumatra (J. Needham, *Science and Civilisation in China*, Vol. 3, Cambridge University Press, 1959). The modern numeric system can therefore be dated to **c. 1320 yr (AD 680)**.

The precursors of printing go back to printed patterns on cloth, and to seals used for making impressions in clay, as marks of ownership or authority. When Europeans eventually adopted printing, long after the Chinese, they had the benefit of an alphanumeric system far simpler to handle than written Chinese. Only late in the twentieth century did computerized systems begin to cure the difficulties of printing and communicating in ideographic languages.

The first electronic digital computer, operational in Britain in AD 1943, was a special-purpose code-breaking machine, Colossus; a general-purpose electronic digital computer, ENIAC, was developed at the University of Pennsylvania, by 1946. Stored programs appeared in various machines in the late 1940s and transistors replaced thermionic tubes in the 1950s. The first microprocessor, developed by M. E. Hoff in **1971**, had 2250 transistors on a chip of silicon of about twelve square millimeters. (C. Evans, *The Mighty Micro*, Gollancz, 1979; National Science Board, *Only One Science*, NSF/GPO, 1980.)

Idiosyncratic data-processing systems such as the quipu knotted cords of the Incas and the European distribution of audio entertainment by telephone are left aside; so are pre-electric telegraph systems, such as the French semaphore of AD 1795. Sources for the timetable, apart from those mentioned above, include R. V. Tooley, *Maps and Mapmakers*, Batsford, 1970; A. G. Sherratt, ed., *Cambridge Encyclopedia of Archaeology*, Cambridge University Press, 1980; P. Phillips, *The Prehistory of Europe*, Allen Lane, 1980; C. Burney, *From Village to Empire*, Phaidon, 1977; P. R. S. Moorey, personal communication, 1981; Hung-hsiang Chou, *Scientific American*, Vol. 240, No. 4, 1979, p. 101; B. Hook, ed., *Cambridge Encyclopedia of China*, Cambridge University Press, 1982; D. Gabor, in D. Fishlock, ed., *A Guide to the Laser*, Macdonald, 1967; and *Encyclopaedia Britannica*, 1981. *See also* arts and crafts.

INFORMATION TECHNOLOGY: TIMETABLE

maps and diagrams: est. 45,000 yr (*see* above). Extant maps include a Mesopotamian town plan c. 4400 yr old. Modern triangulated surveying began in France 256 yr ago (AD 1744).

calendars: **c. 35,000 yr**. France, lunar calendars on bone or stone plaques (Aurignacian period).

accountancy: **10,700 yr**. Iran, symbolic clay tokens (*see* above).

seals: 8500 yr. Anatolia, stamp seals (Çatal Hüyük VIB). Cylinder seals, rolled on clay objects to make impressions, came into use in Mesopotamia 5700 yr (Uruk V).

envelopes: **5600 yr**. Mesopotamia (*see* above).

writing: **5500 yr**. Mesopotamia, pictographs on clay tablets (*see* above).

Egyptian hieroglyphics: c. 5000 yr.

alphabet: **c. 3800 yr**. ?Sinai, phonetic alphabet (Semitic).

Chinese writing: c. 3500 yr. Oracle bones (Shang).

coins: esp. 2600 yr. Anatolia.

paper: **c. 1900 yr (AD 100)**. China, tracing back at least to AD 105, when Ts'ai Lun made paper. It replaced other writing materials in China by c. AD 300.

abacus: **c. 1500 yr (AD 500)**. Eurasia, beads on wires replacing beads on a board. An account in China c. AD 470 is interpreted as a description of an abacus with wires, but the device was also known in Europe c. AD 500.

modern numerals: **c. 1320 yr (AD 680)**. Southeastern Asia, number symbols including zero, with place value (*see* above).

printing: **est. 1300 yr (AD 700)**. China, woodblock printing of written characters, known in Japan by AD 770. Movable type, initially of ceramic, was in use in China c. AD 1040.

paper money: **c. 1090 yr (AD 910)**. China.

printing press: **esp. 545 yr (AD 1455)**. Germany, dated for the first well-printed book in Europe, J. Gutenberg's Bible of AD 1455. Subsequent developments included the high-speed rotary press, **AD 1865**, typewriter, 1867; Linotype and Monotype, 1880s; offset lithography, c. 1904; and high-speed phototypesetting 1949.

photography: **161 yr (AD 1839)**. Experiments in France culminating in daguerreotype, AD 1839, and the modern technique originating with W. H. Fox Talbot in Britain in 1840. Later developments included roll film, 1899, and commercial color film, 1935 (both U.S.).

electric telegraph: esp. 156 yr (AD 1844). U.S., the first public electric telegraph service (S. Morse). The first transatlantic telegraph was completed in 1866.

telephone: **122 yr (AD 1878)**. U.S., dated for A. G. Bell's first commercial telephone exchange, at New Haven, Connecticut, 1878. Automatic telephone exchanges came into service from 1921 onward.

punched-card data processing: esp. 110 yr (AD 1890). U.S., when equipment developed from the Jacquard programmed pattern-weaving loom (France, 1804), handled the U.S. census of 1890 (H. Hollerith).

sound recording: esp. AD 1901. U.S., dated not for the advent of T. A. Edison's phonograph (1877) but for the modern system of spiral grooves in molded disks, 1901 (E. Berliner, et al.). Magnetized plastic tape evolved principally in Germany c. 1938, and found later applications in data storage for computers and video recording.

radio: esp. AD 1901. Transatlantic, dated for G. Marconi of Italy who, defying known physics, transmitted a spark signal from England to Newfoundland.

electronics: **AD 1906**. U.S., dated for the triode, in which L. de Forest improved on the thermionic diode of J. A. Fleming (Britain, 1904), by putting in a grid that regulated the passage of electrons from a heated filament. Later advances in electronic hardware included the magnetron (Britain, 1939) that made modern radar possible by generating intense microwaves, and the transistor (U.S., 1948).

radio broadcasting boom: **AD 1922**. Worldwide. In 1922 the number of stations in the United States increased from 8 to nearly 600, and broadcasting services began in the U.S.S.R., France, and Britain.

computer: **AD 1943**. Britain (*see* above).

television boom: **esp. AD 1950**. U.S., where the number of receiving sets increased tenfold between 1949 and 1951 (from 1 to 10 million), and then climbed to 50 million by 1959. Many countries initiated television broadcasting services during the 1950s.

holography: esp. AD 1963. U.S. (interference patterns in coherent light). Experiments in Britain in 1948 demonstrated the principle using coherent light from a pinhole, but the advent of the laser made holograms useful, as first shown at the University of Michigan, in 1963.

communications satellite: **esp. AD 1965**. U.S., when the first fully commercial communications satellite *Intelsat 1* (alias *Early Bird*) was launched into synchronous orbit, in April 1965. See also map p. 94.

microprocessor: **AD 1971**. U.S. (*see* above). "Personal" computers became available from 1975 onward.

optical-fiber links: AD 1980. Various countries, with revenue-earning installations using hair-thin strands of ultrapure glass, and light-emitting diodes or lasers as the transmitters.

insectivores 80, esp. 40 Myr. *See* mammals.

insects c. 395 Myr; winged **330 Myr**. *See* invertebrates.

interglacials (warm interludes) *See* orbital dating.

internal combustion engine esp. 115 yr (AD 1885). *See* energy; transport.

invertebrates 620 Myr to present. For the origins of the main groups of invertebrate animals 680 to 570 Myr, *see* animal origins, and for the notable reef-building animals, *see* reefs. Sources for the timetable include those cited under animal origins, also A. Hallam, ed., *Planet Earth*, Elsevier Phaidon, 1977, and *Patterns of Evolution*, Elsevier, 1977; W. S. McKerrow, personal communication, 1981, and A. C. Scott, British Association, Liverpool meeting, 1982. For names of periods matching the dates given, *see* geological timescale.

INVERTEBRATES: TIMETABLE

arthropods: **c. 620 Myr**. Notable representatives of these animals with jointed legs appeared **560 Myr**, including trilobites (extinct 256 Myr), pincered chelicerates (later including giant sea scorpions), and crustaceans. Barnacles date to 420 Myr. Arthropods were the first animals ashore, est. 425 Myr, assuming they accompanied the earliest food supplies (*see* plants ashore). Ecosystems with millipedes, insects, spiders, scorpions, and mites are known from c. 394 Myr (Rhynie Chert and Alken), rounded here to 395 Myr. *See* insects, below.

mollusks: **570 Myr**. Famous fossil mollusks include the ammonites, characterized by coiled shells used for floating; from an ammonoid ancestry tracing back to 400 Myr, true ammonites appeared **230 Myr**, crashed 208 Myr, and became abundant thereafter until they were extinguished at the Cretaceous terminal catastrophe, 67 Myr ago.

echinoderms: **560 Myr.** A diverse group of animals related by their fivefold symmetry, most visible in the starfishes, which trace back to 500 Myr. Crinoids or sea-lilies also originated 500 Myr ago, and attained extraordinary abundance 360 to 290 Myr. Sea urchins (echinoids) appeared 470 Myr ago, and since 200 Myr ago they have become the most successful echinoderms.

brachiopods: 560 Myr. Represented nowadays by lampshells. They achieved their greatest diversity 400 Myr ago and have been in overall decline ever since.

foraminifera: 550 Myr. Proto-animals possessing mainly chalky hard parts. They are usually small, although individual forams have ranged up to ten centimeters in diameter. Originally bottom dwellers, forams began living also at the sea surface 216 Myr ago.

radiolaria: 530 Myr. Proto-animals with mainly silica hard parts.

bryozoa: 500 Myr. These moss-like animals that live in colonies and filter food from seawater have had three distinct periods of success: c. 500 to 435 Myr, 96 to 67 Myr, and 55 Myr to present.

insects (arthropods): **395 Myr**. Springtails are plain in the

Rhynie Chert, *see* above. Winged insects, including both fixed-wing (giant dragonflies) and folding-wing forms (beetles, grasshoppers, etc.) were flourishing by **330 Myr**. Butterflies, moths, and early flies appear in the fossil record 270 Myr ago. Explosive evolution among the social insects, including bees, wasps, and ants, occurred **c. 200 Myr** ago.

Iran *North Iran* (microcontinent) *hit Asia* c. 200 Myr; *Arabia hit Iran* (Zagros) **c. 20 Myr**, *see* continents; *goats, sheep, and accountancy* **10,700 yr**, *see* animal domestication; information technology; *Persians and Medes* c. 3350 yr; *Persian empire* **2550 yr** and other movements, *see* breakouts; *Old Persian* (language) by 2000 yr, *see* language; *Sasanian empire* **1776 yr (AD 224)**, *see* historical timescale.

iridium anomalies **67 Myr**, etc. *See* cosmic impacts.

Iron Age **3200, esp. 3050 yr.** *See* archeological timescale; metal industries.

iron boom (banded formations) **esp. 2500 Myr.** *See* ores.

irrigation **8000 yr**. *See* cultivation practice.

Islam 1378 yr (AD 622), **esp. 1368 yr (AD 632)**, *see* breakouts (Arabs); religious surge; *Abbasid caliphate* **1250 yr (AD 750)**, *see* historical timescale; *architecture* esp. 1310 yr (AD 690), *see* building; *science* **esp. 1000 yr (AD 1000)**, *see* science; *Turks busy* 650 yr (AD 1350) etc., *see* breakouts (Mongols and Turks); *Arabian oil* **AD 1932**, *see* energy.

Italy (microcontinent) *quit Africa* c. 110 Myr; *hit Europe* c. 60, **esp. 10 Myr**, *see* continents; Swiss Alps; *early humans* **c. 750,000 yr**, *see* human origins; *Greeks busy* **2800 yr**, *Etruscans* **esp. 2700 yr**, Romans **esp. 2260 yr**, Lombards c. 1442 yr (AD 558), *see* breakouts; *Latin languages* by 2000 yr, *see* language; *Rome fell* **1524 yr (AD 476)**; Renaissance **esp. 500 yr (AD 1500)**; Italians unified **139 yr (AD 1861)**; Fascists **AD 1922 to 1943**, *see* historical timescale.

J

jackals 6 Myr. *See* dogs.

Jacquard loom 196 yr (AD 1804). *See* information technology.

jaguars 1.7 Myr. *See* cats.

Jainism **esp. 2540 yr.** *See* religious surge.

Japan (microcontinent) *quit Eurasia* **c. 30 Myr**, *see* continents; oceans; *pottery* **13,000 yr**, *see* arts and crafts; *deep-sea fishing* **9600 yr**, *see* fishing; *intrusive Yayoi culture* c. 2750 yr, *see* climate 10,300 yr; *Japanese revolution* (Meiji restoration) **132 yr (AD 1868)**; *militarism* **AD 1931**; *Co-Prosperity Sphere* **AD 1942 to 1945**, *see* historical timescale; *breakout* **esp. AD 1905**, *see* breakouts.

Java Man ?1.9 Myr. *See* human origins.

jaws **425 Myr.** *See* vertebrate animals (jawed fishes).

jellyfish **670 Myr.** *See* animal origins.

Judaism **esp. 2450 yr.** *See* religious surge.

Jupiter forming **4550 Myr.** *See* planets.

Jurassic period **208 to 138 Myr**, *see* geological timescale; *terminal extinctions* ?138 Myr, possibly confused with another event (*see* Kimmeridgian turnover 145 Myr; mass extinctions; reefs).

K

kangaroos c. 10 Myr. *See* mammals.

Kenoran Mountains 2700 to 2350 Myr; *supercontinent* **2300 yr**. *See* continents.

Khmer breakout c. 1150 yr (AD 850). *See* breakouts (Indo-chinese peoples).

Kimmeridgian turnover **145 Myr** (Late Jurassic). Changes in life at this time included the disappearance of reefs for 20 million years and the extinction of certain dinosaur groups (e.g., stegosaurs). *See* dinosaurs; mass extinctions; reefs.

king lists Prehistory shades into history with the appearance of writing, and lists of kings prepared by ancient writers. An Egyptian papyrus, the *Turin Royal Canon*, has been the key to early Egyptian history. Such lists have to be anchored by fixing the dates of one or more kings by other means. In the case of Egypt this was done using references from known reigns to the time of year when the bright star Sirius rose just before sunrise. Because the Egyptians used a 365-day calendar, without leap years, the seasons slipped continuously through the calendar, as the centuries passed. From the *Turin Royal Canon* and other sources, the beginning of Dynasty I in Egypt was computed to be 3119 BC, but in 1979 K. Baer (Oriental Institute, Chicago) announced revised reckonings that tended to push this date and others somewhat back in time. For Mesopotamia, king lists span much of Sumerian, Babylonian, and Assyrian history. The king list of Dynasty I of Babylon can be dated astronomically, but with five possible solutions, so that Hammurabi, for example, may have become emperor in 1900 BC, or equally in 1848, 1792, 1728, or 1704 BC, by this method of dating (M. Munn-Rankin, *Antiquity*, Vol. 14, 1980, p. 128).

kings (warrior kings) **5300 yr**. *See* government; weaponry.

Kish (biblical) flood 4900 yr. *See* climate 10,300 yr.

knitting est. 17,000 yr. *See* arts and crafts.

Korea (added to Asia) c. 140 Myr, *see* continents; *Korean language* (links obscure) *see* language; *Korean War* **AD 1950 to 1953**, *see* historical timescale.

L

lamps 17,000 yr. *See* energy.

language est. 45,000 yr. Fully fledged language was probably the trigger for the remarkable success of modern humans from c. 40,000 yr onward (*see* breakouts; mental revolution). If talkative modern humans spread from a common Asian origin (*see* human genetics), they presumably spoke similar languages at first. When populations dispersed their languages mutated rapidly. Linguistic change can be observed in the day-to-day variations of speech (W. Labov, personal communication, 1976) and studied retrospectively by tracing similarities and differences between present languages. Literature gives clues as to how people spoke in the past: apparent illogicalities in English spelling, for instance, trace back to very different pronunciations of vowels at the time when written English was being standardized. One language can split into two mutually unintelligible languages in less than a thousand years.

A method of linguistic dating called glottochronology sets out to fix the lapse of time since the split of two languages, by counting the proportion of important words judged to be "cognate," tracing back to a common root. The age of the split is then said to follow a simple mathematical relationship. This procedure can sometimes lead to split dates that are archeologically plausible. For example, I. Dyen at a conference on lexicostatistics in 1973 offered a date of est. 5200 yr for a split between Akkadian on the one hand and Hebrew, Aramaic, and Arabic on the other. This fits comfortably between a dispersal of dairymen, presumed to be speaking a proto-Semitic language c. 6000 yr (*see* breakouts), and the rise of the Akkadian empire, with its distinctive language, 4400 yr. Dyen put the Hebrew/Arabic split at est. 3800 yr.

Languages ancestral to English and Latvian (both Indo-European languages) split at est. 5700 yr ago by this reckoning, which seems early. R. Jeffers and I. Lehiste (*Principles and Methods for Historical Linguistics*, MIT Press, 1979) pointed out that one word more or fewer in the estimate of cognates results in a change of nearly two hundred years in the split date.

More conventional studies in historical linguistics led to striking discoveries, of which one of the earliest and most important was the kinship of the Indo-European languages from Ireland to Bangladesh. The identity of the Iron Age people of the Hallstatt culture, who had taken over much of Europe 2500 yr ago, was established by the study of place names: they were Celts (T. Bynon, *Historical Linguistics*, Cambridge University Press, 1977). Similarly, ancient pre-Sumerian place names in Mesopotamia implied that the Sumerians were not the earliest inhabitants of Sumer (J. Mellaart, *The Neolithic of the Near East*, Thames & Hudson, 1975). For early China, historians distinguished divergent and interacting cultures, for which they offered linguistic identities: proto-Tungus (northern China, esp. on Yellow River) proto-Turkic (northwestern China), proto-Tibetan (western China), and proto-Thai (southern China, with several distinct cultures).

The connection between genes and languages can range from zero to very high correlations, as noted by geneticists who studied the analogies between genetic and cultural variations (L. L. Cavalli-Sforza and M. W. Feldman, *Cultural Transmission and Evolution*, Princeton University Press, 1981). Demographic and historical events can often drive genes and languages in the same direction, but languages evolve far faster than genes and can change drastically in the course of a few decades, as a result of foreign contacts and impositions. As Cavalli-Sforza and Feldman observe: "Languages express the importance of past political domination."

A genetic-linguistic correlation spanning much of the world associates the milk-drinking mutation and its consequences (*see* livestock revolution) with the language families of Uralic, Indo-European, Afrasian (Semitic and related languages), and Nilo-Saharan. This linkage, together with other dispersals (*see* breakouts), helps to group human languages, as in the first of the accompanying tables. Uralic languages split into Samoyed (e.g., Nenets) and Finno-Ugric (e.g., Finnish, Hungarian) subgroups. Afrasian languages split into Semitic, Berber, Cushitic, Egyptian (Coptic), and Chadic (Hausa) subgroups, and the Semitic subgroup split into Arabic, Hebrew, Aramaic, and Ethiopic branches.

The Indo-European family commands the greatest number of speakers, almost half the world's population. The origins of Indo-European speakers trace back to the Volga basin in eastern Europe, and to the Yamnaya culture there, 5000 yr ago (M. Zvelebil, in A. G. Sherratt, ed., *Cambridge Encyclopedia of Archaeology*, Cambridge University Press, 1980, and personal communication 1982). Their subsequent divisions and evolution are set out in the second table, adapted from data in T. Bynon, *Historical Linguistics*, Cambridge University Press, 1977. See also the maps on pp. 90–95.

APPARENT ROOTS OF SOME MAJOR LANGUAGE FAMILIES.

early scatter and links	modern families
Easternmost Asia Links uncertain; proto-Amerindians spread eastward to Americas more than 11,000 yr ago.	Amerindian (e.g., Maya)
	Eskimo and Aleut
	Paleosiberian (e.g., Ket)
Northeastern Asia Spread westward 760 yr.	Altaic (esp. Turkish)
Southeastern Asia Chinese distinct ?5000 yr, prominent 2200 yr; links with Japanese, Korean, Vietnamese, etc. are cryptic.	Sino-Tibetan (esp. Mandarin Chinese)
	Mon-Khmer (e.g., Khmer)
	other eastern Asian
Australia	Australian aborigine
Indonesia Spread east and west.	Austronesian (e.g., Indonesian)
Southern Asia Shrank southward 3000 yr.	Dravidian (e.g., Telugu)
Western Africa Spread widely 2500 yr on.	Niger-Congo (esp. Bantu)
Southern Africa Shrank southward.	Khoisan (e.g., San Bushman)
North Africa Spread southward 4500 yr.	Nilo-Saharan (e.g., Luo)
Western Eurasia Eurafrasian "dairy" languages: proto-Afrasian spread southward 6000 yr; Uralic northward 4900 yr; Indo-European eastward and westward 4900 yr.	Afrasian (e.g., Arabic)
	Uralic (e.g., Hungarian)
	Indo-European (see next table)

CHRONOLOGY OF SOME INDO-EUROPEAN LANGUAGES.

by 4000 yr	by 3000 yr	by 2000 yr	since 1000 yr ago
eastern group (Indo- Iranian)	Vedic Sanskrit	Classical Sanskrit	Hindi Urdu Bengali Punjabi Marathi Gujerati, etc.
	(proto- Iranian)	Old Persian	Persian Pashto Kurdish Baluchi, etc.
western group (European)	Mycenaean Greek	Classical Greek	Greek
	Hittite	(extinct)	
		(proto- Slavic)	Russian Ukrainian, etc.
			Polish Czech, etc.
			Bulgarian Serbo-Croat, etc.
		Celtic	Gaelic, etc.
		Latin (= proto- Romance)	Spanish Portuguese French Italian Rumanian, etc.
		(proto- Germanic)	English German Dutch, etc.
			Swedish, etc.

laser AD 1960. *See* energy; information technology.

Latin (language) by 2000 yr. *See* language.

Laurasia (Euramerica plus Asia) c. 300 Myr. *See* continents.

law code 3900 yr. *See* government.

lead-lead dating *See* uranium dating.

League of Nations AD 1920. *See* historical timescale.

leather (use of hides) ?1.9 Myr. *See* arts and crafts.

leaves c. 370 Myr. *See* plants ashore.

lemurs (split from human lineage) est. 70 Myr. *See* primates.

leopards 1.8 Myr. *See* cats.

Levallois tools (proto) c. 700,000 yr; (full) c. 160,000 yr. *See* stone tools.

liberalism 310 yr (AD 1690). *See* government.

life on Earth est. 4000 Myr. By a consensus among investigators, life began about 4000 Myr ago (G. Eglinton, et al., *Nature*, Vol. 292, 1981, p. 669). There are reasons for suspecting an earlier origin, but the round figure is preferable at the present state of knowledge. Simple materials abundant in the water and atmosphere of the young Earth supposedly formed a soup of the complex molecules needed for life, and these would have evolved by mutation and natural selection, even before life originated (S. Spiegelman and M. Eigen, personal communications, 1972). In the interval between the appearance of liquid water on the Earth's surface (est. 4450 Myr) and the origin of life, the prebiotic chemical soup was kept simmering by sources of energy that included meteoritic impacts, radioactivity, volcanic heat, thunderclaps, and ultraviolet rays from the sun.

The secret of life on Earth is that nucleic acids carry instructions for the manufacture of protein molecules, and these in turn act to promote the reproduction of the genes. In the hypercycle theory, ribonucleic acid (RNA) chains containing four kinds of units (G, C, A, and U) helped in assembling protein chains of a defined sequence, using the four commonest animo acids in the water of the early Earth. Thus "GGC" specified that glycine should be incorporated in the protein, while "GCC," "GAC," and "GUC" called for alanine, aspartic acid, and valine, respectively. One RNA chain promoted the manufacture of a protein that accelerated the manufacture of another RNA chain, and so on, in a loop that eventually aided the production of the first RNA chain. This set of mutually helpful chemicals was a hypercycle, and concentrated in droplets, it could make living cells that could multiply and take over the world (M. Eigen and P. Schuster, *Naturwissenschaften*, Vol. 64, 1977, p. 541, and Vol. 65, 1978, p. 7 and p. 341). The hypercycle was the putative ancestor of all life on Earth, and all organisms share a similar genetic code for translating genes into proteins.

Studies of molecular evolution revealed the fundamental divisions of living things (C. R. Woese and G. E. Fox, *Proceedings of the National Academy of Sciences*, Vol. 74, 1977, p. 5088, and H. Hori and S. Osawa, in

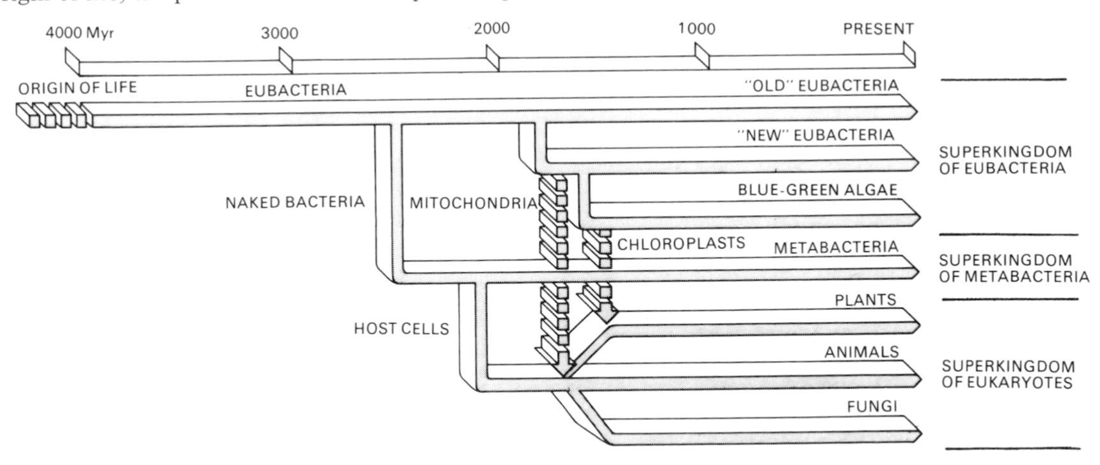

Molecular genetics confirms that human beings, along with other animals, are essentially chimeras of peculiar bacteria. The ancestral organisms were "old" (gram-positive) eubacteria, from which naked metabacteria and "new" (gram-negative) eubacteria evolved. Certain "new" bacteria, taken into a host-cell line of metabacteria, became mitochondria and chloroplasts, working units within modernized, eukaryotic cells. These in turn gave rise to the animal and plant kingdoms.

S. Osawa, et al., eds., *RNA Polymerase, tRNA and Ribosomes*, University of Tokyo Press/Elsevier, 1980). The three superkingdoms are (1) eubacteria, the common bacteria; (2) metabacteria, alias archaebacteria, peculiar methane-producing and salt-resistant organisms; and (3) eukaryotes, organisms that possess "modern" cells with highly organized genetic material. The eukaryotes in turn gave rise to the kingdoms of fungi, animals, and plants (see diagram). Illus. pp. 38, 108, 111, 112.

The origin of eukaryotic organisms, including plants and animals, is to be understood largely in terms of partnership, or symbiosis, between bacteria of different kinds (L. Margulis, *Origin of Eukaryotic Cells*, Yale University Press, 1970, and *Symbiosis in Cell Evolution*, Freeman, 1981). Inside the cells of plants and animals are organelles that resemble bacteria both in appearance and in their molecular systems; they reproduce independently of their hosts, and are inherited only via the mother. The oxygen-handling mitochondria are descended from photosynthetic purple bacteria, while the choloroplasts, green packets inside plant cells, are related to the cyanobacteria (blue-green algae).

The kinship between metabacteria and eukaryotic cells appears in various biochemical tricks. Metabacteria lost their traditional cell walls, although many later reinvented them in idiosyncratic forms. To have been without a wall at the critical stage would have enabled the eukaryotic host bacteria to engulf other bacteria (L. M. Van Valen and V. C. Maiorana, *Nature*, Vol. 287, 1980, p. 248). A modernization of the cells' genetic machinery ensued with genes arranged on chromosomes in a nucleus.

Eukaryotic cells gave rise to protists: fungi, true single-celled algae, and single-celled protozoa (proto-animals). The distinctions between living protists are often elusive, and the earliest modern cells may have had ready access to alternative modes of life. Next came the clubbing together of cells in building metaphytes (true plants) and metazoa (animals). This involved new levels of genetic control to regulate growth and to achieve co-operation and specialization among the participating cells. The dating of sex, and its relationship to the origin of multicelled plants and animals, is obscure. Evolutionary events among animals and higher plants are outlined elsewhere (*see* animal origins; plants ashore).

The timetable that follows reconciles molecular and fossil evidence. Fossils are scanty and often cryptic in the early history of the Earth, and molecules in living organisms provide a better window on the past. Indicative dating can be achieved, for instance, using the cytochrome *c* molecule, much involved in the evolution of respiration (data presented by R. E. Dickerson in various sources, inc. *Scientific American*, Vol. 242, No. 3, 1980, p. 98, and *Nature*, Vol. 283, 1980, p. 210). The procedure followed here is to count differences in corresponding segments of cytochrome *c* (ignoring insertions and deletions), to normalize these to a "standard" molecule of 104 units (amino acids) and correct for overprinting of mutations using a formula due to M. Dayhoff. The inferred date is then $N \times 21$ Myr, where N is the number of mutations and overprintings in the notional standard chain; this is calibrated by taking $N = 20$, for the common fish/land vertebrate split, as being equivalent to 425 Myr, the time of appearance of jawed fishes and a presumed time of the divergence leading to fleshy-finned fishes, ancestors of the land vertebrates. Additional information comes from relationships in 5S RNA, from both bacteria and "higher" organisms, specified in units called $\frac{1}{2}K$, said to be proportional to time (H. Hori and S. Osawa, reference above, and also *Proceedings of the National Academy of Sciences*, Vol. 76, 1979, p. 381). Events are dated by cross calibration at 1500 Myr, from the cytochrome *c* data, $N = 70$, to the 5S RNA data at $\frac{1}{2}K = 0.3$. This also conforms with fossil appearances of apparent green algae.

The fossil data are due to S. M. Awramik, personal communication, 1981, and various of his publications, including a chapter in M. H. Nitecki, ed., *Biotic Crises in Ecological and Evolutionary Time*, Academic Press, 1981, and S. M. Awramik and J. W. Valentine, abstract for Third North American Paleontological Convention, Montreal, 1982. The chief conflict between molecular and fossil evidence concerns the evolution of the blue-green algae. Paleontologists identify fossils from western Australia 2800 Myr old as blue-greens, and by 2000 Myr the fossil blue-greens are said to be unequivocal; they are assigned to modern families. Molecular evidence points to an origin est. 1600 Myr.

LIFE ON EARTH: TIMETABLE

origin of life: **est. 4000 Myr**. Dated for the ancestral split between sulfate-respiring bacteria and other life, 4400 Myr by cytochrome *c*, rounded down to 4000 Myr. No fossil evidence.

photosynthesis: **est. 3900 Myr**. Dated for the ancestral split of green sulfur bacteria and blue-green algae; both approximate to 4100 Myr, by cytochrome *c*, rounded down to 3900 Myr to maintain an evident interval after the origin. Fossil evidence:

chemical peculiarities 3800 Myr in the earliest rocks (Greenland); remains of living reefs built by mats of photosynthetic bacteria 3500 Myr (*see* stromatolites).

naked bacteria (metabacteria): **est. 2500 Myr**. Dated for the metabacteria/eubacteria split, by 5S RNA. No fossil evidence.

host cells: **est. 2100 Myr**. Dated for the metabacteria/eukaryote split, by 5S RNA. No fossil evidence.

oxygen revolution: **1800 Myr**. Free oxygen significant for the first time (*see* oxygen revolution).

purple bacteria: **est. 1800 Myr**. Dated for the gram-negative/gram-positive bacterial split, by 5S RNA.

modern cells (fungi and proto-animals): **est. 1700 Myr**. The oxygen-handling inclusions (mitochondria) diverged from the purple bacteria est. 1700 Myr ago, by cytochrome *c*; ancestors of fungi and animals, both eukaryotic groups, split est. 1700 Myr, by 5S RNA. No fossil evidence.

blue-green algae (cyanobacteria): **est. 1600 Myr**. Dated for the blue-green/*Escherichia coli* split, by 5S RNA. Fossil evidence is discrepant, *see* above.

proto-plants: **1500 Myr**. A calibration age, see above. Fossil evidence: relative large remains resembling those of fossil green algae appeared abundantly c. 1500 Myr ago (e.g. U.S.S.R.).

plants: **c. 1300 Myr**. By fossil evidence. The earliest known multicelled plants, or metaphytes, are microseaweed, millimeter-sized fossil strands from c. 1300 Myr (Greyson Shale, Montana).

seaweed: **c. 1000 Myr**. By fossil evidence. Centimeter-sized multicelled plants from northwestern Canada are loosely dated between 1100 and 800 Myr ago.

sex: **?1000 Myr**. A guess.

animals: **670 Myr**. By fossil evidence (*see* animal origins). Molecular dating gives 720 Myr and 800 Myr for the invertebrate/vertebrate split, by cytochrome *c* and 5S RNA.

brains: **620 Myr**. Fossil evidence of annelid worms and arthropods, possessing relatively elaborate nervous systems.

life ashore: **425 Myr**. By fossil evidence (*see* plants ashore).

vertebrates ashore: **370 Myr**. By fossil evidence (*see* vertebrates).

mental revolution: **45,000 yr**. *See* human origins; mental revolution.

limbs (vertebrates) **370 Myr**. *See* vertebrate animals.

Limpopo Mountains 2600 to 2300 Myr. *See* continents.

lions 1.8 Myr. *See* cats.

literature (oral) est. 45,000 yr; (written) c. 5000 yr. *See* arts and crafts.

Little Ice Age c. 470 yr (AD 1530); *ended* **c. 150 yr (AD 1850)**. *See* climate 10,300 yr.

livestock revolution c. 6000 yr. The use of domesticated animals for purposes beyond supplying meat and skins occasioned a "secondary products revolution" (A. G. Sherratt, in I. Hodder, et al., eds., *Patterns of the Past*, Cambridge University Press, 1981, and personal communication, 1982). Judged by the presence of bones of relatively elderly animals, cattle may have served for traction (and milk?) as early as c. 7800 yr ago in northern Mesopotamia (Middle Halaf), and mutant woolly sheep may have been kept for yarn (and milk?) in Iran from c. 7300 yr ago (Early Ubaid). Early evidence of milking comes in the form of a butter churn from Palestine c. 6200 yr (Ghassulian = Late Ubaid). In Sherratt's view, developments from scattered places were gathered together in northern Mesopotamia (or at any rate on the fringe of the Fertile Crescent) c. 6000 yr ago, as a package in which the plow was a salient invention. Thence it radiated north, south, east, and west, in association with the milk-drinking mutation.

Almost all mammals lose the ability to digest milk sugar (lactose) after infancy, and milk is then harmful to them; the same is true for most human beings. The largest groups of milk drinkers speak Afrasian (Semitic and related languages), Uralic, Indo-European, and Nilo-

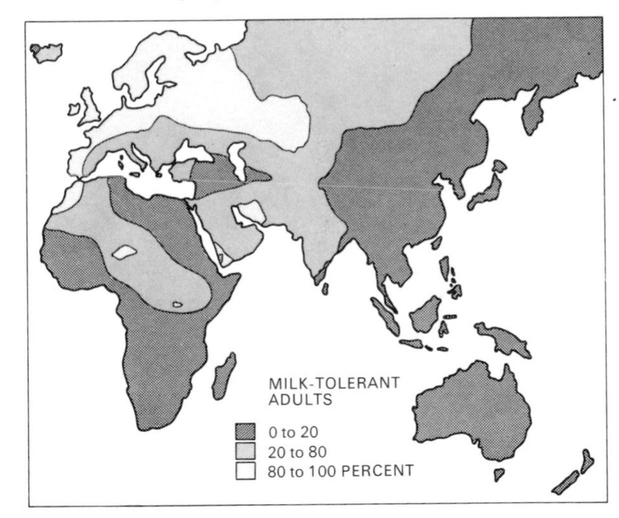

The livestock revolution, involving milking cows, draft oxen, and later tamed horses, helped to spread a mutation that enables a minority of human beings to digest milk after infancy. The map uses provisional data supplied by geneticists at the Stanford University Medical School. It indicates that although milk tolerance is common among aboriginal peoples in Europe, and in parts of Asia and Africa, the trait is rare elsewhere. There is an association between milk drinking and the Afrasian (Semitic), Indo-European, and Uralic languages.

Saharan languages; compare the linguistic maps, pp. 90–95, with milk tolerance as mapped here (p. 251). The onset of milk drinking in adults, preserving a juvenile trait into adulthood, required a genetic mutation. One benefit of milk is that, by carrying calcium, it helps to prevent rickets in bones. Loss of skin pigmentation has a similar effect in dull climates, and there is a correlation between adult milk drinking and fair skins. Several thousand years would be required for the mutation to become general in a population, so where it first occurred is a question open to the widest conjecture. At the time of the livestock revolution, the milk drinkers may have been concentrated between the Volga River and the Caspian Sea. See map p. 90.

The phrase "cow-and-plow revolution" is used here for the core of the livestock revolution: the potent combination of milking cows and ox-drawn plows. Breakouts of milk-drinking "dairymen," "northern dairymen," and "horsemen" distinguish three events and groups in a simple fashion, but note that many "dairymen" later adopted horses, and "horsemen" also kept dairy herds. *See* animal domestication; breakouts; horses; language; plow; weaponry; wheel.

lizards c. 240 Myr. *See* vertebrate animals.

llamas **?500,000 yr**; *domesticated* c. 7000 yr. *See* camelids.

loom 8500 yr. Dated for weaving, *see* arts and crafts.

lungs c. 400 Myr. *See* vertebrate animals.

lystrosaurs (mammal-like reptiles) **245 Myr**. *See* vertebrate animals.

M

Macedonian (Alexander's) **empire** 2334 yr. *See* breakouts (blacksmiths).

machine-gun warfare **esp. 102 yr (AD 1898).** *See* weaponry.

magnetic compass **esp. 885 yr (AD 1115)**, *see* transport; *magnetic tape* esp. AD 1938; *magnetron* AD 1939, *see* information technology.

magnetic stratigraphy Electric currents flowing in the molten iron-nickel core of the Earth generate the magnetism that makes compass needles point roughly north. The planet had repeatedly switched its magnetic poles so that compass needles, had they existed, would often have pointed south. Rocks such as iron-rich sediments and volcanic layers can act like compasses and preserve the direction of magnetism prevailing when they were being formed. Most spectacularly, the ocean floor acts like a tape recorder and shows stripes—alternations of north-pointing and south-pointing magnetization, according to when each section of the ocean floor was formed.

Although B. Brunhes discovered volcanic rocks with reversed magnetism in France in 1906, and M. Matuyama made a similar discovery in Japan in 1929, a later observation that some rocks could reverse their magnetism spontaneously cast doubt on the notion of the Earth's field reversing itself. The concept was fully vindicated when A. Cox and his colleagues in the United States established a firm record of successive reversals for the past few million years, using potassium-argon dating to establish the ages of the rocks (*Science*, Vol. 144, 1964, p. 1537). Since then the timetable of reversals has been extended and refined. For the span of the past 85 million years, when at least 177 switches have a occurred, magnetic stratigraphy dominates the geological timescale.

During a reversal the magnetic poles can vacillate, "flipping" several times between the hemispheres in the course of a thousand years or so, before settling down. A vacillation that was a "failed" reversal occurred 33,000 yr

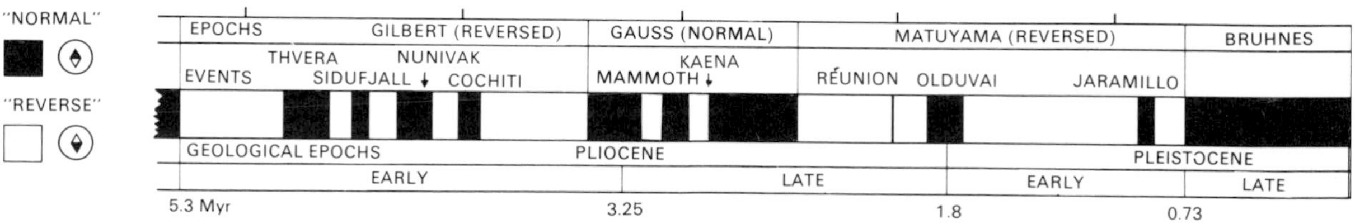

"NORMAL"
"REVERSE"
EPOCHS: GILBERT (REVERSED) — GAUSS (NORMAL) — MATUYAMA (REVERSED) — BRUHNES
EVENTS: THVERA, SIDUFJALL, NUNIVAK, COCHITI — MAMMOTH, KAENA — RÉUNION, OLDUVAI, JARAMILLO
GEOLOGICAL EPOCHS: PLIOCENE (EARLY, LATE) — PLEISTOCENE (EARLY, LATE)
5.3 Myr — 3.25 — 1.8 — 0.73

Magnetic reversals, detectable in many strata, illuminate geological history and give unprecedented precision to dating. The magnetic epochs and events of the Pliocene and Pleistocene geological epochs, shown here, make dating to within about 1 percent of the age a realistic goal. The named events are brief anomalous interludes within the main epochs.

ago (Lake Mungo event). Drastic events affecting living things sometimes coincide with magnetic reversals, but adverse effects on life may have been due to whatever process brought about the reversals, rather than to the reversals themselves.

The frequency of magnetic reversals has changed markedly over time. The high rate of the past 85 million years was equaled or surpassed 400 Myr ago (Early Devonian), but otherwise the past 500 million years have been generally quieter in this respect, and reversals were decidedly rare 430 Myr (Silurian) and 300 Myr ago (Late Carboniferous); 170 Myr (Middle Jurassic) was also a time of magnetic tranquillity (M. W. McElhinny, *Science*, Vol. 172, 1971, p. 157). Fluctuations on shorter timescales are apparent in the record of the past 85 Myr, with minor peaks in the rate of reversals as follows:

BURSTS OF MAGNETIC REVERSALS: TIMETABLE

72 Myr (5 in 1 million years)
54 Myr (10 in 4 million years)
42 Myr (17 in 3 million years)
24 Myr (13 in 3 million years)
15 Myr (51 in 12 million years)

The magnetic timescale used in this book is that of W. Lowrie and W. Alvarez, *Geology*, Vol. 9, 1981, p. 392. Dates for events 3.16 Myr to present are slightly modified to correspond with W. A. Berggren, et al., *Quaternary Research*, Vol. 13, 1980, p. 277, and to note the uncertain Réunion event. The reversals of the past five million years are shown in the diagram.

Magyars (into Hungary) 1105 yr (AD 895). *See* breakouts (northern dairymen).

maize domesticated 7700 yr. *See* crops; grass.

Malagasy peopled c. 1500 (AD 500). *See* breakouts (gardeners).

male dominance esp. 6000 yr. *See* government.

mammal-like reptiles (pelycosaurs) **310 Myr**; (therapsids) **256 Myr**; (cynodonts) **240 Myr**. *See* vertebrate animals.

mammals 216 Myr. The group of vertebrates to which humans belong is usually characterized by milk, warm blood, hair, relatively large brains, and "live" births. The mammals first appeared 216 Myr ago (Norian/Rhaetian boundary; P. E. Olsen and P. M. Galton, *Science*, Vol. 197, 1977, p. 483). The following account draws especially on J. A. Lillegraven, et al., eds., *Mesozoic Mammals*, University of California Press, 1979; W. N. McFarland, et al., *Vertebrate Life*, Macmillan, 1979; P. D. Gingerich, in A. Hallam, ed., *Patterns of Evolution*, Elsevier, 1977; C. Pomerol, *The Cenozoic Era*, Ellis Horwood, 1982, and V. M. Sarich and J. E. Cronin, in R. L. Ciochon and A. B. Chiarelli, eds., *Evolutionary Biology of the New World Monkeys*, Plenum, 1981. Data on the Pleistocene overkill are from J. P. White and J. F. O'Connell, *Science*, Vol. 203, 1979, p. 21, B. Kurtén, *Pleistocene Mammals of Europe*, Weidenfeld & Nicolson, 1968, and P. S. Martin and A. Long, *Science*, Vol. 186, 1974, p. 638. Appearances of selected modern families are mainly from A. W. Gentry, personal communication, 1982; for details and sources *see* families by name (e.g., bears). *See also* American Interchange.

Throughout the period 216 to 67 Myr ago, a variety of small mammals were typically of shrew size or rat size and probably lived mainly on insects or plants. A setback occurred 145 Myr ago, and a resurgence 114 Myr ago is associated with the advent of modern flowering plants. Early specimens may often represent "half-baked" mammals, like the surviving platypus that still lays eggs.

By molecular evidence, pouch-feeding marsupials diverged from the lineage of modern placental mammals **est. 125 Myr** ago. Fossil mammals possibly ancestral to both marsupials and placentals are known from England, 130 Myr ago (Early Cretaceous, Valangian stage; *Aegialodon*). Marsupials dispersed via South America and Antarctica to Australia. Kangaroos evolved c. 10 Myr ago and attained giant size c. 1 Myr ago.

The placental mammals may have originated in Asia, 114 Myr ago, and traveled thence into North and South America. Certain diminutive animals living in Manchuria and Mongolia 114 Myr ago (Early Cretaceous, Aptian/Albian stages; *Prokennalestes*) seem to have been in the mainstream of the evolution of placental mammals. Molecular evidence indicates that the ancestors of hoofed mammals had separated from other mammals by est. 100 Myr ago, preprimates by **95 Myr** ago, and ancestors of carnivores and bats by 90 Myr ago. Rodents and lagomorphs diverged from the primate line est. 85 Myr ago.

In Montana, fossil hunters witness the survival of early mammals during the destruction of the dinosaurs at the Cretaceous terminal catastrophe of 67 Myr ago. Out of 22 placental species, 18 came through, but only 1 out of 13 marsupial species (J. D. Archibald, *Nature*, Vol. 291, 1981, p. 650). Thereafter mammals replaced reptiles as the most conspicuous animals of land and sea, in many different roles.

FOSSIL PLACENTAL MAMMALS: TIMETABLE

early insectivores: **80 Myr**. Extinct 37 Myr ago.

primitive carnivores (creodonts): **70 Myr**.

primates: **69 Myr**. *Purgatorius*. *See* primates.

primitive hoofed mammals (condylarthra): **69 Myr**. Extinct c. 5 Myr ago. Two other extinct orders of herbivorous mammals appeared c. 60 Myr and survived until c. 2 Myr ago.

modern carnivores: **60 Myr**. The modern families of cats and dogs date from **35 Myr**, bears **30 Myr**, and seals **?25 Myr**.

hoofed mammals: **55 Myr**. Odd-toed perissodactyls attained giant size (e.g., *Indricotherium* and titanotheres) c. 30 Myr; the family of horses dates from **55 Myr**, and rhinoceroses from **35 Myr**. Among the very successful even-toed artiodactyls, family origins were: camelids **40 Myr**, pigs **30 Myr**, deer **20 Myr**, bovids **19 Myr**, and hippopotamuses and giraffes 5.5 Myr.

rodents: **55 Myr**. Squirrels trace back to the Miocene (?14 Myr) and the murid family of mice and rats to the Early Pliocene c. 5 Myr. Rats and bobak marmots appeared **c. 650,000 yr** ago.

whales: **55 Myr**. A primitive whale was a superpredator c. 45 Myr ago. The more advanced modern orders (many-toothed and whalebone whales) emerged **40 Myr**.

bats: **55 Myr**.

proboscidians: **c. 50 Myr**. The elephantid family of elephants and mammoths dates from **c. 7 Myr** (Late Miocene).

modern insectivores (shrews, hedgehogs, etc.): **40 Myr**.

lagomorphs (rabbits, hares, etc.): **40 Myr**.

GENERAL EVENTS AMONG MAMMALS: TIMETABLE

new mammals: **c. 55 Myr**. Near-simultaneous debut of modern hoofed mammals, rodents, whales, and bats.

more new mammals: **c. 40 Myr**. Lagomorphs, modern whales and insectivores, and camelids among the artiodactyls.

Eocene terminal turnover: **37 Myr**. Many extinctions (e.g., among whales), followed by the rise, 35 Myr to 30 Myr ago, of modern families of mammals (inc. cats, dogs, rhinos, pigs, and bears); in South America, mylodonts and toxodonts appeared 35 Myr ago, in the peculiar lineages of edentates and notoungulates, respectively.

whales growing: **25 Myr**. Baleen (filter-feeding) whales enlarging progressively as the gap between Australia and Antarctica widened.

Old World Interchange: **20 Myr**. Reunion of fauna of Eurasia and Africa.

Miocene disruption: **15 Myr**. Losses in all major groups of land mammals, amounting to 30 percent of the total number of mammalian genera (species groups).

surge in grazing animals: **11 Myr**. A delayed consequence of the origin of grass, 24 Myr ago.

American Interchange: **3.0 Myr**. A reunion of North and South America resulting in drastic changes in the mammalian fauna.

mammalian peak: **1.0 Myr**. Following a surge in the origination of new mammalian species at the beginning of the Pleistocene epoch 1.8 Myr, a peak of more than 800 genera.

Pleistocene disruption: **730,000 yr**. Loss of about 100 genera, during the late Pleistocene.

Pleistocene overkill: **c. 170,000 yr**. Many species of mammals, especially large ones, dying out 17,000 to 10,000 yr ago, with human hunting at least partly responsible. The starting date given corresponds to the extinction of large marsupials in Australia; the decline of the cave bear and other species in Europe is well recorded in Europe from 12,000 yr onward.

American overkill: **c. 11,300 yr**. Climax of the Pleistocene overkill, in the Americas.

modern overkill: 263 species of the world's mammals listed as "endangered" by the U.S. Department of the Interior (1977).

mammoths 4 Myr; *woolly* **150,000 yr**. *See* elephants.

Manchus in China 364 yr (AD 1636); *Ch'ing empire* **356 yr (AD 1644)**. *See* breakouts, historical timescale.

Manicouagan crater **210 Myr.** *See* cosmic impacts; also map p. 84.

manned spaceflight **AD 1961.** *See* transport.

maps est. 45,000 yr. *See* information technology.

marmots (bobak) **c. 650,000 yr**, *see* mammals; *reservoir of plague* **669 yr (AD 1331)**, *see* disease epidemics.

Mars **c. 4500 Myr**; *defunct* **c. 1000 Myr**. This planet, the next outward from the sun beyond the Earth, has a mass one tenth of the Earth's. Inspections by unmanned spacecraft, especially *Mariner 9* (1971) and the orbiter and lander vehicles of the *Viking* project (1976), revealed a desert planet characterized by heavily cratered landscapes in the south, resembling the lunar highlands, and by younger, lava-covered plains in the north. Mars also possesses extinct volcanoes—notably a conspicuous cluster of four, including Olympus Mons, the largest volcano in the inner solar system. Volcanic activity probably ceased about 1000 Myr ago (P. W. Francis, in D. G. Smith, ed., *Cambridge Encyclopedia of the Earth Sciences*, Cambridge University Press, 1982).

Mars landers (*Viking 1* and *2*) **AD 1976**. *See* spacecraft.

marsupial mammals **est. 125 Myr**. *See* mammals.

Marxism **esp. 133 yr (AD 1867).** *See* government.

mass extinctions 670 Myr to present. The near-simultaneous extinction of large numbers of organisms has repeatedly altered the course of evolution. An early event of this kind coincided with the very severe Varangerian ice ages of 670 Myr ago, when 70 percent of the known kinds of algal plankton disappeared (G. Vidal and A. H. Knoll, *Nature*, Vol. 297, 1982, p. 57). In a thorough analysis of the later fossil record, D. Raup and J. Sepkoski looked for mass extinctions in 3800 families of marine animals (*Science*, Vol. 215, 1982, p. 1501). They pinpointed four events that were statistically significant in their method of analysis, and a fifth (Frasnian) that they considered would be significant but for a probable smearing due to failure to identify precise times of extinction. *See* most of the events mentioned below, described in this index under their names. Illus. p. 120.

MAJOR MARINE EXTINCTIONS ("CATASTROPHES")
(With percentage of families lost; after Raup and Sepkoski.)

Ordovician terminal (12 percent)	440 Myr
Frasnian (Late Devonian, 14 percent)	370 Myr
Permian terminal (52 percent)	245 Myr
Norian (Late Triassic, 12 percent)	216 Myr
Cretaceous terminal (11 percent)	67 Myr

LESSER MARINE EXTINCTIONS ("TURNOVERS")
(Noted by Raup and Sepkoski from other authors but not noticeable in their own data; the Jurassic terminal event may be the same as the Kimmeridgian event in the next table.)

Cambrian disruptions, three to five events	550 to 500 Myr
Toarcian (Early Jurassic)	182 Myr
?Jurassic terminal	?138 Myr
Cenomanian (early Late Cretaceous)	95 Myr
Eocene terminal (Tertiary)	37 Myr

ADDITIONAL EXTINCTIONS OF LAND VERTEBRATES

mid-Permian	256 Myr
dinosaur revolution	esp. 225 Myr
Kimmeridgian (Late Jurassic)	145 Myr
Aptian (mid-Cretaceous)	117 Myr

mastodons c. 20 Myr. *See* elephants.

matter predominant **10,000 years** after origin. *See* Big Bang.

Mauryan empire 2325 yr. *See* breakouts (charioteers); historical timescale.

measles ?1838 yr (AD 162), etc. *See* disease epidemics.

medicine **c. 60,000 yr.** The earliest evidence for systematic use of herbal remedies goes back to the neanderthalers and in particular to the "flower burial" at Shanidar in Iraq, where a medicine man was evidently buried with the flowers of his craft (R. S. Solecki, *Science*, Vol. 190, 1975, p. 880). Medicine remained a neanderthaler art until the twentieth century AD, and folk remedies may have been on the whole safer and more efficacious than magical, astrological, alchemical, and violent treatments devised by learned men. Cowpox vaccination against smallpox, **204 yr (AD 1796)**, led to the eradication of smallpox in **AD 1977**. *See also* disease epidemics.

The germ theory of disease **137 yr (AD 1863)** modernized medicine. Although L. Pasteur in France led the bacteriological revolution of the 1860s, the event is dated for C.-J. Davaine, who in 1863 showed that the blood of sheep dying of anthrax contained rod-like particles, while healthy ones did not. The germ theory gave a boost to hygiene, and the cleanup of modern cities can be dated (by example) to 1866, when New York City administrators assumed responsibility for public health.

Other outstanding medical advances included the discovery of X rays 105 yr (AD 1895), identification of vitamins esp. AD 1912, the isolation of antibiotics **AD 1940** (penicillin 1940, streptomycin 1944), and the pinpointing of a mutation responsible for bladder cancer **AD 1982** (C. J. Tabin, et al., and E. P. Reddy, et al., *Nature*, Vol. 300, 1982, pp. 143, 149).

Mediterranean dryout **6.3 to 5.3 Myr.** Also called the Messinian salinity crisis. Evaporation rates in the Mediterranean today are sufficiently high that, if the connection with the Atlantic were sealed, it would dry in a thousand years. In the interval 6.3 to 5.3 Myr, this occurred not once but perhaps forty times, creating deposits of salt kilometers thick. The account in the narrative follows a summary by J. P. Kennett (*Marine Geology*, Prentice-Hall, 1982), supplemented from other sources, and with dates adjusted.

megalithic stone buildings **6500 yr.** *See* building (stone buildings).

Meiji restoration **132 yr (AD 1868).** *See* historical timescale (Japanese revolution).

mental revolution **est. 45,000 yr.** Animals did not evolve by striving, or by passing on what they had learned to their offspring, as alleged in the Lamarckian theory of evolution, but had to wait for the accidents of mutation, adaptation, and species making. With the advent of modern human beings, a Larmarckian mode of evolution began to operate in human cultures. Purposeful striving multiplied and diversified human skills far more rapidly than biological evolution could ever permit.

The mental revolution was anticipated dimly in neanderthalers and other premodern humans, even in apemen and chimpanzees. But something unusual happened est. 45,000 yr ago that triggered the breakout of fully modern humans c. 40,000 yr ago (*see* breakouts). The "spark" was probably the perfection of language as a social and mental tool (*see* human origins). Biological evolution in human beings continued slowly, but it was not what brought the subspecies to modern knowledge and technology. To suppose that early modern humans were just as bright as their present-day descendants is less likely to be wrong than picturing them as feckless creatures waiting for enlightenment.

The potency of ideas in the life of modern humans was expressed by R. Dawkins (*The Selfish Gene*, Oxford University Press, 1976) in the theory of the "meme," as the mental equivalent of a gene. "It is still . . . drifting clumsily about in its primeval soup, but already it is achieving evolutionary change at a rate which leaves the old gene panting far behind. The new soup is the soup of human culture."

Mercury c. 4500 Myr; *defunct* **c. 3000 Myr**. The mass of this innermost planet is 5 percent of that of the Earth. Inspection of Mercury by an unmanned spacecraft (*Mariner 10*, 1974) showed it to be in some ways similar to the moon, with old, heavily cratered highland areas and younger basins, but the material filling the basins is lighter in color. Large scarps, possibly formed by shocks or shrinkage of the crust, are peculiar to Mercury. The planet was created in the events of 4550 to 4450 Myr ago; *see* solar system forming. A very large cosmic impact est. 4000 Myr created the Caloris basin and shattered the surface even on the far side of the planet. No geological activity has occurred on the surface of Mercury since est. 3000 Myr ago, at the latest (P. W. Francis in D. G. Smith, ed., *Cambridge Encyclopedia of the Earth Sciences*, Cambridge University Press, 1982).

Mesolithic period 18,800 yr. *See* archeological timescale.

meson theory AD 1934. *See* science.

Mesopotamia (innovative center; mainly modern Iraq) **esp. 8000 yr** (irrigation) onward. *See*, e.g., cities; cultivation practice; government; information technology; livestock revolution.

Mesozoic era 245 to 67 Myr. *See* geological timescale.

metabacteria est. 2500 Myr. *See* life on Earth.

metal industries c. 9900 yr to present. Although many archeologists have supposed that metal industries fledged in southwestern Asia, there seems to be no explicit evidence to that effect, if the criterion is production of copper from its ores, by smelting. The following attempt to clarify a muddled subject relies mainly on T. A. Wertime and J. D. Muhly, eds., *The Coming of the Age of Iron*, Yale University Press, 1980; R. F. Tylecote, *A History of Metallurgy*, Metals Society, 1976; J. Needham in Wertime and Muhly, and other publications; and P. Phillips, *The Prehistory of Europe*, Allen Lane, 1980. Dates conform with adjusted or calibrated radiocarbon results.

The chief metals available in native form were nickel-iron from meteorites, and copper and gold from ores reprocessed by natural chemistry. Gold objects were too valuable to be left lying around for archeologists to find, and no history of gold is attempted here. Early Eurasian and North American metalworking benefited from the native copper created by the rapid alternation of wet and dry conditions during the current series of ice ages: in the dry spells, iron compounds were oxidized preferentially, and copper compounds reduced to the metal (J. W. Barnes, cited in Wertime and Muhly).

As early as c. 4100 yr ago, the Laurion mines near Athens were exporting silver and lead to Egypt. The Minoan smiths in Crete obtained most of their copper from the same Laurion mines, and not from Cyprus, as previously supposed (N. H. Gale and Z. A. Stos-Gale, *Science*, Vol. 216, 1982, p. 11). The direction of the tin trade in western Eurasia was westward, and there are suspicions that faraway Thailand may have been the source (B. Landsberger 1965, cited in Wertime and Muhly).

Crucial in many smelting operations is the use of fluxes, materials that remove impurities and make a slag. The first smelting of iron may have occurred on the southern shores of the Black Sea, where a self-fluxing magnetite sand can in principle yield half its weight in metallic iron. The rudiments of successful iron technology seem to have originated only once, in the Black Sea area, and spread from there all over the world. Steel is iron with neither too much nor too little carbon in it. The canonical sequence of stone, copper, bronze, iron, and steel occurred in Europe and western Asia, but in China, bronze remained a precious metal, so that stone gave way to iron for workaday purposes.

There is a literary reference to cast-iron caldrons in China c. 2512 yr; cast-iron molds for bronze implements are known from c. 2350 yr. Cast iron came into widespread use in China, for plows and other agricultural tools, as described in a passage on the iron tax in the Taoist book *Kuan Tzu*, which was put together in the state of Ch'i c. 2300 yr ago. Charcoal and coal served in smelting iron. Cheap steel became available in China c. 2200 yr, produced by decarburization, in an early version of the Kelly-Bessemer process (as in Britain and the United States c. AD 1860). Water-powered blowing engines were in use in 1969 yr (AD 31, Nanyang), and by c. 1500 yr (AD 500) the Chinese were also making cheap steel by the open-hearth process, equivalent to the Siemens-Martin process (as in France, AD 1864). Cast iron was made accidentally in Europe in Roman times, but not routinely until late in the fourteenth century AD, when it met the demand for cannons; priorities are difficult to establish, but early blast furnaces are known from Flanders, Italy, and England. The date of AD 1380 follows Needham.

To list some sixty metallic elements from beryllium to uranium, known to modern chemists and metallurgists, would be misleading. The only "new" metal to rival copper in importance is aluminum, and for all except relatively specialized, small-scale purposes, modern human beings still live in an Iron Age.

METAL INDUSTRIES: TIMETABLE

metalworking with native copper: c. 9900 yr. Anatolia (pins, beads, and a boring reamer, Çayönü); in North America c. 5000 yr (knives, etc., Old Copper Culture, Great Lakes region).

accidental iron: c. 7500 yr. Mesopotamia (Samarra). Thereafter small pieces of accidentally smelted or meteoritic iron appeared in rare and treasured objects, e.g., in Thailand c. 3600 yr (Ban Chang) and China c. 3250 yr (Thai-hsi Thun).

copper smelting and casting: **6500 yr**. Southeastern Europe (ax heads, in Rumania, Bulgaria, and Yugoslavia). Arsenic bronze occurred widely in Europe and southwestern Asia c. 5800 yr (e.g., at Tepe Yahya, Iran).

Amerindian copper smelting: **esp. 3100 yr**. Andes region (Ataura, Peru, and Tiwanaku, Bolivia). Bronze came into widespread use c. 1000 yr ago (AD 1000), with arsenic bronze in the northern Andes and tin bronze in the southern Andes.

tin bronzes: **esp. 5000 yr**. Eurasia, widely. Dates include ?5400 yr for Thailand (Nok-Nok-Tha), 5000 yr for Anatolia (Troy); 4900 yr for the Harrapa culture in northwestern India; 4700 yr for advanced craftsmanship in Mesopotamia (Royal Cemetery at Ur); and c. 4000 yr for Vietnam.

iron in working use: c. 3250 yr. Anatolia (Hittites). Also c. 3200 yr in Iran, the Aegean, and Palestine.

steel: 3100 yr. Cyprus (a knife; carburized, quench-hardened, and tempered, as hard as modern steel).

iron important: **3050 yr**. Eastern Mediterranean region (surpassing bronze). A sudden increase in the use of iron in Etruscan Italy dates from c. 2780 yr. Ironmaking was in progress in western Africa by c. 2750 yr (Nok culture) and in Vietnam, Thailand, and Burma by c. 2500 yr. *See also* breakouts (blacksmiths).

cast iron in China: **c. 2400 yr**. *See* above.

cast iron in Europe: **620 yr (AD 1380)**. *See* above.

aluminum: **112 yr (AD 1888)**. United States and France (Hall-Heroult electrolytic process). Duralumin and other alloys became important in aviation c. AD 1914. Aluminum overtook copper as the world's second metal (by tonnage) c. 1960.

stainless steel: AD 1912. Germany (high chromium steel).

metals from space **c. 1900 Myr.** Vredefort chromite and Sudbury nickel events, *see* cosmic impacts; ores.

meteorites 4550 Myr. *See* solar system forming; *also* cosmic impacts.

methane bacteria esp. 2500 Myr, *see* life on Earth (metabacteria); *methane* (natural gas) *formation* esp. 170 Myr, etc., *see* petroleum.

Mexican prehistory 11,000 yr, etc. *See* archeological timescale; *also* animal domestication, crops.

microprocessor **AD 1971.** *See* information technology.

microscope **c. 340 yr (AD 1660).** *See* science.

microwave background radiation 13,500 Myr. *See* Big Bang, cosmological timescale.

Middle America (innovative region; southern Mexico and Central America) esp. 7700 yr (maize) onward. *See* archeological timescale; crops.

mid-Permian turnover **256 Myr.** At the end of the Early Permian, among land animals, many pelycosaurs and amphibians were annihilated; afterward the therapsids rose to prominence. In the sea, forerunners of the ammonites suffered badly at about the same time, and the ancient trilobites became extinct. This event should not be confused with the far more severe event about 11 million years later (*see* Permian terminal catastrophe).

migrations *See* breakouts; population history.

milder interlude **58,000 yr.** *See* climate 128,000 yr; orbital dating.

milk ?216 Myr, *see* mammals; *milking and milk drinking* **esp. 6000 yr**, *see* livestock revolution.

Milky Way Galaxy **est. 12,500 Myr.** *See* galaxies.

Mill Creek culture (declined) c. 800 yr (AD 1200). *See* climate 10,300 yr.

millets domesticated (various) **c. 8000 yr**. *See* crops.

millipedes **est. 425, esp. 395 Myr.** *See* invertebrates.

Ming empire **632 yr (AD 1368).** *See* historical timescale.

Minoan culture 4000 to 3450 yr. *See* archeological timescale; volcanic events.

Miocene disruption 15 Myr. Extinction of 30 percent of mammalian genera (species groups) occurred, close in time to the Ries impact crater and associated tektite field. *See* cosmic impacts; mammals; radiometric dating.

Miocene epoch **25 to 5.3 Myr.** *See* geological timescale.

missiles (guided and homing) AD 1943; (ballistic) **AD 1944**; (submarine-launched) **AD 1960**; (antiballistic) **AD 1969**; (MIRV) **AD 1970**. *See* weaponry.

Mississippian period **360 to 325 Myr.** *See* geological timescale.

Mitanni culture fl. 3600 yr. *See* breakouts (horsemen).

mites **395 Myr.** *See* invertebrates.

mitochondria **est. 1700 Myr.** *See* life on Earth.

modern amphibians (frogs, etc.) **c. 240 Myr**. *See* vertebrate animals.

modern cells (eukaryotic) **est. 1700 Myr**. *See* life on Earth.

modern humans (talkative *Homo sapiens sapiens*) **est. 45,000 yr**; *breakout* **40,000 yr**. *See* breakouts; human origins; mental revolution.

modern mammals (placental) **114 Myr**. *See* mammals.

modern science formalized **338 yr (AD 1662).** *See* science.

moldboard plow **est. 1500 yr (AD 500).** *See* cultivation practice; plow.

molecular dating Cauliflowers, pork, and rabbit meat all include a protein called cytochrome *c*; its molecules consist of about a hundred chemical units (amino acids) strung together. Between cauliflowers and pigs, forty-seven of the units of cytochrome *c* are different, while between pigs and rabbits there are only four differences. Therefore cauliflowers and pigs once shared a common ancestor, and pigs and rabbits shared a common ancestor more recently. What is more, the molecular clock of cytochrome *c* seems to tick at a rate of about four unit differences per 100 million years, between two divergent lineages.

Thanks to modern methods of reading the entire chemical sequences of nucleic acid (DNA and RNA) and protein molecules, the numbers of mutations in genes, or differences in the protein molecules specified by the genes, are sensitive indicators of the evolutionary distance between two species. Given a little calibration from the fossil record, to fix the dates of selected evolutionary events, the genes and proteins can serve as clocks to date other evolutionary events, assuming that the molecules in question have mutated at a constant rate. Molecular dating already rivals or surpasses fossil evidence in some areas (*see* human genetics; life on Earth; primates).

Rates of change vary widely from molecule to molecule. Histone H4, for example, changes 5 percent of its components in 2500 million years, while fibrinopeptide changes as much in 3 million years. Slow mutation rates, useful for studying evolution over billions of years, become uninformative or misleading about the past few million years. With high rates or very long intervals, early mutations become overprinted by later mutations, so that early changes tend to be understated.

Molecular dating makes sense only within the controversial neutral theory of evolution, according to which molecular changes elude control by natural selection, by neither helping nor harming the organism that owns them (M. Kimura, *Nature*, Vol. 217, 1968, p. 624, and *The Neutral Theory of Evolution*, Cambridge University Press, 1983). For discussion see p. 37. Even in the neutral theory the approximately constant tick rate of molecular clocks seems too good to be true. Why should molecules change just as fast in elephants as in mice? A possible explanation is that, compared with mice, elephants have a long generation time, which slows down the introduction of mutations, but they also have small interbreeding populations, which speed it up.

mollusks **570 Myr.** *See* invertebrates.

money (coins) esp. 2600 yr; (paper) c. 1090 yr (AD 910). *See* information technology (coins).

Mongols' breakout **789 yr (AD 1211)**; *Yuan empire in China* **740 yr (AD 1260)**. *See* breakouts; historical timescale.

Mon-Khmer (languages) origins obscure, *see* language; *Khmer breakout* 1150 yr (AD 850), *see* breakouts (Indochinese peoples).

monkey fossils (New World) 35 Myr; (Old World) 20 Myr. *See* primates.

moon captured **est. 4500 Myr**; *defunct* **c. 3000 Myr**. The moon was a small, embryonic planet, with 1 percent of the mass of the Earth; it was probably captured by the Earth in the midst of the events 4550 to 4450 Myr ago that built both the Earth and the moon (*see* solar system forming). Landings on the moon (AD 1969, etc.) and recovery of rock samples have clarified events of lunar history. In summary: the heavily cratered, light-colored highland areas came into existence before 4000 Myr ago, while the dark maria consist of lava that poured into very large craters from 3800 to 3200 Myr ago. Then the moon died, geologically speaking, although it continued to suffer cratering impacts (P. W. Francis in D. G. Smith, ed., *Cambridge Encyclopedia of the Earth Sciences*, Cambridge University Press, 1982). The date of the moon becoming "defunct" is rounded to 3000 Myr.

moon landings **AD 1969**, etc. *See* transport.

motion pictures 105 yr (AD 1895); *talking* AD 1927. *See* arts and crafts.

mountain-building pulses 2800 Myr to present. *See* continents.

Mousterian technology **c. 120,000 yr.** *See* neanderthalers; stone tools.

Mughals into India **474 yr (AD 1526).** *See* breakouts (Mongols and Turks).

multiple warheads **AD 1970.** *See* weaponry.

music est. 45,000 yr; *formalized* 2600 yr. *See* arts and crafts.

Mycenaean culture **3600 to 3250 yr.** *See* archeological timescale; climate 10,300 yr.

mylodonts **c. 35 Myr.** *See* mammals.

N

naked bacteria (metabacteria) **est. 2500 Myr**. *See* bacteria.

Napoleon's empire **196 to 186 yr (AD 1804 to 1814).** *See* historical timescale.

nationalism (imperial) **c. 360 yr (AD 1640)**; (revolutionary) **224 yr (AD 1776)**. *See* government.

nation-state (evolving) **c. 400 yr (AD 1600)**. *See* government.

natural gas: origins esp. 170 Myr, *see* petroleum; *use* c. 2900 and 179 yr (AD 1821), *see* energy.

natural selection **est. 4450 Myr**, *see* life on Earth; *theory* **142 yr (AD 1858)**, *see* science.

navies c. 2800 yr. *See* weaponry.

navigational compass **885 yr (AD 1115).** *See* transport.

navigators' breakout (Europeans) **508 yr (AD 1492).** *See* breakouts.

neanderthalers **c. 120,000 to c. 34,000 yr.** Before the breakout of fully modern humans (*Homo sapiens sapiens*) c. 45,000 yr ago, near relatives in the subspecies *Homo sapiens neanderthalensis* thrived as hunter-gatherers in southern Europe and in a belt of Asia from the Levant to China (Ma-Pa). They were more heavily built than modern humans. For dating of the origin of neanderthalers *see* human origins. The last known neanderthaler lived near Cognac in France: tools associated with the skeleton show it to be of Chatelperronian times, indicating a date of **c. 34,000 yr**. After the appearance of modern humans in France (nominally c. 35,000 yr ago, *see* breakouts) there was no very long period of coexistence.

The neanderthalers are often associated with Mousterian implements (*see* stone tools), although not all neanderthalers used typical Mousterian tools, nor were all people who had tools of this type necessarily neanderthalers. Another term used by archeologists is Middle Paleolithic, and a rough equation is Middle Paleolithic = Mousterian = mainly neanderthalers = c. 120,000 to 40,000 yr.

Like *Homo erectus* and some modern humans, neanderthalers sometimes ate one another (e.g., at Krapina, Yugoslavia), but they are also credited with care of the sick and aged, with the first known herbal remedies (*see* medicine), and with a few traces of art (*see* arts and crafts). Signs of spirituality appear in ritualized burials of the dead in symbolically decorated graves, and in cults concerning hunted animals. With misgivings both about what constitutes a formal burial, and about the dating of sites, one can only loosely put the onset of burials at **?100,000 yr** ago. The most striking animal cult was the cave-bear cult known from Drachenloch,

Austria, and Regourdou, France. At Regourdou, for example, twenty bears were entombed under a stone slab; the Regourdou site dates from **47,000 yr** (43,550 b.c.). It represents a climax of neanderthaler culture, before these people were replaced by our own kind.

Sources for the foregoing are F. Lévèque and B. Vandermeersch, *Bulletin de la Société Préhistorique Française*, Vol. 77, 1980, p. 35; P. Phillips, *The Prehistory of Europe*, Allen Lane, 1980; L. Freeman in A. G. Sherratt, ed., *Cambridge Encylopedia of Archaeology*, Cambridge University Press, 1980; C. J. Jolly and F. Plog, *Physical Anthropology and Archeology*, Knopf, 1979; and J. G. D. Clark, *World Prehistory*, Cambridge University Press, 1977.

Neolithic period **10,600 yr.** *See* archeological timescale.

Neptune forming **4550 Myr.** *See* solar system forming.

Netherlands *See* Dutch.

neutron-activation analysis Materials inserted into a nuclear reactor become activated, when neutrons change their constituent atoms into radioactive forms. As each kind of radioactive atom emits particles of characteristic energy, this method is very sensitive for detecting elements present in the materials in trace amounts. A notable application in geology has been in detecting the extraterrestrial iridium that arrived on the Earth at the time when the dinosaurs died out (*see* Cretaceous terminal catastrophe). In archeology, trace elements in obsidian, jade, and other materials help to indicate where they came from and hence to establish ancient patterns of trade (*see* metal industries).

neutrinos, loose (decoupled) **1 second** after origin. *See* Big Bang.

neutron discovered **AD 1932.** *See* modern science.

New World monkeys **35 Myr.** *See* primates.

New Zealand (microcontinent): *quit Antarctica* c. 90 Myr, *see* continents; *peopled* by 1000 yr (AD 1000); *colonized by Europeans* 160 yr (AD 1840), *see* breakouts (Polynesians; navigators).

nickel 1840 Myr, etc. *See* ores.

Niger-Congo (esp. Bantu, languages spread) 2500 yr. *See* language.

Nilo-Saharan (languages spread southward) 4500 yr. *See* language.

nitrogen fixation **AD 1913.** *See* cultivation practice.

nomads (mounted pastoralists) by 4000 yr. *See* animal domestication.

Norian catastrophe **216 Myr.** In the Late Triassic, during one of the chief mass extinctions, conspicuous casualties were the marine reptiles, which were almost wiped out, and the reef-building chambered sponges, which were extinguished. On land, the mammal-like reptiles were eclipsed, but they were already suffering from the rise of the dinosaurs. The large impact crater at Manicouagan, dated at c. 210 Myr ago, may correlate with this event. *See* cosmic impacts. Allusions to a Triassic terminal catastrophe (nominally 208 Myr) may refer to the Norian event.

North African desiccation **4500 yr.** *See* climate 10,300 yr.

North America (geology) *Kenoran Mountains* **c. 2700 Myr**, *Grenville* **1150 Myr**, and other mountain-building events; *hit Europe* **460 Myr**; *hit Gondwanaland* **350 Myr**; *quit Africa* **210 Myr**; *quit Europe* **60 Myr**; *early Rockies* 60 Myr, *see* continents; *ores* 2600 Myr, etc., *see* ores; North American seaway **93 to 60 Myr**, *see* continents flooded; *western pileups* c. 300 Myr; *uplifts* **8 Myr**, *see* North American uplifts; *glaciers* (Alaska) **9 Myr**, *see* climate 67 Myr. See maps pp. 81–87. *See also* American Interchange; American intrusive peoples; Amerindians; coal; petroleum.

North American uplifts **8 Myr.** Dramatic scenery of western North America is the product of the continuing interaction between the westward-moving continent and the oceanic plates and rifts of the Pacific. Several island arcs have piled up on the western edge, for example c. 300 Myr ago (Antler event), c. 240 Myr (Golconda thrust), and c. 140 Myr (Pacific border). The Coast Ranges of California are younger shavings of material from a downgoing oceanic plate which also caused volcanic activity in the Sierra Nevada region. Farther inland, an early version of the Rocky Mountains formed c. 60 Myr (Laramide mountain-building event); the mountains eroded, but were rejuvenated by uplifts that occurred across a wide belt of the American West. The cause and the timing of the uplifts are not entirely clear, but their onset roughly coincided with the stretching of the southwestern United States that produced the Basin and Range formation, beginning c. 8 Myr ago (Early Miocene). Uplift seems to have been rapid in the Colorado Plateau, with the Colorado River carving the

Grand Canyon mainly since 2 Myr ago. (P. B. King, *The Evolution of North America*, Princeton University Press, 1977; K. Deffeyes, cited in J. McPhee, *Basin and Range*, Farrar, Straus, Giroux, 1981.) *See also* continents.

northern dairymen's breakout (proto-Uralic speakers) est. 4900 yr. *See* breakouts.

nuclear force 10^{-35} second after origin. *See* Big Bang.

nuclear energy AD 1942, *see* energy; *weapons* AD 1945, etc., *see* weaponry.

nucleic acids (RNA, DNA) **est. 4450 Myr**, *see* life on Earth; *structure* (discovered) **AD 1953**, *see* gene structure. *See also* human genetics; molecular dating.

numerals, modern 1320 yr (AD 680). *See* information technology.

nutrient-film technique AD 1982. *See* cultivation practice.

O

oceans est. 4450, esp. c. 200 Myr to present. Deep water has bathed the Earth ever since the surface was cool enough to allow water to lie on it, est. 4450 Myr ago, yet the floors of the present ocean basins are much younger than continents, because the heavy rock of which they are made continually renews itself, by growth at mid-ocean rifts and destruction at ocean trenches. The oldest part of the Pacific Ocean, at the Challenger Deep in the Marianas Trench, is c. 200 Myr old. It is also the deepest place on the planet.

Traces of former oceans exist in many mountain chains formed by collisions of continents. The ocean that once separated main Eurasia from Italy, Greece, Turkey, Iran, and India is often called Tethys, while Iapetus is a name for a former "Atlantic" destroyed in a collision between North America and Europe. As independent movements of microcontinents become clearer, the patterns of former ocean basins become more complex. Nevertheless, recognition of deep-water features in the past is important for reconstructing marine habitats and climatic conditions. For example, the tropical Atlantic and Pacific were linked until 3.9 Myr ago.

Like the Pacific, the Arctic Ocean is an ancient but diminished feature, renewing itself at the Lomonosov ridge. The Mediterranean is a jumble of shallows and fragmentary, unrenewed ocean basins up to 110 Myr old. Development of new features began approximately as follows: Atlantic Ocean 210 Myr; Indian Ocean 160 Myr; Southern Ocean (Australia/Antarctica) 50 Myr; Gulf of California 35 Myr; Japan Sea 30 Myr; Red Sea 30 Myr; *see* continents.

oil esp. 170 and 93 Myr, *see* petroleum; *industry* 143 yr (AD 1857), *see* energy.

Old Copper Culture 5000 yr. *See* metal industries.

Oldowan technology 2 Myr. *See* stone tools.

Old World Interchange 20 Myr. *See* mammals.

Oligocene epoch 37 to 25 Myr. *See* geological timescale.

Olmecs, rise of 3250 yr. *See* archeological timescale; climate 10,300 yr. Illus. p. 53.

opera (Chinese) c. 1350 yr (AD 650); (Italian) c. 393 yr (AD 1607). *See* arts and crafts.

opossums est. 125 Myr. *See* mammals.

optical-fiber links esp. AD 1980. *See* information technology.

orangutans 10 Myr. *See* primates.

orbital dating A new way of dating events during the past million years relies on the precision with which ice-age cycles follow variations in the Earth's attitude in space and the shape of its orbit. These are computable by well-established astronomical methods, and detectable as climatic rhythms (Milankovitch theory). Rhythms correspond with cycles of about 90,000 years (changes in orbital shape), 41,000 years (tilt of the axis), and 23,000 and 19,000 years ("wobble" of the axis, or precession of the equinoxes). An important dating constraint comes from the last magnetic reversal (730,000 yr ago) detectable in the ocean-bed cores. One can then "tune" the climatic phases, or "core stages," evident from marine fossils (*see* oxygen-18 stratigraphy) to match the orbital variations more closely, and so arrive at a calibrated timescale (J. D. Hays, et al., *Science*, Vol. 194, 1976, p. 1121; J. Imbrie and J. Z. Imbrie, *Science*, Vol. 207, 1980, p. 943; J. J. Morley and J. D. Hays, *Earth and Planetary Science Letters*, Vol. 53, 1981, p. 279; the work is continuing in the SPECMAP Project). Illus. pp. 35, 145, 147, 148.

Dates of glacial maxima given below follow simpler orbital-dating calculations (N. Calder, *Nature*, Vol. 252, 1974, p. 216). Core stages accord with the Morley-Hays (1981) dates, except for core stages 16, 15, and 14, which follow M. A. Kominz, et al., *Earth and Planetary Science Letters*, Vol. 45, 1979, p. 394.

ORBITAL DATING: TIMETABLE

core stage (start)	*age* (yr)	*glacial maximum* (yr)
17	680,000	(warm)
16	649,000	626,000
15	619,000	(warm)
14	551,000	509,000
13	505,000	(warm)
12	475,000	429,000
11	421,000	(warm)
10	347,000	344,000
9	334,000	(warm)
8	279,000	248,000
7	244,000	(warm)
6	188,000	133,000
5	128,000	(warm)
4	72,000	62,000
3	58,000	(warmish)
2	27,000	16,000
1	11,000	(warm)

Note that core stages 4, 3, and 2 cover a single ice age with a mild interlude. Orbital dating also confirms severe but brief cooling episodes 115,000 yr and 95,000 yr ago (*see* false ice ages). Earlier core stages similarity contain a variety of fluctuations, with the warmest spells lasting only 10,000 years. In the timescale and narrative the dates are often rounded.

Although the 90,000-year cycle became emphatic only 800,000 yr ago, a count of twenty-eight ice ages, between 3.25 Myr and the start of an ice age 649,000 yr ago, gives a mean period of 92,900 years. Seven more ice ages can then be counted to the present; the mean period was again 92,900 years.

ice age number	*start (rounded)*
29	650,000 yr
30	550,000 yr
31	475,000 yr
32	350,000 yr
33	280,000 yr
34	188,000 yr
35	72,000 yr
36	about now

Ordovician period 520 to 435 Myr. *See* geological timescale.

Ordovician terminal catastrophe 440 Myr. Many trilobites and early fishes were among the casualties of a disaster to life as severe as that which killed the dinosaurs. The event coincides with ice ages, at the end of the Ordovician period. *See* climate 3800 Myr; mass extinctions.

Oregon eruptions esp. 16 Myr. *See* volcanic events.

ores 3600 Myr to present. Geochemical and biological action have been creating deposits of rich ores since the earliest geological times. By geographical roulette some countries have extraordinarily rich deposits of useful or precious ores (South Africa is an obvious example), while others have very little. Here it is practicable only to offer the broadest outline of a very complex series of processes. Rocks of different ages show different styles of ore deposition, and this summary follows principally J. V. Watson, *Transactions of the Institution of Mining and Metallurgy*, Vol. 82, 1973, p. B107, and *Proceedings of the Royal Society*, Vol. A362, 1978, p. 305. The annotations relating to cosmic impacts are added (*see* cosmic impacts), and economic data are from D. R. Derry, *A Concise World Atlas of Geology and Mineral Deposits*, Mining Journal Books, 1980. (For a remark about the natural production of copper metal during the ice ages, *see* metal industries.)

ORE REGIMES: TIMETABLE

chromite I: 3600 Myr. Chromium, nickel, and copper deposits from molten intrusions through greenstone belts in southwestern Greenland. Similar deposits are known from Zimbabwe and Australia, 2900 to 2700 Myr ago.

gold boom (gold I): **2900 Myr**. Gold deposits in remelted greenstone belts in southeast Africa and a little later in western Australia, down to 1800 Myr ago in western Africa. These deposits are sometimes associated with silver, copper, occasional zinc, and traces of tin.

gold II: 2600 Myr. Gold and uranium deposits in sediments derived from greenstone belts. These began accumulating extravagantly in Witwatersrand, South Africa; similar but lesser deposits formed in Canada, Brazil, Guyana, and western Africa, until 1700 Myr ago.

chromite II: 2600 Myr. Chromium, platinum, and nickel in molten intrusions through continental crust in the Great Dike of Zimbabwe, and later (c. 1950 Myr) in the Bushveld complex of South Africa. More than 90 percent of economically useful chromium and 70 percent of platinum are in these two areas. An association exists with the supposed Vredefort cosmic impact crater in South Africa (1970 Myr).

iron boom: **esp. 2500 Myr**. Iron in banded iron formations (BIF). Vast biochemical deposits settled out of calm water in the U.S.S.R., western Australia, eastern Canada, Brazil, and elsewhere; such formations ceased 1800 Myr ago, but nearly all the world's commercial iron ore now comes from them.

modern-style copper: 2300 Myr. The first major copper-lead-zinc deposits from "modern" volcanic activity and mountain building began in southern Finland; similar deposits formed

c. 1800 Myr in U.S.S.R., Sweden, Australia (Mt. Isa and Broken Hill), and North America.

nickel: 1840 Myr. A rich delivery by the impact of a giant iron-nickel meteorite at Sudbury, Ontario.

diamonds: 1700 Myr. The earliest diamonds in intrusions through a relatively cool upper mantle and crust, in Guyana and southern Africa. Diamond-making events were much commoner 300 to 200 Myr ago, but they still tended to occur in very old crust, as in Siberia, Zaire, South Africa, Botswana, and Namibia.

tin: 900 Myr. The oldest substantial deposits of tin and tungsten formed from remelted granites, in the Kebaran mountain-building event in Africa (Uganda to Namibia), and also in western Brazil. By similar processes, the largest tin deposits formed later (340 to 50 Myr) in southeastern Asia, and tungsten (c. 140 Myr), in southern China.

copper boom: **c. 850 Myr**. Copper-rich sediments derived from the once-molten intrusions of former continental rifts. These formed the great Katanga copper deposits in southern Africa (extending from Zaire to Zambia, etc.); similar but poorer deposits occur in North America. Large copper-ore bodies have accumulated since 300 Myr ago, especially in the United States and Chile, but they are of poorer grade than the Katanga copper.

lead-zinc reconcentration: c. 500 Myr. Lead and zinc derived from older sources. These began to form sediments in the Mississippi Valley and elsewhere in North America, plausibly aided by biological activity.

modern regime: 300 Myr to present. A busy period for the formation of ores of many different kinds. Molten intrusions through continental crust created, for example, copper, nickel, and platinum deposits in Siberia. The remelting of granites concentrated copper, lead, zinc, silver, gold, uranium, tin, and mercury in western North America, western South America, and elsewhere. The uplift of ocean-floor rocks brought fresh copper up into the air in Cyprus, and nickel in New Caledonia, while the destruction of ocean-floor rocks at trenches produced relatively young gold and copper ores in western North America.

bauxite: contemporary. The most useful aluminum ore occurs principally in modern tropical environments, where it is concentrated by the weathering of soil. The largest economic deposits are in Guinea, northern Australia, Brazil, and Jamaica; "fossil" bauxite occurs in southeastern Europe, dating from c. 130 Myr ago, when the region lay in the tropics.

ornithischians 228 Myr. *See* dinosaurs.

outer-planets mission AD 1977. *See* spacecraft (*Voyager 1* and *2*).

overkill (Pleistocene) **c. 17,000 yr**; (in America) **c. 11,300 yr**; (worldwide) at present. *See* mammals.

oxygen (discovered) **229 yr (AD 1771)**. Those who first identified the most active ingredient of the air (C. Scheele in Sweden and independently, perhaps later, J. Priestley in England) thoroughly misunderstood its chemical role; they were put right by A. L. Lavoisier, by AD 1777, and classical chemistry began.

oxygen-18 stratigraphy Oxygen consists mainly of oxygen-16 atoms, mixed with a relatively few oxygen-18 atoms. The proportion of the heavier oxygen-18 in the molecules of ocean water changes when the climate changes. Molecules containing the lighter atoms evaporate more readily. When they fall as snow and become locked up in ice sheets, they leave the oceans relatively depleted of oxygen-16 and enriched in oxygen-18. In short, a high oxygen-18 content is a sign of glacial conditions. Illus. pp. 128, 145, 147, 153, 160.

The proportion of oxygen-18 in the past is recorded in fossils of small marine animals that lived on the seabed and incorporated the oxygen in their carbonate shells. Extracted from a core sample of seabed mud, and analyzed with a mass spectrometer that separates molecules according to their mass, a handful of such fossils is sufficient to determine the volume of the world's ice sheets at the time when they lived. The alternations of glacial and milder conditions can then be used to define "core stages" during the current series of ice ages, recognizable in many seabed cores. They become a method of dating, especially when calibrated by magnetic stratigraphy (N. J. Shackleton and N. D. Opdyke, *Quaternary Research*, Vol. 3, 1973, p. 39). Astronomical calibration has refined the method further (*see* orbital dating).

Oxygen-18 analysis has sampled the snows of yesteryear, preserved in deepfreeze in the ice sheets of Greenland and Antarctica. One of the first quantified descriptions of the course of the most recent ice age came from analysis of a 1390-meter ice core from Greenland (W. Dansgaard, personal communication, 1974). For that work, the University of Copenhagen devised a novel method of dating the layers of ice, by calculating the plastic flow of ice in the ice sheet.

Oxygen-18 stratigraphy works also for nonglacial periods of the more remote past, because the uptake of oxygen-18 by organisms is influenced by the prevailing temperature. The lurches toward increasing cold during the past 50 million years have been analyzed by this method (*see* climate 67 Myr). On a smaller scale, the shells of shellfish reveal variations in oxygen-isotope proportions from season to season, and these provide a method of establishing, for example, at what season a shellfish was eaten by a human being. Variations in the oxygen-18 abundances in stalactites and stalagmites, in caves, are indications of changing climatic conditions on land; so too are changes in the atomic composition of lake water, recorded in the sediments of the lake bed.

oxygen revolution 1800 Myr. Although animals need oxygen to survive, primitive life began in the absence of free oxygen, and when oxygen built up in seawater, it was a deadly poison for organisms unequipped to deal with it. Oxygen originated mainly as a by-product of the growth of plants (*see* atmosphere). For more than half the Earth's history, the formation of sulfates and rich iron ores (banded iron formations) eliminated oxygen almost as fast as it was produced, but eventually these geochemical processes were exhausted. The mobilization of uranium in locally oxygenated conditions created natural reactors c. 2000 Myr ago (*see* uranium reactors). The definitive worldwide change that signaled the appearance of free oxygen was the appearance of red beds, characterized by abundant red oxides of iron (ferric oxides) in sedimentary rocks. The banded iron formation/red bed transition occurred 1800 Myr ago (geological data given in B. F. Windley, *The Evolving Continents*, Wiley, 1977; and H. H. Read and J. V. Watson, *Introduction to Geology*, Vol. 2, Macmillan, 1975). *See also* life on Earth.

P

pack animals esp. 5000 yr. *See* transport.

paintings esp. 20,000 yr. *See* arts and crafts.

Paleocene epoch 67 to 55 Myr. *See* geological timescale.

paleogeography Fossil magnetism in rocks is misaligned with the Earth's present magnetic field, and shows that the continents have moved; it indicates the orientation and latitude of a continent at the time when the rocks were formed. This is the primary source of information about the past locations of continents, but it gives no indication of longitude. Good evidence of recent movements comes from the growth of ocean floors. Traces of ancient oceans, found among mountains, announce that different pieces of present continents were formerly separated, while evidence of rifting along shorelines indicates that continents have split asunder. Edges can be put together again by computer programs that reconcile the coarse shapes of continents with the precise geometry of motions on a sphere, to obtain best fit. A. G. Smith of the University of Cambridge, and his colleagues, have mapped ancient geography strictly by fossil magnetism and by the fitting of continental margins. Present continental outlines are preserved as convenient fiction for finding one's way around the maps, and pieces are not detached unless their positions are well attested. (See base maps pp. 81–85.)

The continents can, though, be dismembered into microcontinents, and maps adjusted by evidence of connections and splits between organisms, climate as indicated by characteristic rocks (e.g., coal, or fossil sand dunes), and geological activity. Global patterns of climate and ocean circulation can be inferred. The most elaborate effort along these lines is that of A. M. Ziegler and his colleagues at the University of Chicago. Data from this school, too, have been incorporated in the maps (pp. 81–85). Continental arrangements before 600 Myr ago are hazy and controversial, although the existence of earlier supercontinents is presumed (*see* continents).

Paleolithic period 2.4 Myr to 10,600 yr. *See* archeological timescale.

Paleozoic era 570 to 245 Myr. *See* geological timescale.

Palestine (innovative region; mainly modern Israel, West Bank, and Jordan): esp. 18,000 yr (animals herded) onward. *See*, e.g., animal domestication; cities; crops.

Pan-African Mountains 650 to 450 Myr. *See* continents; Gondwanaland.

Pangaea esp. 220 Myr. The most recent supercontinent was assembled by the collisions of three main blocks, Gondwanaland, Euramerica, and Asia, 350 to 260 Myr ago (*see* continents). Various smaller blocks, especially in southeastern Asia, were late arrivals. In the initial collision between Gondwanaland and the northern continents, South America butted eastern North America and southern Europe; Spain and central France are former pieces of Venezuela. This first configuration, called Pangaea B, was complete by 260 Myr ago (mid-Permian). A sliding motion then carried Gondwanaland 3500 kilometers westward, relative to the northern continents, until Africa was abutting North America by 220 Myr (Late Triassic), producing the classic configuration of Pangaea A (E. Irving, *Nature*, Vol. 270, 1977, p. 304, dates adjusted). A. G. Smith adopted Irving's concept of Pangaea B for his reconstructions (see maps p. 81), although other paleogeographers dissented from it. The breakup of Pangaea began with the initial rifting at the birth of the North Atlantic ocean c. 210 Myr ago. See map p. 84.

paper 1900 yr (AD 100). *See* information technology.

parliamentary monarchy 312 yr (AD 1688). *See* government.

Parthian empire esp. 2141 yr to 1776 yr (AD 224). *See* historical timescale.

particles (subatomic) **10^{-35} second** after origin. *See* Big Bang.

pebble choppers c. 2.0 Myr, **esp. 750,000 yr**. *See* stone tools.

Peking Man ?650,000 yr. *See* human origins.

pelycosaurs (early mammal-like reptiles) **310 Myr.** *See* vertebrate animals.

penguins est. 40 Myr. *See* birds.

Pennsylvanian period 325 to 290 Myr. *See* geological timescale.

Permian period 290 to 245 Myr. *See* geological timescale; *also* mid-Permian turnover.

Permian terminal catastrophe 245 Myr. This event, which ended the Paleozoic ("old life") era and initiated the Mesozoic ("middle life") era, was the most disastrous event in 670 million years of animal life on Earth (*see* life on Earth). An estimated 96 percent of all species of marine animals were annihilated (D. M. Raup, *Science*, Vol. 206, 1979, p. 218). Reefs disappeared, along with most bottom-dwelling animals; tidal-wave action may be indicated. Among conspicuous land animals, the mammal-like reptiles (therapsids) were destroyed. At the top of the Permian strata, at Nanking, China, a layer of clay is curiously lacking in chromium compared with rocks above and below it (F. Asaro, et al., Conference on Large Body Impacts, Snowbird, 1981). By analogy with other catastrophes, rather than by direct evidence, this event is here ascribed to a head-on collision between the Earth and a comet, producing great violence and relatively little injection of exotic chemicals (*see* cosmic impacts). The Permian terminal event, also called the Permo-Triassic catastrophe, should not be confused with the mid-Permian turnover, 256 Myr ago.

Persians *See* Iran.

perspective (formalized) **c. 585 yr (AD 1415)**. *See* arts and crafts.

pesticides esp. AD 1939. *See* cultivation practice.

petroleum esp. 170 and 93 Myr. Fossil chemicals confirm that petroleum was produced from marine sediments rich in the remains of algae or photosynthetic bacteria, in stagnant water from which the oxygen was removed by the sheer abundance of organic material. The most prolific sources of oil were shallow, landlocked seas on flooded continents, in tropical latitudes. Petroleum trapped underground could move, and as a result the ages of the rocks in which oil now lies are not necessarily the same as the source rocks in which it originated. Nevertheless, source rocks can be traced, and from a study of 148 petroleum zones, a striking pattern emerged. The surviving oil was nearly all made during a few relatively short phases of the Earth's history, generally corresponding to periods of high sea level.

Most early oil was destroyed or converted into natural gas, but important oil fields in the Sahara and western Texas date from 430 to 410 Myr ago (Silurian times). The next notable phase was **170 to 140 Myr** (in Jurassic times), when oil was created in abundance in the North Sea, southwestern Asia, Mexico, Siberia, Australia, and central Asia. A surge in the formation of surviving petroleum, far surpassing anything before or since, occurred 115 to 85 Myr (mid-Cretaceous). This was when most of the petroleum of the great Arabian oil fields was created, along with other fields in Canada, United States, Mexico, Venezuela, Ecuador, and Colombia. When the exceptionally high sea level attained 93 Myr ago is taken into account (*see* continents flooded), the peak in petroleum formation can be set at **93 to 85 Myr** ago (Cenomanian, Turonian, Coniacian, and Santonian stages). The last important episode of oil formation was during a phase of relatively high sea level 15 to 10 Myr ago (mid-Miocene), when oil fields in Indonesia, California, Venezuela, and the Caucasus formed. (Data from B. Tissot, *Nature*, Vol. 277, 1979, p. 465, and *La Recherche*, Vol. 10, 1979, p. 984; dates adjusted.) Two thirds of the world's oil reserves are in the region of the Arabian-Persian gulf; the U.S.S.R., Mexico, United States, Libya, China, Venezuela, Nigeria, and Britain (North Sea), are the next in line, with known endowments of oil (*Oil and Gas Journal*, estimates for 1980).

Phanerozoic era 570 Myr to present. *See* geological timescale.

Phoenician-Carthaginian breakout c. 2800 yr. *See* breakouts (blacksmiths).

phonograph 123 yr (AD 1877). *See* information technology.

photography **161 yr (AD 1839).** *See* arts and crafts; information technology.

photosynthesis **est. 3900 Myr.** *See* bacteria, life on Earth, stromatolites.

physics esp. 396 yr (AD 1604) and AD 1926. *See* science; quantum mechanics.

pigments 1.6 Myr; (multiple) **c. 60,000 yr**. *See* arts and crafts.

pigs **30 Myr.** The family of suids, or pigs, is known from the Late Oligocene and the modern genus *Sus* from **c. 4.5 Myr** (5 or 4 Myr, Early Pliocene). (Data from L. Ginsburg and A. Azzaroli, both 1973, cited by A. W. Gentry, personal communication, 1982.) Modern Eurasian wild hogs (wild boar, *Sus scrofa*) showed up in Europe **c. 1.8 Myr** early in the Pleistocene (C. A. Repenning and O. Fejfar, *Nature*, Vol. 299, 1982, p. 344). Pigs were domesticated **est. 9500 yr** ago, *see* animal domestication.

pineal gland (origin as third eye) est. 510 Myr. *See* vertebrate animals.

planets (forming) **4450 Myr**. *See* solar system form ing; *also* Mars; Mercury; Venus; *outer-planets mission* **AD 1977**, *see* spacecraft.

plant breeding esp. AD 1900. *See* cultivation practice.

plants (multicellular) **c. 1300 Myr**. *See* life on Earth.

plants ashore **esp. 425 Myr** to present. Most plants are prone to total decay, and their fossils are often fragmentary and hard to identify. Leaves are the commonest substantial fossils, but fossils of spores, seeds, and the structures in which they are produced may be the best guide to the evolution and distribution of plants. Land plants evolved from multicelled green algae (*see* life on Earth). Spores of candidate terrestrial or semi-terrestrial ancestors are known from c. 460 Myr (mid-Ordovician), and nematophytales, pioneers apparently unrelated to present land plants, existed c. 430 Myr ago (Early Silurian). The main evolutionary story begins in the mid-Silurian, 425 Myr ago. Liverworts and mosses, which remain "primitive," are known from c. 410 Myr ago (Silurian/Devonian). To venture ashore, plants needed spores that could avoid desiccation, suitable wrappings for the plants themselves, and rigid three-dimensional structures that could support the weight of plants out of water. Plausibly these features evolved in fresh water, in circumstances where rivers or ponds were liable to dry out. J. A. Doyle, in A. Hallam, ed., *Patterns of Evolution*, Elsevier, 1977, and personal communication, 1982, provides the framework for this account. The earliest *Cooksonia* appearance accords with D. Edwards and J. Feehan, *Nature*, Vol. 287, 1980, p. 41.

HIGHER PLANTS: TIMETABLE

pteridophytes (spore-bearing, with eggs fertilized by sperm, via water): 425 Myr. Psilopsids in Ireland dating from 425 Myr (Wenlockian; *Cooksonia*) are the earliest known higher (vascular) plants; they lacked roots and leaves. Club mosses (lycopsids) appeared 395 Myr ago (Siegenian; *Protolepidodendron*; also Emsian; *Asteroxylon*) and horsetails (sphenopsids) c. 370 Myr. Progymnosperms, resembling conifers but still using primitive methods of reproduction, made the first real trees **370 Myr** ago (mid-Late Devonian; e.g., *Archaeopteris*, *Callixylon*). Forerunners of ferns appeared c. 380 Myr ago, and they evolved continuously down to the present.

gymnosperms (ovules pollinated to make naked seeds): 370 Myr. Very primitive seeds are known from c. 370 Myr (latest Devonian; W. H. Gillespie, et al., *Nature*, Vol. 293, 1981, p. 462). They were probably seeds of seed ferns, more complete remains of which date from c. 360 Myr (Lower Carboniferous). Seed ferns survived until c. 67 Myr ago. Cordiatales, c. 350 Myr, attained heights up to 36 meters, but they died out 245 Myr ago. The more durable conifers also appeared **c. 350 Myr** ago. Glossopterids were characteristic of Gondwanaland from 320 to 245 Myr ago. Ginkgoes and cycads appeared c. 280 Myr ago and were conspicuous until c. 100 Myr ago.

bennettitales (early flower-like reproductive structures): **235 Myr.** Flowers attract insects and other animals, thereby obtaining their assistance in cross pollination. Although the angiosperms that evolved later are conventionally called *the* flowering plants, flowers and insect pollination go back much farther; the first known experiments with flowers were made by the bennettitales, gymnosperms that flourished 235 to c. 75 Myr (Middle Triassic to Late Cretaceous).

angiosperms (modern flowering plants): **123 Myr, esp. 114 Myr**. With protected ovules pollinated to make encased seeds, these are the dominant plants of the present era, ranging from minute herbs to great trees. There are good fossil remains of angiosperms from 123 Myr ago (Barremian) and a burst of evolution and migration that brought the angiosperms into prominence worldwide can be dated to 114 Myr ago (*see* floral revolution). *See also* grass.

plastics (wholly synthetic) AD 1910. *See* arts and crafts.

Pleistocene disruption **730,000 yr.** A magnetic reversal, coinciding with a tektite field in Australasia (*see* cosmic impacts), provides a very useful marker for dating. The Arctic acquired permanent summer sea ice soon after (*see* climate 3.25 Myr), and mammalian diversity declined from its earlier peak (*see* mammals).

Pleistocene epoch 1.8 Myr to 10,300 yr. *See* geological timescale.

Pleistocene overkill 17,000 yr. *See* mammals.

Pliocene epoch 5.3 to 1.8 Myr. *See* geological timescale.

plow est. 6000 yr. Plows appear in early pictograms from Sumer (c. 5500 yr), but an informed estimate (A. G. Sherratt) puts their origin in northern Mesopotamia by 6000 yr ago, or earlier. Plow marks dating from 5500 to 5100 yr ago occur widely in Europe (Britain, Denmark, and Poland; 2800 to 2500 BC). The early European plows were simple, being little more than hoes adapted for traction, and perhaps were developed independently of the Middle Eastern plows. Signs of the plow in India date from more than 4000 yr ago (plow marks; Kalibangan). The plow is notable in prehistory for several reasons: it was one of the earliest means of making animal energy (of oxen, in this case) serve human needs; it was an improved method of weed control that multiplied the productivity of farm labor; and it opened wide areas of relatively heavy and difficult soil to cultivation (A. G. Sherratt, in I. Hodder, et al., eds., *Pattern of the Past*, Cambridge University Press, 1981). The plow evolved in many ways. The crook ard, in use in Europe from 5500 yr on, was superseded in northern Europe c. 2500 yr ago by the stave ard and the bow ard. The earliest known turned sod (c. 2000 yr) may have been plowed by a bow ard. Thereafter the European plow evolved into the moldboard plow, properly designed for sod turning, by **est. 1500 yr (AD 500).** Other key innovations included iron plows in southwestern Asia c. 3000 yr ago, cheap cast-iron plows in China c. 2300 yr, and cheap steel plows in North America c. 140 yr (AD 1860). *See also* cultivation practice; livestock revolution.

Polynesian breakout c. 1900 yr (AD 100). *See* breakouts.

population history 10,000 yr to present A median value for estimates of the human population c. 10,000 yr ago is 8 million. Populations began increasing est. 9500 yr ago with the spread of gardeners; they may have grown more rapidly in the period 9500 to 5000 yr than subsequently; as disease tightened its grip, population growth had ceased, or gone into reverse, by 2000 yr ago. A second major transition, ushering in the present population explosion, started in China c. 300 yr ago (AD 1700); it spread to Europe (c. 1800) and then to most of the world (c. 1930). A third demographic transition was evident in Europe c. AD 1900, and later in North America and other industrialized regions: a reduction in birth rates as people adjusted to a fall in death rates. A minor "baby boom" in industrialized countries c. 1950 interrupted this transition. Illus. pp. 175, 179, 187.

All data about past populations are suspect. Unflawed census records are rare, and any other assessment is, at best, informed guesswork. The tactic adopted here is to compare a cautious author, Durand, whose aim was to find overall global trends, with McNeill, who made a formidable case for the importance of disease in history. The chief publications are J. D. Durand, *Historical Estimates of World Population: An Evaluation*, Population Studies Center, University of Pennsylvania, 1974, and W. H. McNeill, *Plagues and People*, Blackwell, 1977. Some theoretical background comes from R. M. May and A. J. Coale, personal communications, 1981; A. J. Coale and K. Davis, *Scientific American*, Vol. 231, No. 3, 1974, p. 40 and p. 22, were also consulted.

The first table shows world and regional population estimates simplified and adjusted from Durand (1974, Table 2), who gave "indifference ranges" of numbers, such that he did not know whether he preferred the higher or lower figure. The middle of his indifference range supplies the number in almost every case, except for *Middle and South America*, AD 1000 and 1500, where the numbers are at the upper end of Durand's range; a

POPULATION ESTIMATES—2000 YR TO PRESENT (MILLIONS)

yr before AD 2000	2000	1000	500	250	100	30
AD	1	1000	1500	1750	1900	1970
World	295	325*	515*	770	1680	3650
China	80	65	125	207	425	800
India[1]	75	75	112	180	180	627
southwestern Asia	35	25	25	30	42	110
Japan	1	5	17	29	44	103
rest of Asia[2]	14	17	22	45	217	422
U.S.S.R. area	7	10	14	35	132	243
Europe[2]	35	35	65	127	297	422
northern Africa	12	7	9	12	54	72
rest of Africa	22	30	45	65	105	280
North America	1	2	2	2	82	228
rest of America[3]	10	50*	60*	15	74	287
Oceania	1	1	1	2	6	19

[1] Territory of present India, Pakistan, and Bangladesh.
[2] Excluding the territory of present U.S.S.R.
[3] Middle and South America.
* Figures doctored as explained above.

plainly wrong figure for Oceania, 1970, is corrected. In relation to world population, these amended numbers are small; nevertheless, the world totals are adjusted.

China's population, in the next table, is the only one in the world for which reasonably rich statistics span a long period of history. The population data relate to political China (not a fixed area) and are mainly from Durand (1974, Table 3), who gives numbers carefully adjusted from census data for the years mentioned. Exceptions are marked "est.": for AD 1500, 1900, and 1970, the figures are from the previous table; and those for AD 1600 and 1700 are from Ping-Ti Ho, cited by McNeill (1977). The census datum for 1982 is added. Discrepancies between this and the previous table are methodological and of no significance. Comments relating to population surges and slumps follow Durand or McNeill.

CHINA'S POPULATION HISTORY—2000 YR TO PRESENT

AD	*millions*	*comments*
2	74	western Han empire
88	45	eastern Han empire
156	64	eastern Han empire
(162)	—	*Eurasian epidemics*
(c. 200)	—	*population crash*
606	53	Sui empire
(c. 700)	—	*population crash*
705	37	T'ang empire
755	53	T'ang empire
(c. 1000)	—	*population surge*
1014	55	Sung empire
1103	123	Sung empire
c. 1194	121	Sung and Chin empires
(1290)	—	*population crash*
(1290)	86	Mongol empire (Yuan)
1393	61	Ming empire
(c. 1450)	—	*population recovery*
1500	est. 125	Ming empire
1600	est. 150	Ming empire
1700	est. 150	Manchu empire (Ch'ing)
(c. 1700)	—	*population surge*
1751	209	Manchu empire
1775	265	Manchu empire
1805	332	Manchu empire
(c. 1850)	—	*population pause*
1851	432	Manchu empire
1900	est. 425	Manchu empire
(c. 1950)	—	*population surge*
1953	583	People's Republic
1970	est. 826	(UN estimate)
1982	1008	People's Republic

Parallels with the Chinese experience can be found in the rest of Eurasia, although estimates are widely divergent. At around 2000 yr ago, when the Chinese census showed a population of 74 million, the population of the Roman empire has been estimated, for example, at 57 million (C. Clark, 1968, cited by Durand). In large areas of the Old World the population fell, or grew little, between 2000 and 1000 yr ago. The Chinese population surge of AD 1000, followed by a crash, was mirrored in Europe.

In Middle America, radar mapping revealed how the Maya cultivated the swamps of the lowlands in raised fields, and created the resources for a large population, estimated at 14 million in AD 800 (R. E. W. Adams, *Science*, Vol. 213, 1981, p. 1457). The pattern of Maya population growth up till then may have been approximately as follows:

2750 yr	1 million
2000 yr	4 million
1200 yr (AD 800)	14 million

The late pre-Columbian history of the Amerindian populations has been transformed by general demographic studies. McNeill cited estimates (S. F. Cook and W. Borah, 1971 to 1973) of about 100 million, with the Mexican and Andean heartlands having 25 to 30 million each. Estimates for the crash from AD 1521 onward, due principally to the epidemics introduced by Europeans, show the population of central Mexico (Aztec territory) shrinking to 3 million by 1568 and 1.6 million by 1620: in other words, a fall on the order of 90 percent.

In the poorer regions of the world, growth remained zero until c. AD 1750, then ran at a moderate rate until c. 1930, when the growth rate overtook that of the industrialized countries (Coale). Growth rates for the world as a whole, and for most developing countries, peaked c. 1970, and eased noticeably in the subsequent decade, except in Africa where they were judged to be still rising toward the unprecedented rate of 3 percent per year (UN statistics, 1980). The diminution in growth rates is called here an inflection.

The excursions that altered the cultural maps of the world (*see* breakouts) did not necessarily involve large movements of populations. Nor did commercial slavery, war, or even genocide have very great effects on the overall sizes of regional populations, except when disruptions of food supplies and war-engendered diseases took their secondary tolls. The chief large-scale movements between regions (following Davis) were European emigration, c. AD 1840 to 1930 (more than 52 million, mainly to North America), and immigration from poor countries 1950 to 1972, into Europe, the United States, Canada, Australia, and New Zealand (20 million people).

POPULATION: SUMMARY TIMETABLE

population growing: c. 9500 yr. From ?8 million to peak, ?300 million, ?2100 yr.

level or falling populations: c. 2000 yr. In the Old World, associated with the increase in disease, and poorly defined crashes, esp. after 1838 yr (AD 162, Eurasian epidemics); populations were still rising in the Americas.

population surge: **c. 1000 yr (AD 1000)**. In China and Europe.

population crash: **c. 790 yr (AD 1290)**. Dated by the Chinese census, perhaps typical of Eurasia.

population recovery: **c. 650 yr (AD 1450)**. Dated for the estimate for China, again perhaps typical for Eurasia.

Amerindian epidemics (population crash): **479 yr (AD 1521)**. A direct consequence of the breakout of the European navigators and their diseases.

population boom: **c. 300 yr (AD 1700)**. First in China (interrupted 1850, resumed 1950); in Europe from c. AD 1800.

global population boom: **c. AD 1930**.

population inflection: **c. AD 1970**. The rate of growth abating in the world as a whole (c. 1900 in Europe).

porcelain c. 1150 yr (AD 650). *See* arts and crafts.

Portuguese: *language* after 1000 yr (AD 1000), *see* language; *overseas empire* (El Mina) **518 yr (AD 1482)**, *see* breakouts (navigators); *carve-up of world* 506 yr (AD 1494), *see* carve-up of continents.

postagricultural British **200 yr (AD 1800)**; *world* **AD 1970**. *See* work.

postindustrial Americans **AD 1960.** *See* work.

potassium-argon dating The common element potassium contains a small proportion of radioactive potassium-40 atoms which decay with a half-life of 1300 millon years; that is to say, half the potassium-40 atoms present in a sample will transform themselves in that period, some of them turning into calcium-40 atoms and others into argon-40, which is a stable, chemically inert gas. The relative abundances of potassium-40 and argon in a sample indicate the time that has elapsed since the argon was last free to escape—typically since the rock was last molten.

This important radiometric technique can span the entire history of the Earth (and the moon) down to about 100,000 yr ago (sometimes less), and it is particularly useful for dating volcanic deposits, for example those in the vicinity of the early humans in east Africa (see p. 24). The possibilities of argon worming its way out of cold rocks, or of argon from the air worming its way in, lent a touch of adventure to the use of potassium-argon dating. In the 1970s a method of testing the integrity of the sample was devised: the sample goes into a reactor for neutron activation, which converts potassium-39 to argon-39, and the investigator then checks whether the proportions of argon-40 and argon-39 change when the sample is heated. Observing yet another isotope, argon-36, provides a check on argon sneaking in from the atmosphere.

Argon-argon dating uses the production of argon-39 in the nuclear reactor to measure the proportion of potassium-39 present in the sample. Melting the sample, either quickly or in stages, then releases the argon-39, together with the argon-40 that had already accumulated in the material by ordinary radioactive decay. The age of the sample is then calculated from the proportions of the two kinds of argon.

A change in the accepted half-life of potassium-40 and in other constants involved in the calculations required a revision of all potassium-argon dates (R. H. Steiger and E. Jager, *Earth and Planetary Science Letters*, Vol. 36, 1977, p. 359). A table calculated by G. B. Dalrymple shows that, 1 Myr ago, corrected ages are 2.7 percent older; the correction diminishes to zero, 1900 Myr ago, and at 4500 Myr the corrected ages are 1.7 percent younger. Given other sources of inaccuracy in dating, the corrections can fairly be regarded as negligible in many older instances, although they become significant for the past 100 million years, for example, in amending the date of the Cretaceous-Tertiary boundary from 65 to 67 Myr.

potatoes domesticated ?10,000 yr. *See* crops.

potter's wheel **6000 yr.** *See* wheel.

pottery **13,000 yr.** *See* arts and crafts.

Precambrian era 3800 to 570 Myr. *See* geological timescale.

premonkeys **est. 70 Myr.** *See* primates.

preprimates **est. 95 Myr.** *See* primates.

presolar supernovas **c. 4650 and 4550 Myr.** Peculiar compositions of certain meteoritic material imply that the material of the solar system was spiked with matter from supernovas (massive exploding stars) shortly before the meteorites formed c. 4550 Myr ago. Xenon atoms made by the fission of plutonium-244, and by the radioactive decay of iodine-129, point to an injection of fresh matter from supernovas 100 million years before the formation of the meteorites. An excess of magnesium-26, formed by the radioactive decay of aluminum-26, which has a half-life of only 720,000 years, requires an injection of newly created aluminum-26 immediately before the formation of the solar system. This late pre-solar supernova probably triggered the creation of the solar system, as evidenced by star formation detected in the vicinity of more recently exploded stars. The radioactivity augmented the sources of heat available in the early stages of planetary formation (D. N. Schramm and R. N. Clayton, *Scientific American*, Vol. 239, No. 4, 1978, p. 98, and other sources).

primates 69 Myr. Humans belong to a group of mammals that live mainly in trees, and possess grasping hands and feet, binocular vision, and relatively large brains. The earliest known primate fossil is *Purgatorius*, **69 Myr**, from Montana (*see* mammals). It is possibly ancestral to the atypical primate *Plesidapsis*, 60 Myr (Paleocene; Europe and North America). Key links 67 to 55 Myr remain to be discovered, but premonkeys with grasping hands and forward-looking eyes appeared **55 Myr** (Eocene).

The interpretation of primate and human evolution has been an area of long-standing conflict between the inferences from fossils and those from molecular evolution. Comparisons of molecules in various species indicated more recent divergences between the various lineages than the fossil evidence seemed to imply. In particular, the separation of humans and apes is set by molecular studies at not more than about 5 Myr ago, while fossils of *Ramapithecus* and *Sivapithecus*, allegedly man-like creatures appearing **15 Myr** ago, suggested much earlier evolutionary events. But *Sivapithecus indicus*, 8 Myr old, turned out to have looked like an orangutan (D. Pilbeam, *Nature*, Vol. 295, 1982, p. 232). The molecular evidence on this point is no longer contradicted by fossil evidence, but fossils are scarce from around the time of human/ape divergence.

The following timetable is written on the assumption that molecular dating and fossil finds are due for reconciliation. In each entry, the divergence date estimated from molecular studies (when available) is given first. The emphasis here, as in many primate studies, is on the divergences leading to humans. Data for the timetable come from W. M. McFarland, et al., *Vertebrate Life*, Collier Macmillan, 1979; J. R. Napier, various sources; E. L. Simons, *Philosophical Transactions of the Royal Society*, Vol. B292, 1981, p. 21; V. Sarich and J. E. Cronin, various sources but especially *Nature*, Vol. 269, 1977, p. 354; D. Pilbeam, various sources; P. Andrews and J. E. Cronin, *Nature*, Vol. 297, 1982, p. 541; and B. Maw, et al., *Nature*, Vol. 282, 1979, p. 65. See also the family trees pp. 126, 137.

PRIMATES: TIMETABLE

preprimates split from other mammals (carnivores): **est. 95 Myr**.

ancestral primates split from tree shrews (Tupaia): est. 85 Myr; *from flying lemur (Cynocephalus):* est. 75 Myr.

earliest primate fossils: **69 Myr** (*Purgatorius*).

premonkeys (anthropoids ancestral to monkeys, apes, etc.) *split from ancestors of present prosimians* (lemurs, tarsiers, etc.): **est. 70 Myr**. If monkeys evolved, as often supposed, from a fossil group of premonkeys 55 to 37 Myr (Eocene omomyids of Eurasia), they had already diverged from the ancestors of present prosimians. *Amphipithecus* and *Pondaungia*, 40 Myr (late Eocene; Burma), and *Parapithecus*, *Apidium*, and *Oligopithecus* 35 Myr (early Oligocene; Fayum, Egypt), appear to be close to the evolutionary line leading to both monkeys and great apes.

ancestors of catarrhine (hang-nose) Old World monkeys and apes split from those of platyrrhine (broad-nose) New World monkeys: est. 36 Myr. New World primate fossils are known from South America **35 Myr** (early Oligocene), but how they reached there is unclear; quite modern-looking catarrhine primates (propliopithecids) are known from **29 Myr** (late Oligocene; Fayum, Egypt).

ancestral apes split from Old World monkeys: **est. 21 Myr**. Despite suggestions that monkeys and apes can be distinguished c. 29 Myr (Egyptian material), surer evidence are a monkey trace fossil c. 20 Myr (early Miocene, Uganda) and dryopithecine ape fossils c. 20 Myr that became widespread in Eurasia and Africa. Some monkeys evolved into distinguishable baboons, living in relatively dry environments, by **c. 4 Myr**.

ancestral great apes split from lesser apes (gibbons and siamang): est. 12 Myr. Here the fossil evidence is discordant with genetic evidence if identifications have been correctly made, namely, *Epipliopithecus*, gibbon, c. 16 Myr (Vindobonian in late Early Miocene; Europe) and *Micropithecus*, siamang, c. 14 Myr (early Middle Miocene; Africa).

orangutans split from ancestral apes: **est. 10 Myr**. The modern orangutan seems to be related to *Sivapithecus*, c. 8 Myr (*see* above).

ancestral apemen (near-human hominids) split from chimpanzees and gorillas: **est. 5 Myr**. African traces of candidate apemen may be somewhat older than 4 Myr, but the prime early apeman fossils are **c. 4.0 Myr** (*see* australopithecines).

humans evolved from apemen: **c. 2.0 Myr** (fossil evidence, Africa) (*see* human origins).

printed textiles 8500 yr, *see* arts and crafts; *printing* **est. 1300 yr (AD 700)**; *printing press* **esp. 545 yr (AD 1455)**, *see* information technology.

prosauropods 225 Myr. *See* dinosaurs.

proteins **est. 4450 Myr.** *See* life on Earth. For applications in dating, *see* human genetics, molecular dating.

Proterozoic era **2500 to 570 Myr.** *See* geological timescale.

Protestantism esp. 479 yr (AD 1521). *See* religious surge.

proto-animals (single-celled, protozoa) **est. 1700 Myr**. *See* life on Earth.

proto-languages *See* languages.

proto-Levallois tools 700,000 yr. *See* stone tools.

proto-modern humans (anatomically *Homo sapiens sapiens*) **c. 100,000 yr**. *See* human origins.

proton era (Hadronic era) **10^{-6} to 10^{-4} second** after origin. *See* Big Bang.

proto-plants (single-celled true algae ancestral to higher plants) **est. 1500 Myr**. *See* life on Earth.

protozoa (single-celled proto-animals) **est. 1700 Myr.** *See* life on Earth.

pterosaurs **225 Myr.** *See* vertebrate animals.

public-health measures **183 yr (AD 1817)**, esp. 134 yr (AD 1866). *See* medicine.

Puchezh-Katunki crater 183 Myr. *See* cósmic impacts; also map p. 84.

Purgatorius (early primate) **69 Myr.** *See* primates.

purple bacteria **est. 1800 Myr.** *See* bacteria; life on Earth.

pyramids, Egyptian 4600 yr; *Amerindian* 2800 yr. *See* building.

Q

quantum mechanics **AD 1926.** Following the foundation of the quantum theory (M. Planck and A. Einstein, 1900 to 1905), the development of Special and General Relativity (A. Einstein, 1905 to 1915), and the discovery of the atomic nucleus and the structure of atoms (E. Rutherford and N. Bohr, 1911 to 1913), a profound shift in thinking about the behavior of matter and energy initiated modern physics. This was the formulation in the late 1920s of quantum mechanics, in which particles were seen to behave in a statistical, chancy manner. The best marker is E. Schrödinger's equation of **1926**, but other outstanding contributors to the development of quantum mechanics, 1924 to 1930, were L. de Broglie, W. Pauli, W. Heisenberg, M. Born, and P. A. M. Dirac (M. Born, *Atomic Physics*, Blackie, 1946).

quark era **10^{-35} to 10^{-6} second** after origin. *See* Big Bang.

quasars **12,200, esp. 11,000 Myr.** *See* galaxies.

Quaternary period **1.8 Myr** to present. *See* geological timescale.

R

rabbits **40 Myr.** *See* mammals.

radar **esp. AD 1939.** *See* information technology.

radio esp. AD 1901; *broadcasting boom* **AD 1922**. *See* information technology.

radioactivity (discovery) **104 yr (AD 1896)**. That uranium emitted rays capable of penetrating paper and fogging a photographic film was discovered by H. Becquerel in Paris in **1896**, but its true import (energy from "nowhere") did not become clear for several years. P. Curie, M. Curie, and others discovered further radioactive elements; E. Rutherford and F. Soddy established the laws of radioactive decay; and A. Einstein cited the phenomenon in 1905 as an example of the equivalence of matter and energy. In one direction, radioactivity led to nuclear physics and the release of nuclear energy (*see* energy; science); in another, it made possible the objective dating of past events (*see* geological timescale; radiocarbon dating; radiometric dating).

radiocarbon dating (carbon-14 dating) Cosmic rays create radioactive carbon-14 atoms in the atmosphere and living things go on absorbing them until the time of their death; thereafter, the carbon-14 atoms decay with a half-life of 5730 years, reducing the ratio of radioactive to normal carbon. (For general principles, *see* radiometric dating.) Conventional procedures make a chemical compound from the carbon in a sample of plant or animal remains, and count the high-energy electrons released by radioactive decay. One new technique developed in the late 1970s enriches the radiocarbon from the sample by diffusion or with lasers; another uses particle accelerators to detect traces of carbon-14 by its mass instead of measuring its abundance by its radioactivity.

The natural supply of radiocarbon is affected by magnetic activity in the sun and the Earth, and by the climate, so radiocarbon dates have to be calibrated, where possible. This was accomplished for the past 7000 years using a very long tree-ring series. The result was a wiggly graph showing that radiocarbon ages could sometimes understate the calendar ages by hundreds of years, and that during certain periods several different calendar ages might give the same radiocarbon age.

Carbon dioxide stored in the sea is somewhat older than in the atmosphere, and bones of fish-eating people will date a few centuries earlier than they should. Trees take time to grow, and constructional wood may have been recycled by prehistoric builders, so that wood-derived dates may be typically a hundred years too old. Contamination of samples by fossil fuels can make them appear older than they really are, while modern contamination may make the dates too young. The age of a sample also depends on the laboratory. In exemplary self-examination, twenty radiocarbon laboratories in eleven countries dated, by almost identical procedures, eight replicate samples of wood of different ages from a drowned forest. The results showed discrepancies going beyond the ranges of quoted statistical errors, and some laboratories gave systematically biased dates (International Study Group, *Nature*, Vol. 298, 1982, p. 619). More generally, a single radiocarbon date may be misleading for any of a number of reasons, so the prudent investigator likes to see it embedded in a series of dates falling in a sensible sequence.

Gratuitous confusions then enter the picture. The least of them is that when a researcher gives a date as so many years "before present" (BP) he means before AD 1950; in this book the reference year is AD 2000. One convention is to write "BP," "AD," or "BC" for a calibrated date and "b.p.," "a.d.," or "b.c." for an uncalibrated one, but it is not always followed. Most troublesome are the altered and realtered estimates of the half-life of radiocarbon. W. F. Libby, who pioneered radiocarbon dating in 1949, gave its half-life as 5720 years; the figure was promptly amended to 5568 years, and in the 1950s and 1960s thousands of dates were computed using that half-life. By the early 1970s it became clear that Libby's original value was nearer the mark; the currently accepted value is 5730 years.

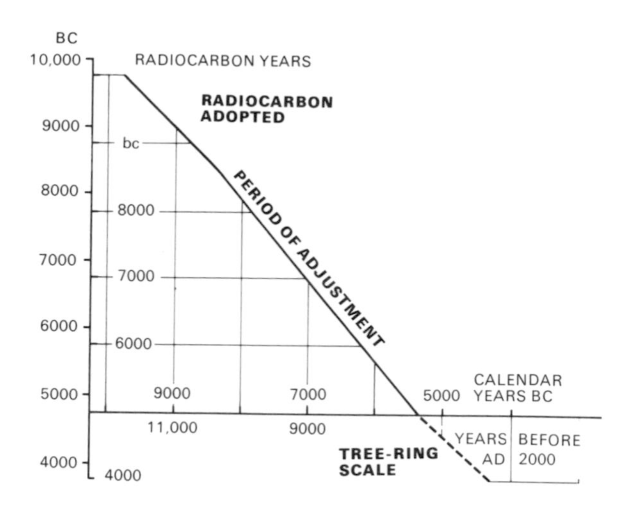

Adjustments to radiocarbon dates are needed between the limit of tree-ring calibration, 7350 yr ago, when a 600-year discrepancy exists, and 10,300 yr ago, when the dates are assumed to be correct. The graph shows how the dates are converted in this book, by joining the two points with a straight line. Radiocarbon dates BC *are those calculated for a 5730-year half-life (the preferred value), while dates b.c. use the conventional 5568-year half-life.*

By convention, raw, uncalibrated radiocarbon dates should always be quoted using the "old" 5568-year half-life even though it is known to be wrong (J. G. D. Clark, *Work Prehistory*, Cambridge University Press,

EXAMPLES OF RADIOCARBON CALIBRATIONS (MASCA SCALE)

radiocarbon dates		*calibrated dates*		
5568-yr half-life	5730-yr half-life	BC/AD	BP years (before AD 1950)	yr ago (before AD 2000)
4560 b.c.	4750 (BC)	5350 BC	7300	7350
3830 b.c.	4000 (BC)	4600 BC	6550	6600
2860 b.c.	3000 (BC)	3640 BC	5590	5640
1890 b.c.	2000 (BC)	2330–2440 BC	c. 4330	c. 4380
920 b.c.	1000 (BC)	1110–1140 BC	c. 3070	c. 3120
a.d. 55	1 (BC)/(AD) 1	AD 70	1880	1930
a.d. 1030	(AD) 1000	AD 1040	910	960

1977). Not everyone adheres to this policy, or makes clear which half-life he or she is using. Various calibration scales are in use, and some authors do not even specify whether they are quoting calibrated or uncalibrated dates. An editorial staff member of a leading scientific journal commented in 1982 that he had "given up" trying to compel authors to be explicit.

For uncalibrated radiocarbon dates back to 4555 b.c. (5568-year half-life) or 6750 yr (5730-year half-life), calibrations in this book follow the MASCA scale issued by the Applied Science Center for Archeology of the University of Pennsylvania's Museum (*MASCA Newsletter*, Vol. 9, No. 1, 1973). The scale is too elaborate to reproduce here, but some sample dates are given in the table.

For the more remote past, one study collated radiocarbon calibrations back to 32,000 yr ago, cross referred them to uranium-series dates, and suggested that the discrepancies were always less than two thousand years (M. Stuiver, *Nature*, Vol. 273, 1978, p. 271). Drastic changes in the Earth's magnetism (Lake Mungo excursion) about 33,000 yr ago influenced the rate of production of carbon-14 (M. Barbetti and K. Flude, *Nature*, Vol. 279, 1979, p. 202), causing possible discrepancies of several thousand years, around 30,000 yr ago. On the other hand, work at the University of Groningen reconciled orbital, deep-sea core, and radiocarbon data back to 70,000 yr ago, with discrepancies remaining less than about four thousand years (G. M. Woillard and W. G. Mook, *Science*, Vol. 215, 1982, p. 159).

In the absence of firmer information, radiocarbon ages of more than 10,300 yr are here assumed to be correct (5730-year half-life) and the gap between that time and the start of the MASCA scale is bridged as shown in the diagram.

radiometric dating The use of radioactivity in dating rocks, archeological remains, and other objects has established new timescales for the Earth, life, and early human activities. For comments, see p. 26. The nucleus of a radioactive atom decays by expelling a piece of atomic radiation and changing its own identity, usually to that of a different chemical element. Uranium-238 decays into lead with a half-life roughly equal to the age of the Earth, meaning that half the atoms orginally present have decayed since the Earth formed. Any radioactive sample diminishes gradually through a succession of half-lives: to one half, one quarter, one eighth, one sixteenth, and so on, in equal intervals. The first "daughters" of uranium-238 are themselves radioactive. Other short-lived radioactive atoms are produced in the atmosphere by natural cosmic rays (atomic particles from outer space) and latterly by tests of nuclear weapons.

The range of ages of samples that each radioactive species can usefully date depends on its half-life: for example, samarium-147 decays to neodymium with a half-life of 110,000 million years, and it is used mainly in dating old rocks, while hydrogen-3 (tritium, half-life 12 years) can detect rainfall supplying underground water, just a few months ago. For an awkward period, roughly 1 Myr to 50,000 yr ago, suitable radiometric elements are scarce, except for some daughters of uranium, as in uranium-series dating. One technique that helps to bridge the gap depends on the natural splitting of uranium nuclei, rather than on radioactive decay (*see* fission-track dating). Cosmic rays in the atmosphere produce several promising materials with half-lives of 130,000 years to 1.5 million years: calcium-41, chlorine-36, aluminum-26, and beryllium-10 (R. E. M. Hedges, *Nature*, Vol. 281, 1979, p. 19). Certain useful radioactive atoms are dealt with separately (*see* potassium-argon dating; radiocarbon dating; rubidium-strontium dating; uranium-series dating; uranium-lead dating).

Figuring out how much of a radioactive constituent has decayed in a sample gives a date for the event that last "set" the radioactive clock. Dating of natural glassy beads, tektites scattered by a cosmic impact in Europe, illustrates the power and scope of radiometric dating. By potassium-argon dating and fission-track dating, the tektites formed 14 Myr ago; by rubidium-strontium dating, the sedimentary rocks from which they were made were laid down 20 Myr ago; by samarium-neodymium dating, the material from which the sedimentary rocks were themselves derived dates from 900 Myr ago. Comparisons with ages of sediments and basement rocks near the Ries crater in Germany make it highly probable that the crater and the tektites record the same event.

railroads esp. 175 yr (AD 1825). *See* transport.

ramapithecines 15 Myr. *See* primates.

rats c. 650,000 yr, *see* mammals; *agents of plague* **669 yr (AD 1331)**, *see* disease epidemics.

recooling 10,900 yr. *See* climate 128,000 yr (Younger Dryas).

Red Sea rifting 30 Myr; *ocean-like* **3.5 Myr**. *See* continents.

red shift *See* cosmological timescale.

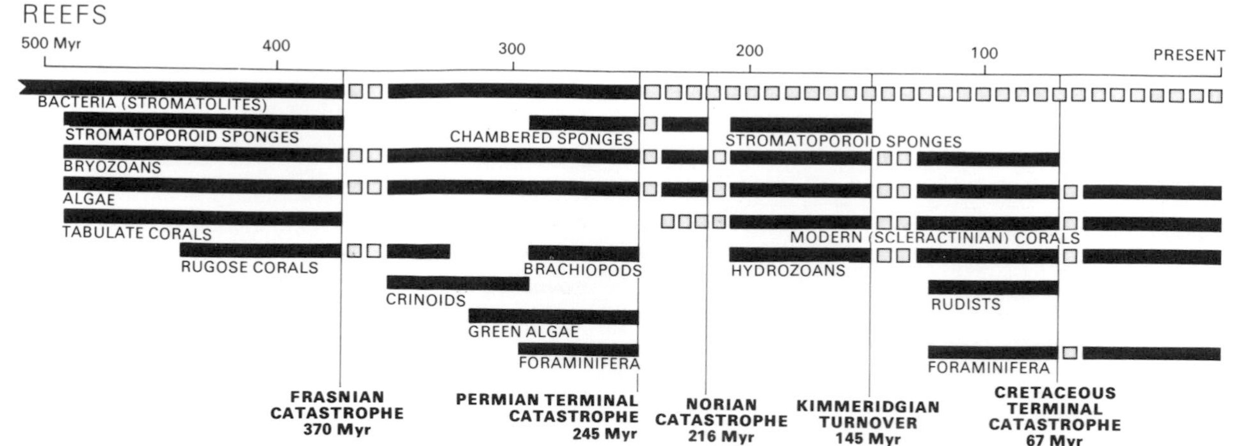

Changing fortunes of organisms that built banks and reefs in the sea help to identify major crises in life on Earth. See *mass extinctions; and individual crises by name.*

reefs (living) **3500 Myr** to present. Coral reefs, which alarm mariners in tropical waters and accommodate extraordinary varieties of plants and animals, are the most familiar living reefs of our own period. The Great Barrier Reef, two thousand kilometers long, has been growing on the northeast continental shelf of Australia since c. 20 Myr ago (Miocene times), while that continent has nosed northward into the tropics. Fossil reefs made by quite different reef builders are found, high and dry, in many continental strata. Each indicates a tropical or subtropical climate prevailing at the time and place of its formation. The earliest living reefs were built by photosynthetic bacteria **3500 Myr** ago (*see* stromatolites); such reefs have been scarce since 245 Myr ago (end of Permian). Archaeocyathids, cup-shaped, sponge-like animals, flourished in reefs 570 Myr ago (Early Cambrian) but the Cambrian disruptions, beginning 550 Myr ago, eclipsed the archaeocyathids; most were gone by 530 Myr (end of Middle Cambrian). Rugose corals built the first true coral reefs, 435 to 325 Myr (early Silurian to end of Mississippian in Carboniferous). Various other organisms have been conspicuous as builders of banks and reefs in various intervals, during the past 500 million years. (*See* the diagram: data adapted, with adjustment of dates, from N. D. Newell, *Scientific American*, Vol. 224, No. 6, 1971, p. 64.)

Reformation (Protestant) 479 yr (AD 1521). *See* religious surge.

reindeer **est. 200,000 yr.** *See* deer.

relativity **AD 1905.** The Special Theory (A. Einstein, 1905) established that high-speed motion could cause discrepancies between clocks; in the same year Einstein showed that matter was frozen energy ($E = mc^2$; for applications to origins, *see* Big Bang). The General Theory of Relativity (1915) is the current theory of gravity, in which gravity deflects light and makes clocks run slow (N. Calder, *Einstein's Universe*, Viking, 1979).

religious surge **2550 yr.** Hinduism is the oldest of the classic religions widely followed today; it pooled elements of proto-Indo-European and indigenous beliefs in India, est. 3600 yr ago. The religious surge 2550 yr ago is not well understood but even a short list of teachers of that time is striking (data adapted from G. Barraclough, ed., *The Times Atlas of World History*, Times Books, 1978).

Lao-tzu fl. 2550 yr (Taoism)
Zoroaster fl. 2550 yr (Zoroastrianism)
Deutero-Isaiah fl. 2540 yr (Judaism)
Mahariva fl. 2540 yr (Jainism)
Pythagoras fl. 2530 yr (numerology)
Buddha fl. 2500 yr (Buddhism)
Confucius fl. 2500 yr (Confucianism)

Christianity is here dated by the birth of Christ **2007 yr (7 BC)**, because of its calendrical significance; the date adopted follows a persuasive astronomical analysis by D. W. Hughes (*The Star of Bethlehem Mystery*, Dent, 1979). Later key events were the legalization of Christianity in the Roman empire 1387 yr (AD 313), and the Protestant Reformation initiated by Luther 479 yr (AD 1521). The Muslim calendar (which does not conform to astronomical years), starts from the Hegira of Mohammed 1378 yr (AD 622); here Islam is dated geopolitically from the military breakout of the Arabs **1368 yr (AD 632)**.

Renaissance (in Italy) **esp. 500 yr (AD 1500).** *See* historical timescale.

reptiles **313 Myr.** *See* vertebrate animals.

repulsion, grand **10^{-38} second** after origin. *See* Big Bang.

revolutionary nationalism **224 yr (AD 1776)**; revolutionary one-party government AD 1917. *See* government.

rhinoceroses **35 Myr.** *See* mammals.

rice domesticated **est. 9500 yr.** *See* crops.

Riphean period **1700 to 680 Myr.** *See* geological timescale.

RNA (ribonucleic acid) est. 4450 yr. *See* life on Earth; molecular dating.

roads (highways) c. 5500 yr. *See* transport.

robust apemen **c. 2.2 Myr.** *See* australopithecines.

rockets (liquid-fuel) AD 1926. *See* spacecraft.

rocks, oldest **3800 Myr.** *See* continents.

Rocky Mountains c. 60, **esp. 8 Myr**. *See* continents; North American uplifts.

rodents **55 Myr.** *See* mammals.

Roman breakout **2260 yr**, *see* breakouts; *empire* **esp. 2027 yr** (Augustus); western empire fell **1524 yr (AD 476)**. *See* historical timescale; language.

Romance languages (origins) by 2000 yr. *See* language.

ropes **17,000 yr.** *See* arts and crafts.

rubber domesticated 128 yr (AD 1872). *See* crops.

rubidium-strontium dating Rubidium-87's half-life of 53,000 million years is much longer than the span of the universe so far. This radioactive species changes at its leisurely pace, into strontium-87. The relative abundances of these, and of other forms of rubidium and strontium, indicate the ages of rocks. Although the technique is most suitable for rocks billions of years old, improved methods have brought the lower end of the useful span to less than a million years. Comparisons of the ages of meteorites given by uranium-lead and rubidium-strontium methods suggest that the half-life of rubidium-87 ought to be increased by 1 percent (J.-F. Minster, et al., *Nature*, Vol. 300, 1982, p. 414).

Russians *proto-Slavic language* by 2000 yr, *see* language; *regional population estimates* 2000 yr, *see* population history; *state founded* 1118 yr (AD 882); *breakout* **414 yr (AD 1586)**, *see* breakouts (Vikings; Russians); *Tsars* **522 yr (AD 1478)**; *revolutions* **AD 1917**; *Stalin* **AD 1924 to 1953**, *see* historical timescale; *carve-up of Asia* **esp. AD 1727**; *of Europe* AD 1945, *see* carve-up of continents.

S

Sangamonian warm interlude 128,000 yr. *See* climate 128,000 yr.

Sanskrit by 3000 yr. *See* language.

Sasanian empire (Iran) **1776 yr (AD 224).** *See* historical timescale.

satellites, man-made AD 1957. *See* spacecraft; transport.

sauropods 225 Myr. *See* dinosaurs.

scavenging **2.0 Myr.** *See* human origins.

science **2600 yr** etc. Inquiries into natural phenomena may have begun with the neanderthalers (*see* medicine), and intuitive and systematic exploration of natural laws and processes must go back at least to the origin of modern humans est. 45,000 yr ago. Premodern science prospered in the Greek and Hellenic cultures from **2600 yr** (Thales) to 1850 yr (Ptolemy), notably in geometry **c. 2300 yr**. Chinese science can be dated for an assembly of experts at Ch'ang-an **1996 yr (AD 4)**, and it flowered in Sung times 1040 to 721 yr (AD 960 to 1279). Islamic science entered its peak phase **c. 1000 yr (AD 1000)**, when Avicenna was at work and optical studies were leading to Alhazen's treatise on mirrors and lenses (c. AD 1038).

By modern-minded experimentation and reasoning, Galileo in Italy articulated the law of equal and uniform acceleration under gravity, c. AD 1604, contradicting Greek ideas on motion and launching classical physics. Telescopes of indifferent quality appeared in Europe c. 1607, but Galileo made an improved one and turned it on the sky **391 yr (AD 1609)**, discovering the ruggedness of the moon, the moons of Jupiter, and the phases of Venus. His telescope is a tangible symbol of the new philosophical order. Other early European discoveries included the circulation of the blood **372 yr (AD 1628)** at the start of modern physiology; the pressure of the air **357 yr (AD 1643)** initiating modern meteorology; vacuum **346 yr (AD 1654)** launching vacuum physics and technology; and microscopic observations **c. 340 yr (AD 1660)** at the start of micro-anatomy.

Modern science was formalized by the granting of a royal charter **338 yr (AD 1662)** to the Royal Society of London, the first durable scientific society. The experimental philosophers had invented a social system for announcing and criticizing discoveries, and the charter endorsed a special relationship between scientists and rulers. This relationship was always a delicate one: while seemingly secure in northwestern Europe, it was repeatedly compromised elsewhere.

For a limited selection of subsequent scientific discoveries (as noted on the timescale) *see* gene structure; germ theory of disease; gravity, theory of; medicine (and cross references therein); oxygen discovered; quantum mechanics; radioactivity; relativity; spacecraft. Other items mentioned in the narrative include the theory of evolution by natural selection (C. Darwin and A. R. Wallace, **1858**), electromagnetic theory (J. C. Maxwell, **1862**), atomic nucleus (E. Rutherford, **1911**), neutron (J. Chadwick, **1932**), antimatter (C. Anderson, **1932**), meson theory (H. Yukawa, **1932**), charm confirmed (G. Goldhaber, et al., **1976**), and the W particle discovered (CERN, **1983**).

scorpions **395 Myr.** *See* invertebrates.

sculpture **29,000 yr.** *See* arts and crafts.

seafaring **40,000 yr.** *See* breakouts; transport.

sea-level fall **29 Myr**: the most pronounced drop of the Tertiary period. *See* climate 67 Myr; sea-level stratigraphy.

sea-level stratigraphy The oil prospectors' technique of seismic sounding uses explosives, air guns, or thumping devices that generate waves in the Earth's crust, revealing characteristic surfaces (unconformities) within deep-lying deposits, produced by worldwide retreats of the sea. The surfaces can be dated by samples recovered by drilling. This information is then useful in exploration for oil (P. R. Vail and R. M. Mitchum, Jr., *Proceedings of the Tenth World Petroleum Congress*, Vol. 2, 1979, p. 95).

For most of the Earth's history the sea level has been higher than at present, with the flooded areas of continental shelf being typically twice as large as they are now. Indeed, the only known phases in the past 500 million years when the sea has been persistently as low as it is now were c. 330 Myr (mid-Carboniferous), c. 280 Myr (Early Permian), c. 200 Myr (Early Jurassic), and episodically since 30 Myr ago (mid-Oligocene).

Gradual variations in sea level make good sense in relation to the waxing and waning of the ridges at mid-ocean rifts, which displace a great deal of water. High or rising sea levels 430 to 360 Myr and 170 to 70 Myr ago coincided with periods when oceans were growing and ocean rifts were very active; during the latter case, the sea level may have risen 200 or 300 meters above the present level. (*See also* continents flooded.) Very recent falls in sea level, recorded for the current series of ice ages, are also understandable, in view of the buildup of ice on land, which deprives the ocean of much water. Altogether more puzzling is the discovery of a large number of relatively quick falls in sea level, for which there is no obvious correlation with other trends or events; for comments see p. 34. A skeptical view of the rate and severity of the changes reported by Vail is expressed by A. Hallam in a study of the Early Jurassic (*Journal of the Geological Society*, Vol. 138, 1981, p. 735).

Major sea-level falls occurred about 400, 330, and 280 Myr ago. Falling sea-level events since 220 Myr ago are shown in the diagram. The most extraordinary fall

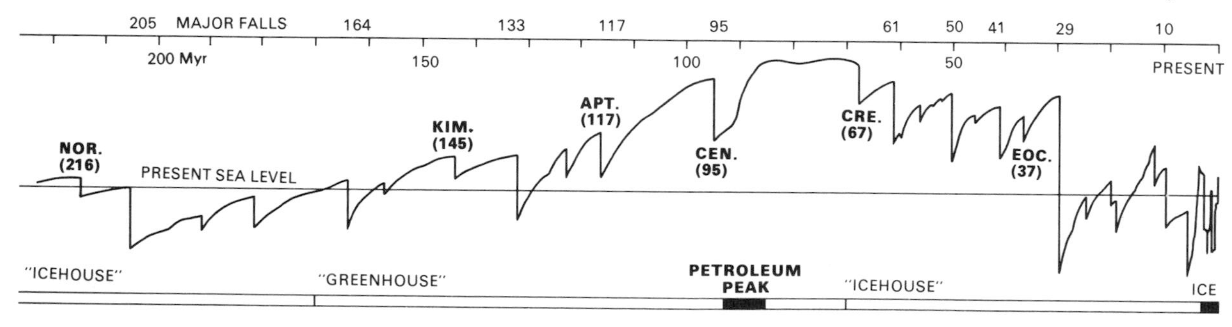

Sea level is an indicator of how the global environment has changed during the past 200 Myr, with sharp falls in sea level superimposed on a gradual rising and falling. During phases of high sea level, large areas of the continents were flooded. The events marked with abbreviations are catastrophes or turnovers in life, namely the Norian, Kimmeridgian, Aptian, Cenomanian, Cretaceous terminal, and Eocene terminal; note that they were typically associated with minor falls in sea level. Data are from Exxon (P. R. Vail, et. al) with dates adjusted to conform with the geological timescale adopted here. The 123 Myr event is added (Gulf Oil). The heights of the sea are not precisely specified, but they ranged about two hundred meters above and below the present sea level.

occurred 29 Myr ago, apparently associated with the onset of cold conditions in Antarctica.

seals (animals) **?25 Myr**. *See* mammals.

seals (stamp) 8500 yr; (cylinder) 5700 yr. *See* information technology.

sea pens 670 Myr. *See* animal origins.

sea urchins 470 Myr. *See* invertebrates.

seaweed c. 1000 Myr. *See* life on Earth.

Semitic (languages origin) **c. 6000 yr**. *See* breakouts (dairymen); languages.

sex ?1000 Myr. *See* life on Earth.

Shang culture esp. 3800 to 3100 yr. *See* archeological timescale; breakouts (charioteers).

sharks 380 Myr, *giant* c. 4 Myr. *See* vertebrate animals.

sheep c. 1.8 Myr, *see* bovids; *domesticated* **10,700 yr**, *see* animal domestication; *woolly* **c. 7300 yr**, *see* livestock revolution.

shellfish collecting ?420,000, esp. 128,000 yr. *See* fishing.

shelly (hard) animals 600 Myr. *See* animal origins.

ships *See* transport.

Siberia hit Europe (Euramerica) c. 300 Myr, *see* continents; *peopled* (modern humans) c. 32,000 yr, *see* breakouts.

silk 6000 yr. *See* animal domestication (silkworm).

Silk Road c. 2106 yr. The opening of the first transcontinental silk road across Asia was marked by "through" caravans from China to Persia; this event was preceded by the Chinese establishment of fortified posts strung westwards (2120 yr); the route then passed through the lands of the Huns before reaching Persian-controlled territory. (J. Needham, *Science and Civilisation in China*, Vol. 1, Cambridge University Press, 1956.) Illus. p. 45.

Silurian period 435 to 410 Myr. *See* geological timescale.

Sinhalese (Aryan speakers to Sri Lanka) c. 2500 yr. *See* breakouts (charioteers); languages.

sivapithecines fl. 10 Myr. *See* primates.

slash-and-burn cultivation ?2500 yr. *See* cultivation practice.

slavery esp. 2800 yr. *See* work.

slaves-and-sugar trade c. 440 yr (AD 1560). The Portuguese invented the triangular Atlantic trade that took miscellaneous goods (notably textiles and trinkets) from Europe to Africa, carried slaves from Africa to the Americas (Brazil, in the first instance), and brought sugar from the Americas to Europe. From AD 1625 onward, first Dutch and then British and French traders and settlers (notably in the Caribbean) exploited the same markets. The total flow of slaves arriving in the Americas from Africa between 1600 and 1870 was about 8 to 10 million (mainly from west Africa), of which more than 80 percent went to Brazil or the West Indies; about 5 percent were imported into North America, where conditions were less deadly for the slaves. (P. D. Curtin, personal communication, 1981; A. Calder, *Revolutionary Empire*, Cape, 1981.)

Slavic (languages origins) by 2000 yr, *see* language; *breakouts* c. 1450 yr (AD 550), etc., *see* breakouts (Slavs; Russians).

sleds (animal-drawn) ?7800 yr. *See* livestock revolution; transport.

smallpox epidemics ?1749 yr (AD 251), etc., *see* disease epidemics; *cowpox vaccination* **204 yr (AD 1793)**; *smallpox eradicated* **AD 1977**, *see* medicine.

snakes esp. 30 Myr. *See* vertebrate animals.

social behavior (evolving) **?400,000 yr**, *see* human origins; social systems 29,000 yr, etc.; *social stratification* (class distinctions) **esp. 5600 yr**, *see* government.

solar elements 10,000 Myr. Average age of elements heavier than hydrogen or helium, in the solar system and on Earth. *See* elements.

solar system forming 4550 Myr. This date is the same as the age of most meteorites, which give the only direct indication of when the solar system began (*see* Earth forming). The assumption is that the sun, its family of planets, and the meteorites all came into existence at about the same time. Support comes from evidence that the Earth's uranium was gathered together approximately 4600 Myr ago and that no known rocks from the moon are as old as 4600 Myr.

A few million years before the meteorites formed, a massive star exploded nearby (*see* presolar supernovas) and presumably triggered the collapse of the cloud of gas, dust, and ice that made the solar system. Completing the sun may have taken 10 million years (to **4540 Myr**), and the concurrent construction of the Earth and other planets 100 million years (to **4450 Myr**). Some meteorites look as if they were formerly embodied in small planets at least a hundred kilometers wide, so the implication is that objects of at least that size were orbiting the embryonic sun within a few million years of the start of the process, together with icy comets.

The general picture favored by astronomers has the sun bursting into light in the midst of a swirling disk of gas and dust, the solar nebula, which was flattened by the combined effects of gravity, centrifugal action, and the drag of the gas on the dust. Out of this disk the planets formed. It is possible that each planet formed directly in one episode of gravitational collapse within the solar nebula, but cited for preference is the theory of planetisimal growth in which the inner rocky planets (including the Earth) are supposed to have formed by the gradual coalescence of swarms of smaller bodies, like those in the asteroid belt between Mars and Jupiter. The theory was developed by V. Safronov in Moscow in the 1960s and pursued in computer models in the United States and Japan in the 1970s.

The astronomers visualize dust grains sticking together by simple attraction (Van de Waals force) to make thimble-sized particles which then grow by collision into small boulders a meter wide. These in turn gather together with the residual dust to make microplanets (planetisimals) a kilometer or so in diameter. The computational problem is then to check whether further growth is possible, despite the chances that the planetisimals scatter when near misses occur, or smash themselves into smaller pieces when they collide. The answer is that they can grow, and tend to form a small number of massive bodies at different distances from the sun (G. W. Wetherill, *Annual Reviews of Astronomy and Astrophysics*, Vol. 18, 1980, p. 77).

sound recording esp. AD 1901. *See* information technology.

soup (prebiotic) **est. 4450 Myr.** *See* life on Earth.

South African gold **esp. 2600 Myr**, *see* ores; *gold rush* 114 yr (AD 1886), *see* gold rushes.

South America (geology) *ores* 2600 Myr, etc., *see* ores. *Transamazonian Mountains* **c. 2000 Myr** and other mountain-building events; *part of Gondwanaland* 500 Myr; *quit Africa* 130, **esp. 85 Myr**, *see* continents; *Andes* esp. 100 and **4.5 Myr**, *see* Andes; *peculiar mammals* 85 Myr, *see* mammals; *glaciers* **6.6 Myr**, *see* climate 67 Myr. See maps pp. 81–87. *See also* American Interchange; American intrusive peoples; Amerindians; petroleum.

Soviet Union *See* Asia; Russians.

spacecraft **AD 1957.** *Sputnik 1* (U.S.S.R.) went into orbit in October 1957. R. H. Goddard (1926, U.S.) had launched the first small rocket using liquid propellants, and he dreamed of space travel. The first large liquid-propellant rocket, the German military V-2, was operational in 1944 with a range of about 300 kilometers. After 1945, the V-2 became the basis for developments in other countries, especially the United States and the U.S.S.R., where ballistic missiles for carrying nuclear weapons were under development in the 1950s. Meanwhile, preliminary space research began, using V-2s and other vehicles as sounding rockets to probe the outer atmosphere. The possibility of launching artificial satellites to orbit the Earth was plain to all concerned; the Americans began the Vanguard project in 1954, but the Russians used a converted intercontinental ballistic missile to launch *Sputnik 1* first. Other countries to fly satellites (using their own launchers) were then as follows: United States (*Explorer 1*), 1958; France (*Asterix*), 1965; Japan (*Osumi*), 1970; China (*China*), 1970; Britain (*Prospero*), 1971; India (*Rohini*), 1980; European Space Agency (joint venture: *Meteosat 2*), 1981.

Applications for unmanned spacecraft included intelligence gathering, missile detection, navigation, meteorology, communications, earth survey, antisatellite systems, astronomy, and general space research. Long-range scientific space probes for the exploration of the solar system began with *Luna 1* (U.S.S.R.) 1959 and led, for example, to the American *Viking 1* and *2* Mars missions **1976**. Each of the *Viking* craft consisted of an orbiter and a lander; the orbiter generated survey pictures of the planet's surface and moons, while the landers sent pictures from the surface and carried out a variety of experiments on the soil and the atmosphere—showing, in particular, that there was no sign of life. The *Voyager 1* and *2* missions to the outer planets, launched **1977**, sent back pictures and data from Jupiter and Saturn, and *Voyager 2* was intended to go on to Uranus and Neptune (1990). *See also* information technology (communications satellite); weaponry (military satellites).

spaceflight, manned **AD 1961**, etc. *See* transport.

space shuttle **esp. AD 1982.** *See* transport.

spacetime **13,500 Myr, esp. 10^{-38} second** after origin. *See* Big Bang (inflation era); cosmological timescale.

Spanish *language* after 1000 yr (AD 1000), *see* language; *overseas empire* (Hispaniola) **507 yr (AD 1493)**, *see* breakouts (navigators); *carve-up of world* 506 yr (AD 1494), *see* carve-up of continents; *Latin American revolutions* AD 1808, etc., *see* historical timescale.

speech est. 45,000 yr. *See* human origins; mental revolution.

spiders 395 Myr. *See* invertebrates.

spiral-arm elements est. 9000 Myr. Average age of elements heavier than hydrogen or helium, in interstellar clouds of the spiral arms of the Milky Way Galaxy. *See* elements.

sponges 580 Myr, *see* animal origins.

springtail (first known insect) **395 Myr**. *See* invertebrates.

***Sputnik* AD 1957.** *See* spacecraft.

squashes domesticated ?8000 yr. *See* crops.

stars ?13,500 **esp. 12,500 Myr**. Understanding cosmic history depends largely on modern conceptions of the life cycles of stars, the commonplace entities of the universe. Stars are massive aggregations of gas and dust ignited by gravity and capable of burning nuclear fuels—initially hydrogen, making helium. The oldest stars visible to astronomers are in globular star clusters orbiting around and through the Milky Way Galaxy. Estimates of their ages based on the astrophysical theory of stars range from 18,000 to 13,000 Myr. With an adopted age for the universe of 13,500 Myr (*see* cosmological timescale), consistency requires that the globular star clusters appeared quite quickly after the origin of the universe.

The making of stars relevant to the existence of the Earth, and of life, began in earnest with the formation of the Milky Way Galaxy, **est. 12,500 Myr** (*see* galaxies). Stars less massive than the sun can survive for far longer than 12,500 million years. On the other hand, more massive stars burn intensely and lead much shorter lives: for example, a star twenty times as massive as the sun survives for only a million years before destroying itself. The brightest stars noticeable in the sky are mainly very young, massive stars. The Orion nebula is a well-known nursery of stars, where the four bright stars called the Trapezium are thought to be only about 100,000 years old. Some regions of present star formation are characterized by unusual emissions of infrared radiation—a notable example is again in the Orion nebula, beyond the Trapezium. Very young stars can appear as dusty and unstable objects that flare up vigorously (as T-Tauri stars) at intervals of, say, ten thousand years: FU Orionis in Orion, a star seen for the first time by astronomers in AD 1936, is such an object.

Aged stars, on the other hand, run out of hydrogen fuel at their cores, and swell into red giant stars like Betelgeuse (also in Orion). Stars of the mass of the sun or smaller then puff off much of their substance in a planetary nebula and collapse to make white dwarf stars; more massive stars come to a dramatic end in a supernova explosion that manufactures heavy elements and flings many of them out into space, while the heart of the star collapses to make a neutron star (pulsar) or a black hole. (*See also* elements; galaxies; presolar supernovas.)

state 5100 yr. *See* government.

steam energy esp. 288 yr (AD 1712); *fossil-energy revolution* **175 yr (AD 1825)**. *See* energy; transport.

steel 3100 yr; *stainless* AD 1912. *See* metal industries.

stegosaurs 164 to 145 Myr. *See* dinosaurs.

Steinheim Man c. 230,000 yr. *See* human origins.

sterilization of galaxies (gas stripping) **esp. 6000 Myr**. *See* galaxies.

stirrups c. 2100 yr. *See* transport.

Stone Age 2.4 Myr to 6000 yr. *See* archeological timescale (Paleolithic, Neolithic, and Chalcolithic).

stone buildings 6500 yr; *temples* 5000 yr. *See* building.

stone tools 2.4 Myr. Careful searches for tools in the vicinity of early apemen fossils drew a blank (M. D. Leakey, *Philosophical Transactions of the Royal Society*, Vol. B292, 1981, p. 95). Man-made stone tools are, though, known from **2.4 Myr** ago (Omo, Ethiopia; G. H. Curtis, same volume, p. 7). These tools antedate fossil appearances of the earliest humans (*see* human origins). Marks on the cutting edges of ancient stone tools can be compared with marks in similar newly made implements applied to various tasks, such as slicing meat, scything plants, or cutting wood. Such microwear analysis established that the earliest humans ate meat (L. J. Kelley and N. Toth, *Nature*, Vol. 293, 1981, p. 464).

Subsequent improvements in stone tools depended less on experimentation than on evolutionary changes in the toolmakers themselves. Precursors can be found: for example, Levallois techniques, in which the face of a tool was struck whole in one bold action from a prepared stone, are detectable in African tools of c. 700,000 yr ago ("proto-proto-Levallois"), and more clearly in tools of c. 230,000 yr ago ("proto-Levallois"). Many experiments with wooden handles and shafts may have preceded the appearance of the refined composite tools of the Gravettians. The table notes widespread adoption of more-or-less perfected technologies by identifiable human groups. (Data from G. L. Isaac and L. G. Freeman, personal communications, 1981; J. A. J. Gowlett and A. G. Sherratt, personal communications, 1982; C. J. Jolly and F. Plog, *Physical Anthropology and Archeology*, Knopf, 1979; C. Bonsall, et al., *Man Before Metals*, British Museum, 1979; and other sources.)

EARLY STONE TECHNOLOGIES: TIMETABLE

early stone tools: **2.4 Myr**. Omo, Ethiopia, Oldowan type.

Oldowan: **2.0 Myr**. Pebble choppers and knife-like flakes (*Homo habilis*). Illus. p. 140.

Acheulian: **1.9 Myr**. Characterized by hand axes (early *Homo erectus* and archaic *Homo sapiens*, in southwestern Asia, Africa, southern Europe, and India. Illus. p. 140.

Developed Oldowan: **c. 1.6 Myr**. Pebble choppers and flakes (*Homo erectus* in Africa, but esp. in eastern and central Asia; also in Europe, c. 750,000 yr, in Molise). Illus. p. 144.

Levallois: **esp. 160,000 yr**. Characterized by prepared cores (archaic *Homo sapiens* and neanderthalers). Dated from Biache-Saint-Vaast, Europe. Illus. p. 152.

Mousterian: **c. 120,000 yr**. Toolkits of relatively high quality (neanderthalers and their contemporaries).

blades: **esp. 40,000 yr**. Delicate blades and flakes made using punches and other novel methods (modern humans). Illus. p. 158.

composite tools: **esp. 29,000 yr**. Handle fitting and spear tipping (Gravettian culture). Dated from Dolní Věstonice, Europe.

stratification, social **esp. 5600 yr.** *See* government (class distinctions).

streetcars (electric tramways) 119 yr (AD 1881). *See* transport.

stromatolites **3500 Myr.** As the oldest easily visible form of life on Earth, stromatolites consist of heaps of mats of primitive photosynthesizing bacteria growing as reefs on the seashore. Their fossils were impressive enough to be mistaken for animals (*Cryptozoon*) when first discovered in the nineteenth century. Even more impressive are reefs of stromatolites kilometers wide and hundreds of meters thick. The earliest fossil stromatolites occur in northwestern Australia c. 3500 Myr; they are conical in shape and contain filamentary bacteria, possibly resembling *Chloroflexus*, a green filamentous bacteria that build stromatolites in hot springs today (D. R. Lowe, *Nature*, Vol. 284, 1980, p. 441, and M. R. Walter, same volume, p. 443). Stromatolites of similar age have been found in Zimbabwe. Blue-green algae are known as builders of modern stromatolites, but there is no proof that the earliest stromatolites were built by them; on the contrary, molecular evidence suggests that the blue-greens evolved later (est. 1600 Myr, *see* life on Earth).

Stromatolites became very widespread as continental platforms began to stabilize c. 2500 Myr ago, and column-like forms developed. These increased in diversity from c. 2000 Myr onward until, by c. 850 Myr, there were almost a hundred known kinds of columnar stromatolites, probably reflecting diversity among their bacterial architects. This diversity diminished after 680 Myr ago, perhaps because early animals grazed on them and burrowed into them. The oldest (conical) forms of stromatolites were wiped out, and the variety of columnar forms was greatly reduced (S. M. Awramik, in M. H. Nitecki, ed., *Biotic Crises in Ecological and Evolutionary Time*, Academic Press, 1981). Stromatolites nevertheless continued as notable reef builders in intertidal regions until the Permian terminal event 245 Myr ago; since then, they have been scarce.

subatomic particles **10^{-35} second** after origin. *See* Big Bang.

submarines (warfare) esp. AD 1914; (nuclear-powered) AD 1955; (with ballistic missiles) **AD 1960**. *See* weaponry.

sugar cane domesticated **?9000 yr.** *See* crops.

Sui empire **1419 yr (AD 581).** *See* historical timescale.

Sumer (innovative center; southern Mesopotamia, modern Iraq) esp. 5800 yr (urbanization) onward, *see* cities; government; information technology; transport; *Sumerian trading colonies* esp. 5500 yr, *see* trading.

sun forming **4550 to 4540 Myr.** *See* solar system forming.

Sung empire **1040 yr (AD 960).** *See* historical timescale.

supercontinents **2300, 1500, 800, and 220 Myr**. *See* continents.

supercycle 430 Myr to present. *See* climate 3800 Myr.

supernovas **esp. 12,500, 4650, and 4550 Myr.** *See* elements; presolar supernovas; stars.

Swanscombe Man c. 230,000 yr. *See* human origins.

Swedish empire (Baltic) **esp. 342 yr (AD 1658).** *See* historical timescale.

sweet potatoes ?10,000 yr. *See* crops.

Swiss Alps c. 10 Myr. The term "Alpine" is often applied, somewhat confusingly, to a worldwide mountain-building phase 240 or 100 Myr to the present. Moreover, the activities that built the Alps of Europe have a long and complex history, involving collisions between various microcontinents and the European margin starting c. 60 Myr ago, in some cases resurrecting much older mountains. High mountains reportedly existed in Switzerland in Oligocene times (c. 30 Myr ago). The microcontinent including Italy pushed, tugged, and pushed again at Europe over a long period. The present conspicuous mountains of Switzerland and adjacent provinces began to rise c. 10 Myr ago, in the Middle Miocene or early Late Miocene (K. J. Hsü, et al., *Nature*, Vol. 267, 1977, p. 399). Italy had been caught in a convergence of Africa and Europe beginning c. 18 Myr ago, and was driven like a ramrod into Switzerland. Old faults were reactivated across Europe, raising minor mountains and hills, e.g., in southern England. (*See also* continents.) Illus. p. 134.

synthetic dyes 135 yr (AD 1865); *plastics* esp. AD 1910; *fibers* esp. AD 1935. *See* arts and crafts.

syphilis 506 yr (AD 1494). *See* disease epidemics.

T

tailored clothing 25,000 yr. *See* arts and crafts.

Tairona culture fl. 500 yr (AD 1500). Contemporaries of Incas in Colombia (*New Scientist*, Vol. 94, 1982, p. 512).

taming of animals *See* animal domestication.

T'ang empire 1382 yr (AD 618). *See* historical timescale.

Taoism c. 2550 yr. *See* religious surge.

taxes 5600 yr. *See* government; information technology.

teeth 425 Myr. *See* vertebrate animals.

tektites *See* cosmic impacts.

telegraph (electric) **esp. 156 yr (AD 1844).** *See* information technology.

telephone esp. 122 yr (AD 1878). *See* information technology.

telescope, astronomical 391 yr (AD 1609). *See* science.

television boom AD 1950. *See* information technology.

temples 7800 yr, *see* buildings; *temple-market communities* 7800 yr, *see* government.

Tertiary period 67 to 1.8 Myr. *See* geological timescale.

textiles (knitted) est. 17,000 yr; (woven) c. 8500 yr; (wholly synthetic) esp. AD 1935. *See* arts and crafts.

Thais' breakout c. 850 yr (AD 1150). *See* breakouts.

thaw started 14,000 yr. *See* climate 128,000 yr (Bølling).

theater esp. 2500 yr. *See* arts and crafts (formal drama).

thecodonts 240 Myr. *See* vertebrate animals.

therapsids (mammal-like reptiles) **256 Myr.** *See* vertebrate animals.

tigers 600,000 yr. *See* cats.

time *zero* at 13,500 Myr. *See* Big Bang, cosmological timescale.

tin ores 900 Myr, *see* ores; *tin bronze* **esp. 5000 yr**, *see* metal industries.

Toarcian turnover 182 Myr. In the Early Jurassic, a hesitation in the increase in diversity of marine animals after the Permian terminal catastrophe, with small declines in some groups, for example ammonites. It coincides with the date given for an impact crater, Puchezh-Katunki, c. 183 Myr. *See* cosmic impacts.

Toba explosion 73,000 yr. *See* volcanic events.

town 10,600 yr. *See* cities.

toxodonts 35 Myr. *See* mammals.

tractors esp. AD 1915. *See* cultivation practice.

trading c. 30,000 yr. Dated for a regular trade in flint from mines in Poland and Czechoslovakia, covering a wide area. Other early trading items were obsidian for toolmaking (going by sea in the Mediterranean by 10,000 yr ago), salt, bitumen, and ornamental shells. Lapis lazuli from Afghanistan showed up in Egypt 6000 yr ago, and Sumerians from southern Mesopotamia established trading colonies in northern Syria 5500 yr ago, exporting mass-produced pottery. (P. Phillips, *The Prehistory of Europe*, Allen Lane, 1980; J. Mellaart, *Antiquity*, Vol. 53, 1979, p. 6.) For selected aspects of later trading, *see* British Bengal; carve-up of Africa; Chinese naval ventures; Dutch spice monopoly; imperial companies, Silk Road, slaves-and-sugar trade.

Transamazonian Mountains c. 2000 Myr. *See* continents.

transatlantic communication (telegraph) **134 yr (AD 1866)**; (radio) AD 1901. *See* information technology.

transistor AD 1948. *See* information technology.

transparent universe **1,000,000 years** after origin. *See* Big Bang.

transport 40,000 yr to present. Seafarers crossed from Indonesia to Australia c. 40,000 yr ago (*see* breakouts) and by inference water transport was available from very early times (N. C. Flemming and P. M. Masters, eds., *Quaternary Shorelines and Marine Archeology*, Academic Press, 1983). It is difficult to believe that sailing craft were not invented until 6900 yr ago, when the first evidence exists. Homer's *Odyssey* (c. 2750 yr) was a mnemonic for pilotage data for the Mediterranean (E. Bradford, *Ulysses Found*, Hodder & Stoughton, 1963). In land transport, a technological scandal surrounds the early harnessing of horses as draft animals; for almost 2000 years no one seemed to notice that throat straps choked the horses and reduced their effectiveness.

Sources for the timetable included N. C. Flemming, personal communication, 1982; J. Needham, *Science and Civilization in China*, Cambridge University Press, various volumes, 1954 to 1966; J. Needham and C. A. Ronan, in B. Hook, ed., *Cambridge Encyclopedia of China*, Cambridge University Press, 1982; C. Burney, *From Village to Empire*, Phaidon, 1977; J. G. D. Clark, *World Prehistory*, Cambridge University Press, 1977; R. M. Adams, *Heartland of Cities*, University of Chicago Press, 1981; and *Encyclopaedia Britannica*, 1981.

TRANSPORT: TIMETABLE

seafaring: **40,000 yr**. *See above.*

animal-drawn sleds: ?7800 yr. Mesopotamia (*see* livestock revolution).

sailing craft: 6900 yr. Mesopotamia (model at Eridu), but *see* comment above. Chinese advances included the sternpost rudder c. 1950 yr, and the fore-and-aft lugsail c. 1750 yr.

animal-drawn wagon: **5500 yr**. Mesopotamia (*see* wheel).

highways: esp. 5500 yr. Mesopotamia to Anatolia. Note also the llama-train roads of the Incas through the Andes c. 550 yr (AD 1450), and expressways for motor traffic conceived in the AD 1920s (U.S., Italy, Germany).

pack animals: **esp. 5000 yr**. Asia (donkey, camels, elephant; *see* animal domestication). The use of llamas as pack animals in South America may be older.

horses: c. 5000 yr. Riding in eastern Europe (*see* horses). Chariots appeared c. 4100 yr in Cappadocia (*see* weaponry), and horse bits c. 3900 yr in Hungary. Improved Scythian horses had been bred by 2850 yr. In China a rational breast strap was introduced c. 2250 yr ago, and the horse collar c. 2000 yr ago. A big-toe stirrup was in use by riders in India c. 2100 yr ago, and foot stirrups in China by c. 1700 yr (AD 300).

canals: esp. 4200 yr. Mesopotamia, dated for a boom in canal building for irrigation (Third Dynasty of Ur), assuming secondary use for transport. The Maya in Middle America, from c. 2750 yr ago, carried goods by canal. A Chinese imperial network starting c. 2220 yr ago (Ch'in dynasty) culminated in the opening of the Grand Canal, **1389 yr (AD 611)**; it incorporated gates, sluices, and slipways. Oceans were linked by the Suez Canal in AD 1869 and the Panama Canal in 1914.

harbors, man-made: c. 3200 yr. Syria and Egypt, "protoharbors" made by modifying reefs. Piers date from c. 2800 yr (Phoenicians and Greeks).

navigational compass: **c. 885 yr (AD 1115)**. China. The earliest well-dated description of the use of the magnetic compass at sea falls between AD 1111 and 1117.

global navigation: **478 yr (AD 1522)**. The ship *Vittoria* completed the first circumnavigation (Magellan/Elcano). Map p. 92.

manned balloons: **217 yr (AD 1783)**. France (Montgolfier brothers).

powered ships: esp. 193 yr (AD 1807). United States, dating from the steamboat *Clermont* (J. Fulton). Other variants of powered ships included the hydrofoil (Italy, c. AD 1900) and the hovercraft air-cushion vehicle (Britain, 1959). For submarines, *see* weapons.

railroads: **esp. 175 yr (AD 1825)**. Britain, first passenger steam train service. The first subway (urban underground railroad) operated in London in AD 1863.

electric streetcars (trams): 119 yr (AD 1881). Germany (Berlin).

automobiles: **esp. 115 yr (AD 1885)**. Germany (C. Benz). The motor cycle dates from the same year (G. Daimler, Germany). Automobiles came into general use with the mass production of cheap machines, esp. AD 1908 (H. Ford, U.S.).

bicycle: **esp. 112 yr (AD 1888)**. Britain (pneumatic rubber tires, Dunlop).

airship: esp. AD 1900. Germany (F. von Zeppelin).

aircraft: **AD 1903**. United States (Wright brothers). The first international scheduled passenger flights may have been those between Vienna and Kiev, 1918. Experimental helicopters appeared from AD 1924 onward, but were not perfected until 1939 (I. Sikorsky, U.S.). Jet engines fitted to passenger and cargo aircraft from the 1950s made air transport safer, faster, and cheaper.

manned spaceflight: **AD 1961**. U.S.S.R. (circumnavigation by *Vostok 1*, Y. A. Gagarin, April 1961). Manned flights to the surface of the moon followed (*Apollo 11*, U.S., N. A. Armstrong and others, July **1969**, and five subsequent landings); also semipermanent space stations (*Salyut 1*, U.S.S.R., 1971, and *Skylab*, U.S., 1973). A reusable space shuttle proved itself in its first commercial flight (*Columbia*, U.S., November **1982**). (For unmanned spacecraft, *see* spacecraft.)

tree-ring dating *See* dendrochronology; radiocarbon dating.

trees 370 Myr. *See* plants ashore.

Triassic period 245 to 208 Myr. *See* geological timescale.

Triassic terminal event ?208 Myr. *See* mass extinctions; Norian catastrophe.

trilobites 560 to 256 Myr. *See* invertebrates.

triode (thermionic tube) **AD 1906**. *See* information technology.

Trojan War 3250 yr. *See* climate 10,300 yr.

tungsten esp. 140 Myr. *See* ores (tin).

Turgai Sea 160 to 29 Myr. *See* continents flooded.

Turkish *breakouts* **789 yr (AD 1211)** etc., *see* breakouts (Mongols and Turks); *language* (spread westward) esp. 760 yr (AD 1240), *see* language.

turnovers (of animals) **550 Myr**, etc. *See* mass extinctions.

turtles c. 230 Myr. *See* vertebrate animals.

type (movable) c. 960 yr (AD 1040). *See* information technology.

typewriter AD 1867. *See* information technology.

tyrannosaurs esp. 84 Myr. *See* dinosaurs.

U

United Nations AD 1945. *See* historical timescale.

United States *See* American intrusive peoples; North America (geology); North American uplifts.

universe 13,500 Myr; *quarter-size* **est. 6700 Myr**; *half-size* **est. 4000 Myr**, *see* cosmological timescale; *transparent* **1,000,000 years** after origin, *see* Big Bang.

Ural Mountains 300 to 260 Myr. *See* continents.

Uralic (languages origin) est. 4900 yr ago. *See* breakouts (northern dairymen); language.

uranium-lead dating Half of the uranium-238 atoms in a given sample will decay into lead-206 in 4510 million years, while the equivalent half-life of uranium-235 atoms is 713 million years; the decay product is lead-207. Careful interpretations of the relative abundances of different uranium and lead atoms in a rock permit these overall transformations to yield the ages of rocks. The uranium-lead method is suitable for very old rocks, but not for rocks less than about 10 Myr old. When attention focuses on the lead isotopes the method is called lead-lead dating. As uranium decays into lead, it forms intermediate materials with relatively short half-lives, including uranium-234 (247,000 years), protoactinium-231 (32,000 years), and thorium-230 (75,000 years). These serve in uranium-series (otherwise, uranium-disequilibrium) dating, measuring relatively young ages.

uranium reactors (natural) **c. 2000 Myr**. At Oklo in Gabon, Africa, some compact bodies of rich uranium ore engaged in spontaneous chain reactions which consumed at least six tons of uranium-235 and released the usual mess of fission products (R. Bodu, et al., *Comptes Rendus*, Vol. 275, 1972, p. 275; J. R. DeLaeter, et al., *Earth and Planetary Science Letters*, Vol. 50, 1980, p. 238).

Uranus forming 4550 Myr. *See* solar system forming.

urbanization 5800 yr. *See* cities.

U.S. *See* American intrusive peoples; North America (geology); North American uplifts.

U.S.S.R. *See* Asia; Russians.

V

vaccination esp. 204 yr (AD 1796). *See* medicine.

Varangerian ice ages 670 Myr. *See* climate 3800 Myr.

Vendian period 680 to 570 Myr. *See* geological timescale.

Venus c. 4500 Myr. This planet is almost the same size as the Earth, but 20 percent less massive, and it orbits closer to the sun. Although wrapped entirely in cloud, it has been explored by several Soviet spacecraft landing on the surface and by radar survey from an American spacecraft. The surface possesses raised continents and sunken (but dry) ocean basins, implying that geological processes on Venus may be similar in some respects to the Earth's, and still continuing. (P. W. Francis, in D. G. Smith, ed., *Cambridge Encyclopedia of the Earth Sciences*, Cambridge University Press, 1982.) *See also* solar system forming.

vertebrate animals 510 Myr to present. An army of relatively large animals, including human beings, possesses an internal skeleton that includes a brain case and a backbone through which passes a nerve cord. The first traces of vertebrates are bony fragments from the mid-Late Cambrian of Wyoming (J. E. Repetski, *Science*, Vol. 200, 1978, p. 529, and personal communication, 1982). Because of ambiguities in the dating of the Cambrian-Ordovician boundary (520 to 500 Myr), a cautious date of 510 Myr ago is adopted here. General information for the following account is from W. N. McFarland, et al., *Vertebrate Life*, Macmillan, 1979, and A. Hallam, ed., *Patterns of Evolution*, Elsevier, 1977. Data on early reptiles are from T. Kemp, *New Scientist*, Vol. 93, 1982, p. 381, and personal communication, 1982; dates adjusted.

The distinctive body plan of vertebrates may have evolved from that of the tadpole-like larvae of invertebrate animals; in theory, some of the young of these unpromising-looking ancestors became precociously capable of reproduction without first undergoing metamorphosis into the adult form. This evolutionary event occurred 550 Myr ago, with the appearance of the first chordates, animals with spinal columns but lacking bones (*see* animal origins). Vertebrates constitute a subdivision of the phylum of chordates.

The first vertebrates were fishy animals lacking jaws but using bone for external armor. They were typically twenty to thirty centimeters long and they fed by sucking in water and whatever it contained. The invention of jaws was the next landmark in vertebrate evolution, and jawed fishes gave rise to early amphibians. Amphibians begat reptiles, the first of a broad group of "amniotes" characterized by active membranes wrapped around the embryo, as in a bird's egg. From reptiles, mammals and birds arose, and with the evolution of the birds, the main categories of modern vertebrates were complete. See the family tree p. 122.

VERTEBRATES: TIMETABLE

jawless fishes: **c. 510 Myr**. Bony protective scales, already elaborate in structure, from the jawless fish *Anatolepsis*, are the oldest fossil remains of bone and of vertebrates. Vertebrates' eyes evolved at about this time; better preserved jawless fishes (425 Myr ago) had three eyes, and so do the jawed vertebrates that are descended from them; in humans, the third eye has become the pineal gland deep inside the head.

hemoglobin: est. 480 Myr. The characteristic red blood pigment of the vertebrates evolved in jawless fishes about 480 Myr ago.

jawed fishes: **425 Myr**. The first vertebrate jaws, complete with teeth, are those of acanthodians, with large mouths and a vertebral column (mid-Silurian).

ray-finned fishes: **415 Myr**. The actinopterygians are the characteristic fishes of modern seas and inland waters, and the most abundant of all vertebrates; the first uncertain fragments of ray-finned fishes appear late in the Silurian period.

lungs: **c. 400 Myr.** These may have originated as buoyancy bags in early fishes, formed by balloons of gut, which then lent themselves to oxygen intake from the air when fishes were stranded or found themselves in stagnant water.

fleshy-finned fishes: **c. 400 Myr.** Notable chiefly as ancestors of the vertebrates that settled the land, the sarcopterygians flourished from the Early Devonian onward but declined in the Carboniferous c. 360 Myr ago. Today only a few species remain, including lungfishes and coelacanths.

sharks: **380 Myr**. The earliest known shark-like elasmobranchs appeared in Middle Devonian times. A giant shark was a superpredator, c. 4 Myr ago.

early amphibians: **370 Myr**. The main land-dwelling vertebrates (but not the modern amphibians) are descended from fleshy-finned fishes that came ashore, possessing lungs, and also bones within their fins, as potential limb bones. The oldest known fully fledged amphibians, various ichthyostegids from Greenland and Australia 370 Myr ago (Famennian, Late Devonian), already sported well-developed limbs. The diversity of early amphibians reached a peak 310 Myr (Late Carboniferous), but the reptiles displaced them from the land by c. 256 Myr (mid-Permian turnover), and from the water by 208 Myr (end Triassic).

reptiles: **313 Myr**. The earliest primitive reptiles (*Hylonomus*, etc.; Nova Scotia, Westphalian B, in Late Carboniferous) seem to have evolved from relatively unsuccessful amphibians. Their chief distinguishing features were skins resistant to desiccation, more powerful, less trap-like jaws, and the invention of amniotic eggs. In addition to spectacular reptiles listed below, lizards appeared 240 Myr ago. Snakes are legless lizards, and the chief family of snakes, the colubrids, evolved c. 30 Myr ago.

mammal-like reptiles: **310 Myr**. Deriving from an early reptile with a superior jaw (*Archaeothyris*; Nova Scotia, Westphalian C), these very successful animals evolved through three dynasties, as pelycosaurs, 310 Myr; therapsids, 256 Myr (start of Late Permian); and cynodonts esp. 240 Myr (Early Triassic). In each phase they radiated into a variety of carnivores and large and small herbivores. (For comments on the events that ended the reigns of the pelycosaurs and the therapsids, *see* mid-Permian turnover; Permian terminal catastrophe.) During an interregnum at the start of the Triassic, 245 to 240 Myr, a single genus (species group) of herbivores, *Lystrosaurus*, had the world almost to itself, and large predators were lacking. The mammal-like reptiles were eclipsed by the dinosaurs by 216 Myr ago.

waterborne reptiles: **278 Myr**. Notable reptilian fish eaters, including the nothosaurs, emerged in the Early Permian. Large fish-like ichthyosaurs were superpredators, 230 Myr ago. The long-necked plesiosaurs appeared 216 Myr ago, and giant pliosaurs flourished 159 Myr ago. Seagoing lizards, the mosasaurs, evolved 89 Myr, and became superpredators, 72 Myr ago. Turtles, together with the nearest surviving reptilian relatives of the dinosaurs, the freshwater crocodiles, evolved c. 230 Myr ago.

modern amphibians: **c. 240 Myr**. Frogs and salamanders are sufficiently different from early amphibians to indicate a fresh, independent origin from fishes, with a date of est. 250 Myr implied by molecular dating (*see* life on Earth). The earliest fossil resembling a frog dates from c. 240 Myr (*Triadobatrachus*; Malagasy, Lower Triassic). Frogs and salamanders surged c. 170 Myr ago (Middle Jurassic).

dinosaurs: **235 Myr**. Related large reptiles, the thecodonts, flourished 240 Myr ago. The earliest dinosaurs of 235 Myr ago were relatively small; the first giant dinosaurs appeared 225 Myr ago and displaced the mammal-like reptiles; they became extinct themselves 67 Myr ago (*see* dinosaurs).

airborne reptiles: **225 Myr**. Pterosaurs, the winged reptiles, evolved 225 Myr ago (dimorphodontid; Italy, Norian stage, Upper Triassic) and flourished until they met their doom alongside the dinosaurs 67 Myr ago; the most remarkable specimen known is *Quetzalcoatlus*, with a span of 15 meters.

mammals: **216 Myr**. Rapid evolution among late mammal-like reptiles gave rise to new forms, including the ictidosaurs and tritylodonts, some of which survived for about 50 million years. But the most successful of the novelties were the true mammals, descendants of mammal-like reptiles that became very small (*see* mammals).

birds: **150 Myr**. Some small dinosaurs evolved into birds; after *Archaeopteryx*, 150 Myr, true birds akin to modern birds appeared c. 133 Myr ago (*see* birds).

Vietnamese breakout 529 yr (AD 1471), *see* breakouts (Indochinese peoples); *Americans in Vietnam* **esp. AD 1961 to 1973**, *see* historical timescale. *See also* Indochina.

Vikings' breakout **1207 yr (AD 793).** *See* breakouts; climate 10,300 yr.

viruses *See* disease epidemics.

vitamins esp. AD 1912. *See* medicine.

volcanic events **esp. 16 Myr** to present. The most voluminous production of lava occurs at ocean rifts, the most violent, beside ocean trenches, and the most incongruous, in the middle of plates that are running over hot spots. In the dating technique called tephrochronology, layers of ash from the same events can be matched in different places. Direct dating of lavas in volcanic areas on land, and the detection of volcanic dust and acid fallout preserved in ice sheets, extend the overview of global volcanic history. Compilations of ash-layer records reveal, contrary to expectation, that volcanism is not merely intermittent but episodic, with marked increases in volcanism occurring globally in particular intervals. Around 32 Myr ago, Indonesia and the Philippines were racked by volcanism. Close study of the past twenty million years has linked volcanism with other global events, notably cooling climates and falling sea levels, which stimulated volcanic action. (J. P. Kennett in C. Emiliani, ed., *The Sea*, Vol. 7, Wiley, 1981.) Adapting Kennett's data, major volcanic episodes were as follows:

VOLCANIC EVENTS: TIMETABLE

Oregon eruptions: **c. 16 Myr**. A pronounced maximum of ash ages for central Oregon, scene of the largest recent outpourings on land.

chorus: **c. 14 Myr**. Global ash reached a maximum c. 15 Myr, but exceptional outbursts occurred, in the Hawaiian chain (Gardner Pinnacles) and elsewhere, c. 14 Myr.

lull: 10 to 6 Myr. Oregon was active.

chorus: c. 4 Myr. Upturn in global ash c. 5 Myr, reaching a peak c. 4 to 3 Myr.

lull: c. 3 Myr. Also in Oregon.

chorus: **c. 1 Myr to present**. Activity began to increase markedly about 2 Myr ago, and since 1 Myr has been especially high (Oregon active).

The largest single eruption of the past 2 million years was probably the Toba explosion in Sumatra **73,000 yr** ago. The event has been analyzed in detail (D. Ninkovitch, et al., *Nature*, Vol. 276, 1978, p. 574). It probably put 100 times as much dust into the stratosphere as the Krakatoa eruption (AD 1883). Comparison with the European pollen record (*see* climate 128,000 yr) indicates that Toba was responsible for a drastic cooling before the Brørup mild spell. The largest events of the past ten thousand years were probably an unidentified eruption c. 9640 yr ago, and the explosion that made Crater Lake, Oregon c. 6400 yr ago (C. U. Hammer, et al., *Nature*, Vol. 288, 1980, p. 230). Although the explosion of Santorini (Thera) in the Aegean Sea c. 3470 yr ago covered nearby Crete in five centimeters of ash, it did not cause the immediate fall of the Minoan culture there (N. D. Watkins, et al., *Nature*, Vol. 271, 1978, p. 122). During the past two hundred years, the main explosion of Tambora, Indonesia, 185 yr ago AD 1815) caused a marked, though transient, cooling of the world;

similar effects have been noted in the aftermath of lesser eruptions (H. H. Lamb, *Climate: Present, Past and Future*, Vol. 1, Methuen, 1972). Although disasters such as Pompeii, 1921 yr (AD 79), Krakatoa, 117 yr (AD 1883), and Mt. Pelée, AD 1902, were locally awesome, individual volcanic eruptions have had little lasting effect on human history, except perhaps in Middle America (*see* archeological timescale).

W

walls (town) 10,500 yr; (military) c. 4170 yr; *Great Wall of China* **2214 yr**. *See* weaponry.

warfare, chronic **5300 yr.** *See* government; weaponry.

warmings (global) **esp. 4900 and 1070 yr (AD 930).** *See* climate 10,300 yr.

warm interludes (interglacials) **noted** **500,000 yr** onward. *See* orbital dating.

warrior kings **5300 yr.** *See* government.

water (in liquid form) est. 4450 yr. *See* Earth forming.

water buffalo c. 1.8 Myr, *see* bovids; *domesticated* est. 9500 yr, *see* animal domestication.

water energy (waterwheel) c. 2030 yr. *See* energy.

weak force (active) **10^{-35} second**; (weakened) **10^{-11} second** after origin. *See* Big Bang.

weaponry 20,000 yr to present. Axes, clubs, spears, slings, and fire, employed by hunter-gatherers against animals, were always liable to be used for homicide, or in skirmishes between bands. The same can be said of the bow and arrow, dated here for arrowheads (Tunisia and Spain, Aterian and Solutrean cultures) at 20,000 yr. Threats from robbers are evident in the construction of the walls of Jericho, 10,500 yr ago. Fortifications began to appear in western Asia **c. 8000 yr** (between levels I and III at Tell-es-Sawwan). Fighting remained intermittent until the present era of chronic warfare, involving professional warriors, which began in Egypt 5300 yr ago (Nakada III). The most consequential invention of early warriors was the chariot, known from Cappadocia, in modern Turkey, c. 4100 yr ("late third millennium"). Navies can be dated from the building of special-purpose war galleys, 2800 yr ago, and air forces, from those of the Germans and French, AD 1913.

General sources for the timetable include J. G. D. Clark, *World Prehistory*, Cambridge University Press, 1977, W. H. McNeill, *World History*, Oxford University Press, 1979; A. G. Sherratt, ed., *Cambridge Encyclopedia of Archaeology*, Cambridge University Press, 1980; R. M. Adams, *Heartland of Cities*, University of Chicago Press, 1981; *Encyclopaedia Britannica*, 1981; and Imperial War Museum, London, personal communications. Accounts of the galley and the catapult follow W. Soedel and V. Foley, *Scientific American*, Vol. 240, No. 4, 1979, p. 120, and Vol. 224, No. 4, 1981. The stated origins of gun warfare reconcile H. Thomas, *An Unfinished History of the World*, Hamish Hamilton, 1979, with J. Needham, in R. Dawson, ed., *The Legacy of China*, Oxford University Press, 1964, and other sources. Accounts of recent weaponry draw on N. Calder, ed., *Unless Peace Comes*, Viking, 1968, and *Nuclear Nightmares*, Viking, 1979.

WEAPONRY: TIMETABLE

bow and arrow: **c. 20,000 yr**. Western Mediterranean (*see* above).
fortifications: **c. 8000 yr**. Western Asia (*see* above).
slings: esp. 7300 yr. Western Asia (e.g., Mersin VII).
bronze weapons: c. 4800 yr. Western Asia (e.g., Alaca Hüyük).
battle wagon: c. 4700 yr. Western Asia (Ur).
long wall: c. 4170 yr. Western Asia (Euphrates to Tigris).
chariot: **c. 4100 yr**. Western Asia (*see* above).
iron weapons: c. 3050 yr. Europe (Greece).
war galley: esp. 2800 yr. Eastern Mediterranean.
cavalry: esp. 2800 yr. Europe (Scyths).
heavy cavalry: c. 2400 yr. Europe (Sarmatians).
catapult: 2399 yr. Europe (Sicily).
compound bow: (mounted archers) c. 2300 yr. Asia (Huns).
crossbow: c. 2300 yr. Southern China.
Great Wall of China: **2214 yr**.
knight-in-armor: **c. 1280 yr (AD 720)**. Europe (Franks).
gunpowder: 1100 yr (AD 900). China.
longbow: c. 800 yr (AD 1200). Europe (Wales).
guns: by 725 yr (AD 1275). China and India.
gun warfare: **esp. 644 yr (AD 1356)**. China.
naval gun warfare: **623 yr (AD 1377)**. Spain.
hand guns: esp. 475 yr (AD 1525). Spain.
armored ship: 141 yr (AD 1859). France.
machine-gun warfare: **esp. 102 yr (AD 1898)**. Britain.
high explosive: esp. AD 1904. Europe (TNT).
military aircraft: AD 1908. United States.
submarine warfare: esp. AD 1914. Germany.
gas warfare: esp. AD 1915. Germany.
battle tank: AD 1916. Britain and France.
aircraft carrier: AD 1918. Britain.
bomber warfare: **esp. AD 1936**. Europe (Madrid).
nerve gases: AD 1939. Germany (not used).
napalm: AD 1941. United States.
guided and homing missiles: AD 1943. Germany.
cruise missile: AD 1944. Germany (V-1).
ballistic missile: **AD 1944**. Germany (V-2).

A-bomb: **AD 1945**. United States (fission devices).
H-bomb: **AD 1953**. U.S., U.S.S.R. (fusion devices, median date).
nuclear-powered submarine: AD 1955. United States.
military satellite: **AD 1960**. United States (photographic).
submarine-launched ballistic missile: **AD 1960**. United States.
antisatellite missile: AD 1963. United States.
antisatellite satellite: AD 1968. U.S.S.R.
antiballistic missile: AD 1969. U.S.S.R.
multiple warheads: **esp. AD 1970**. United States (MIRV).

weaving 8500 yr. *See* arts and crafts.

West Antarctic deep freeze 6.6 Myr. *See* climate 67 Myr.

whales, primitive 55 Myr; *modern* **40 Myr**; *growing* **25 Myr**, *see* mammals.

wheat domesticated 10,600 yr. *See* crops.

wheel 5500 yr. Early use of the wheel, in pottery making in Mesopotamia, was associated particularly with mass-produced Sumerian pottery, from **6000 yr** ago (Uruk times). The earliest evidence for the wagon wheel comes in a Sumerian pictogram of a wagon **5500 yr** ago (Uruk IVA), where it is clearly seen to be a sled fitted with wheels, and the pictograms capture a technological change better than one is entitled to expect in archeology. The wagon wheel was not necessarily invented by the Sumerians, and people with both plows and carts showed up in Europe from 5300 yr ago (P. R. S. Moorey and A. G. Sherratt, personal communications, 1981). Its relatively late appearance from some single origin in the Old World suggests that the wheel required a rare imaginative leap; it was also known in the ancient New World, but only in models, and the lack of suitable draft animals may have inhibited its exploitation. True spoked wheels appeared c. 4100 yr ago in Cappadocia (modern Turkey; end of the third millennium BC). *See also* transport; weaponry.

wildcats c. 120,000 yr. *See* cats.

wild hogs c. 1.8 Myr, *see* pigs; *domesticated* **c. 9500 yr**, *see* animal domestication.

windmills est. 1400 yr (AD 600). *See* energy.

wine (grape) est. 6000 yr. *See* alcohol.

winged insects c. 330 Myr. *See* invertebrates.

winged reptiles 225 Myr. *See* vertebrate animals.

Wisconsin ice age 72,000 yr. *See* orbital dating.

wolves 650,000 yr, *see* dogs; *tamed* **12,000 yr**, *see* animal domestication.

woolly mammoth 150,000 yr. *See* elephants.

woolly sheep c. 7300, **esp. 6000 yr**. *See* livestock revolution.

work est. 9000 yr. Surviving hunter-gatherers of recent times, driven into the least bountiful margins of the world, were able to obtain plenty of food for everyone from about four hours' work-like activity a day. "Probably, with the advent of agriculture, people had to work harder" (M. P. Sahlins, *Essays in Stone Age Economics*, Aldine, 1972). Early horticulture was probably fairly effortless, and work in the sense of regular hard labor may have begun only with the advent of the plow, est. 6000 yr ago.

Taxation dates back to c. 5600 yr ago, and punishment for farmers who did not work hard enough was prescribed in early laws c. 3900 yr ago. Fully organized commercial slavery seems to have begun with the Greeks and Phoenicians c. 2800 yr ago. In the past, work typically meant work on the land; around AD 1970, work in industry and information processing outstripped it, worldwide. In countries with high productivity, industrial work began to join agriculture as a minority pursuit while electronic techniques made inroads on the information-processing sector. Four business cycles of 40 to 60 years' duration (Kondratieff cycles) have occurred in Western industrialized countries during the past two centuries, characterized by high unemployment during phases of contraction starting in AD 1815, 1873, 1920, and 1980; this last contraction may prove to be a terminal event after 9000 years of worldwide work. Data in the timetable follow C. M. Cipolla, *The Economic History of World Population*, Penguin, 1978, various authors in *Scientific American*, Vol. 247, No. 3, 1982, and other sources.

WORK: TIMETABLE

posthunting world: est. 9000 yr.

postagricultural British: **c. 200 yr (AD 1800)**. The cotton industry boomed and the proportion of the population dependent on agriculture for their income fell to less than 40 percent; similar milestones were passed in Germany and the United States by AD 1890 or 1900, and in Japan c. 1920.

trade unions ceased to be illegal: 176 yr (AD 1824). Britain.

assembly line: **esp. AD 1913**. United States (automotive industry).

forty-hour week: c. AD 1947. United States.

postindustrial Americans: **c. AD 1960**. The fraction of the

American labor force engaged in farming, mining, construction, and manufacture fell to less than 40 percent of the total.

postagricultural world: **c. AD 1970**. By this date, less than half the world's working population was engaged in agriculture.

World War I AD 1914 to 1918; *II* **AD 1939 to 1945**. *See* historical timescale.

worms 620 Myr. *See* animal origins.

writing 5500 yr. *See* information technology.

Würm ice age 72,000 yr. *See* orbital dating.

X

X force c. 10^{-35} second after origin. *See* Big Bang.

X rays discovered 105 yr (AD 1895). *See* medicine.

Y

yaks c. 400,000 yr, *see* bovids; *tamed* c. 4500 yr, *see* animal domestication.

Yamnaya culture 5000 yr. *See* breakouts (horsemen); language.

yams domesticated ?9000 yr. *See* crops.

Yayoi culture c. 2750 yr. *See* climate 10,300 yr.

yellow fever esp. 352 yr (AD 1648). *See* disease epidemics.

Yuan (Mongol) empire 740 yr (AD 1260). *See* historical timescale.

Z

Zagros Mountains c. 20 Myr. *See* continents.

zebras (arrival in Africa) 1.9 Myr. *See* horses.

zebu cattle domesticated c. 5500 yr. *See* animal domestication.

zero ("Arabic") **c. 1317 yr (AD 683).** *See* information technology.

Zimbabwe, Great 800 yr (AD 1200). *See* breakouts (blacksmiths).

Zoroastrianism c. 2550 yr. *See* religious surge.